Introduction to Lattice Algebra

Introduction to Lattice Algebra

With Applications in AI, Pattern Recognition, Image Analysis, and Biomimetic Neural Networks

Gerhard X. Ritter
University of Florida
Gonzalo Urcid
INAOE

CRC Press
Taylor & Francis Group
Boca Raton London New York

CRC Press is an imprint of the
Taylor & Francis Group, an **informa** business
A CHAPMAN & HALL BOOK

First edition published 2022
by CRC Press
6000 Broken Sound Parkway NW, Suite 300, Boca Raton, FL 33487-2742

and by CRC Press
2 Park Square, Milton Park, Abingdon, Oxon, OX14 4RN

CRC Press is an imprint of Taylor & Francis Group, LLC

Library of Congress Cataloging-in-Publication Data

Names: Ritter, G. X., author. | Urcid, Gonzalo, author.
Title: Introduction to lattice algebra : with applications in AI, pattern recognition, image analysis, and biomimetic neural networks / Gerhard X. Ritter, Gonzalo Urcid.
Description: First edition. | Boca Raton : C&H\CRC Press, 2021. | Includes bibliographical references and index.
Identifiers: LCCN 2021007716 (print) | LCCN 2021007717 (ebook) | ISBN 9780367720292 (hardback) | ISBN 9780367722951 (paperback) | ISBN 9781003154242 (ebook)
Subjects: LCSH: Lattice theory. | Computer science--Mathematics. | Artificial intelligence--Mathematical models.
Classification: LCC QA76.9.L38 R58 2021 (print) | LCC QA76.9.L38 (ebook) | DDC 004.01/5113--dc23
LC record available at https://lccn.loc.gov/2021007716
LC ebook record available at https://lccn.loc.gov/2021007717

ISBN: 978-0-367-72029-2 (hbk)
ISBN: 978-0-367-72295-1 (pbk)
ISBN: 978-1-003-15424-2 (ebk)

DOI: 10.1201/9781003154242

To Cheri and my daughters: Andrea and Erika

In loving memory of my parents, Gustavo and Elena,
and to my wife, María Teresa, for her patience and understanding

Contents

Preface

Background. Lattice theory extends into virtually every branch of mathematics, ranging from measure theory and convex geometry to probability theory and topology. A more recent development has been the rapid escalation of employing lattice theory for various applications outside the domain of pure mathematics. These applications range from electronic communication theory and gate array devices that implement Boolean logic, to artificial intelligence and computer science in general. An excellent summary of lattice theory in practice can be found in Vassilis G. Kaburlasos's book *Towards a Unified Modeling and Knowledge-Representation based on Lattice Theory: Computational Intelligence and Soft Computing Applications* [131].

The early development of a *generalized* lattice algebra came about when H. Heijmans and J. Davidson completed the mathematical foundation for *mathematical morphology* by, respectively, formulating its embedding into lattice theory [112] and an algebra known as *mini-max algebra* [65, 66, 63]. At that time it was known that mathematical morphology, as formulated by Matheron and Serra [184, 246], grew out of the early work of Minkowski and Hadwiger [191, 107]. The definitions of *Minkowski addition*, also known as Minkowski *sum* or *dilation*, and *Minkowski subtraction*, also known as the Minkowski *difference*, are directly related to the notions of *dilation* and *erosion* which form the foundation of mathematical morphology. Further details of this relationship can be found in Chapter 7 of [220]. Mathematical morphology became a useful member in the toolbox for image processing and image analysis. As such, it caught the attention of researchers involved in the development of *image algebra*. The main goal of the image algebra project was the establishment of a comprehensive and unifying theory of image transformations, image analysis, and image understanding in the discrete as well as the continuous domain [220]. It became clear that for some image transforms there was a need for lattice-based techniques that mimicked linear transformations and other techniques from the linear algebra domain. The development of these techniques resulted in the expansion of lattice theory into a larger algebraic theory referred to as *lattice algebra*. The lattice algebra defined in this treatise refers to the mathematical structure derived by merging basic concepts from lattice theory, abstract algebra, and linear algebra. A consequence of this amalgamation has resulted in a branch of mathematics that deals with concepts such as lattice-based vector spaces, lattice independence, and lattice products of matrices. These concepts are based on the theory of lattice semirings and lattice semifields discussed in this book. Although G. Birkhoff also used the term lattice vector space in his classic book on lattice theory [30], the lattice vector spaces introduced in Chapter 4 are drastically different. The same holds true for concepts such as lattice products of matrices and lattice independence.

The emphasis of this book is on two subjects, the first being lattice algebra and the second the practical applications of that algebra. As the title indicates, this is an introduction and not a definitive all-encompassing text of the two subjects. The immense breadth of lattice algebra and its applications limits the choice of topics. The book's epilogue lists several additional topics for future consideration.

The Audience. This book is intended for use as a textbook for a special topics course in artificial intelligence with focus on pattern recognition, multispectral image analysis, and biomimetic artificial neural networks. Since a large part of this book deals with applications in artificial intelligence, pattern recognition, and artificial neural networks, we assume that the reader has some familiarity with these subjects.

The book is self-contained and–depending on the student's major–can be used as a senior undergraduate-level or a first-year graduate-level course. Since lattice theory is not a required course for computer scientists, the book is an ideal self-study guide for researchers and professionals in the above mentioned disciplines.

Organization. The book consists of nine chapters. Most chapter sections contain exercises for the student and interested reader. The first two chapters are introductory chapters that can be skipped by graduate mathematics students. The focus of the first chapter is on abstract algebra, a subject generally not taught to engineering students. Since comprehension of abstract algebra is necessary for understanding lattice algebra, we found it necessary to include it. Chapter 2 introduces the reader to the basic topological properties of n-dimensional Euclidean space. These properties are used in later chapters and also help in the understanding of proofs of theorems that deal with the capabilities of biomimetic neural networks. Chapter 3 provides a brief review of pertinent concepts of lattice theory and its relationships with other branches of mathematics. Chapter 4 is dedicated to the edifice of lattice algebra. The remaining chapters deal with practical applications of lattice algebra. Chapter 5 presents applications of minimax matrix products in the area of associative memories. The chapter is interesting from a historic perspective. As far as we know, it was the first associative memory completely specified in lattice algebra [216, 217, 218, 219]. Its success in outperforming such well-known matrix-based associative memories as the Hopfield memory gave impetus for further research in lattice algebra and artificial neural networks. Chapter 6 takes a sidestep into convex set theory with a focus on lattice polytopes and the computation of extreme points of lattice polytopes. The results obtained in Chapter 6 are relevant for Chapter 7 which discusses applications in hyperspectral image unmixing and color image segmentation. Chapter 8 explains the rationale for biomimetic neural networks and presents a model of such a network. Chapter 9 provides various learning methods for biomimetic neural networks and compares the performance of biomimetic neural networks versus several current artificial neural networks on a variety of different data sets.

Acknowledgments. We are deeply indebted to the many people who have contributed in different ways to the process of writing this book. We offer our sincere appreciation to Professors Ralph Selfridge, David Wilson, Panagiotis E. Nastou, Anand Rangarajian, and Alireza Entezari. Gonzalo Urcid, my co-author, thanks the Mexican National System of Researchers (SNI-CONACYT) for partial funding (grant no. 22036) during the realization of the book project.

Special thanks go to Joseph Sadeq for proofreading the entire manuscript, correcting typographical and other errors, and creating the Index. A very special thanks goes to our editors, Callum Fraser and Mansi Kabra, at CRC Press/Taylor & Francis Group. Callum, for believing in our project from the beginning, and Mansi and Callum for their patience, support, and guidance in all matters.

Gerhard X. Ritter
Gainesville, Florida, December 2020

CHAPTER 1

Elements of Algebra

To gain a thorough understanding of the subject of this book, the reader will have to learn the language used in it. We will try to keep the number of technical terms as small as possible, but there is a certain minimum vocabulary that is essential. Much of the language used is taken from the theory of abstract algebra, geometry, and topology, subjects which are pertinent but not the focus of this book. These subjects are, indeed, independent branches of mathematics. The aim of this chapter is to present the mathematical notation as well as pertinent facts from algebra that are used throughout this study of lattice algebra and its applications in various areas of artificial intelligence.

The initial focus is on the foundation of abstract algebra, namely operations on sets and the concepts of groups, rings, and fields. These concepts play a major role in lattice algebra. The chapter concludes with a brief discourse of linear algebra with emphasis on vector spaces. Although the reader of this manuscript is most likely very familiar with linear algebra, the addition of vector spaces will provide for a quick comparison with the concept of vector spaces based on lattice algebra (Chapter 4).

1.1 SETS, FUNCTIONS, AND NOTATION

Throughout this book we try to use common mathematical notation and terminology. We adopt the convention of denoting sets by capital letters and the elements of the set by lower case letters. Intuitively, we think of a set as a collection of objects that satisfy some given condition or property such as the set of integers, the pages in this book, or the set of objects named in a list. These objects are called the *elements* of the set. Given a set A, then $x \in A$ means that x is an element of A and the symbol $\in$ reads "is an element of" or "is a member of," while the symbol $\notin$ corresponds to "is not a member of" or

DOI: 10.1201/9781003154242-1

"is not an element of." There is exactly one set with no elements. This set is called the *empty set* and is denoted by $\varnothing$.

All through this text we will use certain notation from symbolic logic in order to shorten statements. For example, if p and q represent mathematical propositions, then the statement $p \Rightarrow q$ means that "p implies q" or, equivalently, "if p is true, then q is true." The statement $p \Leftrightarrow q$ reads "p if and only if q" and means that p and q are logically equivalent, i.e., $p \Rightarrow q$ and $q \Rightarrow p$.

An expression $p(x)$ that becomes a proposition whenever values from a specified domain of discourse are substituted for x is called a *propositional function* or, equivalently, *a condition* on x, and p is called a *property* or *predicate*. The assertion "x has property p" means that $p(x)$ is true. For instant, if $p(x)$ is the propositional function "x is an integer," then p denotes the property "is an integer," and $p(2)$ is true, whereas $p(\frac{1}{2})$ is false.

A set may be described by giving a characterizing property or by listing its elements. It is common practice to enclose either description in braces. When using a characterizing property, then the set assumes the form $\{x : p(x)\}$, which translates into the "set of all elements x such that $p(x)$ is true.". For example, the set of positive integers 1 through 5 could be expressed as $\{x :$ x is a positive integer less than $6\}$. Here $p(x)$ is the proposition "x is a positive integer less than 6" and p denotes the property "is a positive integer less than 6." This set can also be specified by listing the elements: $\{1,2,3,4,5\}$. For our purposes, we shall assume that the construct $X = \{x : p(x)\}$ is a set if and only if given an object x, the proposition $p(x)$ is either true or false, but not both (i.e., true *and* false). This enables us to decide whether or not an object under discussion is an element of X.

The existential quantifier "there exists" is denoted by $\exists$ while the symbol $\nexists$ means "there does not exist." The existential quantifier "there exists one and only one" or, equivalently, "there exists a unique" is denoted by $\exists!$. The universal quantifier "for all" or "for every" is denoted by $\forall$. The common symbol for the connective "such that" is $\ni$. Thus, the assertion $\forall x \, \exists y \ni \forall z :$ $p(x,y,z)$ reads: "for every x there exists a y such that for all z, $p(x,y,z)$ is true."

If X and Y are sets, then $X = Y$ means that they have the same elements; that is, $\forall x : x \in X \Leftrightarrow x \in Y$. The notation $X \subset Y$ reads "X is a *subset* of Y," and signifies that each element of X is an element of Y. More precisely, $\forall x : x \in X \Rightarrow x \in Y$. Equality is not excluded. Thus, whenever necessary, we shall say that X is a *proper subset* of Y if $X \subset Y$ and $X \neq Y$. The set whose elements are all the subset of a given set X is called the *power set* of X and is denoted by 2^X. The following subset relationships are a direct consequence from the aforementioned definitions:

S_1. $X \subset X$ for every set X;

S_2. if $X \subset Y$ and $Y \subset Z$, then $X \subset Z$;

S_3. $X = Y$ if and only if $X \subset Y$ and $Y \subset X$;

S_4. $\varnothing \subset X$ for every set X;

S_5. $\varnothing \in 2^X$ and $x \in X \Leftrightarrow \{x\} \in 2^X$.

Although these subset relations are self-evident, they are of fundamental importance as we will discuss many other types of relationships that are analogous to those stated in S_1 through S_4.

In addition to the subset relations, there are also three elementary operations on sets that provide for an algebra of a collection of sets. For such an algebra, it is customary to view the sets under consideration as subsets of some larger set U, called a *universal* set or the *universe of discourse*. For example, in plane geometry the universal set consists of all points in the plane.

Definition 1.1 Let X and Y be two sets. The *union* of X and Y, denoted by $X \cup Y$, is defined as the set whose elements are either in X or in Y, or in both X and Y. Thus,

$$X \cup Y = \{z : z \in X \textit{ or } z \in Y\}.$$

The *intersection* of X and Y, denoted by $X \cap Y$, is defined as the set of all elements that are in both X and Y. Thus,

$$X \cap Y = \{z : z \in X \textit{ and } z \in Y\}.$$

If $X \subset U$ then the *complement* of X (with respect to U) is denoted by X' and is defined by $X' = \{x : x \in U \textit{ and } x \notin X\}$.

Notwithstanding that the notation X^c is also commonly used to denote the complement of X, the notation X' will be used throughout this book to denote the complement.

The complement X' should not be confused with the *difference* of X and Y, also known as the *subtraction* of Y from X as well as the *complement* of X with respect to Y. The *difference* of X and Y, is denoted by $X \setminus Y$ and defined by $X \setminus Y = \{x : x \in X \textit{ and } x \notin Y\}$. Consequently, $X' = U \setminus X$. The relationship between $\cup$, $\cap$, and $\subset$ are given by the following theorem:

Theorem 1.1 *The following statements are all equivalent:*

(i) $X \subset Y$, *(ii)* $X = X \cap Y$, *(iii)* $Y = X \cup Y$, *(iv)* $Y' \subset X'$, *and* (v) $X \cap Y' = \varnothing$.

The laws covering the basic operations of union, intersection, and subtraction of sets are as follows:

S_6. $X \cup U = U$ and $X \cap U = X$; $X \cup \varnothing = X$ and $X \cap \varnothing = \varnothing$ (identity laws);

S_7. $X \cup X = X$ and $X \cap X = X$ (idempotent laws);

S_8. $(X')' = X$, $X \cup X' = U$ and $\varnothing' = U, U' = \varnothing$, $X \cap X' = \varnothing$ (complement laws);

S_9. $(X \cup Y) \cup Z = X \cup (Y \cup Z)$ and $(X \cap Y) \cap Z = X \cap (Y \cap Z)$ (associative laws);

S_{10}. $X \cup Y = Y \cup X$ and $X \cap Y = Y \cap X$ (commutative laws);

S_{11}. $X \cup (Y \cap Z) = (X \cup Y) \cap (X \cup Z)$ and $X \cap (Y \cup Z) = (X \cap Y) \cup (X \cap Z)$ (distributive laws);

S_{12}. $(X \cup Y)' = X' \cap Y'$ and $(X \cap Y)' = X' \cup Y'$ (De Morgan's laws);

S_{13}. $\varnothing = A \setminus A$ and $A \cap B = A \setminus (A \setminus B)$ (equivalency laws).

In addition to the set operations of union and intersection, the *Cartesian product* of sets is one of the most important construction of combining two sets as it enables us to express many concepts in terms of sets.

Definition 1.2 Suppose X and Y are sets. The *Cartesian product* of X and Y, denoted by $X \times Y$, is the set of all ordered pairs $\{(x, y) : x \in X and y \in Y\}$.

The Cartesian product is also known as the *cross product*. The element (x, y) of the set $X \times Y$ indicates the order in that the first component x of the pair must be an element of X while the second component y must be an element of Y. Additionally, these ordered pairs are subject to the condition: $(a, b) = (c, d) \Leftrightarrow a = c$ and $b = d$. The components of the ordered pair are also known as *coordinates*.

The notion of Cartesian product extends the set of elementary set-theoretic operations. However, in comparison to $\cup$ and $\cap$, the Cartesian product is neither commutative nor associative. When $X \neq Y$, then $X \times Y \neq Y \times X$. It follows from the definition of the product that $(X \times Y) \times Z \neq X \times (Y \times Z)$. For instance, the first component of the element $((x, y), z)$ of $(X \times Y) \times Z$ may correspond to a location (x, y) and the second component z to the temperature at that location. In such a case, $(x, (y, z)) \in X \times (Y \times Z)$ has a totally different meaning and in addition an equality $((x, y), z) = (x, (y, z))$ would mean that $x = (x, y)$ and $z = (y, z)$. Also, $X \times Y = \varnothing \Leftrightarrow X = \varnothing$ or $Y = \varnothing$ (or both). The relation between the Cartesian product and the operation of union and intersection can be summarized as follows:

Theorem 1.2
$X \times (Y \cup Z) = (X \times Y) \cup (X \times Z)$ *and* $X \times (Y \cap Z) = (X \times Y) \cap (X \times Z)$.

Exercises 1.1

1. Suppose $A = \{a, b, c\}$, $B = \{b, d\}$, and $C = \{a, c, e\}$, Determine $A \cup B \cup C$, $A \cap B \cap C$, $A \times B \times C$, and $(A \times B) \cap (B \times A)$.

2. Prove Theorem 1.1.

3. Prove the laws S_6 through S_{13}.

4. Prove Theorem 1.2.

1.1.1 Special Sets and Families of Sets

Certain sets of numbers occur naturally in this treatise and we reserve the notation $\mathbb{R}$ to denote the set of real numbers, $\mathbb{R}^+$ for the set of positive real numbers, $\mathbb{R}^-$ for the set of negative real numbers, and $\mathbb{R}_{[0,\infty)}$ for the set $\{r \in \mathbb{R} : r \geq 0\}$ of non-negative real numbers. The symbol $\mathbb{Z}$ denotes the set of integers $\{\ldots, -3, -2, -1, 0, 1, 2, 3, \ldots\}$, $\mathbb{N}$ the set $\{1, 2, 3, \ldots\}$ of natural numbers, $\mathbb{N}_{[0,\infty)} = \mathbb{N} \cup \{0\} = \{0, 1, 2, \ldots\}$, $\mathbb{N}_n = \{1, 2, 3, \ldots, n\}$, and $\mathbb{Z}_n = \{0, 1, 2, \ldots, n-1\}$. Whenever a and b are real numbers, then the relations $a < b$, $a \leq b$, $b > a$, and $b \geq a$ mean, respectively, that a is strictly less than b, a is less than or equal to b, b is strictly greater than a, and b is greater than or equal to a. The relation $\leq$ (or $\geq$) is an *order relation* on $\mathbb{R}$ called the *natural* order for the set of real numbers. This order relation has the additional property that for any two real numbers a and b, either $a \leq b$ or $b \leq a$. This type of order is called a *total* order since any two elements are related. A precise definition of order relations on sets is given in Chapter 3. There are some similarities between laws of subsets and that of the order for integers. For instance, when considering the set $\mathbb{N}_{[0,\infty)}$, we see that the statements

s_1. $x \leq x \ \forall x \in \mathbb{N}_{[0,\infty)}$,

s_2. if $x \leq y$ and $y \leq z$, then $x \leq z$,

s_3. $x = y$ if and only if $x \leq y$ and $y \leq x$, and

s_4. $0 \leq x \ \forall x \in \mathbb{N}_{[0,\infty)}$.

with the exception of the use of different symbols, these laws are a mirror image of S_1 through S_4. These similarities do not stop here. Considering the identity laws, the associative laws, and the distributive laws for unions and intersections of sets, it becomes obvious that they resemble the laws of addition and multiplication for elements of $\mathbb{N}_{[0,\infty)}$:

$$identity: \; x+0=x \text{ and } x\times 0=0;$$

$$associativity: \; (x+y)+z=x+(y+z) \text{ and } (x\times y)\times z=x\times(y\times z);$$

$$commutativity: \; x+y=y+x \text{ and } x\times y=y\times x;$$

$$distributivity: \; x\times(y+z)=(x\times y)+(x\times z).$$

These types of similarities will occur throughout this book.

The set of real numbers can be extended by employing the concepts of largest element and smallest element. The symbols ∞ and $-\infty$, called *positive infinity* and *negative infinity*, respectively, can be used to extend $\mathbb{R}$ in three ways by setting $\mathbb{R}_\infty = \mathbb{R}\cup\{\infty\}$, $\mathbb{R}_{-\infty} = \mathbb{R}\cup\{-\infty\}$, $\mathbb{R}_{\pm\infty} = \mathbb{R}_\infty \cup \mathbb{R}_{-\infty} = \mathbb{R}\cup\{\infty,-\infty\}$, and $\mathbb{R}_{[0,\infty]} = \mathbb{R}_{[0,\infty)}\cup\{\infty\}$. By declaring ∞ to be larger than any real number and $-\infty$ smaller than any real number, these sets become ordered sets. The subset $\mathbb{N}_{[0,\infty]}$ of $\mathbb{R}_{[0,\infty]}$ is defined by setting $\mathbb{N}_{[0,\infty]} = \mathbb{N}_{[0,\infty)}\cup\{\infty\}$. Finally, the sets of rational numbers and complex numbers will be denoted by $\mathbb{Q}$ and $\mathbb{C}$, respectively.

If for each element λ of some non-empty set Λ there corresponds a set X_λ, then the collection $\{X_\lambda : \lambda\in\Lambda\}$ is called a *family* of sets, and Λ is called an *indexing set* for the family. Such indexed family is also commonly denoted by $\{X_\lambda\}_{\lambda\in\Lambda}$. If the indexing set $\Lambda=\mathbb{N}$, then the indexed family $\{X_i\}_{i\in\mathbb{N}}$ is called a *sequence* (of sets) and may also be denoted by $\{X_i\}_{i=1}^{\infty}$.

The notion of union and intersection can be generalized to any arbitrary indexed family of subsets of some universal set U.

Definition 1.3 Let $\{X_\lambda\}_{\lambda\in\Lambda}$ be a family of subsets of a universal set U. The *union* of this family is denoted by $\bigcup_{\lambda\in\Lambda} X_\lambda$ and is the set

$$\{x\in U : x\in X_\lambda \text{ for at least one } \lambda\in\Lambda\}.$$

The *intersection* is denoted by $\bigcap_{\lambda\in\Lambda} X_\lambda$ and is the set

$$\{x\in U : x\in X_\lambda \text{ for every } \lambda\in\Lambda\}.$$

For finite and infinite sequences $\{X_i\}_{i=1}^n$ and $\{X_i\}_{i=1}^\infty$ of sets we will also use notations such as

$$\bigcup_{i=1}^{n} X_i = X_1 \cup \ldots \cup X_n, \quad \bigcup_{i=1}^{\infty} X_i = X_1 \cup X_2 \cup \ldots, \text{ and } \bigcap_{i=1}^{\infty} X_i = X_1 \cap X_2 \cap \ldots.$$

Example 1.1 Let $\Lambda = \{\lambda : \lambda \in \mathbb{R} \text{ and } 0 \le \lambda \le 1\}$. For each $\lambda \in \Lambda$, let $X_\lambda = \{r : r \in \mathbb{R},\ 0 \le r \le \lambda\}$. Then

$$\bigcup_{\lambda\in\Lambda} X_\lambda = \{r : r \in \mathbb{R},\ 0 \le r \le 1\} = \Lambda, \text{ and } \bigcap_{\lambda\in\Lambda} X_\lambda = \{0\}.$$

This example can be generalized as follows: let X be any set and for each $x \in X$, let X_x be a subset of X such that $x \in X_x \subset X$. Then $X = \bigcup_{x\in X} X_x$.

Suppose $\{A_\lambda\}_{\lambda\in\Lambda}$ is a family of sets and X is any set. Then the union and intersection obey the following respective equivalency law:

S_{14}. $X \setminus (\bigcup_{\lambda\in\Lambda} A_\lambda) = \bigcap_{\lambda\in\Lambda}(X \setminus A_\lambda)$ and $X \setminus (\bigcap_{\lambda\in\Lambda} A_\lambda) = \bigcup_{\lambda\in\Lambda}(X \setminus A_\lambda)$.

These laws play a key role in our discourse on measure theory.

The n-fold Cartesian product of a family $\{X_i\}_{i=1}^n$ of sets can be defined inductively, yielding the form

$$\prod_{i=1}^{n} X_i = X_1 \times X_2 \times \cdots \times X_n = \{(x_1, x_2, \ldots, x_n) : x_i \in X_i \text{ for } i \in \mathbb{N}_n\}, \tag{1.1}$$

where x_i is called the *ith coordinate*. If $X = X_i$ for $i = 1, \ldots, n$, then we define $X^n = \prod_{i=1}^n X_i$. In particular, if $\mathbb{R} = X_i$ for $i = 1, \ldots, n$, then n-dimensional Euclidean space is given by the n-fold Cartesian product $\mathbb{R}^n = \prod_{i=1}^n \mathbb{R}$. Similarly, $\mathbb{R}^n_\infty = \prod_{i=1}^n \mathbb{R}_\infty$, $\mathbb{R}^n_{-\infty} = \prod_{i=1}^n \mathbb{R}_{-\infty}$, and $\mathbb{R}^n_{\pm\infty} = \prod_{i=1}^n \mathbb{R}_{\pm\infty}$.

Exercises 1.1.1

1. Suppose $A = \{(x, y) : x, y \in \mathbb{R} \text{ and } 3x + y = 4\}$, and $B = \{(x, y) : x, y \in \mathbb{R} \text{ and } x - 2y = 2\}$. Determine the graphs of A, B, $A \cap B$, and $A \times B$.

2. Prove the equivalency law S_{14}.

1.1.2 Functions

The notion of a function or map is basic in all mathematics. Intuitively, a function f from a set X to a set Y, expressed as $f : X \to Y$, is a rule which assigns to each $x \in X$ some element y of Y, where the assignment of x to y by the rule f is denoted by $f(x) = y$. However, we shall define the notion of a function formally in terms of a *set*.

Definition 1.4 Let X and Y be two sets. A *function* f from X to Y, denoted by $f : X \to Y$, is a subset $f \subset X \times Y$ with the property: for each $x \in X$ $\exists!$ $y \in Y \ni (x,y) \in f$. The set of all functions from X to Y will be denoted by Y^X. Thus, $Y^X = \{f : f \text{ is a function from } X \text{ to } Y\}$. Functions will also be referred to as *mappings*.

We write $f(x) = y$ for $(x,y) \in f$ and say that y is the *value* f assumes at x, or that y is the *evaluation* of f at x. For example, defining $X = Y = \mathbb{N}$, then the set $f \subset X \times Y$ defined by

$$f = \{(x,y) : y = 2x+1,\, x \in X\} \quad \text{or by} \quad f = \{(x, 2x+1) : x \in X\}$$

represents a function $f : X \to Y$. This function is completely specified by the rule $f(x) = 2x+1$ or, equivalently, by $y = 2x+1$.

We will periodically refer to certain special properties and types of functions. In particular, it will be important to distinguish between the following types of functions:

f_1. a function $f : X \to Y$ is said to be *injective* or *one-to-one* if and only if $x \neq z \Rightarrow f(x) \neq f(z)$ or, equivalently, $f(x) = f(z) \Rightarrow x = z$;

f_2. a function $f : X \to Y$ is said to be *surjective* or *onto* if and only if $\forall y \in Y$, $\exists x \in X \ni f(x) = y$;

f_3. a function $f : X \to Y$ is said to be a *bijection* if and only if it is both injective and surjective;

f_4. a function $f : X \to X$ with the property $f(x) = x \;\forall x \in X$ is called the *identity* function;

f_5. a function $f : X \to Y$ is said to be a *constant* function if and only if for some $y \in Y$, $f(x) = y \;\forall x \in X$;

f_6. given a function $f : X \to Y$ and $W \subset X$, then the function $f|_W : W \to Y$ defined by $f|_W = f \cap (W \times Y) = \{(x, f(x)) : x \in W\}$ is called the *restriction* of f to W;

f$_7$. given sets $X_1, \ldots, X_n$, the function $p_j : \prod_{i=1}^{n} X_i \to X_j$, where $1 \le j \le n$, defined by $p_j(x_1, \ldots, x_j, \ldots, x_n) = x_j$, is called the *projection* into the jth coordinate;

f$_8$. for $X \subset U$, the function $\chi_X : U \to \{0,1\}$, defined by

$$\chi_X(x) = \begin{cases} 1 & \text{if } x \in X \\ 0 & \text{if } x \notin X, \end{cases}$$

is called the *characteristic* function of X.

The identity function defined in f$_4$. will generally be denoted by 1_X.

Recall that the *composite* of two functions $f : X \to Y$ and $g : Y \to Z$ is a function $g \circ f : X \to Z$ defined by $(g \circ f)(x) = g[f(x)]\ \forall x \in X$. The following theorem indicates a simple method for establishing that a given function f (respectively g) is one-to-one (respectively onto).

Theorem 1.3 *Let $f : X \to Y$, $g : Y \to X$, and $h : Y \to Z$.*

1. *If $g \circ f = 1_X$, then f is injective and g is surjective.*
2. *If f and h are surjective, then $h \circ f : X \to Z$ is surjective.*
3. *If f and h are injective, then $h \circ f : X \to Z$ is injective.*

Proof. We only prove part 1 of the theorem and leave the remaining parts as exercises.

Since $f(x) = f(z) \Rightarrow x = (g \circ f)(x) = g(f(x)) = g(f(z)) = (g \circ f)(z) = z$, we have that f is injective. The function g is surjective since for any $x \in X\ \exists y \in Y$, namely $y = f(x)$, such that $x = (g \circ f)(x) = g(f(x)) = g(y)$. □

Following the last sentence in the proof of Theorem 1.3 is the Halmos symbol □. This symbol stands for "Q.E.D.," the abbreviation of the Latin phrase *quod erat demonstrandum* (which was to be proven) and will be used throughout this text at the end of proofs of theorems.

As a simple illustration of Theorem 1.3 we show that for any function $h : X \to Z$, the function $f : X \to Y$, where $Y = X \times Z$ and $f(x) = (x, h(x))$ is injective. Let $p_1 : X \times Y \to X$ be the projection onto the first coordinate. Then $p_1 \circ f : X \to X$ is 1_X. Hence, by Theorem 1.3, f is injective (and p_1 is surjective).

Definition 1.5 Suppose $f : X \to Y$, $A \subset X$, and $B \subset Y$. Then

the *range* of f is the set $R(f) = \{f(x) : x \in X\}$;

the *image* of A is the set $f(A) = \{y : y = f(x) \text{ for some } x \in A\}$;

the *pre-image* of B is the set $f^{-1}(B) = \{x : f(x) \in B\}$.

The *domain* of $f : X \to Y$ is the set X. The pre-image is also known as the *inverse image*. A direct consequence of these definitions is that $f^{-1}(R(f)) = X$, $R(f) \subset Y$, and $R(f) = Y \Leftrightarrow f$ is onto. In addition, $f^{-1}(B)$ may be empty even when $B \neq \emptyset$.

The concept of the pre-image induces a function $f^{-1} : 2^Y \to 2^X$ defined by $f^{-1}(B) \in 2^X \; \forall B \in 2^Y$. Similarly, the function f induces a function $2^X \to 2^Y$, again denoted by f, and defined by $f(A) \in 2^Y \; \forall A \in 2^X$, where $f(\emptyset) = \emptyset$. These two induced functions are called *set* functions since they map sets to sets. We conclude this section by listing some essential properties of set functions. Many of the properties listed will occur periodically in similar form in latter chapters.

Theorem 1.4 *Let $f : X \to Y$, $A \subset B \subset X$, and $\{A_\lambda\}_{\lambda \in \Lambda}$ be a family of subsets of X. Then*

1. $f(A) \subset f(B)$
2. $f(\bigcup_{\lambda \in \Lambda} A_\lambda) = \bigcup_{\lambda \in \Lambda} f(A_\lambda)$
3. $f(\bigcap_{\lambda \in \Lambda} A_\lambda) \subset \bigcap_{\lambda \in \Lambda} f(A_\lambda)$

If $A \subset B \subset Y$, and $\{B_\lambda\}_{\lambda \in \Lambda}$ be a family of subsets of Y. Then

4. $f^{-1}(A) \subset f^{-1}(B)$
5. $f^{-1}(B') \subset (f^{-1}(B))'$
6. $f^{-1}(\bigcup_{\lambda \in \Lambda} B_\lambda) = \bigcup_{\lambda \in \Lambda} f^{-1}(B_\lambda)$
7. $f^{-1}(\bigcap_{\lambda \in \Lambda} B_\lambda) = \bigcap_{\lambda \in \Lambda} f^{-1}(B_\lambda)$

Proof. We only prove property *7* as the proofs of *1* through *6* are just as simple. Observe that

$$x \in f(\bigcap_{\lambda \in \Lambda} B_\lambda) \Leftrightarrow f(x) \in \bigcap_{\lambda \in \Lambda} B_\lambda \Leftrightarrow f(x) \in B_\lambda \; \forall \lambda \in \Lambda$$

$$\Leftrightarrow x \in f^{-1}(B_\lambda)\ \forall \lambda \in \Lambda \Leftrightarrow x \in \bigcap_{\lambda \in \Lambda} f^{-1}(B_\lambda). \qquad \square$$

The following important relationship between the two set functions is also easy to verify:

Theorem 1.5 *Let $f : X \to Y$, $A \subset X$, and $B \subset Y$. Then*

1. *$A \subset f^{-1} \circ f(A)$ and*
2. *$f \circ f^{-1}(B) \subset B$.*

The following example shows that the inclusion (*3*) of Theorem 1.4 and the inclusions of Theorem 1.5 cannot, in general, be replaced by equalities.

Example 1.2 Let $A = \{(x, y) : 1 \le x \le 2, 1 \le y \le 2\}$, $B = \{(x, y) : 1 \le x \le 2, 3 \le y \le 4,\}$, and $p_1 : \mathbb{R} \times \mathbb{R} \to \mathbb{R}$ be the projection into the first coordinate (i.e., the x-axis). Since $A \cap B = \varnothing$, $p_1(A \cap B) = \varnothing \ne \{x : 1 \le x \le 2\} = p_1(A) \cap p_1(B)$. Also,

$$A \ne p_1^{-1}(p_1(A)) = p_1^{-1}(\{x : 1 \le x \le 2\}) = \{(x, y) : 1 \le x \le 2, -\infty < y < \infty\}.$$

If $f : X \to Y$ is a bijection, then for all $y \in Y$, the set $f^{-1}(\{y\})$ consists of a single element $x \in X$ so that $\{x\} = f^{-1}(\{y\})$. Thus, the set function f^{-1} defines a bijection from Y to X, again denoted by f^{-1}, and defined by $f^{-1}(y) = f^{-1}(\{y\})$. In this case, we have $(f^{-1})^{-1} = f$ and equality holds for both properties (*1*) and (2) of Theorem 1.5. Furthermore, if $g : Y \to Z$ is also a bijection, then both $f \circ g$ and $(g \circ f)^{-1}$ are bijections. This follows from Theorem 1.3 and the fact that $(g \circ f)^{-1} = f^{-1} \circ g^{-1}$.

Exercises 1.1.2

1. Prove part *2* and part *3* of Theorem 1.3.
2. Prove parts *1*, *2*, and *3* of Theorem 1.4.
3. Prove parts *4*, *5*, and *6* of Theorem 1.4.
4. Prove Theorem 1.5.

1.1.3 Finite, Countable, and Uncountable Sets

Two sets are said to be *equivalent* if there exists a bijection $f : X \to Y$. Hence the idea that two sets X and Y are equivalent means that they are identical except for the names of the elements. That is, we can view Y as being obtained from X by renaming an element x in X with the name of a certain unique element y in Y, namely $y = f(x)$. If the two sets are finite, then they are equivalent if and only if they contain the same number of elements. Indeed, the idea of a finite set X is the same as saying that X is equivalent to a subset $\mathbb{N}_n \subset \mathbb{N}$ for some integer $n \in \mathbb{N}$. To make this and related ideas more precise, we list the following definitions involving a set X and the set $\mathbb{N}$.

Definition 1.6 A set X is called

$finite$ if and only if either $X = \varnothing$ or $\exists n \in \mathbb{N}$ such that X is equivalent to $\mathbb{N}_n$;

$infinite$ if it is not finite;

denumerable if and only if it is equivalent to $\mathbb{N}$;

countable if and only if it is finite or denumerable;

uncountable if it is not countable.

Some properties of countable sets are given in the next example.

Example 1.3

1. $X \subset \mathbb{N} \Rightarrow X$ is countable.
2. X is countable $\Leftrightarrow \exists$ a set $Y \subset \mathbb{N}$ such that X is equivalent to Y.
3. If X is equivalent to Y and Y is countable, then X is countable.
4. If $X \subset Y$ and Y is countable, then X is countable.
5. If $f : X \to Y$ is onto (surjective) and X is countable, then Y is countable.
6. If $f : X \to Y$ is one-to-one (injective) and Y is countable, then X is countable.
7. If Λ is countable and $\{X_\lambda : \lambda \in \Lambda\}$ is a collection of countable sets, then $\bigcup_{\lambda\in\Lambda} X_\lambda$ is countable. In other words, the countable union of countable sets is countable.

Assuming Example 1.3 as a fact, it is not difficult to show that the set $\mathbb{Z}^2$ and the set $\mathbb{Q}$ of rational numbers are countable sets. First note that the function $f : \mathbb{N} \to \mathbb{Z}$, defined by $f(2n) = n$ and $f(2n-1) = -n+1$ is a bijection. Thus, $\mathbb{Z}$ is equivalent to $\mathbb{N}$ and therefore countable. For each $i \in \mathbb{Z}$ the set $X_i = \{(i,n) : n \in \mathbb{Z}\}$ is countable. But then since $\mathbb{Z}$ is countable and $\{X_i : i \in \mathbb{Z}\}$ is a countable collection of countable sets, it follows from Example 1.3(7) that $\mathbb{Z}^2 = \bigcup_{i\in\mathbb{Z}} X_i$ is countable.

To show that $\mathbb{Q}$ is countable, define $f : \mathbb{Z}^2 \to \mathbb{Q}$ by

$$f(i,j) = \begin{cases} \frac{i}{j} & \text{if } j \neq 0 \\ 0 & \text{if } j = 0 \end{cases} \tag{1.2}$$

Then f is obviously onto. It now follows from Example 1.3(5) that $\mathbb{Q}$ is countable.

We attach a label *card*(X) to each set X, called the *cardinality* of X, which provides a measure of the *size* of X. In particular, the label should distinguish in some way if one of two given sets has more members than the other. We shall say that two sets X and Y have the same cardinality if and only if they are equivalent. By assigning $card(X) = n$ whenever X is equivalent to $\mathbb{N}_n$ and using the convention that $card(X) = 0$ if and only if $X = \emptyset$, satisfies this requirement for finite sets. Thus, if X and Y are finite sets, then $card(X) < card(Y)$ if and only if X has fewer elements than Y. In particular, if X is a strict subset of Y, then $card(X) < card(Y)$. However, for infinite countable sets this no longer holds true since we have the sequence of strict subsets $\mathbb{N} \subset \mathbb{Z} \subset \mathbb{Q}$ while $card(\mathbb{N}) = card(\mathbb{Z}) = card(\mathbb{Q})$. Although these three sets contain different elements, their sizes are all the same in the sense that for each element of $\mathbb{Q}$ there exists a unique integer $n \in \mathbb{N}$, which means that each rational number can be labeled with a unique integer. This demonstrates that there are just as many integers as there are rational numbers. Thus, they are all of the *same* size. What about sets that are not countable? Are there sets of different infinite sizes? The answer to this question is indeed yes and the labeling of different sizes is accomplished via transfinite numbers called *cardinal numbers*. The cardinal number of a set X is commonly denoted by $\aleph(X)$ and is really a set defined in terms of ordinal numbers. Since the subject of ordinal and cardinal numbers are outside the scope of our discussion, we will only list two pertinent properties of cardinal numbers and their relationship to the definition of the label *card*.

C_1. $\quad card(X) \leq card(Y) \Leftrightarrow \aleph(X) \leq \aleph(Y)$.

C_2. $\quad card(X) = card(Y) \Leftrightarrow \aleph(X) = \aleph(Y)$.

Georg Cantor proved that there does not exist a bijection $f : \mathbb{N} \to \mathbb{R}$ [37, 38]. Since $\mathbb{N} \subset \mathbb{R}$ is a strict subset and in view of Cantor's result, we must have $card(\mathbb{N}) < card(\mathbb{R})$. The cardinal number of $\mathbb{N}$ is commonly denoted by $\aleph(\mathbb{N}) = \aleph_0$ while the cardinality of $\mathbb{R}$ is denoted by $\aleph(\mathbb{R}) = \mathfrak{c}$, where $\mathfrak{c}$ is called the cardinality of the *continuum*. We note that any set X with $\aleph(X) \leq \aleph_0$ is countable, while any set X with $\aleph(X) > \aleph_0$ is uncountable.

The reader needs to remember that cardinal numbers are *sets*, even in the finite case, but are closely related to the concept of *card* due to the properties C_1 and C_2. For this reason we can associate unique integers to finite sets, but need transfinite numbers for labeling infinite sets. For example, defining the symbols $\bar{0} = \varnothing$, $\bar{1} = \{\varnothing\}$, $\bar{2} = \{\varnothing, \{\varnothing\}\}$, $\bar{3} = \{\varnothing, \{\varnothing\}, \{\varnothing, \{\varnothing\}\}$, and so on, then $\aleph(\mathbb{N}_n) = \bar{n}$ is a set, while $card(\mathbb{N}_n) = n$. It also follows that $\aleph(\aleph(\mathbb{N}_n)) = \aleph(\bar{n}) = \bar{n}$.

Another consequence of Cantor's work is that $card(X) < card(2^X)$ for any set X. If $\aleph(X) = \aleph_0$, then the cardinal number of 2^X is denoted by $\aleph(2^X) = \aleph_1$. Thus, $\aleph_0 < \aleph_1$ and $\aleph_0 < \mathfrak{c}$. The question as to whether or not $\aleph_1 = \mathfrak{c}$ is known as the *continuum hypothesis*. Kurt Gödel proved that it is impossible to *disprove* the continuum hypothesis based on the Zermelo-Fraenckel axioms, with or without the axiom of choice, while Paul J. Cohen proved that it is impossible to *prove* the continuum hypothesis based on the Zermelo-Fraenckel axioms, with or without the axiom of choice [97, 56]. For our purposes we shall simply identify any set X with $\aleph(X) \geq \aleph_0$ with the infinity symbol ∞. This simplifies the problem of *measuring* the size of sets.

Definition 1.7 If U denotes the universal set of discourse, then the *counting function* $|\cdot| : 2^U \to \mathbb{N}_{[0,\infty]}$ is defined by

$$|A| = \begin{cases} card(A) & \text{if } A \text{ is finite} \\ \infty & \text{if } \aleph(A) \geq \aleph_0, \end{cases}$$

Consequently $0 \leq |A| \leq \infty \ \forall A \in 2^U$, and $|A| = 0 \Leftrightarrow A = \varnothing$, and all infinite sets map to the same symbol ∞ representing infinity.

If a set X is denumerable, then there exists a bijection $f : \mathbb{N} \to X$. Such a bijection provides a convenient tool for *counting* the elements of X by identifying each element of X with a unique numerical label determined by f. Specifically, for $x \in X$ we set $x = x_i \Leftrightarrow f(i) = x$. Since f is one-to-one, $x_i = x_j \Leftrightarrow i = j$. Since f is also onto, the set X can now be expressed as a sequence $X = \{x_1, x_2, x_3, \ldots\}$.

Example 1.4 Let $g : \mathbb{N}^2 \to \mathbb{N}$ be defined by

$$g(i, j) = \frac{1}{2}(i + j - 1)(i + j - 2) + j \quad \forall (i, j) \in \mathbb{N}^2.$$

By the definition of g we have that $g(1,1) = 1$, $g(2,1) = 2$, $g(1,2) = 3$, $g(3,1) = 4$, and so on. When $j = 1$ or $i = 1$ the function assumes the respective simple expression $g(i,1) = \frac{i}{2}(i-1)+1$ and $g(1, j) = \frac{j}{2}(j-1)+j$. These observations also make it an easy exercise to show that g is one-to-one and onto. Thus, g has an inverse $g^{-1} : \mathbb{N} \to \mathbb{N}^2$. This means that we have two equivalent ways of labeling the points $\mathbf{x} = (i, j)$ of $\mathbb{N}^2$. More specifically, we define $\mathbf{x}_\ell = \mathbf{x} \in \mathbb{N}^2 \Leftrightarrow g^{-1}(\ell) = \mathbf{x}$, or equivalently $\mathbf{x}_\ell = \mathbf{x} \Leftrightarrow g(\mathbf{x}) = \ell$. Figure 1.1 illustrates this particular labeling of the elements of $\mathbb{N}^2$.

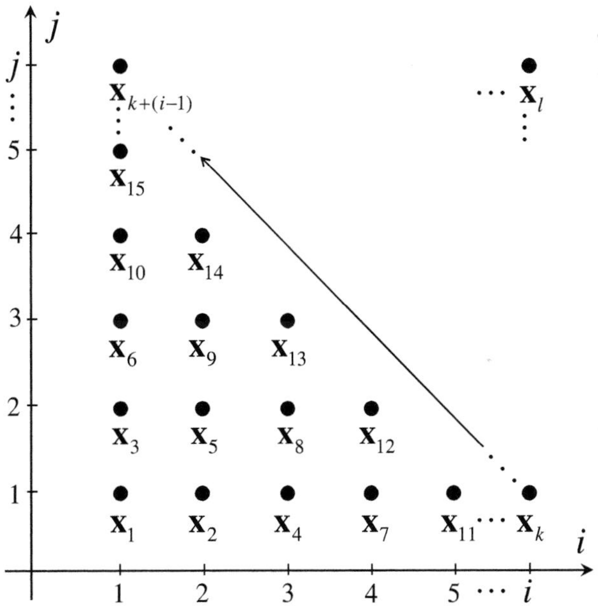

Figure 1.1 The sequential labeling of elements of $\mathbb{N}^2$. The ordering of elements of $\mathbb{N}^2$ is along the diagonals of slope -1 passing through the points of $\mathbb{N}^2$ as indicated by the arrow. Here $k = g(i,1)$, $k + (i-1) = g(1, j)$ with $i = j$, and $\ell = g(i, j)$.

Exercises 1.1.3

1. The solutions of equations of form $x^2 + bx + c = 0$, where $b, c \in \mathbb{Z}$, are called *quadratic integers*. Prove that the set of quadratic integers is countable.

2. Assuming Example 1.3 as fact, show that the interval $(0,1) = \{x \in \mathbb{R} : 0 < x < 1\}$ is uncountable.

1.2 ALGEBRAIC STRUCTURES

If one surveys the subjects of arithmetic, elementary algebra, or matrix theory, certain features stand out. One notes that these subjects deal with some given or derived set of objects, usually numbers or symbolic expressions, and with rules for combining these objects. Examples of these are the set of real numbers, the set of real valued functions on a set X, and the set of complex valued $n \times n$ square matrices with the usual rules of addition, subtraction, and multiplication. Moreover, one finds that there are some properties which these combining operations have in common: e.g. adding zero to any real number, adding the zero function to a function, or adding the zero matrix to a matrix does not change the value of the real number, the function, or the matrix, respectively. Other properties such as commutativity, do not always hold. Multiplication of square matrices is, in general, not a commutative operation.

The field of algebra aims at providing a fuller understanding of these subjects through a systematic study of typical algebraic structures. Such a study has the advantage of economy in that many superficially distinct structures are found to be basically the same, and hence open to a unified treatment. We begin our survey of algebraic structures by considering rules for combining elements of sets.

1.2.1 Operations on Sets

A *binary operation* on a set X can be thought of as a rule for combining any two elements of the set in order to produce another element of the set. More precisely, a *binary operation* on X is simply a function $\circ : X \times X \to X$. The evaluation $\circ(x,y)$ is commonly denoted by $x \circ y$ and is called the *resultant* of the operation. Accordingly, if $(x,y) \in X \times X$, then $x \circ y \in X$.

If $\circ$ is a binary operation on X and $Y \subset X$, then $\circ$ is a binary operation on Y if and only if for every pair $x, y \in Y$, $x \circ y \in Y$. Thus, if $\circ$ is also a binary operation on $Y \subset X$, then we say that Y is *closed* under the operation $\circ$. For instance, the operation $\circ = +$ is a binary operation on $\mathbb{Q} \subset \mathbb{R}$. However the operation $x \circ y = x^y$ is a binary operation on $\mathbb{R}$ but not on $\mathbb{Q}$. Therefore $\mathbb{Q}$ is not closed under the operation of exponentiation.

Some binary operations may satisfy special properties. Commutativity and associativity are the most important of these special properties. A binary operation $\circ$ on X is called *commutative* whenever $x \circ y = y \circ x \; \forall x, y \in X$, and *associative* whenever $(x \circ y) \circ z = x \circ (y \circ z) \; \forall x, y \in X$. For example, addition and multiplication are commutative and associative binary operations on $\mathbb{R}$ as well as on $\mathbb{R}^+$. However, the binary operation of division is not commutative on $\mathbb{R}^+$.

A set X is said to have an *identity* element with respect to a binary operation $\circ$ on X if there exists an element $e \in X$ with the property that $x \circ e = x = e \circ x \; \forall x \in X$.

The identity element for $\mathbb{R}$ with respect to addition is 0 since $0 + x = x + 0 = x \; \forall x \in \mathbb{R}$. Note that addition has no identity in the set $\mathbb{R}^+$. However, the identity with respect to the binary operation of multiplication on $\mathbb{R}^+$ is 1 since $1 \cdot x = x \cdot 1 = x \; \forall x \in \mathbb{R}^+$. The next theorem shows that identity elements are unique.

Theorem 1.6 *An identity element, if it exists, of a set X with respect to a binary operation $\circ$ on X is unique.*

The proof of this theorem is given in most introductory level books on modern algebra (e.g., [124, 293]).

If a set X has an identity element e with respect to a binary operation $\circ$, then an element $y \in X$ is called an *inverse* of $x \in X$ provided that $x \circ y = y \circ x = e$. Note that the set of all real-valued $n \times n$ matrices under matrix multiplication has a multiplicative identity, namely the $n \times n$ identity matrix. Nonetheless, not every $n \times n$ matrix has a multiplicative inverse.

The proof of the next theorem is similar to the proof of Theorem 1.6 and can also be found in [124, 293].

Theorem 1.7 *Let $\circ$ be a binary operation on a set X. The inverse with respect to the operation $\circ$ of $x \in X$, if it exists, is unique.*

Although not every binary operation on a set X provides for inverse elements, many operations provide for elements that behave almost like inverses. Obviously, if y is the inverse of $x \in X$ with respect to the operation $\circ$, then $x \circ y \circ x = x$ and $y \circ x \circ y = y$. Any element y satisfying the two conditions

$$x \circ y \circ x = x \quad \text{and} \quad y \circ x \circ y = y \tag{1.3}$$

is called a *pseudoinverse* of $x \in X$. Thus, every inverse is a pseudoinverse, but the converse does not necessarily hold.

Suppose that X is a set with two binary operations $\diamond$ and $\circ$. The operation $\circ$ is said to be *left distributive* with respect to $\diamond$ if

$$x \circ (y \diamond z) = (x \circ y) \diamond (x \circ z) \ \forall x, y, z \in X \tag{1.4}$$

and *right distributive* if

$$(y \diamond z) \circ x = (y \circ x) \diamond (z \circ x) \ \forall x, y, z \in X. \tag{1.5}$$

When both equations 1.4 and 1.5 hold, we simply say that $\circ$ is distributive with respect to $\diamond$. Note that the right members of equations 1.4 and 1.5 are equal whenever $\circ$ is commutative. Obviously, on $\mathbb{R}$, multiplication is distributive with respect to addition. However, division on $\mathbb{R}^+$ is not left distributive over addition; $(y+z)/x = (y/x)+(z/x)$ but $x/(y+z) \neq (x/y)+(x/z)$.

The remainder of this chapter consists of a brief survey of algebraic structures in terms of listing features of special abstract algebraic systems that are relevant in the study of lattice structures employed in constructing lattice-based neural networks.

Exercises 1.2.1

1. Prove Theorem 1.6 without using the listed references.
2. Prove Theorem 1.7 without using the listed references.
3. For each binary operation $\circ$ on $\mathbb{Q}$, decide whether the operation is commutative, associative, or both. Suppose (a) $x \circ y = (x \cdot y)/2$, (b) $x \circ y = x \cdot y + 1$, (c) $x \circ y = x - y$.

1.2.2 Semigroups and Groups

We begin our discussion with the simplest algebraic structure, namely that of a groupoid.

Definition 1.8 A *groupoid* is a set X together with a binary operation $\circ$ on X. If the operation $\circ$ is also associative, then the groupoid is called a *semigroup*.

To be completely precise in denoting a semigroup, we should use some symbolism such as $(X, \circ, =)$, which specifies the set of elements, the binary relation, and the equality relation used to specify the equality of elements; i.e., $x \circ (y \circ z) = (x \circ y) \circ z$. However, it is customary to use either the pair $(X, \circ)$ or simply the letter designation of the set of elements, in this case X, as a designation of the groupoid or semigroup, provided there is no danger of confusion as to the notation being used for binary composition. Furthermore, algebraists, as a rule, do not use a special symbol such as $\circ$ to denote a binary operation different from the usual addition or multiplication. They stick with the conventional additive or multiplicative notation and even call these operations *addition* and *multiplication*. We follow this convention to some extent by using the notation $x+y$ when viewing $x \circ y$ as an additive operation and $x \cdot y$ or $x \times y$, or even xy, when viewing $x \circ y$ as a multiplicative operation. There is a sort of gentlemen's agreement that the zero symbol 0 is used to denote an additive identity and the symbol 1 to denote a multiplicative identity, even though they may not actually be denoting the integers 0 and 1. Of course, if a person is also talking about numbers at the same time, other symbols are used to denote these identities in order to avoid confusion.

To the uninitiated, semigroups may seem too poor in properties to be of much interest. However, the set of $n \times n$ square matrices under matrix multiplication forms a semigroup. Anyone who has had experience with matrix theory is well-aware that this system, far from being too poor in properties to be of interest, is, indeed, extremely rich in properties. Research into the fascinating ramifications of matrix theory has provided the stimulus to a great deal of mathematical development and remains an active and growing branch of mathematics.

The set of $n \times n$ square matrices under matrix multiplication has the additional property of having a multiplicative identity. This leads to the next definition:

Definition 1.9 A *monoid* is a semigroup with identity.

Example 1.5 The set of positive integers, $\mathbb{N}$, together with the operation of addition is not a monoid. However, $(\mathbb{N}, +)$ is a semigroup. The system $(\mathbb{N}, \cdot)$ is a monoid with identity the integer 1.

Of the various possible algebraic systems having a single associative operation, the system known as a *group* has been by far the most extensively studied. Also, the theory of groups is one of the oldest parts of abstract algebra, as well as one rich in applications.

Definition 1.10 A *group* is a monoid with the property that each element has an inverse.

It is customary to denote the inverse of an element x in a group G by x^{-1} if multiplicative notation is used, and by $-x$ if additive notation is used.

Recalling the definition of a monoid, one can define a group alternatively as a set G together with a binary operation, say $(G, \circ)$, such that:

G_1. the operation $\circ$ is associative, i.e., $\forall x, y, z \in G$, $x \circ (y \circ z) = (x \circ y) \circ z$,

G_2. There exists an identity $1 \in G$ such that $\forall x \in G$, $x \circ 1 = 1 \circ x = x$, and

G_3. $\forall x \in G$, $\exists$ an inverse element $x^{-1} \in G$, such that $x \circ x^{-1} = x^{-1} \circ x = 1$.

If $(G, \circ)$ is a group and $H \subset G$, then H is called a *subgroup* of G if and only if $(H, \circ)$ is a group. Also, in addition to the above listed three properties the operation is commutative, then the group G is called an *abelian* group.

Example 1.6 The set $\mathbb{N}_0$ with the operation of addition is not a group. There is an identity element 0, but no integers greater than 0 have inverses in $\mathbb{N}_0$. However, $(\mathbb{Z}, +)$ is a group and this group is abelian.

A common activity among scientists and engineers is to solve problems. Often these problems lead to equations involving some unknown number or quantity x which is to be determined. The simplest equations are the linear ones of the forms $a + x = b$ for the operation of addition, and $a \cdot x = b$ for multiplication. Equations of the form $a \cdot x = b$ are in general not solvable in the monoid $(\mathbb{N}, \cdot)$. For instance, $2 \cdot x = 3$ has a solution $x = 3/2$, which is not an integer. However, the equations of form $a \cdot x = b$ are always solvable in the structure $(\mathbb{R}^+, \cdot)$. The reason for this is that $(\mathbb{R}^+, \cdot)$ is a group. As the next theorem shows, the properties necessary to solve linear equations within a system are precisely the properties of a group.

Theorem 1.8 *If $(G, \circ)$ is a group and $a, b \in G$, then the linear equations $a \circ x = b$ and $y \circ a = b$ have unique solutions in G.*

Proofs of this theorem can be found in most textbooks on abstract algebra; e.g., [293, 124]. It is important to note that the solutions $x = a^{-1} \circ b$ and $y = b \circ a^{-1}$ need not be the same unless the group is abelian. However, the equation $a \circ x \circ b = c$ has a unique solution in any group $(G, \circ)$, namely $x = a^{-1} \circ c \circ b^{-1}$. The equation $a \circ x \circ b = c$ is called a *group translation* of x. The terminology stems from the following definition:

Definition 1.11 If $(G, \circ)$ is a group, then a function $\psi : G \to G$ is called a *group translation* if $\psi(x) = a \circ x \circ b \; \forall x \in G$, where $a, b \in G$ are constants. If $(G, \circ)$ is only a semigroup, then ψ is called a *semigroup translation.*

Exercises 1.2.2

1. Give examples other than those in the text of semigroups and monoids that are not groups.
2. Prove the following Proposition: If G is a semigroup, then G is a group $\iff$ the following two conditions are satisfied, (1) there exists an element $i \in G \ni i \circ x = x \; \forall x \in G$ and (2) $\forall x \in G \; \exists x^{-1} \in G \ni x^{-1} \circ x = i$.
3. Prove Theorem 1.8.
4. Suppose G is a group and $\emptyset \neq H \subset G$. Prove that H is a subgroup of G if and only if $\forall x, y \in H$, $x^{-1} \circ y \in H$.
5. Prove that $(\mathbb{R}^n, +)$ is an abelian group.

1.2.3 Rings and Fields

The structures we have considered thus far have been concerned with sets with a single binary operation. Our earliest experience with arithmetic, however, has taught us the use of two distinct binary operations on sets of numbers, namely addition and multiplication. This early and important experience should indicate that a study of sets on which two binary operations have been defined is of great importance. Modeling our definition on properties common to these number systems, as well as such structures as the set of all $n \times n$ matrices with elements in one of the number systems, or the set of all polynomials with coefficients in, say, the set of all integers, we now define a type of algebraic structure known as a ring.

Definition 1.12 A *ring* $(R, \diamond, \circ)$ is a set R together with two binary operations $\diamond$ and $\circ$ of addition and multiplication, respectively, defined on R such that the following three conditions are satisfied:

R_1. $(R, \diamond)$ is an abelian group.

R_2. $(R, \circ)$ is a semigroup.

R_3. $\forall a,b,c \in R$, $a \circ (b \diamond c) = (a \circ b) \diamond (a \circ c)$ and $(a \diamond b) \circ c = (a \circ c) \diamond (b \circ c)$.

If condition R_1 of this definition is weakened to $(R, \diamond)$ is a commutative semigroup, then R is called a *semiring*.

Of the many examples of rings which come readily to mind from experience with common systems of elementary mathematics, the most natural is, perhaps, the ring $\mathbb{Z}$ of integers with the usual addition and multiplication. However, if we examine the properties of the ring $(\mathbb{Z}, +, \times)$ of integers, where $+$ and $\times$ correspond to $\diamond$ and $\circ$, we note that it has properties not enjoyed by rings in general. Among these properties are:

R_4. The existence of an identity element for the operation $\circ$, called the *unit* element and usually denoted by 1 or I.

R_5. The commutativity of the operation $\circ$.

R_6. The nonexistence of an element $a \neq 0$ such that for some positive integer n, $na := \underbrace{a \diamond a \cdots \diamond a}_{n} = 0$.

On the other hand, the integers themselves fail to possess a most useful property, namely that:

R_7. For every nonzero $a \in R$ there is an element in R, denoted by a^{-1}, such that $a \circ a^{-1} = a^{-1} \circ a = 1$; i.e., $(R \setminus \{0\}, \circ)$ is a group.

In fact, $(\mathbb{Z}, +, \times)$ also fails to have the slightly weaker property:

R_8. For every nonzero $a \in R$ there is an element in R, denoted by $\widetilde{a}$, such that $a \circ \widetilde{a} \circ a = a$ and $\widetilde{a} \circ a \circ \widetilde{a} = \widetilde{a}$; i.e., every nonzero element has a pseudoinverse.

These properties lead to some further definitions.

Definition 1.13 If a ring satisfies property R_4, then it is called a *ring with unity*. If a ring satisfies property R_5, it is called *commutative* or *abelian*. If a ring satisfies property R_6, it is said to have *characteristic zero*. If it satisfies R_7, it is called a *division ring* or a *quasi-field*. A *field* is a commutative division ring. A ring which satisfies property R_8 is called a *von Neumann ring*. A ring R is called a *Boolean ring* if for all $a \in R$, $a^2 = a$.

As we noted, $(\mathbb{Z}, +, \times)$ is a commutative ring but not a division ring. On the other hand, $(\mathbb{R}, +, \times)$ is an example of a commutative division ring and, hence, a field. It is also an example of a ring with unity and characteristic zero as well as a *totally ordered field*. This means that not only is the set $\mathbb{R}$ totally ordered by the natural order of $\leq$, but also that the field operations $+$ and $\times$ are compatible with the order in that

1. $\forall x, y, z \in \mathbb{R}$, if $x \leq y$, then $x + z \leq y + z$ (preservation of order under $+$)
2. $\forall x, y \in \mathbb{R}$, if $0 \leq x$ and $0 \leq y$, then $0 \leq x \times y$ (preservation of order under $\times$)

Other well-known examples of fields are $(\mathbb{Q}, +, \times)$ and $(\mathbb{C}, +, \times)$. Not all rings are necessarily rings of different sets of numbers.

Example 1.7 Consider the set $\mathbb{R}^X$ of real valued functions on a set X. Addition and multiplication of two functions $f, g \in \mathbb{R}^X$ are defined by

1. $(f + g) : X \to \mathbb{R}$ by $(f + g)(x) = f(x) + g(x)\ \forall x \in X$, and
2. $(f \times g) : X \to \mathbb{R}$ by $(f \times g)(x) = f(x) \times g(x)\ \forall x \in X$.

The *zero* function $\phi \in \mathbb{R}^X$ is defined by $\phi(x) = 0\ \forall x \in X$, and the *unit* function $I \in \mathbb{R}^X$ by $I(x) = 1\ \forall x \in X$. For each $f \in \mathbb{R}^X$, define its inverse $-f \in \mathbb{R}^X$ by $(-f)(x) = -f(x)$. Then

$$(f + (-f))(x) = f(x) + (-f)(x) = f(x) - f(x) = 0 = \phi(x)\ \forall x \in X,$$

and, hence, $(f + (-f)) = \phi$. Also,

$$(f + g)(x) = f(x) + g(x) = g(x) + f(x) = (g + f)(x)\ \forall x \in X.$$

Therefore, addition on $\mathbb{R}^X$ is commutative. It is just as easy to show that $f \times g = g \times f$ and that

$$(f + g) + h = f + (g + h), \quad (f \times g) \times h = f \times (g \times h),$$
$$f \times (g + h) = (f \times g) + (f \times h), \quad \text{and} \quad (f + g) \times h = (f \times h) + (g \times h).$$

Thus, $(\mathbb{R}^X, +)$ is an abelian group, $(\mathbb{R}^X, \times)$ is a commutative semigroup with unity, and axiom R_3 is also satisfied. Therefore, $(\mathbb{R}^X, +, \times)$ is an abelian ring with unity. However, $(\mathbb{R}^X, +, \times)$ is not a field, since not every $f \in \mathbb{R}^X$ with $f \neq \phi$ has a multiplicative inverse. We cannot define $f^{-1}(x) = 1/f(x)$ since it may be possible that $f(x) = 0$ for *some* $x \in X$ but not all $x \in X$. Nevertheless, if $f \neq \phi$, then the function $\widetilde{f}$, defined by $\widetilde{f}(x) = 1/f(x)$ whenever $f(x) \neq 0$ and $\widetilde{f}(x) = f(x)$ when $f(x) = 0$, is a pseudoinverse since $f \times \widetilde{f} \times f = f$ and $\widetilde{f} \times f \times \widetilde{f} = \widetilde{f}$. Therefore, $(\mathbb{R}^X, +, \times)$ is a von Neumann ring.

All the rings considered thus far were commutative rings. However, non-commutative rings play an important role in a wide variety of applications. We present the most pertinent example of such a ring.

Example 1.8 Let $\mathbb{F}$ be any field, say $\mathbb{Q}$, $\mathbb{R}$, or $\mathbb{C}$, and consider the set $M_{2\times2}(\mathbb{F})$ of all 2×2 matrices of form

$$(a_{ij}) = \begin{pmatrix} a_{11} & a_{12} \\ a_{21} & a_{22} \end{pmatrix},$$

where the a_{ij}'s are all in $\mathbb{F}$. The set $M_{2\times2}(\mathbb{F})$ of all $n \times n$ matrices over $\mathbb{F}$ is similarly defined.

Matrix addition on $M_{2\times2}(\mathbb{F})$ is defined by

$$\begin{pmatrix} a_{11} & a_{12} \\ a_{21} & a_{22} \end{pmatrix} + \begin{pmatrix} b_{11} & b_{12} \\ b_{21} & b_{22} \end{pmatrix} = \begin{pmatrix} a_{11}+b_{11} & a_{12}+b_{12} \\ a_{21}+b_{21} & a_{22}+b_{22} \end{pmatrix},$$

that is, by the corresponding entries using addition in $\mathbb{F}$. Obviously, $(M_{2\times2}(\mathbb{F}), +)$ is an abelian group with additive identity

$$\begin{pmatrix} 0 & 0 \\ 0 & 0 \end{pmatrix},$$

and with additive inverse

$$-\begin{pmatrix} a_{11} & a_{12} \\ a_{21} & a_{22} \end{pmatrix} = \begin{pmatrix} -a_{11} & -a_{12} \\ -a_{21} & -a_{22} \end{pmatrix}.$$

Matrix multiplication on $M_{2\times2}(\mathbb{F})$ is defined by

$$\begin{pmatrix} a_{11} & a_{12} \\ a_{21} & a_{22} \end{pmatrix} \times \begin{pmatrix} b_{11} & b_{12} \\ b_{21} & b_{22} \end{pmatrix} = \begin{pmatrix} a_{11}\cdot b_{11}+a_{12}\cdot b_{21} & a_{11}\cdot b_{12}+a_{12}\cdot b_{22} \\ a_{21}\cdot b_{11}+a_{22}\cdot b_{21} & a_{21}\cdot b_{12}+a_{22}\cdot b_{22} \end{pmatrix}.$$

If $\mathbb{F}$ equals $\mathbb{R}$ or $\mathbb{C}$, then this multiplication corresponds, of course, to the common matrix products over these fields and can best be remembered by $(a_{ij})(b_{ij}) = (c_{ij})$, where $c_{ij} = \sum_{k=1}^{2} a_{ik}b_{kj}$. Of course, the analogous definition holds for matrix multiplication in which the sum goes from $k = 1$ to n. In short, everything we said about the system $(M_{2\times 2}(\mathbb{F}), +, \times)$ is also valid for the system $(M_{n\times n}(\mathbb{F}), +, \times)$.

To show that $(M_{n\times n}(\mathbb{F}), +, \times)$ is a ring, the associative and distributive laws remain to be proven. Using the field properties of $\mathbb{F}$ and the definition of matrix multiplication in the structure $(M_{n\times n}(\mathbb{F}), +, \times)$, then if d_{rs} denotes the entry in the rth row and sth column of $(a_{ij})[(b_{ij})(c_{ij})]$, we have

$$d_{rs} = \sum_{k=1}^{n} a_{rk}\Big(\sum_{j=1}^{n} b_{kj}c_{js}\Big) = \sum_{j=1}^{n}\Big(\sum_{k=1}^{n} a_{rk}b_{kj}\Big)c_{js} = e_{rs},$$

where e_{rs} denotes the entry in the rth row and sth column of $[(a_{ij})(b_{ij})](c_{ij})$. The distributive property is proved in a similar fashion.

This example proves the following theorem:

Theorem 1.9 *If $\mathbb{F}$ is a field, then the set $M_{n\times n}(\mathbb{F})$ of all $n\times n$ matrices with entries from $\mathbb{F}$ forms a ring under matrix addition and matrix multiplication.*

The rings of matrices over a field $\mathbb{F}$ are an important tool in the theory and practice of various scientific and engineering endeavors. On the downside, we need to point out that the ring $(M_{n\times n}(\mathbb{F}), +, \times)$ lacks some important algebraic properties. Since matrix multiplication is not commutative, $M_{n\times n}(\mathbb{F})$ is not a commutative ring. Additionally, one of the most important properties of the real number system is that the product of two numbers can only be zero if at least one of the factors is zero. The working engineer or scientist uses this fact constantly, perhaps without realizing it. Suppose for example that one needs to solve the equation $2x^2 + 9x - 5 = 0$. The first thing one usually does is to factor the left side: $2x^2 + 9x - 5 = (2x-1)(x+5)$. One then concludes that the only possible values for x are $1/2$ and -5 because the resulting product is zero if and only if one of the factors $2x-1$ or $x+5$ is zero. The property that if a product equals zero, then at least one of the factors must also equal zero does not hold for rings in general. For instance, the definition of the matrix product in $M_{2\times 2}(\mathbb{F})$ shows that

$$\begin{pmatrix} 0 & 0 \\ 0 & 1 \end{pmatrix} \times \begin{pmatrix} 0 & 1 \\ 0 & 0 \end{pmatrix} = \begin{pmatrix} 0 & 0 \\ 0 & 0 \end{pmatrix}.$$

Exercises 1.2.3

1. Prove whether or not $X = \{2x : x \in \mathbb{Z}\}$ is a ring.

2. Prove whether or not $X = \{2x+1 : x \in \mathbb{Z}\}$ is a ring.

3. Prove that $(\mathbb{Q}, +, \times)$ is a ring. Is it a division ring?

4. Suppose $X = \{(w,x,y,z) : w,x,y,z \in \mathbb{Q}\}$ is endowed with a binary operation of addition and a binary operation of multiplication defined by

$$(w,x,y,z)+(s,t,u,v) = (w+s, x+t, y+u, z+v) \text{ and}$$

$$(w,x,y,z)\cdot(s,t,u,v) = (w\cdot s + x\cdot u, w\cdot t + x\cdot v, y\cdot s + z\cdot u, y\cdot t + z\cdot v)$$

$\forall\ (w,x,y,z),(s,t,u,v) \in X$. Prove that X is a ring.

5. Prove that the set $X = \{x + y\cdot 3^{1/3} + z\cdot 9^{1/3} : x,y,z \in \mathbb{Q}\}$ is a ring.

6. Determine if the set X of Exercise 5 is a field.

7. Prove that $(\mathbb{R}, +, \times)$ is a field.

8. Prove that $(-1)\cdot(-1) = 1$.

9. Show that every Boolean ring is commutative.

10. Let 2^X denote the set of all subset of a set X. Define two binary operations $+$ and $\cdot$ on 2^X by $A+B = (A\cup B)\setminus(A\cap B)$ and $A\cdot B = A\cap B$ for $A,B \in 2^X$. Prove that $(2^X,+,\cdot)$ is a Boolean ring.

1.2.4 Vector Spaces

The theory of solutions of systems of linear equations is part of a more inclusive theory of an algebraic structure known as a *vector space*. As we assume the reader's acquaintance with this topic, our treatment of vector spaces will be very brief indeed, designed only as a recall of a few basic concepts and theorems. These basic concepts and theorems have their counterpart in the algebraic structure of lattice vector spaces.

Although vector space theory, as covered in elementary linear algebra courses, is usually concerned with the Euclidean vector spaces $\mathbb{R}^n$, the operations of vector addition and scalar multiplication are used in many diverse contexts in mathematics and engineering. Regardless of the context, however, these operations obey the same set of arithmetic rules. In the general case, the scalars are elements of some field, which may be different than the real numbers.

Definition 1.14 A *vector space* $\mathbb{V}$ *over a field* $(\mathbb{F}, \diamond, \circ)$, denoted by $\mathbb{V}(\mathbb{F})$, is an "additive" abelian group $\mathbb{V}$ together with an operation called *scalar multiplication* of each element of $\mathbb{V}$ by each element of $\mathbb{F}$ on the left, such that $\forall \alpha, \beta \in \mathbb{F}$ and $\mathbf{v}, \mathbf{w} \in \mathbb{V}$ the following five conditions are satisfied:

V_1. $\alpha \circ \mathbf{v} \in \mathbb{V}$

V_2. $\alpha \circ (\beta \circ \mathbf{v}) = (\alpha \circ \beta) \circ \mathbf{v}$

V_3. $(\alpha \diamond \beta) \circ \mathbf{v} = (\alpha \circ \mathbf{v}) \diamond (\beta \circ \mathbf{v})$

V_4. $\alpha \circ (\mathbf{v} \diamond \mathbf{w}) = (\alpha \circ \mathbf{v}) \diamond (\alpha \circ \mathbf{w})$

V_5. $1 \circ \mathbf{v} = \mathbf{v}$

The elements of $\mathbb{V}$ are called *vectors* and the elements of $\mathbb{F}$ are called *scalars*.

If the field $\mathbb{F}$ of scalars is clear from the context of discussion, then it is customary to use the symbol $\mathbb{V}$ instead of $\mathbb{V}(\mathbb{F})$. We also note that in comparison to the algebraic systems discussed thus far, scalar multiplication for a vector space is not a binary operation on a set, but a rule which associates an element α from $\mathbb{F}$ and an element $\mathbf{v}$ from $\mathbb{V}$ with an element $\alpha \circ \mathbf{v}$ of $\mathbb{V}$.

Example 1.9 It follows from Example 1.8 that for any field $\mathbb{F}$, $M_{n \times n}(\mathbb{F})$ is an additive abelian group. It is also easy to see that if one defines scalar multiplication by

$$\alpha \circ (a_{ij}) = (\alpha \circ a_{ij}) \quad \forall \alpha \in \mathbb{F} \quad \text{and} \quad \forall (a_{ij}) \in M_{n \times n}(\mathbb{F}).$$

Obviously, scalar multiplication is well-defined since both α and a_{ij} are elements of $\mathbb{F}$ and, hence, $\alpha \circ a_{ij} \in \mathbb{F}$. Thus, axioms V_1 through V_5 of Definition 1.14 will hold, proving that $M_{n \times n}(\mathbb{F})$ is a vector space over $\mathbb{F}$.

One of the most important concepts in the theory of vector spaces is the notion of linear independence. To be in accord with the standard notation used in vector space theory, we shall let $+ = \diamond$ and $\times = \circ$ for the remainder of this chapter

Definition 1.15 Let $\mathbb{V}(\mathbb{F})$ be a vector space and let $S = \{\mathbf{v}_1, \mathbf{v}_2, \dots, \mathbf{v}_k\}$ be a subset of $\mathbb{V}$. If for every combination of scalars $\alpha_1, \alpha_2, \dots, \alpha_k$

$$\alpha_1 \cdot \mathbf{v}_1 + \alpha_2 \cdot \mathbf{v}_2 + \cdots + \alpha_k \cdot \mathbf{v}_k = \mathbf{0} \quad \Rightarrow \quad \alpha_i = 0 \quad \text{for} \quad i = 1, 2, \dots, k, \quad (1.6)$$

then the vectors in S are said to be *linearly independent* over $\mathbb{F}$. In this definition, $\mathbf{0}$ denotes the zero in $\mathbb{V}$ and 0 denotes the zero in $\mathbb{F}$. If the vectors are not linearly independent over $\mathbb{F}$, then they are *linearly dependent* over $\mathbb{F}$.

Note that if the vectors are linearly dependent over $\mathbb{F}$, then for some combination of scalars $\alpha_1, \alpha_2, \ldots, \alpha_k$, $\exists\ \alpha_i \neq 0$ for at least one $i \in \{1, 2, \ldots, k\}$ such that

$$\alpha_1 \cdot \mathbf{v}_1 + \alpha_2 \cdot \mathbf{v}_2 + \cdots + \alpha_k \cdot \mathbf{v}_k = \mathbf{0}. \tag{1.7}$$

In this case we can solve for $\mathbf{v}_i$:

$$\mathbf{v}_i = \sum_{j \neq i} \beta_j \mathbf{v}_j, \quad \text{where} \quad \beta_j = -\frac{\alpha_j}{\alpha_i}. \tag{1.8}$$

That is, $\mathbf{v}_i$ can be expressed as a linear combination of the remaining vectors; i.e., $\mathbf{v}_i$ *depends* on the remaining $\mathbf{v}_j$'s.

If $\mathbb{V}$ is a vector space, then certain subsets of $\mathbb{V}$ themselves form vector spaces under the vector addition and scalar multiplication defined on $\mathbb{V}$. These vector spaces are called *subspaces* of $\mathbb{V}$. For example, the set $\mathbb{W} = \{(x,y) \in \mathbb{R}^2 : y = 2x\}$ is a subspace of $\mathbb{R}^2$. Obviously, if $(x,y) \in \mathbb{W}$, then defining $u = \alpha x$ and $v = \alpha y$, we see that $v = \alpha y = \alpha(2x) = 2(\alpha x) = 2u$ for any real number α. Thus, $\alpha(x,y) = (\alpha x, \alpha y) = (u, 2u) \in \mathbb{W}$ and axiom V_1 of Definition 1.20 is satisfied. The remaining four vector space axioms are just as easily verified. However, in order to show that a subset $\mathbb{W}$ is a subspace of a vector space $\mathbb{V}$, it is not necessary to verify axioms V_2 through V_5. The following theorem, which we state without proof, shows that in addition axiom V_1, all we need to show is that $\mathbb{W}$ is closed under vector addition.

Theorem 1.10 *If $\mathbb{W}$ is a non-empty subset of a vector space $\mathbb{V}$ over $\mathbb{F}$, then $\mathbb{W}$ is a subspace of $\mathbb{V}$ $\Leftrightarrow$ the following conditions hold:*

1. *if $\mathbf{v}, \mathbf{w} \in \mathbb{W}$, then $\mathbf{v} + \mathbf{w} \in \mathbb{W}$, and*

2. *if $\alpha \in \mathbb{F}$ and $\mathbf{w} \in \mathbb{W}$, then $\alpha\mathbf{w} \in \mathbb{W}$.*

Let $\mathbb{W} = \{\mathbf{v}_1, \mathbf{v}_2, \ldots, \mathbf{v}_k\}$ be a subset of a vector space $\mathbb{V}$ and let $S(\mathbb{W})$ denote the set of all linear combinations of the vectors $\mathbf{v}_1, \mathbf{v}_2, \ldots, \mathbf{v}_k$. That is,

$$S(\mathbb{W}) = \{\mathbf{v} : \mathbf{v} = \sum_{i=1}^{k} \alpha_i \mathbf{v}_i\}. \tag{1.9}$$

Thus, if $\mathbf{v}, \mathbf{w} \in S(\mathbb{W})$, with $\mathbf{v} = \sum_{i=1}^{k} \alpha_i \mathbf{v}_i$ and $\mathbf{w} = \sum_{i=1}^{k} \beta_i \mathbf{v}_i$, then $\forall \alpha \in \mathbb{F}$, $\alpha\mathbf{v} = \sum_{i=1}^{k} \gamma_i \mathbf{v}_i$, where $\gamma_i = \alpha\alpha_i$, and $\mathbf{v} + \mathbf{w} = \sum_{i=1}^{k} \delta_i \mathbf{v}_i$, where $\delta_i = \alpha_i + \beta_i$. Therefore, $\alpha\mathbf{v}$ and $\mathbf{v} + \mathbf{w}$ are in $S(\mathbb{W})$ and by Theorem 1.10, $S(\mathbb{W})$ is a subspace of $\mathbb{V}$. The subspace $S(\mathbb{W})$ is a of $\mathbb{V}$ is said to be *spanned* by $\mathbb{W}$ and the vectors of $\mathbb{W}$ are called the *generators* of the vector space $S(\mathbb{W})$. If $S(\mathbb{W}) = \mathbb{V}$, then we say that $\mathbb{W}$ is a *span* for $\mathbb{V}$. In particular, if $\mathbb{W}$ is a span for $\mathbb{V}$, then every vector $\mathbf{v} \in \mathbb{V}$ is a linear combination of the vectors in $\mathbb{W}$.

Closely related to the notion of a span are the concepts of *basis* and *dimension*. To students of science, the concept of dimension is a natural one. They usually think of a line as one-dimensional, a plane as two-dimensional, and the space around them as three-dimensional. The following definition makes these concepts more precise.

Definition 1.16 A *basis* for a vector space $\mathbb{V}$ is a linearly independent set of vectors of $\mathbb{V}$ which spans $\mathbb{V}$.

We remind the reader that a basis for a vector space is merely one of many kinds of bases encountered in the study of various mathematical systems. A basis for a vector space portrays the algebraic properties of the vector space and is intimately connected to the *linear* algebraic properties of the space while a basis for a topological space provides for the basic geometric properties of the space. Once a basis for a mathematical system has been established, we may proceed to describe the properties of the system under investigation relative to that basis. The most important facts about a vector space basis can be summarized as follows. Let $\mathbb{V}$ be a vector space and $B = \{\mathbf{v}_1, \mathbf{v}_2, \ldots, \mathbf{v}_k\} \subset \mathbb{V}$.

1. If B is a basis for $\mathbb{V}$, then every vector of $\mathbb{V}$ can be *uniquely* expressed as a linear combination of the elements of B.

2. If every vector of $\mathbb{V}$ can be uniquely expressed as a linear combination of the elements of B, then B is a basis for $\mathbb{V}$.

3. If B is a basis for $\mathbb{V}$ and $C \subset \mathbb{V}$ with $card(C) > card(B)$, then C is not linearly independent.

Statement 2 follows immediately from the observation that the zero vector has a unique representation in terms of the $\mathbf{v}_i$'s. Thus, the $\mathbf{v}_i$'s must be linearly independent. Statement 3 implies that the number of elements in a basis of a vector space must be unique; that is, *every* basis of a given vector space has the same number of elements. Thus, if B is a basis and $card(C) > card(B)$, then C will also span $\mathbb{V}$ but C cannot be a basis for $\mathbb{V}$. With this in mind, we define the *dimension* of a vector space $\mathbb{V}$ as the cardinality of any basis for $\mathbb{V}$ and use the notation $dim(\mathbb{V})$ to denote the dimension of $\mathbb{V}$.

We conclude this section with the following pertinent examples of vector spaces.

Example 1.10

1. For $i = 1, 2, \ldots, n$, let $\mathbf{e}^i = (e^i_1, e^i_2, \ldots, e^i_n) \in \mathbb{R}^n$ be defined by $e^i_j = 1$ if $j = i$, and $e^i_j = 0$ if $j \neq i$. Then $\mathfrak{E} = \{\mathbf{e}^i : i = 1, 2, \ldots, n\}$ is a basis for $\mathbb{R}^n$ since any vector $\mathbf{x} = (x_1, x_2, \ldots, x_n) \in \mathbb{R}^n$ can be uniquely expressed as $\mathbf{x} = x_1\mathbf{e}^1 + x_2\mathbf{e}^2 + \cdots + x_n\mathbf{e}^n$. The set $\mathfrak{E}$ is called the *standard basis* for $\mathbb{R}^n$. The dimension of $\mathbb{R}^n$ is n.

2. Let

$$\mathbf{v}_1 = \begin{pmatrix} 1 & 0 \\ 0 & 0 \end{pmatrix}, \mathbf{v}_2 = \begin{pmatrix} 0 & 1 \\ 0 & 0 \end{pmatrix}, \mathbf{v}_3 = \begin{pmatrix} 0 & 0 \\ 1 & 0 \end{pmatrix}, \text{ and } \mathbf{v}_4 = \begin{pmatrix} 0 & 0 \\ 0 & 1 \end{pmatrix}.$$

Then $B = \{\mathbf{v}_1, \mathbf{v}_2, \mathbf{v}_3, \mathbf{v}_4\}$ is a basis for $M_{2\times2}(\mathbb{R})$. Any vector $\mathbf{v} = \begin{pmatrix} a & b \\ c & d \end{pmatrix}$ can be uniquely written as $\mathbf{v} = a\mathbf{v}_1 + b\mathbf{v}_2 + c\mathbf{v}_3 + d\mathbf{v}_4$. Thus, $dim(M_{2\times2}(\mathbb{R})) = 4$.

Exercises 1.2.4

1. Show that if $B = \{\mathbf{v}_1, \mathbf{v}_2, \ldots, \mathbf{v}_n\}$ is a basis of the vector space $\mathbb{V}$ over the field $\mathbb{F}$ and $C = \{\mathbf{w}_1, \mathbf{w}_2, \ldots, \mathbf{w}_k\}$ is a linearly independent set of vectors of $\mathbb{V}$, then $k \leq n$.

2. Show that if $B = \{\mathbf{v}_1, \mathbf{v}_2, \ldots, \mathbf{v}_n\}$ is a basis of the vector space $\mathbb{V}$ over the field $\mathbb{F}$, then any other basis for $\mathbb{V}$ has the same number of vectors as B.

3. Suppose $dim(\mathbb{V}) = n$ and $U = \{\mathbf{w}_1, \mathbf{w}_2, \ldots, \mathbf{w}_k\}$ is a basis of $\mathbb{W} \subset \mathbb{V}$, where $k < n$. Prove that if $\mathbf{v} \in \mathbb{V}$ and $\mathbf{v} \notin \mathbb{W}$, then $U \cup \{\mathbf{v}\}$ is a linearly independent set.

1.2.5 Homomorphisms and Linear Transforms

Throughout this book we deal with various kind of mathematical systems. The names "abstract mathematical system" or "algebraic system" are used to describe any well-defined collection of mathematical objects consisting, for example, of a set together with relations and operations on the set, and a collection of postulates, definitions, and theorems describing various properties of the structure.

It is a fundamentally important fact that even when these systems have very little structure, such as semigroups or groups, it is often possible to classify them according to whether or not they are mathematically similar or equivalent. These notions are made mathematically precise by the *morphism* relationship between abstract systems.

Definition 1.17 Let $G = (G, \star)$ and $\mathcal{G} = (\mathcal{G}, \circ)$ denote two algebraic systems with binary operations $\star$ and $\circ$, respectively. A *homomorphism* from G to $\mathcal{G}$ is a function $\psi : G \to \mathcal{G}$ such that for each $g, h \in G$,

$$\psi(g \star h) = \psi(g) \circ \psi(h). \tag{1.10}$$

Thus, a homomorphism is required to preserve the operations of the systems; i.e., performing the operation $g \star h$ in G and then applying the function ψ to the result is the same as first applying ψ to each g and h and then performing the operation $\psi(g) \circ \psi(h)$ in $\mathcal{G}$. If such a function exists, then the two systems are said to be *homomorphic*.

By definition, a homomorphism need not be a one-to-one correspondence (injection) between the elements of G and $\mathcal{G}$. Onto functions (surjections) that are also one-to-one are called *bijections* and lead to the extremely important concept of an isomorphism.

Definition 1.18 Let $G = (G, \star)$ and $\mathcal{G} = (\mathcal{G}, \circ)$ denote two systems. An *isomorphism* from G to $\mathcal{G}$ is a homomorphism $\psi : G \to \mathcal{G}$ that is a bijection.

If such a homomorphism exists, then we say that the two systems are *isomorphic* and denote this by $G \approx \mathcal{G}$. Hence the idea that the two systems G and $\mathcal{G}$ are isomorphic means that they are identical except for the names of the elements and operations. That is, we can obtain $\mathcal{G}$ from G by renaming an element g in G with a certain element $\mathfrak{g}$ in $\mathcal{G}$, namely $\mathfrak{g} = \psi(g)$, and by renaming the operation $\star$ as $\circ$. Then the counterpart of $g \star h$ will be $\mathfrak{g} \circ \mathfrak{h}$. The next theorem, proved in [293], is very obvious if we consider an isomorphism to be a renaming of one system so that it is just like another.

Theorem 1.11 *Let G and $\mathcal{G}$ be two groups and suppose that e is the identity of G. If $\psi : G \to \mathcal{G}$ is an isomorphism, then $\psi(e)$ is the identity of $\mathcal{G}$. Moreover,*

$$\psi(g^{-1}) = [\psi(g)]^{-1} \; \forall\, g \in G. \tag{1.11}$$

The essence of the theorem is that isomorphisms map identities onto identities and inverses onto inverses.

It is immediate from our discussion that every system is isomorphic to itself; we simply let ψ be the identity function. To show whether two different systems are isomorphic can be a difficult task. Proceeding from the definition, the following algorithm can be used to show that two systems $G = (G, \star)$ and $\mathcal{G} = (\mathcal{G}, \circ)$ are isomorphic:

Step 1. Define the function ψ from G to $\mathcal{G}$ which is proposed as a candidate for the isomorphism.

Step 2. Show that ψ is a one-to-one function.

Step 3. Show that ψ is an onto function.

Step 4. Show that $\psi(g \star h) = \psi(g) \circ \psi(h)$.

Step 4 is usually just a question of computation. One computes both sides of the equation and checks out whether or not they are the same. We illustrate this procedure with an example.

Example 1.11 We want to show that $(\mathbb{R}, +)$ is isomorphic to $(\mathbb{R}^+, \cdot)$.

Step 1. Define the function $\psi : \mathbb{R} \to \mathbb{R}^+$ by $\psi(x) = e^x \; \forall\, x \in \mathbb{R}^+$.

Step 2. If $\psi(x) = \psi(y)$, then $e^x = e^y$ and by taking the natural log we obtain $x = y$. Thus, ψ is one-to-one.

Step 3. If $x \in \mathbb{R}^+$, then $\psi(\ln x) = e^{\ln x} = x$. Thus, for every $x \in \mathbb{R}^+$, $\exists\, y \in \mathbb{R}$, namely $y = \ln x$, such that $\psi(y) = x$. Therefore, ψ is onto.

Step 4. For $x, y \in \mathbb{R}$, we have

$$\psi(x+y) = e^{x+y} = e^x \cdot e^y = \psi(x) \cdot \psi(y).$$

In algebra, the concept of being identical is always called isomorphic. To show that two systems are not isomorphic means to show that there cannot exist a one-to-one correspondence which preserves the algebraic structure of the systems. This is a trivial problem whenever the two system have a different number of elements. For example, since $card(\mathbb{Z}) = \aleph_0 < \mathfrak{c} = card(\mathbb{R})$, they can never be isomorphic as algebraic structures.

Thus far, we have restricted our discussion to systems with one operation. However, the concept of isomorphism easily extends to systems with more than one operation.

Definition 1.19 Two rings R and $\mathfrak{R}$ are said to be *isomorphic*, denoted by $R \approx \mathfrak{R}$, if and only if there exists a one-to-one and onto function $\psi : R \to \mathfrak{R}$ such that

$$\psi(r + s) = \psi(r) + \psi(s), \tag{1.12}$$

$$\psi(r \cdot s) = \psi(r) \cdot \psi(s). \tag{1.13}$$

If such a function ψ exists, then ψ is called a *ring isomorphism.*

Example 1.12 Consider the ring $(\mathbb{R}^n, +, \cdot)$, where addition corresponds to vector addition and multiplication is defined by multiplying the corresponding vector components; i.e.,

$$(x_1, x_2, \ldots, x_n) + (y_1, y_2, \ldots, y_n) = (x_1 + y_1, x_2 + y_2, \ldots, x_n + y_n) \tag{1.14}$$

and *Hadamard multiplication*

$$(x_1, x_2, \ldots, x_n) \cdot (y_1, y_2, \ldots, y_n) = (x_1 \cdot y_1, x_2 \cdot y_2, \ldots, x_n \cdot y_n). \tag{1.15}$$

We leave it up to the reader to convince herself that $(\mathbb{R}^n, +, \cdot)$ is a commutative ring with unity. Suppose that X is a finite set with n elements, say $X = \{x_1, x_2, \ldots, x_n\}$ and let $\psi : \mathbb{R}^X \to \mathbb{R}^n$ be defined by

$$\psi(f) = (f(x_1), f(x_2), \ldots, f(x_n)).$$

If $(y_1, y_2, \ldots, y_n) \in \mathbb{R}^n$ and $g \in \mathbb{R}^X$ is the function defined by $g(x_i) = y_i$ for $i = 1, 2, \ldots, n$, then $\psi(g) = (y_1, y_2, \ldots, y_n)$ Therefore, ψ is onto. To show that ψ is one-to-one is just as easy. Furthermore,

$$\begin{aligned}
\psi(f + h) &= ((f + h)(x_1), (f + h)(x_2), \ldots, (f + h)(x_n)) \\
&= (f(x_1) + h(x_1), f(x_2) + h(x_2), \ldots, f(x_n) + h(x_n)) \qquad (1.16) \\
&= (f(x_1), f(x_2), \ldots, f(x_n)) + (h(x_1), h(x_2), \ldots, h(x_n)) \\
&= \psi(f) + \psi(h).
\end{aligned}$$

An analogous argument show that $\psi(f \cdot h) = \psi(f) \cdot \psi(h)$. This shows that the rings $(\mathbb{R}^n, +, \cdot)$ and $(\mathbb{R}^X, +, \cdot)$ are isomorphic. Of course, by arguing in an analogous fashion, we can prove that for any field $\mathbb{F}$, the corresponding rings $(\mathbb{F}^n, +, \cdot)$ and $(\mathbb{F}^X, +, \cdot)$ are isomorphic.

The final class of functions considered in this section are linear transformations. These are the most important functions in the study of vector spaces.

Definition 1.20 A *linear transformation* or *linear operator* of a vector space $\mathbb{V}(\mathbb{F})$ into a vector space $\mathbb{W}(\mathbb{F})$ is a function $L : \mathbb{V}(\mathbb{F}) \to \mathbb{W}(\mathbb{F})$ which satisfies the equation

$$L(\alpha \cdot \mathbf{u} + \beta \cdot \mathbf{v}) = \alpha \cdot L(\mathbf{u}) + \beta \cdot L(\mathbf{v}) \tag{1.17}$$

for all $\mathbf{u}, \mathbf{v} \in \mathbb{V}(\mathbb{F})$ and for all scalars $\alpha, \beta \in \mathbb{F}$.
If in addition L is both one-to-one and onto, then L is called a *vector space isomorphism* and the vector spaces $\mathbb{V}$ and $\mathbb{W}$ are called *isomorphic* vector spaces. As before, the notation $\mathbb{V} \approx \mathbb{W}$ means that $\mathbb{V}$ and $\mathbb{W}$ are isomorphic.

.

An equivalent definition of a linear transformation which can be found in many text books is given by the following theorem (see also references [10, 85, 163]).

Theorem 1.12 *A function $L : \mathbb{V}(\mathbb{F}) \to \mathbb{W}(\mathbb{F})$ is a linear transformation $\Leftrightarrow$ $\forall \alpha \in \mathbb{F}$ and $\forall \mathbf{u}, \mathbf{v} \in \mathbb{V}$ the following conditions hold:*

1. $L(\alpha \cdot \mathbf{u}) = \alpha \cdot L(\mathbf{u})$ *and*
2. $L(\mathbf{u} + \mathbf{v}) = L(\mathbf{u}) + L(\mathbf{v})$.

If L is a linear transformation mapping a vector space $\mathbb{V}(\mathbb{F})$ into a vector space $\mathbb{W}(\mathbb{F})$, then

1. $L(\mathbf{0}_{\mathbb{V}}) = \mathbf{0}_{\mathbb{W}}$, where $\mathbf{0}_{\mathbb{V}}$ and $\mathbf{0}_{\mathbb{W}}$ are the zero vectors in $\mathbb{V}$ and $\mathbb{W}$, respectively, and
2. if $\mathbf{v}_1, \ldots, \mathbf{v}_n$ are elements of $\mathbb{V}$ and $\alpha_1, \ldots, \alpha_n$ are scalars from $\mathbb{F}$, then

$$L(\alpha_1 \cdot \mathbf{v}_1 + \alpha_2 \cdot \mathbf{v}_2 + \ldots + \alpha_n \cdot \mathbf{v}_n) = \alpha_1 \cdot L(\mathbf{v}_1) + \alpha_2 \cdot L(\mathbf{v}_2) + \ldots + \alpha_n \cdot L(\mathbf{v}_n).$$

The first statement (1) follows from the condition $L(\alpha \cdot \mathbf{u} + \beta \cdot \mathbf{v}) = \alpha \cdot L(\mathbf{u}) + \beta \cdot L(\mathbf{v})$ with $\alpha = 0$ and $\beta = 0$. The second statement (2) can be easily proven by mathematical induction.

Example 1.13 Recall that the function $\psi : \mathbb{R}^X \to \mathbb{R}^n$ defined in Example 1.12 is one-to-one and onto whenever $card(X) = n$. Now suppose that $k \in \mathbb{R}$ and $f, h \in \mathbb{R}^X$. By defining the function $(k \cdot f) : X \to \mathbb{R}$ by $(k \cdot f)(x) = k \cdot f(x)$, we have that

$$\begin{aligned}
\psi(k \cdot f) &= ((k \cdot f)(x_1), (k \cdot f)(x_2), \ldots, (k \cdot f)(x_n)) \\
&= (k \cdot f(x_1), k \cdot f(x_2), \ldots, k \cdot f(x_n)) \\
&= k \cdot (f(x_1), f(x_2), \ldots, f(x_n)) \\
&= k \cdot \psi(f)
\end{aligned}$$

The fact that $\psi(f + h) = \psi(f) + \psi(h)$ was proven in Example 1.12. Thus, according to Theorem 1.12, ψ is a vector space isomorphism, and $\mathbb{R}^X$ and $\mathbb{R}^n$ are isomorphic vector spaces.

Exercises 1.2.5

1. Consider the groups $(\mathbb{Z}, +)$ and $(3\mathbb{Z}, +)$, where $3\mathbb{Z} = \{3n : n \in \mathbb{Z}\}$. Determine whether or not the function $\psi : \mathbb{Z} \to 3\mathbb{Z}$ defined by $\psi(n) = 3n$ is an isomorphism.

2. Let $G = \mathbb{R} \setminus \{0\}$. Determine whether or not the two groups $(G, \times)$ and $(\mathbb{R}, +)$ are isomorphic.

3. Determine whether or not the rings $(2\mathbb{Z}, +, \times)$ and $(3\mathbb{Z}, +, \times)$ are isomorphic.

4. Suppose $\mathbb{V}$, $\mathbb{W}$, and $\mathbb{U}$ are three vector spaces and the two vector spaces $\mathbb{V}$ and $\mathbb{W}$ are both isomorphic to $\mathbb{U}$. Show how these two isomorphisms can be combined in order to generate an isomorphism from $\mathbb{V}$ onto $\mathbb{W}$.

CHAPTER 2

Pertinent Properties of Euclidean Space

THE objective of this chapter is to familiarize the reader with several pertinent properties of Euclidean space, including relevant concepts from topology such as bounded sets, closed and open sets, connected sets, compactness, and metrics. These concepts are essential tools in various AI areas such as pattern recognition, image analysis and understanding, and artificial neural networks.

2.1 ELEMENTARY PROPERTIES OF $\mathbb{R}$

Since n-dimensional Euclidean space $\mathbb{R}^n$ is derived from the n-fold cross product of the set of real numbers, it inherits many of its basic properties from $\mathbb{R}$. For this reason we begin our discussion by recalling some of the foundational concepts associated with the real numbers.

2.1.1 Foundations

Although the axiomatic foundation of modern mathematics had its start with Euclid's *Elements* around 300 BC, it took another 2000 years before the axiomatic approach became the foundation for all branches of mathematics. In this study we will not cover the Zermelo-Fraenkel axioms of set theory nor the axioms for totally ordered rings used by Dedekind in his construction of the set of real numbers. We will, however, use basic theorems that are directly derived from these axioms.

Definition 2.1 Suppose $X \subset \mathbb{R}$.

DOI: 10.1201/9781003154242-2

1. If $\exists y \in \mathbb{R} \ni x \leq y \, \forall x \in X$, then X is said to be *bounded above*, and y is called an *upper bound* for X.

2. If $\exists z \in \mathbb{R} \ni z \leq x \, \forall x \in X$, then X is said to be *bounded below*, and z is called a *lower bound* for X.

3. If X is bounded above and below, then X is said to be *bounded.*

4. Suppose y is an upper bound of X. If $x < y$ implies that x is not an upper bound, then y is called the *least upper bound* or *supremum* of X.

5. Suppose z is a lower bound of X. If $z < x$ implies that x is not a lower bound, then z is called the *greatest lower bound* or *infimum* of X.

We shall abbreviate the least upper bound of a set X by $lub(X)$ or $sup(X)$ and the greatest lower bound by $glb(X)$ or $inf(X)$.

Example 2.1

1. The set $Y = \{y : y = e^x, x \in \mathbb{R}\}$ has 0 as its *glb*, but Y is not bounded above.

2. The set $X = \{\frac{1}{n} : n \in \mathbb{N}\}$ is a bounded set with $glb(X) = 0$ and $lub(X) = 1$.

Suppose $\mathbb{F}$ is a totally ordered field and $P = \{A \in 2^{\mathbb{F}} : A \neq \varnothing$ and A is bounded above$\}$. If $\mathbb{F}$ also satisfies the property

D_1. $\quad A \in P \Rightarrow lub(A)$ exists,

then the order is said to be *Dedekind complete.*

Note that $(\mathbb{Q}, +, \times)$ is a totally ordered field with respect to the natural order $\leq$. However, the order is not Dedekind complete. For example, the set $A = \{x \in \mathbb{Q} : x \times x \leq 2\}$ is bounded above by such upper bounds as 2 and $\frac{71}{50} \times \frac{71}{50}$ but there does not exist an element x of $\mathbb{Q}$ such that $x \times x = lub(A)$. We also know (see Section 1.3) that the field $(\mathbb{R}, +, \times)$ is a totally ordered field. Thus, the question arises as to whether or not the natural order $\leq$ of $\mathbb{R}$ is Dedekind complete. Here the answer is yes. This very fundamental property of the real numbers, known as *the completeness theorem for real numbers*, is due to Richard Dedekind [71] and a simple proof of this theorem is given in [239].

Theorem 2.1 *(Dedekind) Let A and B be subsets of* $\mathbb{R}$ *such that*

1. *if* $x \in \mathbb{R}$, *then either* $x \in A$ *or* $x \in B$;

2. $A \cap B = \varnothing$;

3. $A \neq \varnothing \neq B$;

4. *if* $x \in A$ *and* $y \in B$, *then* $x < y$.

Then $\exists! r \in \mathbb{R} \ni x \leq r \; \forall x \in A$ *and* $r \leq y \; \forall y \in B$.

Any pair of sets A, B that satisfies properties 1 through 4 of Theorem 2.1 is called a *Dedekind cut*. As can be inferred from the completeness theorem, any number $r \in \mathbb{R}$ can be used to create a cut by setting $A = \{x \in \mathbb{R} : x < r\}$ and $B = \{x \in \mathbb{R} : r \leq x\}$. Dedekind used cuts of rational numbers in order to construct the real numbers by filling in the *gaps* that exists in the totally ordered ring $\mathbb{Q}$. These gaps exist despite the fact that between any pair of rational numbers there exist another rational number.

The following fact is an easy Corollary of Theorem 2.1.

Corollary 2.2 *If* A *and* B *are as in Theorem 2.1, then either* A *contains a largest number or* B *contains a smallest number.*

Neither Theorem 2.1 or its corollary mentioned the *lub* property for subsets of $\mathbb{R}$ that are bounded above. This property is established by the next theorem.

Theorem 2.3 *If* $X \subset \mathbb{R}$ *is bounded above and* $X \neq \varnothing$, *then* $lub(X)$ *exists.*

Proof. Let $A = \{a \in \mathbb{R} : \exists x \in X \ni a < x\}$ and $B = \{b \in \mathbb{R} : b \notin A\}$. We shall demonstrate that the sets A and B satisfy properties 1 through 4 of Theorem 2.1.

Clearly, $A \cap B = \varnothing$. Since $X \neq \varnothing$, we have that for $x \in X$ the set $C(x) = \{a \in \mathbb{R} : a < x\} \neq \varnothing$. Hence $A = \bigcup_{x \in X} C(x) \neq \varnothing$. By construction of A, no element of A is an upper bound for X. But X is bounded above. Thus, $\exists b \in \mathbb{R} \ni b$ is an upper bound of X with $b \notin A$. Hence $b \in B$. This proves that $A \neq \varnothing \neq B$. It also follows from the construction of A and B that for $x \in \mathbb{R}$, either $x \in A$ or $x \in B$, which means that property 1 is also satisfied. Finally, for any $a \in A$ and $b \in B \; \exists x \in X \ni a < x \leq b$ so that $a < b$, which shows that the sets A and B also satisfy property 4.

We are now able to apply Corollary 2.2, which implies that either A contains a largest element or B contains a smallest element. Let us suppose that A contains a largest element, say a_0. Since $a_0 \in A$, $\exists x \in X \ni a_0 < x$. Set

$a = a_0 + \frac{1}{2}(x - a_0)$. Since $\mathbb{R}$ is a totally ordered field, we now have $a_0 < a < x$ which contradicts the assumption that A contains a largest element. Therefore B must contain a smallest element b_0. Hence $x \leq b_0 \; \forall \, x \in X$ or, equivalently, $lub(X) = b_0$. □

Since $\mathbb{R}$ consists of negative and non-negative numbers, the notion of the magnitude of a real number, also known as its absolute value, is of basic importance in mathematical theory and applications.

Definition 2.2 If $x \in \mathbb{R}$, then the *absolute value* of x is defined as

$$|x| = \begin{cases} x & \text{if } 0 \leq x \\ -x & \text{if } x < 0 \end{cases}$$

Although the vertical delimiters $|\cdot|$ are also being used in defining the counting function for sets, there is no chance for confusion since the context of discussion will clarify whether or not we are dealing with a number or a set.

For a simple but fundamental application of the absolute value, consider the *distance* between two numbers x and y. This distance is defined as the absolute value of the number $z = x + (-y)$, namely $|z| = |x - y|$, where $-y$ denotes the additive inverse of y.

The absolute value also possesses the following simple properties:

1. $|x| \geq 0$ and $|x| = |-x| \; \forall \, x \in \mathbb{R}$.

2. $|r \cdot x| = |r| \cdot |x| \; \forall \, r, x \in \mathbb{R}$ and $|r \cdot x| = r \cdot |x|$ if $r \geq 0$.

3. $|x - y| = |y - x|$ and $|x + y| \leq |x| + |y| \; \forall \, x, y \in \mathbb{R}$.

Exercises 2.1.1

1. Prove Corollary2.2.

2. Prove that $|x + y|^2 + |x - y|^2 = 2|x|^2 + 2|y|^2$

3. Suppose $x, y \in \mathbb{R}$ and $x < y$. Prove that $\exists r \in \mathbb{Q}$ such that $x < r < y$.

4. Suppose $x, y \in \mathbb{R}$ and $x < y$. Prove that $\exists s \in \mathbb{I}$, where $\mathbb{I} = \mathbb{R} \setminus \mathbb{Q}$ denotes the set of irrational numbers, such that $x < s < y$.

2.1.2 Topological Properties of $\mathbb{R}$

A basic understanding of the topological properties of subsets of $\mathbb{R}$ expedites the comprehension of these properties for subsets of $\mathbb{R}^n$ as well as sets in general topological spaces. We will use conventional topological terminology and call the elements of $\mathbb{R}$ *points* when discussing topological concepts.

Definition 2.3 Let $\mathbf{y} \in \mathbb{R}$ and $r \in \mathbb{R}^+$. The set

1. $N_r(y) = \{x \in \mathbb{R} : y - r < x < y + r\}$ is called an *open neighborhood of* y with *radius* r.

2. The set $\bar{N}_r(y) = \{x \in \mathbb{R} : y - r \leq x \leq y + r\}$ is called a *closed neighborhood of* y with *radius* r.

3. Any open or closed neighborhood of y with an unspecified radius will be denoted by $N(y)$ or $\bar{N}(y)$, respectively.

4. The set $N(y')$ defined by $N(y') = N(y) \setminus \{y\}$ is called the *center deleted* neighborhood of y.

We shall also use the notation $(y - r, y + r)$ and $[y - r, y + r]$ to denote the respective neighborhoods $N_r(y)$ and $\bar{N}_r(y)$. Consequently, an open neighborhood of zero will always be of form $N_r(0) = (-r, r)$ for some $r > 0$. The sets $\{x \in \mathbb{R} : a < x < b\}$ and $\{x \in \mathbb{R} : a \leq x \leq b\}$, where $a, b \in \mathbb{R}$ with $a < b$ are respectively denoted by (a, b) and $[a, b]$ and called *open* and *closed* intervals. Here the center and radius are not explicitly specified but implicitly given by $c = \frac{a+b}{2}$ and $r = \frac{b-a}{2}$, respectively.

With respect to a given set $X \subset \mathbb{R}$, each point $y \in \mathbb{R}$ has one of three relations and for each we use a familiar word in a precise way:

1. y is an *interior* point of X if there exists a neighborhood $N(y)$ such that $N(y) \subset X$;

2. y is an *exterior* point of X if there exists a neighborhood $N(y)$ such that $N(y) \cap X = \varnothing$; and

3. y is a *boundary* point of X if y is neither an exterior point or an interior point of X.

The set of all interior points of X is called the *interior* of X and is denoted by *int*X. The set of all boundary points of X is called the *boundary* of X and is denoted by ∂X. Note that a boundary point may or may not belong to X: If we let $X = N_r(y)$ and $Y = \bar{N}_r(y)$, then $\partial X = \{x \in \mathbb{R} : |x-y| = r\}$. Therefore $\partial X \cap X = \varnothing$ while $\partial Y = \partial X \subset Y$.

Beginning students of calculus often confuse the two distinct notions of limit points and boundary points. Limit points of sets in $\mathbb{R}$ are defined as follows:

Definition 2.4 A point $y \in \mathbb{R}$ is a *limit point* of $X \subset \mathbb{R}$ if and only if for every open neighborhood $N(y)$ of y, $N(y') \cap X \neq \varnothing$.

It follows from this definition that every interior point of X is a limit point of X and every point of $[y-r, y+r]$ is a limit point of $(y-r, y+r)$. This means that limit points need not be boundary points. The next example shows that the converse is also true.

Example 2.2 Let $X = \{x \in \mathbb{R} : 0 < |x| \leq 1\} \cup \{2\}$. The boundary of X consists of three separate pieces: The points x satisfying the equation $|x| = 1$, and the two points 0 and 2. The interior of X is the set of all points x that satisfy the condition $0 < |x| < 1$, while the set of all limit points of X is the set $\bar{N}_1(0)$. In particular, 2 is a boundary point but not a limit point. A boundary point that is not a limit point of a set X is called an *isolated* point of X. Thus, a point x is an isolated point if there exist a neighborhood $N(x)$ such that $X \cap N(x') = \varnothing$.

Definition 2.5 Suppose $X \subset \mathbb{R}$. Then

1. X is an *open* set if and only if $X = intX$;
2. X is a *closed* set if and only if every limit point of X is a point of X;
3. X is a *perfect* set if and only if X is closed and every point of X is a limit point of X.

It follows that open sets are the union of neighborhoods, and closed sets are sets that contain all their limit points. We wish to stress the difference between having a limit point and containing one. The set $X = \{\frac{1}{n} : n \in \mathbb{N}\} \subset \mathbb{R}$ has a limit point, namely $x = 0$, but no point of X is a limit point of X. In every day usage, "open" and "closed" are antonyms; this is not true for the technical meaning. The set X just described is neither open or closed in $\mathbb{R}$. The following theorem summarizes several important facts about open and closed sets.

Theorem 2.4

1. *Every neighborhood $N(x)$ is an open set and every closed neighborhood $\bar{N}(x)$ is a closed and perfect set.*

2. *The union of any collection of open sets is open and the intersection of any finite collection of open sets is open.*

3. *The intersection of any collection of closed sets is closed and the union of any finite collection of closed sets is closed.*

4. *A set is open if and only if its complement is closed.*

5. *A point* $\mathbf{x}$ *is a boundary point of* X *if and only if every neighborhood of* $\mathbf{x}$ *contains a point of* X *and a point of* X'.

6. $X \cup \partial X$ *is a closed set.*

7. *Any open set can be expressed as a union of a countable number of neighborhood.*

The proofs of the properties listed are left to the reader since they follow almost immediately from the definitions. It is important to realize that the intersection of an infinite collection of open sets need not be open: $\{0\} = \bigcap_{n=1}^{\infty}(-\frac{1}{n}, \frac{1}{n})$ is closed. Similarly, the union of an infinite number of closed sets need not to be closed: $(-1,1) = \bigcup_{n=1}^{\infty}[\frac{1}{n} - 1, 1 - \frac{1}{n}]$ is open even though $[\frac{1}{n} - 1, 1 - \frac{1}{n}] = \bar{N}_{\frac{n-1}{n}}(0)$ is closed for each $n \in \mathbb{N}$.

The set $\bar{X} = X \cup \{x : x \text{ is a limit point of } X\}$ is called the *closure* of X and is a closed set by definition of the concept "closed." In particular, X is closed if and only if $\bar{X} = X$. Also, by property (6.) of Theorem 2.4, $\bar{X} = X \cup \partial X$. Another distinguishing feature concerning limit points and boundary points is provided by the next theorem and its corollary.

Theorem 2.5 *If p is a limit point of X, then every neighborhood of p contains infinitely many points of X.*

.

Proof. Suppose there is a neighborhood $N(p)$ which contains only a finite number of points of X. Let $x_1, x_2, \ldots, x_n$ denote the points of $N(p) \cap X$ which are distinct from p and let

$$r = inf\{|p - x_i| : x_i \in N(p') \cap X\}.$$

Obviously every finite set in $\mathbb{R}$ has a greatest lower bound (as well as a least upper bound) which means that r exists and since $p \neq x_i \ \forall i \in \mathbb{N}_n$, $r > 0$. But then for any number ρ with $0 < \rho < r$ we obtain $N_\rho(p') \cap X = \varnothing$ which contradicts the hypothesis that p is a limit point of X. Thus our assumption that there is a neighborhood $N(p)$ which contains only a finite number of points of X must be false. □

Corollary 2.6 *A finite subset of* $\mathbb{R}$ *has no limit points.*

According to the corollary, finite sets are always closed sets. However, there are also infinite subsets of $\mathbb{R}$ that have no limit points. For instance, the set $\mathbb{Z} \subset \mathbb{R}$ has no limit points and all its elements are isolated points. Sets without limit points and whose elements are all isolated points are called *discrete sets*.

In ordinary usage, the words "finite" and "bounded" are sometimes synonymous. In mathematics they are used to describe quite different aspects of a set. It follows from definition 2.1(3) that if a set X is bounded, then there exists a number r such that $|x| < r \ \forall x \in X$. Geometrically, this means that no point of X is farther than a distance r from the origin; that is $X \subset N_r(0)$. As our next theorem shows, closed bounded sets in $\mathbb{R}$ have the property that they contain their least upper bound and greatest lower bound.

Theorem 2.7 *If* $X \subset \mathbb{R}$ *is closed and bounded with* $a = glb(X)$ *and* $b = lub(X)$, *then* $a, b \in X$.

Proof. Suppose $b \notin X$. For every $r > 0 \ \exists x \in X \ni b - r \leq x \leq b$, for otherwise $b - r$ would be an upper bound less than $lub(X)$. Thus every neighborhood $N_r(b)$ contains a point x of X with $x \neq b$, since $b \notin X$. It follows that b is a limit point of X which is not in X. But this contradicts the hypothesis that X is closed. To show that $a \in X$ is similar. □

Another basic property of closed and bounded sets is the *nested set property* established by the following theorem:

Theorem 2.8 *(Nested set property) Let* $\{A_n\}_{n\in\mathbb{N}}$ *be a family of non-empty closed and bounded sets in* $\mathbb{R}$. *If* $A_1 \supset A_2 \supset A_3 \supset \cdots$, *then* $\bigcap_{n=1}^{\infty} A_n \neq \varnothing$.

Proof. Let $X = \bigcap_{n=1}^{\infty} A_n$. Applying Theorem 2.7 let $a_n = glb(A_n)$ and $b_n = lub(A_n)$ for $n = 1, 2, \ldots$. The set of greatest lower bounds $S = \{a_n : n \in \mathbb{N}\}$ has the property that for any $j \in \mathbb{N}$, $a_n \leq b_j \ \forall n \in \mathbb{N}$. Thus each b_n is an upper bound of S. By Theorem 2.3 there exists a number a such that $a = lub(S)$. Since a is an upper bound for S, $a_n \leq a \ \forall n \in \mathbb{N}$ and since a is the least upper bound, $a \leq b_n \ \forall n \in \mathbb{N}$. Now if $a_i = a$, then $a_n = a \ \forall n \geq i$ and, therefore, $a \in X$. A similar argument holds if $a = b_i$ for some $i \in \mathbb{N}$. if $a_n \neq a \neq b_n \ \forall n \in \mathbb{N}$, then $a_n < a < b_n \ \forall n \in \mathbb{N}$. Now suppose that $a \notin X$. We claim that for any $r > 0$, $N_r(a') \cap X \neq \varnothing$. Suppose otherwise. Then there exists $x \in N_r(a')$, with $x \notin X$ and $a - r < x < a$. But this makes x an upper bound of S less than the $lubS$. Hence we must have $N_r(a') \cap X \neq \varnothing$. Hence a is a limit point of X which is not in X. But this is impossible since by Theorem 2.4(3) X is closed. Therefore, $a \in X$ and $X \neq \varnothing$. □

The proof of the next theorem is an excellent example of the application of the nested set property.

Theorem 2.9 *(Bolzano-Weierstrass) If $X \subset \mathbb{R}$ is bounded and $card(X) \geq \aleph_0$, then X has at least one limit point.*

Proof. Since X is bounded, there exists a closed interval $I_1 = [a_1, b_1]$ such that $X \subset I_1$. Divide this interval into two intervals $[a_1, \frac{a_1+b_1}{2}]$ and $[\frac{a_1+b_1}{2}, b_1]$. Since X contains infinitely many points, at least one of these intervals contains infinitely many points of X. Choose the one containing infinitely many points and denote it by I_2. If both of the two subintervals contain infinitely many points, then it does not matter which one is chosen. Now repeat this process with the interval I_2, keeping the subinterval I_3 containing infinitely many points of X. This process generates a sequence of closed intervals

$$I_1 \supset I_2 \supset I_3 \supset \cdots \supset I_n \supset \cdots$$

with each containing an infinite number of points of X. According to the nested set property, there exists a point $a \in \bigcap_{n=1}^{\infty} I_n$. If $\ell = b_1 - a_1$ denotes the diameter (or length) of I_1, then the diameter of I_n will be $\frac{\ell}{2^{n-1}}$. Thus, given a number $r > 0$, we can choose $n \in \mathbb{N} \ni \frac{\ell}{2^{n-1}} < \frac{r}{2}$. Then $I_n \subset N_{\frac{r}{2}}(a)$ and $I_n \cap X \subset N_{\frac{r}{2}}(a) \cap X$ contains infinitely many points of X. Therefore, $N_{\frac{r}{2}}(a') \cap X \neq \varnothing$ which means that a is a limit point of X. □

Closely associated with the notion of bounded sets is the concept of compactness. Compactness is usually defined in terms of open covers. By an *open cover* of X we mean a family $\{Y_\lambda\}_{\lambda \in \Lambda}$ of open sets such that $X \subset \bigcup_{\lambda \in \Lambda}$.

Definition 2.6 A subset X of $\mathbb{R}$ is *compact* if and only if every open cover of X contains a finite subcover of X.

More explicitly, the requirement is that if $\{Y_\lambda\}_{\lambda\in\Lambda}$ is an open cover of X, then there exists finitely many indices $\{\lambda_1, \dots, \lambda_n\} \subset \Lambda$ such that $X \subset \bigcup_{i=1}^{n} Y_{\lambda_i}$. Obviously, every finite subset of $\mathbb{R}$ is compact. On the other hand, $\mathfrak{N} = \{(\frac{1}{n}, 1 + \frac{1}{n})\}_{n\in\mathbb{N}}$ is an open cover of the set $X = \{x \in \mathbb{R} : 0 < x \leq 1\}$ for which there does not exist a finite subcollection of $\mathfrak{N}$ which covers X.

Although compactness is defined in terms of open sets, the relationships between the three topological concepts of closed, bounded, and compact sets are of particular importance. For instance, replacing boundedness for compactness in the hypothesis of the Bolzano-Weierstrass Theorem one obtains the following result:

Theorem 2.10 *Suppose X is compact and $Y \subset X$. If Y is infinite, then Y has a limit point in X.*

Proof. If no point of X is a limit point of Y, then each $x \in X$ has a neighborhood $N(x)$ such that $N(x') \cap Y = \emptyset$ and either $N(x) \cap Y = \emptyset$ or $N(x) \cap Y = \{x\}$. Obviously, the collection $\mathfrak{N} = \{N(x)\}_{x\in X}$ of these neighborhoods is infinite since Y is infinite and no finite subcollection of $\mathfrak{N}$ can cover Y. But $Y \subset X \subset \bigcup_{x\in X} N(x)$, which now implies that no finite subcollection of $\mathfrak{N}$ can cover X. This contradicts the hypothesis that X is compact. Thus our assumption that no point of X is a limit point is false. □

Another useful property concerns closed subsets of compact sets.

Theorem 2.11 *Suppose X is compact and $Y \subset X$. If Y is closed, then Y is compact.*

Proof. Let $\mathfrak{V} = \{V_\lambda\}_{\lambda\in\Lambda}$ be an open cover of Y. Since Y is closed, Y' is open (Theorem 2.4(3)). But then the collection $\mathfrak{U} = \{U : U \in \mathfrak{V} \text{ or } U = Y'\}$ is an open cover of X. Since X is compact, there exists a finite subcollection $\{U_i\}_{i=1}^{n} \subset \mathfrak{U}$ which covers X and since $Y \subset X$, it also covers Y. To obtain a finite subcollection of $\mathfrak{V}$ which covers Y, simply throw out Y' from the collection $\{U_i\}_{i=1}^{n}$ if $Y' \in \{U_i\}_{i=1}^{n}$; i.e., the finite subcover of Y will be $\{U_i\}_{i=1}^{n} \setminus \{y'\}$. This will not affect the finite subcover of Y since $Y \cap Y' = \emptyset$. □

The proof of the next theorem provides another good example of the varied application of the nested set property.

Theorem 2.12 *Every closed interval is compact.*

Proof. Let $I = [a, b] \subset \mathbb{R}$ for some $a, b \in \mathbb{R}$ with $a < b$ since the case $a = b$

is trivial. Suppose $\mathfrak{V} = \{V_\lambda\}_{\lambda\in\Lambda}$ is an open cover for I for which there does not exist a finite subcover for I. In this case let $c = \frac{a+b}{2}$ and consider the two subintervals $[a,c]$ and $[c,b]$ of I. Since $[a,c] \cup [c,b] = I$, at least one of these two subintervals cannot be covered by a finite subcollection of $\mathfrak{V}$ for otherwise I could be covered by a finite subcollection. Let I_1 denote this subinterval and let a_1, b_1 denote the boundary points of I_1 with $a_1 < b_1$. Now subdivide I_1 into two intervals $[a_1,c]$ and $[c,b_1]$, where $c = \frac{a_1+b_1}{2}$. Again, at least one of these subintervals of I_1 can not be covered by a finite subcollection of $\mathfrak{V}$. Continuing in this fashion, one obtains a sequence of closed intervals

$$I \supset I_1 \supset I_2 \supset I_3 \supset \cdots \supset I_n \supset \cdots$$

By the nested set property there exists a point $p \in \bigcup_{n=1}^{\infty} I_n$. Thus $p \in I_n \; \forall\, n \in \mathbb{N}$ and $p \in I$. Since $\mathfrak{V}$ is an open cover for I, $p \in V_\lambda$ for some $\lambda \in \Lambda$. Also, since V_λ is an open set, there exists a number $r > 0$ such that $N_r(p) \subset V_\lambda$. By construction of I_n, the diameter of $I_n = [a_n, b_n]$ is $\frac{b-a}{2^n}$. Thus, choosing n sufficiently large so that $0 < \frac{b-a}{2^n} < r$ means that $p \in I_n = [a_n, b_n] \subset N_r(p) \subset V_\lambda$. But this is absurd since by construction I_n cannot be covered by a finite number of elements of $\mathfrak{V}$. Therefore our assumption that there exists an open cover $\mathfrak{V}$ that has no finite subcollection covering I is false. □

Recall also that every finite set is also closed and bounded. The next theorem shows that this property is shared by all compact sets.

Theorem 2.13 *(Heine-Borel) $X \subset \mathbb{R}$ is compact if and only if X is closed and bounded.*

Proof. Suppose X is compact. Let $p \in \mathbb{R} \setminus X = X'$. For each $x \in X$, consider the neighborhood $N_r(x)$, where $0 < r < \frac{|p-x|}{2}$. Then the collection $\{N_r(x)\}_{x\in X}$ covers X. Since X is compact, there exists a finite number of points $\{x_1, x_2, x_3, \ldots, x_n\} \subset X$ such that $X \subset \bigcup_{i=1}^{n} N_{r_i}(x_i)$, where $0 < r_i < \frac{|p-x_i|}{2}$. Letting $\rho = min\{r_1, r_2, \ldots, r_n\}$, then the neighborhood $N_\rho(p)$ has the property that $N_\rho(p) \cap [\bigcup_{i=1}^{n} N_{r_i}(x_i)] = \varnothing$. Thus p is an interior point of X'. Since p was arbitrarily chosen, every point of X' is an interior point of X'. Therefore X' is open and by Theorem 2.4(*4*) X is closed.

Next assume that X is not bounded. Then for each $n \in \mathbb{N} \; \exists\, x_n \in X \ni n < |x_n|$. It follows that the set $Y = \{x_n \in X : n < |x_n|, n \in \mathbb{N}\}$ is finite and no point of $\mathbb{R}$ is a limit point of Y. Since $X \subset \mathbb{R}$, no point of X can be a limit point of Y. But this contradicts Theorem 2.10. Thus, our assumption that X is not bounded must be false. This proves that X is also bounded.

Conversely, suppose that X is closed and bounded. Then there exists an interval $I = [a,b]$ such that $X \subset I$. By Theorem 2.12 I is compact. Since X is closed and I is compact with $X \subset I$, it follows from Theorem 2.11 that X is compact. □

Suppose $Y \subset \mathbb{R}$ is closed. If $X \subset \mathbb{R}$ is compact, then by the Heine-Borel theorem X is also closed and, hence, $Y \cap X$ is closed (see Theorem 2.4(3)). According to Theorem 2.11 this means that $Y \cap X$ is compact and proves the following theorem:

Theorem 2.14 *If Y is closed and X is compact, then $Y \cap X$ is compact.*

Given an interval $[a,b]$ and $r \in \mathbb{R}$, then we define $r + [a,b] = [r+a, r+b]$ and $r[a,b] = [ra, rb]$. More generally, if $X = \bigcup_i [a_i, b_i]$ is a disjoint union of intervals, then we define $r + X = \bigcup_i [r + a_i, r + b_i]$ and $rX = \bigcup_i [ra_i, rb_i]$.

Thus far we have considered open, closed, and compact sets. Considering the definition of a perfect set, it may be surprising that there exist perfect compact sets that contain no open intervals.

Example 2.3 (G. Cantor) Let $I = [0,1]$ and remove the middle third of I by setting $C_1 = I \setminus (\frac{1}{3}, \frac{2}{3}) = [0, \frac{1}{3}] \cup [\frac{2}{3}, 1]$. Next remove the middle third from each of the two intervals $[0, \frac{1}{3}]$ and $[\frac{2}{3}, 1]$ and set $C_2 = [0, \frac{1}{9}] \cup [\frac{2}{9}, \frac{3}{9}] \cup [\frac{6}{9}, \frac{7}{9}] \cup [\frac{8}{9}, 1]$. Continue this process in order to obtain the sequence

$$C_1 \supset C_2 \supset C_3 \supset \cdots \supset C_n \supset \cdots$$

with the property that for $n \in \mathbb{N}$, the set C_n is the union of 2^n closed intervals and the diameter of each interval is of length $\frac{1}{3^n}$. The set $C = \bigcap_{n=1}^{\infty} C_n$ is known as the *Cantor set*. An explicit inductive formulation of C_n is given by the equation

$$C_n = \frac{1}{3}C_{n-1} \cup (\frac{2}{3} + \frac{1}{3}C_{n-1}).$$

Since each C_n is the finite union of closed intervals, it follows from Theorem 2.4(3) that C_n and C are closed. By the nested set property $C \neq \varnothing$ and since I is compact (Theorem 2.12), the compactness of C now follows from Theorem 2.11. That $C \neq \varnothing$ also follows from the fact that the endpoints of the intervals constituting C_n are also points of C_m $\forall m \in \mathbb{N}$ and, hence, are elements of C. Thus C is infinite and in fact uncountable. It can also be shown that C contains elements that are not endpoints of any of the interval components of C_n for any integer n. One such element is the number $\frac{1}{4}$.

Now let $x \in C$ and consider a neighborhood $N_r(x)$ for some $r > 0$. Let

$n \in \mathbb{N}$ be sufficiently large so that $\frac{1}{3^n} < \frac{r}{2}$, and let I_1 denote the interval of C_n containing x. Since I_1 has two endpoints, we can always choose an endpoint that is not equal to x in case x is one of the endpoints, and let x_1 denote the chosen endpoint of I_1. We now have $x_1 \in (I_1 \setminus \{x\}) \cap C \subset N_r(x') \cap C$. Repeat this process in order to obtain an interval $I_2 \subset C_{n+1}$ having the property that $x \in I_2$ and x_2 is an endpoint of I_2 with $x_2 \neq x$. Continuing in this fashion shows that there exist an infinite number of points $\{x_i : i = 1, 2, \ldots\}$ contained in $N_r(x') \cap C$. Thus x is a limit point of C. Since x was arbitrary, every point of C is a limit point of C. This proves that C is perfect.

Suppose that $(a,b) \subset C$. Then $(a,b) \subset C_n\ \forall n \in \mathbb{N}$. Since C_n is a disjoint union of intervals, (a,b) must be a subset of an interval of C_n. But for n sufficiently large we have that $\frac{1}{3^n} < b-a$, which means that C_n has no intervals large enough to contain (a,b). Therefore, (a,b) cannot be a subset of C_n. Since $C_n \supset C$, our assumption that $(a,b) \subset C$ must be wrong. This shows that C contains no open intervals.

Definition 2.7 A function $f : \mathbb{R} \to \mathbb{R}$ is said to be *continuous* if and only if for every open set $U \subset \mathbb{R}$, $f^{-1}(U)$ is open in $\mathbb{R}$.

If U is open in $\mathbb{R}$ and $U \cap f(\mathbb{R}) = \varnothing$, then $f^{-1}(U) = \varnothing$, which is open. If $U \cap f(\mathbb{R}) \neq \varnothing$, then there exists $y \in U \cap f(\mathbb{R})$ and $x \in \mathbb{R} \ni f(x) = y$. Since any open subset of $\mathbb{R}$ is a union of open intervals, there exists $r > 0$ such that $N_r(y) \subset U$. Hence for any number $\epsilon > 0$, $f^{-1}(N_\epsilon(y))$ is open and $x \in f^{-1}(N_\epsilon(y))$. Thus there exists a neighborhood $N_\delta(x) \subset f^{-1}(N_\epsilon(y))$. It now follows from Theorem 1.4(1) that $f(N_\delta(x)) \subset N_\epsilon(f(x))$. This shows that the concept of continuity defined in terms of open sets is equivalent to the ϵ, δ definition used in beginning calculus courses. Continuity expressed in terms of open sets is more general and applies to functions defined on topological spaces and not just to real valued functions on $\mathbb{R}$.

Example 2.4 Suppose $f : \mathbb{R} \to \mathbb{R}$ is defined by $f(x) = 2$ if $x \leq 0$ and $f(x) = 5$ if $x > 0$. Then $N_1(2)$ is open and does not contain 5. If f is continuous, then $f^{-1}(N_1(2))$ is open and contains 0. Therefore 0 is an interior point of $f^{-1}(N_1(2))$. Thus there exists $\delta > 0$ such that $N_\delta(0) \subset f^{-1}(N_1(2))$. Note that the set $S = \{x \in \mathbb{R} : 0 < x < \delta\}$ is a subset of $N_\delta(0)$. Since $f(N_\delta(0)) \subset N_1(2)$ all the elements of S map to 2. But this is impossible by the definition of f. Thus, our assumption that f is continuous is incorrect.

Note that f is almost a constant function as it maps all of $\mathbb{R}$ to only two points. Any constant function $c : \mathbb{R} \to \mathbb{R}$, however, is always continuous.

The notion of continuity was defined for functions $f : \mathbb{R} \to \mathbb{R}$. We now turn our attention to functions $f: X \to \mathbb{R}$, where $X \subset \mathbb{R}$. For this we employ the idea of what it means to be an open set in X.

Definition 2.8 Given the sets $X \subset \mathbb{R}$ and $Y \subset X$, then Y is called a *relatively open* set in X if and only if there exists an open set $V \subset \mathbb{R}$ such that $V \cap X = Y$.

Obviously, if X an open set, then any relatively open subset Y of X is also an open set in $\mathbb{R}$. The same does not hold for sets that are not open. For instance, given the closed interval $[a,b]$ with $a < b$, then for any number c with $a < c < b$ the sets $\{x \in \mathbb{R} : a \leq x < c\}$ and $\{x \in \mathbb{R} : c < x \leq b\}$ as well as $[a,b]$ are all relatively open in $[a,b]$ but not open in $\mathbb{R}$.

Another equivalent definition of the concept of "relatively open" is as follows:

Theorem 2.15 *The set $Y \subset X$ is relatively open in X if and only if for each $y \in Y$ there exists a number $r > 0$ such that $N_r(y) \subset \mathbb{R}$ with $N_r(y) \cap X \subset Y$.*

Proof. Suppose that Y is relatively open in $X \subset \mathbb{R}$. By definition there exists an open set $V \subset \mathbb{R}$ such that $Y = V \cap X$. Since V is open in $\mathbb{R}$, $V = \bigcup_{v \in V} N_{r(v)}(v)$, where the notation $r(v)$ indicates that the radius depends on v. Thus for each $y \in Y \subset V$ there exists a radius $r(y)$ such that $N_{r(y)}(y) \cap X \subset V \cap X = Y$.

The proof of the converse is just as easy and left as an exercise. □

Definition 2.9 Suppose X is a subset of $\mathbb{R}$ and $f: X \to \mathbb{R}$. Then f is said to be *continuous* if and only if for every open set $V \subset \mathbb{R}$, $f^{-1}(V) \cap X$ is relatively open in X.

Any constant function $f: X \to \mathbb{R}$ is continuous since the inverse image of any open set V is either $\varnothing$ or X, both of which are relatively open in X. The proofs of the following elementary properties are left as an exercise.

2.1 If X and Y are two subsets of $\mathbb{R}$ and the two functions $f: X \to Y$ and $g: Y \to \mathbb{R}$ are continuous, then $g \circ f: X \to \mathbb{R}$ is continuous.

2.2 If $f: X \to \mathbb{R}$ is continuous and $Y \subset X$, then $f|_Y: Y \to \mathbb{R}$ is continuous.

Theorem 2.16 *Suppose X is a subsets of $\mathbb{R}$ and $f: X \to \mathbb{R}$ is continuous. If X is compact, then $f(X)$ is compact.*

Proof. Let $\{V_\lambda\}_{\lambda \in \Lambda}$ be an open cover for $f(X)$. Since f is continuous, $f^{-1}(V_\lambda) \cap X$ is relatively open in X and $X = \bigcup_{\lambda \in \Lambda} f^{-1}(V_\lambda) \cap X$. Thus for each λ there exists an open set $U_\lambda \subset \mathbb{R}$ such that $U_\lambda \cap X = f^{-1}(V_\lambda) \cap X$. The family of open sets $\{U_\lambda\}_{\lambda \in \Lambda}$ covers X and since X is compact, there exists a finite subcover $U_{\lambda_1}, U_{\lambda_2}, \ldots, U_{\lambda_n}$ such that $X \subset \bigcup_{i=1}^{n} U_{\lambda_i}$. It follows that the finite sub-collection $\{V_{\lambda_i}\}_{i=1}^{n}$ covers $f(X)$. □

The theorem asserts that compactness of sets is preserved by continuous functions. The same holds for connected sets. Intuitively, a set $X \subset \mathbb{R}$ is connected if it does not consist of two or more separated pieces. This intuitive notion can be rigorously defined in terms of open sets.

Definition 2.10 A set $X \subset \mathbb{R}$ is *connected* if and only if there do not exist two disjoint open sets U and V in $\mathbb{R}$ such that $U \cap X \neq \varnothing \neq V \cap X$ and $X \subset U \cup V$.

As an example, consider the set $\mathbb{Q} \subset \mathbb{R}$, and let $U = (-\infty, \sqrt{2})$ and $V = (\sqrt{2}, \infty)$. Then U and V are open in $\mathbb{R}$ with $\mathbb{Q} \subset U \cap V$ but $U \cap V = \varnothing$. Thus $\mathbb{Q}$ is not connected. Obviously, if $X = \varnothing$ or $X = \{p\}$, where $p \in \mathbb{R}$, then X is connected. For $X \subset \mathbb{R}$ with $|X| > 1$ we can employ the following fact:

Theorem 2.17 *A set $X \subset \mathbb{R}$ is connected if and only if for any triple $\{x, y, z\} \subset \mathbb{R}$, we have that if $x < z < y$ and $x, y \in X$, then $z \in X$.*

Proof. Suppose that X is connected but assume that for some triple $\{x, y, z\}$ with $x < z < y$ and $x, y \in X$, $z \notin X$. Let $U = \{u \in \mathbb{R} : u < z\}$ and $V = \{v \in \mathbb{R} : z < v\}$. Then U and V are disjoint open sets with $X \subset U \cup V$. This contradicts the hypothesis that X is connected. Hence our assumption that $z \notin X$ is false.

To prove the converse, the hypothesis is that for any triple $\{x, y, z\} \subset \mathbb{R}$ with $x < z < y$ and $x, y \in X$, then $z \in X$. Now assume that X is not connected. Then there exist two disjoint open sets U and V such that $X \subset U \cup V$ and $U \cap X \neq \varnothing \neq V \cap X$. Thus there exist $u \in U \cap X$ and $v \in V \cap X$. If $u < v$, then let $A = \{a \in U \cap X : u \leq a < v\}$ and $z = lub(A)$.

Since V is open and $v \in V$, we have that $z < v$. If $z \in U$, then since U is open, z cannot be an upper bound of A. Therefore $z \notin U$.

On the other hand, if $z \in V$, then since V is open we have that z is not the least upper bound of A. Thus, $z \notin V$. Also, since U is open and $x \in U$, we must have $x < z$. Therefore $x < z < y$ and $z \notin X$, which contradicts the hypothesis. □

As an immediate consequence we have the following classification of connected sets in $\mathbb{R}$:

Corollary 2.18 *The only connected subsets of* $\mathbb{R}$ *are* $\mathbb{R}$, *any open or closed interval, and the sets* $\{x \in \mathbb{R} : a \le x < b\}$, $\{x \in \mathbb{R} : a < x \le b\}$, $\{x \in \mathbb{R} : -\infty < x < a\}$, $\{x \in \mathbb{R} : -\infty < x \le a\}$, $\{x \in \mathbb{R} : a \le x < \infty\}$, *and* $\{x \in \mathbb{R} : a < x < \infty\}$, *where a and b are real numbers with* $a \le b$.

The next theorem provides another way of identifying connected sets.

Theorem 2.19 *X is connected if and only if no continuous function* $f : X \to \mathbb{Z}_2$ *is surjective.*

Proof. Suppose that X is connected and assume that there exists a continuous surjection $f : X \to \mathbb{Z}_2$. Then the sets $f^{-1}(N_r(0))$ and $f^{-1}(N_r(1))$ are open for $0 < r < 0.5$ and $X = f^{-1}(N_r(0)) \cup f^{-1}(N_r(1))$ with $f^{-1}(N_r(0)) \cap X \neq \varnothing \neq f^{-1}(N_r(1)) \cap X$. Since X is connected, $f^{-1}(N_r(0)) \cap f^{-1}(N_r(1)) \neq \varnothing$. This implies that there exists $z \in f^{-1}(N_r(0)) \cap f^{-1}(N_r(1)) \neq \varnothing$. But then $f(z)$ assumes the two values 0 and 1, which means that f is not a function.

To prove the converse, suppose that no continuous function $f : X \to \mathbb{Z}_2$ is surjective and assume that X is not connected. Then there exist disjoint open sets U and V such that $X \subset U \cup V$ and $X \cap U \neq \varnothing \neq X \cap V$. But then the function $\chi_U : X \to \mathbb{Z}_2$ is a continuous bijection. This contradicts the hypothesis. □

The following two theorems are now easy consequences of Theorem 2.19.

Theorem 2.20 *If X is connected and* $f : X \to \mathbb{R}$ *is continuous, then the image* $f(X)$ *is connected.*

Proof. Suppose $f(X)$ is not connected. Then it follows from Theorem 2.19 that there must exist a continuous surjection $g : f(X) \to \mathbb{Z}_2$. But then $g \circ f : X \to \mathbb{Z}_2$ is a continuous surjection. This contradicts the connectedness of X. □

Theorem 2.21 *Suppose* $\{A_\lambda\}_{\lambda \in \Lambda}$ *is a family of connected subsets of* $\mathbb{R}$ *and* $A = \bigcup_{\lambda \in \Lambda} A_\lambda$. *If* $\bigcap_{\lambda \in \Lambda} A_\lambda \neq \varnothing$, *then A is connected.*

Proof. Suppose that A is not connected. Then there exists a continuous surjection $f : A \to \mathbb{Z}_2$. Since $\bigcap_{\lambda \in \Lambda} A_\lambda \neq \varnothing$, there exists $a \in \bigcap_{\lambda \in \Lambda} A_\lambda$. But since A_λ is connected for each $\lambda \in \Lambda$ and $f|_{A_\lambda} : A_\lambda \to \mathbb{Z}_2$ is continuous, $f|_{A_\lambda}$ cannot be surjective. Thus for each λ and $x \in A_\lambda$, $f|_{A_\lambda}(x) = f(a)$. It follows that given $x \in A$, then $x \in A_\lambda$ for some $\lambda \in \Lambda$ and $f(x) = f|_{A_\lambda}(x) = f(a)$. Therefore f is not surjective. Hence, the assumption that A is not connected must be incorrect. □

If X is not connected, then it can be *uniquely* decomposed into connected subsets called *components*.

Definition 2.11 Suppose that X is a subset of $\mathbb{R}$. If $x \in X$, then the *component* $C(x)$ *of* x *in* X is the union of all connected subsets of X containing x.

In view of Theorem 2.21, $C(x)$ is connected. Considering the set $X = \mathbb{Q} \subset \mathbb{R}$, we have $C(q) = \{q\}\ \forall\, q \in \mathbb{Q}$. Any set X with the property $C(x) = \{x\}\ \forall\, x \in X$ is called *totally disconnected*.

Exercises 2.1.2

1. Determine the interior points, exterior points, boundary points, and limit points of the following subsets of $\mathbb{R}$: (1) $\mathbb{Z}$, (2) $\mathbb{Q}$, (3) $\mathbb{I} = \mathbb{R} \setminus \mathbb{Q}$, and (4) $[a,b) = \{x \in \mathbb{R} : a \leq x < b\}$.

2. Prove statements *1* through *4* of Theorem 2.4.

3. Prove statements *5* through *7* of Theorem 2.4.

4. Prove the converse of Theorem 2.15.

5. Prove statements **2.1** and **2.2** preceding Theorem 2.16.

6. Prove Corollary 2.18.

2.2 ELEMENTARY PROPERTIES OF EUCLIDEAN SPACES

The elementary properties considered in Section 2.1 can be effortlessly applied to define corresponding properties for the n-fold Cartesian product space $\mathbb{R}^n$ of $\mathbb{R}$. Although the notion of *distance* was not dealt with in Section 2.1.2, it was indirectly used in defining the neighborhood $N_r(y)$. Specifically, the neighborhood $N_r(y) = \{x \in \mathbb{R} : y - r < x < y + r\} = \{x \in \mathbb{R} : |x - y| < r\}$, where the absolute value $|x - y|$ corresponds to the distance between x and y. The notion of distance is mathematically well-defined and the starting point of our discussion.

2.2.1 Metrics on $\mathbb{R}^n$

A type of real-valued function of particular importance in image processing, pattern analysis, and machine learning is the distance function also known as a *metric*. Distance functions induce geometric structures on sets through the notion of nearness of one element to another. The general definition of a metric on a set X is as follows.

Definition 2.12 Let X be a nonempty set. A *metric* d on X is a function $d : X \times X \to \mathbb{R}$ satisfying the following four conditions:

1. $d(x,y) \geq 0 \; \forall \; x,y \in X$. (non-negativity)
2. $d(x,y) = 0 \Leftrightarrow x = y$. (identity)
3. $d(x,y) = d(y,x) \; \forall \; x,y \in X$. (symmetry)
4. $d(x,z) \leq d(x,y) + d(y,z) \; \forall \; x,y,z \in X$. (triangle inequality)

If d is a metric on X, then the pair (X, d) is called a *metric space*. Calling a set X as a metric space assumes that there is some metric d defined on X. The elements of a metric space are generally referred to as *points* and the functional value $d(x,y)$ is called *the distance between the points x and y*. Although the non-negativity is always listed as a condition for a metric, it is actually a consequence of the remaining three conditions. More specifically, conditions 2 through 4 imply non-negativity through the following observation:

$$0 = d(x,x) \leq d(x,y) + d(y,x) = d(x,y) + d(x,y) = 2d(x,y) \Rightarrow 0 \leq d(x,y)$$

The reason for listing all four properties is to emphasize the difference between metrics and *similarity measures*. Metrics are often used as measures of similarity on sets of objects or patterns, but not every similarity measure is a metric. In fact, some authors even use the term *distance* measure even though these distance measures are not metrics. For example, the cosine distance between two vectors $\mathbf{x}, \mathbf{y} \in \mathbb{R}^n$ is defined as

$$d_{\cos}(\mathbf{x},\mathbf{y}) = 1 - \left(\frac{\sum_{i=1}^{n} x_i y_i}{||\mathbf{x}||\,||\mathbf{y}||} \right),$$

where $||\mathbf{x}||$ denotes the *norm* of $\mathbf{x}$ as defined below (Definition 2.13). It is not difficult to show that $d_{\cos}$ does not satisfy condition 2 by simply setting $\mathbf{y} = c\mathbf{x}$, where $c > 0$, results in $d_{\cos}(\mathbf{x},\mathbf{y}) = 0$. However, $d_{\cos}$ does satisfy all the remaining conditions of a metric. Such similarity measures are called *pseudometrics*. Similarity measures that do not satisfy the symmetry condition but all remaining conditions of a metric are called *quasimetrics*. This does not mean that such *distance* functions are poor similarity measures. They have been successful in solving various classification problems. For instance, the cosine distance measures the angle between vectors **x** and **y** and is concerned about their directions, not their sizes. As such, this function has proven very useful when vectors represent data sets of different sizes and one is interested in comparing the similarity of distributions but not sizes.

Another interesting and useful metric is the *Jaccard* distance between finite subsets of some universal set U of discourse. This distance is defined as

$$d(X,Y) = 1 - \frac{|X \cap Y|}{|X \cup Y|},$$

whenever $X \cup Y \neq \varnothing$, and $d(\varnothing,\varnothing) = 0$. If $X = Y$, then $d(X,Y) = 1 - 1 = 0$. This proves that axiom 2 of the metric definition holds. Proving the remaining metric axioms is just as easy and left as an exercise.

The most common metrics on the set $\mathbb{R}^n$ and other vector spaces are the ℓ_p distance functions, where $1 \leq p \leq \infty$. For pairs $\mathbf{x}, \mathbf{y} \in \mathbb{R}^n$ and $p \in [1,\infty)$, the ℓ_p metric is defined as

$$d_p(\mathbf{x},\mathbf{y}) = \Big(\sum_{i=1}^{n} |x_i - y_i|^p\Big)^{1/p}. \tag{2.1}$$

Three ℓ_p distances of particular interest in this treatise are the ℓ_1, ℓ_2, and ℓ_∞ metrics given by $d_1(\mathbf{x},\mathbf{y}) = \sum_{i=1}^{n} |x_i - y_i|$, $d_2(\mathbf{x},\mathbf{y}) = \big(\sum_{i=1}^{n}(x_i - y_i)^2\big)^{1/2}$, and $d_\infty(\mathbf{x},\mathbf{y}) = \lim_{p\to\infty}\big(\sum_{i=1}^{n} |x_i - y_i|^p\big)^{1/p} = \bigvee_{i=1}^{n} |x_i - y_i|$, respectively. The ℓ_1 metric is also known as the *Manhattan distance* or *taxicab distance*, while the ℓ_∞ metric is known as the *Chebyshev distance* or the *checkerboard distance*. The ℓ_2 distance is commonly known as the *Euclidean distance* since it measures the distance between two points along the straight (Euclidean) line determined by the two points. The ℓ_p distance illuminates a major difference between $\mathbb{R}$ and $\mathbb{R}^n$ for $n > 1$ in that for any $p \in [1,\infty]$, $d_p(x,y) = |x - y| \ \forall\, x,y \in \mathbb{R}$.

In general, there are no algebraic operations defined on a metric space, only a distance function. However, the use of vector space theory in such fields as physics and engineering, the *length* of vectors play a major role in the solution of problems. These lengths are usually expressed in terms of norms. In the following definition we assume that $\mathbb{V}$ is a vector space over the field $\mathbb{R}$.

Definition 2.13 A *normed vector space* $(\mathbb{V}, ||\cdot||)$ is a vector space $\mathbb{V}$ together with a function $||\cdot|| : \mathbb{V} \to \mathbb{R}$, called a *norm* for $\mathbb{V}$ such that for all pairs of vectors $\mathbf{v}, \mathbf{w} \in \mathbb{V}$ and any $c \in \mathbb{R}$:

1. $0 \leq ||\mathbf{v}|| < \infty$
2. $||\mathbf{v}|| = 0 \Leftrightarrow \mathbf{v} = \mathbf{0}$.
3. $||c\mathbf{v}|| = |c|\,||\mathbf{v}||$.
4. $||\mathbf{v} + \mathbf{w}|| \leq ||\mathbf{v}|| + ||\mathbf{w}||$.

Each ℓ_p distance defines an ℓ_p norm, namely $||\mathbf{v}|| = d_p(\mathbf{v}, \mathbf{0})$, where $\mathbf{0}$ denotes the zero vector. The converse can be stated as a theorem.

Theorem 2.22 *If* $(\mathbb{V}, ||\cdot||)$ *is a normed vector space, then the function* $d : \mathbb{V} \times \mathbb{V} \to \mathbb{R}$ *defined by* $d(\mathbf{w}, \mathbf{v}) = ||\mathbf{w} - \mathbf{v}||$ *is a metric on* $\mathbb{V}$.

The proof of this theorem is an elementary exercise while the proof of the next theorem, known as the *Cauchy-Schwartz inequality*, is not quite as easy (see Exercise 5).

Theorem 2.23 *(Cauchy-Schwartz) If* $\mathbf{x}, \mathbf{y} \in \mathbb{R}^n$, *then*

$$\left|\sum_{k=1}^{n} x_k y_k\right| \leq ||\mathbf{x}||\,||\mathbf{y}||.$$

The norm in the conclusion of the theorem is usually the ℓ_2 norm, also called the *Euclidean norm*. To avoid confusion, the notation $||\cdot||_p$ is often used to denote the ℓ_p norm.

Exercises 2.2.1

1. Suppose $U = \{V \subset \mathbb{R}^n : \exists k \in \mathbb{N} \ni |V| = k\}$. Prove that the Jaccard distance is a metric on U.
2. Suppose X is a non-empty set and d is a metric on X. Prove that

$$\delta(x, y) = \frac{d(x, y)}{1 + d(x, y)} \text{ is a metric.}$$

3. Let X be a non-empty set. Show that the function

$$d(x,y) = \begin{cases} 1 & \text{if } x \neq y \\ 0 & \text{if } x = y \end{cases} \quad \text{is a metric on } X.$$

4. Prove Theorem 2.22 for $\mathbb{V} = \mathbb{R}^n$ and $\ell_p = \ell_2$.

5. Prove Theorem 2.23 for $p = 2$.

6. Prove the Minkowski inequality $\|\mathbf{x} + \mathbf{y}\|_2 \leq \|\mathbf{x}\|_2 + \|\mathbf{y}\|_2$.

2.2.2 Topological Spaces

In Section 2.1.2, we discussed such notions as continuity, compactness, limit points, and boundary points. These notions are all *topological* concepts and a careful look at these concepts reveals that the basic ingredient in all of them is the idea of an *open* set. Continuity of functions was defined purely in terms of inverse images of open sets; closed sets are merely complements of open sets (see Theorem 2.4 for a summary of open and closed sets); the concept of compactness requires little more than the idea of open sets. Nevertheless, open sets in $\mathbb{R}$ and in $\mathbb{R}^n$ are really just examples of open sets in more general spaces known as topological spaces.

Definition 2.14 Let X be a set. A set $\mathcal{T} \subset 2^X$ is called a *topology* on X if and only if $\mathcal{T}$ satisfies the following axioms:

$\mathfrak{T}_1$. X and $\varnothing$ are elements of $\mathcal{T}$.

$\mathfrak{T}_2$. The union of any number of elements of $\mathcal{T}$ is an element of $\mathcal{T}$.

$\mathfrak{T}_3$. The intersection of any finite number of elements of $\mathcal{T}$ is an element of $\mathcal{T}$.

The pair $(X,\mathcal{T})$, consisting of a set X and a topology $\mathcal{T}$ on X is called a *topological space*.

Whenever it is not necessary to specify $\mathcal{T}$ explicitly we simply let X denote the topological space $(X,\mathcal{T})$. The elements of a topological space are called *points*. The elements of $\mathcal{T}$ are called *open sets* of the topological space X. There is no preconceived idea of what "open" means other than that the sets called open in any discussion satisfy the three axioms. Exactly what sets qualify as open sets depends entirely on the topology $\mathcal{T}$ on X; a set open with respect to one topology on X may be closed with respect to another topology on X.

Definition 2.15 Suppose (X, τ) is a topological space.

1. If $x \in X$ and N is an open set ($N \in \tau$) with $x \in N$, then N is called a *neighborhood* of x and denoted by $N(x)$ whenever necessary.

2. $A \subset X$ is called *closed* if $X \setminus A$ is an open set.

3. if $A \subset X$, then $y \in X$ is called a *limit point* of A if and only if $\forall N(y)$, $N(y') \cap A \neq \varnothing$.

4. if $A \subset X$, then the set $\bar{A} = \{x \in X : x \text{ is a limit point of } A\} \cup A$ is called the *closure* of A.

The points of $N(x)$ are *neighboring* points of x, sometimes called N*-close* to x. Thus, a topology τ organizes X into regions of neighboring points. In this way topology provides us with a rigorous general working definition of the concept of nearness for two distinct points or two other objects in the space X.

Example 2.5

1. Let X be any set and $\tau = \{\varnothing, X\}$. This topology, in which no set other than $\varnothing$ and X is open, is called the *indiscrete* topology on X. There are no "small" neighborhoods.

2. Let X be any set and $\tau = 2^X$. This topology, in which every subset of X is an open set, is called the *discrete* topology on X, and X together with this topology is called a *discrete* space. Comparing this with example (1) above indicates the sense in which different topologies on X give different organizations of the points of X.

3. Recall from Section 2.1.2 that a subset W of $\mathbb{R}$ is an "open" set in $\mathbb{R}$ if and only if $W = intW$ or, equivalently, every point of W is an interior point. It is not difficult to verify that the collection of all sets satisfying this definition of "open in $\mathbb{R}$" determines a topology τ on $\mathbb{R}$. Axiom $\mathfrak{T}_1$ is trivial. Axiom $\mathfrak{T}_2$ is also obvious: If $\tau = \{W \subset \mathbb{R} : W = intW\}$ and $\{W_\lambda : W_\lambda \in \tau \ \forall \lambda \in \Lambda\}$, where Λ is some set of indices, then so is $(\bigcup_{\lambda \in \Lambda} W_\lambda) \in \tau$ since given a point

$$\mathbf{x} \in \bigcup_{\lambda \in \Lambda} W_\lambda \Rightarrow \exists \lambda \in \Lambda \ni \mathbf{x} \in W_\lambda \Rightarrow \exists r > 0 \ni N_r(\mathbf{x}) \subset W_\lambda \subset \bigcup_{\lambda \in \Lambda} W_\lambda.$$

Thus, $\mathbf{x}$ is an interior point and since $\mathbf{x}$ was arbitrary, every point of $\bigcup_{\lambda\in\Lambda} W_\lambda$ is an interior point. Therefore $\bigcup_{\lambda\in\Lambda} W_\lambda = int(\bigcup_{\lambda\in\Lambda} W_\lambda) \in \tau$.

To prove that axiom $\mathfrak{T}_3$ holds is similar and left as an exercise.

4. Let $X = \mathbb{Z}\times\mathbb{Z} = \mathbb{Z}^2$. We call a point $\mathbf{x} = (x_1, x_2) \in X$ *even* whenever $x_1 + x_2$ is even and *odd* when $x_1 + x_2$ is odd, and let

$$N(\mathbf{x}) = \begin{cases} \{\mathbf{x}\} \\ \{(x_1,x_2),(x_1-1,x_2),(x_1+1,x_2),(x_1,x_2-1),(x_1,x_2+1)\} \end{cases},$$

where the first value applies if $\mathbf{x}$ is odd, and the second value is assigned if $\mathbf{x}$ is even. Define $\tau = \{V \subset \mathbb{Z}^2 : V = \bigcup_\lambda N(\mathbf{x}_\lambda)$ is an arbitrary union$\}$. Then τ is a topology on X, also known as the *Khalimsky topology* or *4-connected topology*. This topology belongs to a class of *digital topologies* and plays a major role in classifying such objects as digital curves, the connectivity of objects, and the Euler characteristic of digital objects in computer vision, image processing, and pattern recognition [153].

It is also of interest to note that functions that map sets onto topological spaces induce topologies on their domain sets.

Theorem 2.24 *Suppose X is a set, Y is a topological space, and $f : X \to Y$. If f is onto (surjective) and $\tau = \{V : \exists U \subset Y$ with U open in Y and $V = f^{-1}(U)\}$, then τ is a topology for X.*

Proof. Since Y is open in Y and f is onto, we have $f^{-1}(Y) = X \in \tau$. Furthermore, $f^{-1}(\varnothing) = \varnothing \in \tau$. Thus axiom $\mathfrak{T}_1$ is satisfied.

Now let $\{V_\lambda\}$ be an arbitrary collection of elements of τ. Then for each λ there exists an open set U_λ in Y such that $f^{-1}(U_\lambda) = V_\lambda$. It now follows from Theorem 1.4(6) that

$$\bigcup_\lambda V_\lambda = \bigcup_\lambda f^{-1}(U_\lambda) = f^{-1}(\bigcup_\lambda U_\lambda),$$

and since U_λ is open in Y for each λ, $\bigcup_\lambda U_\lambda$ is open in Y. Thus, $\bigcup_\lambda V_\lambda \in \tau$, which proves that $\mathfrak{T}_2$ is satisfied.

Finally, suppose that $V = f^{-1}(U)$ and $W = f^{-1}(O)$, where U and O are open sets in Y. Then $U \cap O$ is open in Y and according to Theorem 1.4(7)

$$V \cap W = f^{-1}(U) \cap f^{-1}(O) = f^{-1}(U \cap O),$$

which proves that $V \cap O \in \tau$. Using induction shows that every finite intersection of elements of τ is again an element of τ. □

In Section 2.1.2, we developed the topological notions of open and closed sets in terms of neighborhoods. We now extend these notions to any metric space (M,d).

Definition 2.16 Let $x \in M$ and $r \in \mathbb{R}^+$. The set

1. $N_r(x) = \{y \in M : d(y,x) < r\}$ is called an *open neighborhood of* x with *radius* r.
2. The set $\bar{N}_r(x) = \{y \in M : d(y,x) \leq r\}$ is called a *closed neighborhood of* x with *radius* r.
3. Any open or closed neighborhood of x with an unspecified radius will be denoted by $N(x)$ or $\bar{N}(x)$, respectively.
4. The set $N(x')$ defined by $N(x') = N(x) \setminus \{x\}$ is called the *center deleted* neighborhood of x.

The definitions of an interior point, exterior point, boundary point, and an isolated point with respect to a set $X \subset M$ are identical to those given in Section 2.1.2. The same holds true for the definition of a limit point when replacing $\mathbb{R}$ with M in Definition 2.4. Similarly, replacing $X \subset \mathbb{R}$ with $X \subset M$ in Definition 2.5 one obtains the definitions for an open, closed, and perfect set in M. Because of these identical formulations, Theorems 2.4, 2.5, and Corollary 2.6 also hold when replacing $\mathbb{R}$ with M. Specifically, we now have that a set $V \subset M$ is open in M if and only if for every $\mathbf{x} \in V$ there exists a number $r > 0$ (depending on $\mathbf{x}$) such that $N_r(\mathbf{x}) \subset V$. Thus, neighborhoods are the basic building blocks of open sets in that every open set can be expressed as a union of neighborhoods (see also Theorem 2.4(7)). This observation has the following precise formulation.

Definition 2.17 Let (X,τ) be a topological space. A class B of open subsets of X (i.e., $B \subset \tau$) is a *basis* for the topology τ if and only if every non-empty set $U \in \tau$ is the union of elements of B. Equivalently, $B \subset \tau$ is a basis for τ if and only if for any open set U and any $x \in U$, there exists a set $V \in B$ such that $x \in V \subset U$.

If B is a basis for a topology τ, then we say that B *generates* τ and an element of B is called a *basic open set* of X.

Example 2.6

1. τ is a basis for τ.

2. Let τ be the discrete topology on a set X. Then $B = \{\{x\} : x \in X\}$ is a basis for τ.

3. The set $B_1 = \{N_r(\mathbf{x}) : \mathbf{x} \in \mathbb{R}^n, r \in \mathbb{R}^+\}$, where $N_r(\mathbf{x}) = \{\mathbf{y} \in \mathbb{R}^n : d_2(\mathbf{x}, \mathbf{y}) < r\}$, is a basis for a topology on $\mathbb{R}^n$ called the *Euclidean* topology.

4. If $B_2 = \{N_r(\mathbf{x}) : \mathbf{x} \in \mathbb{Q}^n, r \in \mathbb{Q}^+\}$, then B_2 is a countable basis for the Euclidean topology on $\mathbb{R}^n$. For if $\mathbf{x} \in U \in \tau$, where τ denotes the Euclidean topology, then $\exists r > 0$ such that $N_r(\mathbf{x}) \subset U$. Now, if $\mathbf{x} \notin \mathbb{Q}^n$ and $r \notin \mathbb{Q}^+$, pick $r' \in \mathbb{Q}^+$ such that $r' < \frac{1}{2}r$ and choose a point $\mathbf{y} \in \mathbb{Q}^n$ such that $\|\mathbf{x} - \mathbf{y}\|_2 < r'$. Then $\mathbf{x} \in N_{r'}(\mathbf{y}) \subset N_r(\mathbf{x}) \subset U$. A slight modification of this argument will verify the case where $r \in \mathbb{Q}^+$ and $\mathbf{x} \notin \mathbb{Q}^n$ The case where $\mathbf{x} \in \mathbb{Q}^n$, but $r' \notin \mathbb{Q}^+$ is trivial. Thus B_2 is a basis for τ with $B_2 \subset B_1$. The fact that B_2 is countable follows from Example 1.3(7).

5. Let (X, d) be a metric space. Define an open ball of radius $r \in \mathbb{R}^+$ about the point $x \in X$ by

$$N_r(x) = \{y \in X : d(x, y) < r\}.$$

   Then $B = \{N_r(x) : x \in X, r \in \mathbb{R}^+\}$ is a basis for a topology on X. This topology is called the *metric* topology τ_d induced by the metric d. The topological space (X, τ_d) is called a *metric* space and is customary to use the simpler notation (X, d) for (X, τ_d).

In regards to example 3, the term *Euclidean space*, or *Euclidean n-space*, refers to the topological space $\mathbb{R}^n$ generated by the d_2 metric.

A consequence of example 5 is that every metric space is also a topological space with the topology induced by the metric. The converse is, in general, not true. There exist topological spaces X with topology τ that are not metrizable; i.e., it is impossible to define a metric d on the set X such d induces τ. Furthermore, in view of these examples it is easy to see that a given topology may have many different bases that will generate it. This is analogous to the concept of a basis for a vector space: Different bases can generate the same vector space. Any linearly independent set of n vectors in $\mathbb{R}^n$ can be used as a basis for the vector space $\mathbb{R}^n$.

We now ask the following question: Given $B \subset 2^X$, when will B be a basis for some topology on X? Clearly, $X = \bigcup_{V \in B} V$ is necessary since X is open in every topology on X. The next example shows that other conditions are also needed.

Example 2.7 Let $X = \{1,2,3\}$. The set $B = \{\{1,2\},\{2,3\}\}$ cannot be a basis for any topology on X. For otherwise the sets $\{1,2\}$ and $\{2,3\}$ would themselves be open sets and therefore their intersection $\{1,2\} \cap \{2,3\} = \{2\}$ would also be an open set; but the set $\{2\}$ is not equal to a union of elements of B.

The following theorem gives both necessary and sufficient condition for a class of sets to be a basis for some topology.

Theorem 2.25 *Let B be a collection of subsets of X. Then B is a basis for some topology on X if and only if it possesses the following two properties:*

1. $X = \bigcup_{V \in B} V$.
2. *If for any U, $V \in B$, $x \in U \cap V$, then $\exists W \in B$ such that $x \in W \subset U \cap V$, or equivalently, $U \cap V$ is the union of elements of B.*

Proof. Suppose that B is a basis for a topology τ on X. Since X is open, X is the union of elements of B. Thus $X = \bigcup_{V \in B} V$ and, therefore, property (*1.*) of the theorem is satisfied. Now if U, $V \in B$, then, in particular, U and V are open sets. Hence $U \cap V$ is also open; that is, $U \cap V \in \tau$. Since B is a basis for τ, $U \cap V = \bigcup_{W \in B} W$. Thus, if $x \in U \cap V = \bigcup_{W \in B} W$, then $x \in W \subset U \cap V$ for some $W \in B$. This satisfies property (2).

Conversely, suppose that B is a collection of subsets of X which satisfies the two properties of the theorem. Let $\tau(B) = \{U : U = \varnothing$ or U is the union of elements of $B\}$, i.e., $\tau(B)$ is a collection of all possible subsets of X which can be formed from unions of elements of B. Then, obviously, $\tau(B)$ contains both X and $\varnothing$. Therefore, Axiom $\mathfrak{T}_1$ holds.

If $\{U_\lambda\}$ is a collection of elements of $\tau(B)$, then each U_λ is the union of elements of B; thus the union $\bigcup_\lambda U_\lambda$ is also a union of elements of B. Therefore $\bigcup_\lambda U_\lambda \in \tau(B)$. This shows that Axiom $\mathfrak{T}_2$ is satisfied.

The proof for verifying Axiom $\mathfrak{T}_3$ is similar and left an exercise. □

If X is a set and B is a collection of subsets of X satisfying properties (*1*) and (2) of Theorem 2.25, then we say that $\tau(B)$ is the *topology on X generated*

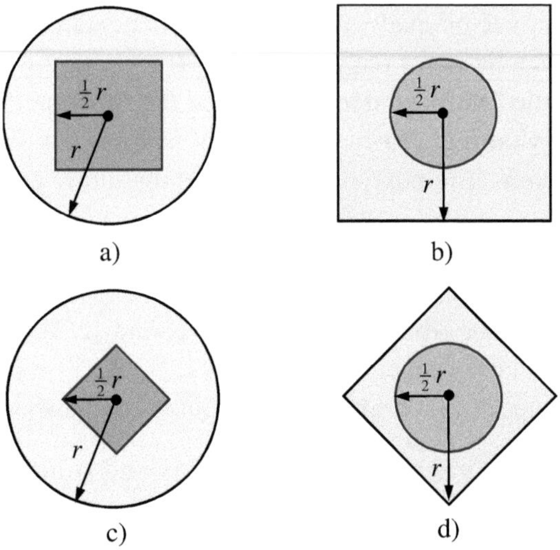

Figure 2.1 (a) and (b): subset relationships between $N_{2,r}(\mathbf{x})$ and $N_{\infty,r}(\mathbf{x})$; (c) and (d): subset relationships between $N_{2,r}(\mathbf{x})$ and $N_{1,r}(\mathbf{x})$.

by B. If B_1 and B_2 are two bases for some topologies on X, then it is possible that $\tau(B_1) = \tau(B_2)$ even though $B_1 \neq B_2$. The two bases defined in Example 2.6(3) and 2.6(4) illustrate this case. If $\tau(B_1) = \tau(B_2)$, then we say that the two bases B_1 and B_2 are *equivalent*. A necessary and sufficient condition that two bases B_1 and B_2 are equivalent is that both of the following two conditions hold:

1. For each $U \in B_1$ and each $x \in U$, there exists $V \in B_2$ such that $x \in V \subset U$.

2. For each $V \in B_2$ and each $x \in V$, there exists $U \in B_1$ such that $x \in U \subset V$.

Example 2.8 For $p \in \{1, 2, \infty\}$ let d_p denote the ℓ_p metric on $\mathbb{R}^n$. Then the three bases B_1, B_2, and B_∞ defined by $B_p = \{N_{p,r}(\mathbf{x}) : \mathbf{x} \in \mathbb{R}^2,\ r \in \mathbb{R}^+\}$, where $N_{p,r}(\mathbf{x}) = \{\mathbf{y} \in \mathbb{R}^2 : d_p(\mathbf{x}, \mathbf{y}) < r\}$, are equivalent bases. The equivalence follows from the fact that $N_{p,\frac{r}{2}}(\mathbf{x}) \subset N_{q,r}$ for any pair $p, q \in \{1, 2, \infty\}$. Two of the three different subset relationships are illustrated in Figure 2.1. Consequently, the Euclidean topological space $\mathbb{R}^2$, as well as $\mathbb{R}^n$, can be generated by any one of these three bases. The various neighborhoods $N_{p,r}(\mathbf{x})$ are also called *open discs* even though their shapes do not conform to the traditional shape of discs for $p \in \{1, \infty\}$.

The specification of a topology in terms of a basis is generally accomplished by specifying for each $x \in X$ a family of neighborhoods $\{N_\lambda(x) :$

$\lambda \in \Lambda(x)\}$, called a *neighborhood basis* at x, and verifying that the family $B = \{N_\lambda(x) : x \in X, \lambda \in \Lambda(x)\}$ satisfies the two conditions of Theorem 2.25. If the two conditions of the theorem are met, then each member $N_\lambda(x)$ is called a *basic* neighborhood of x. For example, the set $\{N_r(\mathbf{x}) : \mathbf{x} \in \mathbb{R}^2, r \in \mathbb{R}^+\}$ (generated by an ℓ_p metric) is a neighborhood basis for the Euclidean topology on $\mathbb{R}^2$ and each open disc $N_r(\mathbf{x})$ is a basic neighborhood of $\mathbf{x}$.

We need to point out that if (X, τ) is a topological space and $Y \subset X$, then Y can be viewed as a subspace of X by defining the topology τ_Y for Y by $\tau_Y = \{Y \cap U : U \in \tau\}$. Then the topological space (Y, τ_Y) is called a *subspace* of X and τ_Y is called the *induced topology*, i.e., induced by τ. It follows that if B is a basis for X, then $B_Y = \{Y \cap U : U \in B\}$ is a basis for Y.

A topological space X that satisfies the property that $\forall\, x, y \in X$ with $x \neq y$, $\exists N(x)$ and $N(y) \ni N(x) \cap N(y) = \varnothing$ is called a *Hausdorff space*. Most topological spaces used in applications, especially in physics and engineering, are Hausdorff spaces. Euclidean n-space is obviously a Hausdorff space.

Exercises 2.2.2

1. Complete the proof of Theorem 2.24 using induction.

2. Define the metric topology τ_δ, where δ denotes the metric defined in Exercise 2 of Exercises 2.2.1.

3. Define the metric topology τ_d, where d denotes the metric defined in Exercise 3 of Exercises 2.2.1.

4. Prove that Axiom $\mathfrak{T}_3$ holds in Example 2.5(3).

5. Verify Axiom $\mathfrak{T}_3$ in the proof of Theorem 2.25.

6. Prove that A is closed if and only if $A = \bar{A}$.

7. Is the Khalimsky topological space $\mathbb{Z}^2$ (Example 2.5(4)) a Hausdorff space?

2.2.3 Topological Properties of $\mathbb{R}^n$

Considering Example 2.8 it becomes obvious that for any $p \in \mathbb{N} \cup \{\infty\}$ the metric d_p can be used to define a neighborhood basis $B_p = \{N_{p,r}(\mathbf{x}) : \mathbf{x} \in \mathbb{R}^n, r \in \mathbb{R}^+\}$ which generates the Euclidean topology on $\mathbb{R}^n$. Another equivalent basis can be derived from the fact that $\mathbb{R}^n$ is the Cartesian product of n

copies of $\mathbb{R}$. This basis proves useful when generalizing the topological properties established in Section 2.1.2. In order to define this basis we need the following Lemma:

Lemma 2.26 *If* $\{A_i\}_{i=1}^n$ *and* $\{B_i\}_{i=1}^n$ *are two collections of sets, then*

$$(\Pi_{i=1}^n A_i) \cap (\Pi_{i=1}^n B_i) = \Pi_{i=1}^n (A_i \cap B_i).$$

Proof.

$$\begin{aligned} x = (x_1, x_2, \ldots, x_n) \in (\Pi_{i=1}^n A_i) \cap (\Pi_{i=1}^n B_i) \\ &\Leftrightarrow x \in \Pi_{i=1}^n A_i \quad \text{and} \quad x \in \Pi_{i=1}^n B_i \\ &\Leftrightarrow x_i \in A_i \quad \text{and} \quad x_i \in B_i \quad \text{for } i = 1, 2, \ldots, n \\ &\Leftrightarrow x_i \in A_i \cap B_i \quad \text{for } i = 1, 2, \ldots, n \\ &\Leftrightarrow x \in \Pi_{i=1}^n (A_i \cap B_i). \quad \square \end{aligned}$$

Theorem 2.27 *For* $i = 1, 2, \ldots, n$ *let* $\mathbb{R}_i = \mathbb{R}$ *denote the topological space with the Euclidean topology* τ *defined in Example 2.5(3). If* $B = \{\Pi_{i=1}^n U_i : U_i$ *is open in* $\mathbb{R}_i\}$*, then* B *is a basis for a topology for* $\mathbb{R}^n$*.*

Proof. Suppose U and V are elements of B. Then $U = \Pi_{i=1}^n U_i$ and $V = \Pi_{i=1}^n V_i$ and $U \cap V = (\Pi_{i=1}^n U_i) \cap (\Pi_{i=1}^n V_i) = \Pi_{i=1}^n (U_i \cap V_i)$, where the last equation follows from Lemma 2.26. By Axiom $\mathfrak{T}_3$, $U_i \cap V_i$ is open in $\mathbb{R}_i$ for $i = 1, 2, \ldots, n$. Note that if $\mathbf{x} = (x_1, x_2, \ldots, x_n) \in U \cap V$, then $x_i \in U_i \cap V_i$ for $i = 1, 2, \ldots, n$. Thus, there exist open neighborhoods $N_{r_i}(x_i) \subset U_i \cap V_i$ for $i = 1, 2, \ldots, n$ so that $\mathbf{x} \in \Pi_{i=1}^n (N_{r_i}(x_i)) \subset U \cap V$ with $\Pi_{i=1}^n (N_{r_i}(x_i)) \in B$. This shows that property (2.) of Theorem 2.25 is satisfied. The proof of property (*1.*) is trivial. $\square$

The topology on $\mathbb{R}^n$ generated by the basis B defined in Theorem 2.27 is called the *product topology* on $\mathbb{R}^n$. Although the Cartesian product of open subsets of $\mathbb{R}$ is open in $\mathbb{R}^n$ for both the product topology as well as the Euclidean topology, an open set in the Euclidean topology is generally not the product of open sets in $\mathbb{R}$. For instance, there do not exist two open sets U and V in $\mathbb{R}$ such that $U \times V = N_{d_2,r}(\mathbf{x}) \subset \mathbb{R}^2$. However, if $\mathbf{y} = (y_1, y_2) \in N_{d_2,r}(\mathbf{x})$, then there exist open sets $N_{r_i}(y_i) \subset \mathbb{R}_i$ for $i = 1, 2$ such that $W(\mathbf{y}) = N_{r_1}(y_1) \times N_{r_2}(y_2) \subset N_{d_2,r}(\mathbf{x})$. Simplifying notation by letting $N(\mathbf{x}) = N_{d_2,r}(\mathbf{x})$ we can now write $N(\mathbf{x}) = \bigcup_{\mathbf{y} \in N(\mathbf{x})} W(\mathbf{y})$, where each $W(\mathbf{y}) \in B$.

More generally, identical reasoning shows that any basic neighborhood for the Euclidean space $\mathbb{R}^n$ can be expressed as $N(\mathbf{x}) = N_{d_2,r}(\mathbf{x}) = \bigcup_{\mathbf{y}\in N(\mathbf{x})} W(\mathbf{y})$, where $W(\mathbf{y}) = \Pi_{i=1}^n N_{r_i}(y_i)$. Thus, if $W \subset \mathbb{R}^n$ is open in the Euclidean topology, then

$$W = \bigcup_{\mathbf{x}\in W} N(\mathbf{x}) = \bigcup_{\mathbf{x}\in W} [\bigcup_{\mathbf{y}\in N(\mathbf{x})} W(\mathbf{y})].$$

Therefore W is also open in the product topology. This shows that the product topology is the same as the Euclidean topology.

Viewing $\mathbb{R}^n$ as a product of the one-dimensional Euclidean space $\mathbb{R}$ expedites the generalization of the concepts and properties established in Section 2.1.2. Intervals played a key role in defining and determining various properties of subsets of $\mathbb{R}$. The definition of an interval in $\mathbb{R}^n$ is a simple generalization of the one-dimensional case.

Definition 2.18 Let $\mathbf{a} = (a_1, a_2, \ldots, a_n)$ and $\mathbf{b} = (b_1, b_2, \ldots, b_n)$ be two points in $\mathbb{R}^n$ with $a_i \leq b_i$ for $i = 1, 2, \ldots, n$.

1. The set $[\mathbf{a}, \mathbf{b}] = \{(x_1, x_2, \ldots, x_n) \in \mathbb{R}^n : a_i \leq x_i \leq b_i \ \forall i \in \mathbb{N}_n\}$ is called a *closed interval*. The order relation $a_i \leq x_i \leq b_i$ characterizing $[\mathbf{a}, \mathbf{b}]$ allows for any or all of the inequalities ($\leq$) to be strict equalities ($=$) or strict inequalities ($<$). By definition $[\mathbf{a}, \mathbf{b}] = \Pi_{i=1}^n [a_i, b_i]$.

2. If k denotes the number of integers i for which $a_i = b_i$, then $[\mathbf{a}, \mathbf{b}]$ is also called an *$(n-k)$-dimensional* closed interval, and if $k = 0$, then $[\mathbf{a}, \mathbf{b}]$ is often referred to as an *n-dimensional hyperbox* or *n-dimensional rectangle* as well as a *non-degenerate* closed interval. If $k = n$, then $[\mathbf{a}, \mathbf{b}]$ is a point.

3. The sets $[\mathbf{a}, \mathbf{b}) = \Pi_{i=1}^n [a_i, b_i)$ and $(\mathbf{a}, \mathbf{b}] = \Pi_{i=1}^n (a_i, b_i]$ are called *half-open intervals*.

4. The set $(\mathbf{a}, \mathbf{b}) = \Pi_{i=1}^n (a_i, b_i)$ is called an *open interval*.

It follows that $(\mathbf{a}, \mathbf{b}] = \{\mathbf{x} \in \mathbb{R}^n : a_i < x_i \leq b_i \text{ for } i = 1, 2, \ldots, n\}$. Furthermore, if $a_j = b_j$ for some $j \in \mathbb{N}_n$, then $(a_j, b_j) = \varnothing$ so that the Cartesian product $(\mathbf{a}, \mathbf{b}) = \Pi_{i=1}^n (a_i, b_i) = \varnothing$. Thus, an open interval in $\mathbb{R}^n$ is either empty or a nonempty open set. Since the empty set is an open set, every open interval is an open set in $\mathbb{R}^n$. This is in contrast with closed intervals which are nonempty for any given pair of points $\mathbf{a}$, $\mathbf{b} \in \mathbb{R}^n$.

Theorem 2.28 *Every closed interval* $[\mathbf{a},\mathbf{b}] \subset \mathbb{R}^n$ *is a closed set in the Euclidean topology.*

The proof is simple and left as an exercise.

Also, as a consequence of Theorem 2.27 and the discussion following its proof, we have the additional result:

Theorem 2.29 *If* $\mathcal{V}$ *is the set of all open intervals in* $\mathbb{R}^n$*, then* $\mathcal{V}$ *is a basis for a topology on* $\mathbb{R}^n$*. Furthermore, there exists a countable set* $B \subset \mathcal{V}$ *which is a basis for a topology on* $\mathbb{R}^n$ *which is equivalent to the Euclidean basis.*

Definition 2.19 If $X \subset \mathbb{R}^n$, then X is said to be *bounded* if and only if there exists a closed non-degenerate interval $[\mathbf{a},\mathbf{b}]$ such that $X \subset [\mathbf{a},\mathbf{b}]$.

Suppose X is a bounded set. Defining $X_i = \{p_i(\mathbf{x}) : \mathbf{x} \in X\} \subset \mathbb{R}$, where p_i denotes the projection of X into the ith coordinate axis and $i = 1, 2, \ldots, n$, it follows that $X_i \subset [a_i, b_i]$ is bounded for every $i \in \mathbb{N}_n$ and $X \subset \Pi_{i=1}^n [a_i, b_i]$. This observation in conjunction with the notions of *intervals* and *bounded* sets form the basis for proving the Bolzano-Weierstrass theorem for $\mathbb{R}^n$.

Theorem 2.30 *(Bolzano-Weierstrass) If* $X \subset \mathbb{R}^n$ *is bounded and contains infinitely many points, then* X *has at least one limit point.*

Note that the formulation of the Bolzano-Weierstrass Theorem is identical to the formulation of Theorem 2.9. In order to prove Theorem 2.9 we created a sequence of intervals $I_1 \supset I_2 \supset I_3 \supset \cdots \supset I_n \supset \cdots$ with each interval containing an infinite number of points. The proof of the n-dimensional case uses the same reasoning.

Proof. For $a \in \mathbb{R}^+$, define the interval $J_1 = [\mathbf{a}_1, \mathbf{b}_1]$, where $\mathbf{a}_1 = (a_{1,1}, a_{1,2}, \ldots, a_{1,n})$, $\mathbf{b}_1 = (b_{1,1}, b_{1,2}, \ldots, b_{1,n})$, $a_{1,i} = -a$, and $b_{1,i} = a$ for $i = 1, 2, \ldots, n$. Since X is bounded we can choose a sufficiently large so that $X \subset J_1$. Expressed as a Cartesian product J_1 has the form

$$J_1 = \prod_{i=1}^{n} [a_{1,i}, b_{1,i}] = [a_{1,1}, b_{1,1}] \times [a_{1,2}, b_{1,2}] \times \cdots \times [a_{1,n}, b_{1,n}]$$
$$= \{(x_1, x_2, \ldots, x_n) : x_i \in [a_{1,i}, b_{1,i}]\},$$

where $b_{1,i} - a_{1,i} = 2a$.

Next, divide each interval $[a_{1,i}, b_{1,i}]$ into two intervals $[a_{1,i_1}, b_{1,i_1})]$ and $[a_{1,i_2}, b_{1,i_2}]$ of equal length. This is achieved by simply setting

$$[a_{1,i_1}, b_{1,i_1})] = \{x \in \mathbb{R} : a_{1,i} \leq x \leq 0\} \text{ and } [a_{1,i_2}, b_{1,i_2}] = \{x \in \mathbb{R} : 0 \leq x \leq b_{1,i}\}.$$

For each $i \in \mathbb{N}_n$ let $k(i) \in \{i_1, i_2\}$ and consider all possible Cartesian products of form

$$\prod_{i=1}^{n}[a_{1,k(i)}, b_{1,k(i)}] = [a_{1,k(1)}, b_{1,k(1)}] \times [a_{1.k(2)}, b_{1,k(2)}] \times \cdots \times [a_{1,k(n)}, b_{1,k(n)}].$$

Each product is an n-dimensional closed interval and there are 2^n such distinct intervals. Since the union of these 2^n intervals is J_1, there must be at least one n-dimensional interval among these 2^n intervals that contains infinitely many points of X. Let J_2 denote one of the n-dimensional intervals containing infinitely many points of X and let $[a_{2,i}, b_{2,i}]$ denote the ith component of the Cartesian product defining J_2 so that

$$J_2 = \prod_{i=1}^{n}[a_{2,i}, b_{2,i}],\ b_{2,i} - a_{2,i} = a, \text{ and } J_2 \subset J_1.$$

This procedure is now repeated ad infinitum, resulting in a sequence of n-dimensional closed intervals $\{J_j\}_{j=1}^{\infty}$, where

$$J_j = \prod_{i=1}^{n}[a_{j,i}, b_{j,i}],\ b_{j,i} - a_{j,i} = \frac{a}{2^{j-2}},\ J_j \subset J_{j-1},$$

and J_j contains infinitely many points of X. According to Theorem 2.8, $\bigcap_{j=1}^{\infty}[a_{j,i}, b_{j,i}] \neq \emptyset$. In addition we also have that $\lim_{j\to\infty} \frac{a}{2^{j-2}} = 0$. Thus for each integer $i = 1, 2, \ldots, n$ there exists a point $p_i = lub\{b_{j,i} : j \in \mathbb{N}\} = glb\{a_{j,i} : j \in \mathbb{N}\}$. We now claim that the point $\mathbf{p} = (p_1, p_2, \ldots, p_n)$ is a limit point of X.

Given a number $r > 0$, choose an integer j sufficiently large such that $\frac{a}{2^{j-2}} < \frac{r}{2}$. Then $J_j \subset N_r(\mathbf{p})$ and since J_j contains infinitely many points of X, $N_r(\mathbf{p}') \cap X \neq \emptyset$. Since $r > 0$ was arbitrary, $\mathbf{p}$ is a limit point of X. □

Note that in the proof of the Bolzano-Weierstrass theorem for n-dimensional space $\mathbb{R}^n$ we used the nested set property (Theorem 2.8) for subsets of $\mathbb{R}$. We now use the Bolzano-Weierstrass theorem in order to prove the nested set property for subsets of $\mathbb{R}^n$

Theorem 2.31 *(Nested set property for $\mathbb{R}^n$) Let $\{A_n\}_{n\in\mathbb{N}}$ be a family of non-empty closed and bounded sets in $\mathbb{R}^n$. If $A_1 \supset A_2 \supset A_3 \supset \cdots$, then the set $A = \bigcap_{n=1}^{\infty} A_n$ is closed and non-empty.*

Proof. Since A_n is closed for every $n \in \mathbb{N}$, it follows from Theorem 2.4(*3*) that A is closed. We may assume that each A_n contains infinitely many points since otherwise the proof is trivial. Let $S = \{\mathbf{x}_k : \mathbf{x}_k \in A_k \ni \mathbf{x}_\ell \neq \mathbf{x}_k$ whenever $\ell \neq k\}$. The set S is bounded since $S \subset A_1$ and A_1 is bounded. Since S is bounded and infinite it follows from the Bolzano-Weierstrass theorem that S has at least one limit point, say $\mathbf{x}$. Thus $N_r(\mathbf{x}') \cap S$ contains infinitely many points. Since all but a finite number of points of S are points of A_k, $N_r(\mathbf{x}') \cap A_k \neq \emptyset$. Hence $\mathbf{x}$ is a limit point of A_k and since A_k is closed, $\mathbf{x} \in A_k$. Since k was arbitrarily chosen, $\mathbf{x} \in A_k \; \forall k \in \mathbb{N}$. Therefore $\mathbf{x} \in A = \bigcap_{n=1}^{\infty} A_n \neq \emptyset$. □

The definition of a compact subset of a topological space remains the same as Definition 2.6.

Definition 2.20 Let Y be a topological space and $X \subset Y$. The set X is said to be *compact* if and only if every open cover of X contains a finite subcover of X.

Lemma 2.32 *Let $\mathfrak{N} = \{N_r(\mathbf{x}) : r \in \mathbb{Q}$ and $\mathbf{x} \in \mathbf{Q}^n\}$ and let U be an open subset of $\mathbb{R}^n$. If $\mathbf{p} \in U$, then there exists an element $N_r(\mathbf{x})$ of $\mathfrak{N}$ such that $\mathbf{p} \in N_r(\mathbf{x}) \subset U$.*

The proof of the lemma is a consequence of the definition of an open set in the Euclidean topological space $\mathbb{R}^n$ and left as an exercise.

Theorem 2.33 *Suppose $X \subset \mathbb{R}^n$ and $V = \{V_\lambda : \lambda \in \Lambda\}$ is an open cover of X. Then there exists a countable subcollection of V which also covers X.*

Proof. Let $\mathfrak{N} = \{N_r(\mathbf{x}) : r \in \mathbb{Q}$ and $\mathbf{x} \in \mathbf{Q}^n\}$. Since $\mathfrak{N}$ is countable (why?), we shall denote the elements of $\mathfrak{N}$ by $\{N_1, N_2, \ldots\}$. Suppose that $\mathbf{p} \in X \subset \bigcup_{\lambda \in \Lambda} V_\lambda \subset \mathbb{R}^n = \bigcup_{i=1}^{\infty} N_i$. Then there exists an open set $V_\lambda \in V$ such that $\mathbf{p} \in V_\lambda$. By Lemma 2.32 there exists a neighborhood $N_j \in \mathfrak{N}$ such that $\mathbf{p} \in N_j \subset V_\lambda$. Since there may be many elements of $\mathfrak{N}$ that contain $\mathbf{p}$ and are subsets of V_λ, let $k(\mathbf{p}) = \inf\{j \in \mathbb{N} : \mathbf{p} \in N_j \subset V_\lambda\}$ and set $\lambda_{k(\mathbf{p})} = \lambda$. Then $N_{k(\mathbf{p})} \in \mathfrak{N}$ and $\mathbf{p} \in N_{k(\mathbf{p})} \subset V_\lambda$. The set $\{N_{k(\mathbf{p})} : \mathbf{p} \in X\}$ covers X and is countable. Since $N_{k(\mathbf{p})} \subset V_{\lambda_k(\mathbf{p})}$, the set $\{V_{\lambda_k(\mathbf{p})} : \mathbf{p} \in X\}$ also covers X and is a countable subcollection of V. □

Theorem 2.34 *(Heine-Borel) $X \subset \mathbb{R}^n$ is compact if and only if X is closed and bounded.*

Proof. Suppose X is compact. Let $\mathbf{p} \in \mathbb{R}^n \setminus X = X'$. For each $\mathbf{x} \in X$, consider the neighborhood $N_r(\mathbf{x})$, where $0 < r < \frac{\|\mathbf{p}-\mathbf{x}\|_2}{2}$. Then the collection

$\{N_r(\mathbf{x})\}_{\mathbf{x}\in X}$ covers X. Since X is compact, there exists a finite number of points $\{\mathbf{x}_1, \mathbf{x}_2, \mathbf{x}_3, \ldots, \mathbf{x}_k\} \subset X$ such that $X \subset \bigcup_{i=1}^{k} N_{r_i}(\mathbf{x}_i)$, where $0 < r_i < \frac{\|\mathbf{p}-\mathbf{x}_i\|_2}{2}$. Letting $\rho = min\{r_1, r_2, \ldots, r_k\}$, then the neighborhood $N_\rho(\mathbf{p})$ has the property that $N_\rho(\mathbf{p}) \cap [\bigcup_{i=1}^{n} N_{r_i}(\mathbf{x}_i)] = \varnothing$. Thus $\mathbf{p}$ is an interior point of X'. Since $\mathbf{p}$ was arbitrarily chosen, every point of X' is an interior point of X'. Therefore X' is open and by Theorem 2.4(*4*) X is closed.

For each integer $k \in \mathbb{N}$ let $\mathbf{k}$ and $-\mathbf{k}$ denote two points of $\mathbb{R}^n$ defined by $\mathbf{k} = (k, k, \ldots, k)$ and $-\mathbf{k} = (-k, -k, \ldots, -k)$. Then the open intervals $\{(-\mathbf{k}, \mathbf{k})\}_{k\in\mathbb{N}}$ cover $\mathbb{R}^n$ (why?) and, in particular, $X \subset \bigcup_{k=1}^{\infty}(-\mathbf{k}, \mathbf{k})$. Since X is compact, a finite subcollection $\{(-\mathbf{k}, \mathbf{k})\}_{k=1}^{m}$ covers X and since $(-\mathbf{k}, \mathbf{k}) \subset (-\mathbf{m}, \mathbf{m})$ whenever $k < m$, we have $X \subset [-\mathbf{m}, \mathbf{m}]$. Thus X is bounded.

Conversely, suppose that X is closed and bounded and $V = \{V_\lambda\}_{\lambda\in\Lambda}$ is an open cover for X. According to Theorem 2.33 there exists a countable subcollection $\{V_1, V_2, \ldots\} \subset V \ni X \subset \bigcup_{k=1}^{\infty} V_k$. For each $m \in \mathbb{N}$ define $U_m = \bigcup_{k=1}^{m} V_k$. Then for each positive integer m, U_m is open and U'_m is closed. Now set $A_1 = X$ and for $m > 1$ set $A_m = X \cap U'_m$. Since $U_m \subset U_{m+1}$ we have $U'_m \supset U'_{m+1}$ $\forall m \in \mathbb{N}$. Thus $A_1 \supset A_2 \supset A_3 \supset \cdots$, were each A_m is closed and bounded since each U'_m is closed, and X is closed and bounded. Now suppose that $A_m \neq \varnothing$ $\forall m$, then by Theorem 2.31, $\bigcap_{m=1}^{\infty} A_m \neq \varnothing$. Therefore there must be a point $\mathbf{x} \in X \ni \mathbf{x} \in U'_m$ $\forall m \in \mathbb{N}$. But this means that $\mathbf{x} \in X$ and $\mathbf{x} \notin U_m = \bigcup_{k=1}^{m} V_k$ $\forall m \in \mathbb{N}$. This contradicts the fact that $\mathbf{x} \in X \subset \bigcup_{k=1}^{\infty} V_k = \bigcup_{m=1}^{\infty} U_m$. Therefore our assumption that $A_m \neq \varnothing$ $\forall m$ must be incorrect. This implies that for some positive integer $m > 1$, $A_k = \varnothing$ $\forall k \geq m$. Equivalently we have that $X \cap U'_m = X \setminus U_m = \varnothing$ so that $X \subset U_m = \bigcup_{k=1}^{m} V_k$. This proves that X is compact. □

The final theorem of this chapter is a valuable tool for deciding if a set is uncountable.

Theorem 2.35 *Suppose* $\varnothing \neq P \subset \mathbb{R}^n$. *If* P *is perfect, then* P *is uncountable.*

Proof. Since P has limit points $|P| = \infty$. Let us suppose that P is countable so that the points of P can be labeled with the elements of $\mathbb{N}$, namely $P = \{\mathbf{x}_n \in P : n \in \mathbb{N}\} = \{\mathbf{x}_1, \mathbf{x}_2, \mathbf{x}_3, \ldots\}$. Next construct a sequence of neighborhoods as follows. Let $N_1 = N_r(\mathbf{x}_1) = \{\mathbf{y} \in \mathbb{R}^n : |\mathbf{y} - \mathbf{x}_1| < r\}$. Note that $\bar{N}_1 = \{\mathbf{y} \in \mathbb{R}^n : |\mathbf{y} - \mathbf{x}_1| \leq r\}$ and $\bar{N}_1 \cap P \neq \varnothing$. Since every point of P is a limit point, there exists a neighborhood N_2 such that $\bar{N}_2 \subset N_1$, $\mathbf{x}_1 \notin \bar{N}_2$, $\bar{N}_2 \cap P \neq \varnothing$. Now continue this process so that (1) $\bar{N}_{n+1} \subset N_n$ (2) $\mathbf{x}_n \notin \bar{N}_{n+1}$, and (3) $\bar{N}_{n+1} \cap P \neq \varnothing$.

Next set $A_n = \bar{N}_n \cap P$ $\forall n \in \mathbb{N}$. Since $\mathbf{x}_n \notin A_{n+1}$, no point of P is a point of $\bigcap_{n=1}^{\infty} A_n$. However $A_n \subset P$ which implies that $\bigcap_{n=1}^{\infty} A_n = \varnothing$, which contradicts

Theorem 2.31. Thus our assumption that P is countable is false. □

Consequently, the Cantor set is uncountable since the Cantor set is perfect (see Example 2.3).

Theorems 2.10, 2.11, and 2.14 as well as their proofs remain the same for any topological space. For example, the definitions of a continuous function (Definition 2.7) and the definition of a connected set (Definition 2.10) remain the same for general topological spaces:

Definition 2.21 Suppose X and Y are topological spaces and $f : X \to Y$. Then f is continuous on X if and only if $f^{-1}(U)$ is open in X for every open set U in Y.

and

Definition 2.22 Suppose Y is a topological space and $X \subset Y$. Then X is said to be *connected* if and only if there do not exist two disjoint open sets U and V in Y such that $U \cap X \neq \varnothing \neq V \cap X$ and $X \subset U \cup V$. X is said to be *disconnected* if X is not connected.

In concert with these definitions, theorems such as 2.15, 2.16 2.19, 2.20, and 2.21 have the same formulation and proofs when $\mathbb{R}$ is replaced by $\mathbb{R}^n$ or by any topological space. However, a note of caution is necessary. Although most of the theorems and definitions listed in Section 2.1.2 remain unchanged when replacing the set $\mathbb{R}$ with the set $\mathbb{R}^n$, there are various exceptions. For instance, according to Corollary 2.18, the only connected subsets of $\mathbb{R}$ are the different types of intervals. Since intervals play an important role in various types of lattice-based artificial neural networks it is useful to know that they are also connected objects in $\mathbb{R}^n$. However, in contrast to the space $\mathbb{R}$, intervals are not the only connected subsets of $\mathbb{R}^n$.

Example 2.9 Let $X = A \cup B$ denote the set shown in Figure 2.2, where

$$A = \{(0,y) \in \mathbb{R}^2 : -1 \leq y \leq 1\} \text{ and } B = \{(x,y) \in \mathbb{R}^2 : y = \sin(\frac{1}{x}), 0 < x \leq 1\}.$$

That X is connected follows from the observation that each of the sets A and B are connected and every point of A is a limit point of B, but A and B are not disconnected.

Note that in the example we have $A \cap B = \varnothing$ but A and B are not disconnected. Another way of approaching the topic of connected sets in $\mathbb{R}^n$

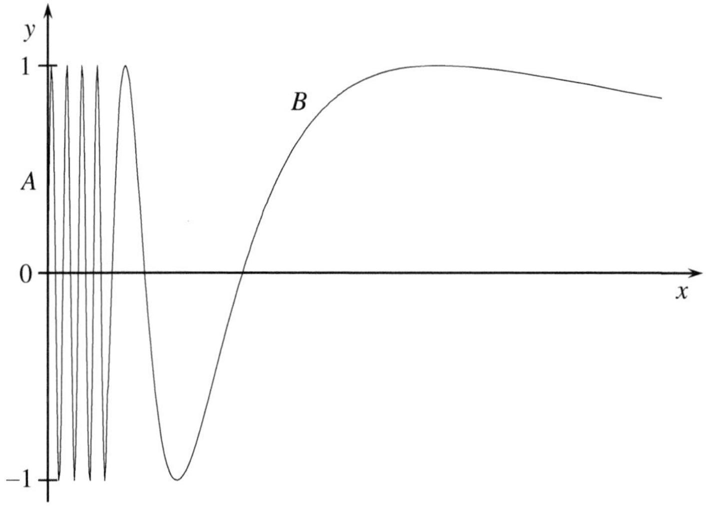

Figure 2.2 The set $X = A \cup B$.

or topological spaces in general is by defining the concept of separated sets. More specifically, two subsets A and B of a topological space X are said to be *separated* if and only if $A \cap \bar{B} = \varnothing$ and $\bar{A} \cap B = \varnothing$. Using this definition of separation results into the following theorem:

Theorem 2.36 *If A and B are connected sets which are not separated, then $A \cup B$ is connected.*

Exercises 2.2.3

1. Prove Theorem 2.28.

2. Prove Theorem 2.29.

3. Prove Lemma 2.32.

4. Prove that for $n \in \mathbb{N}$, the Euclidean topology for $\mathbb{R}^n$ has a countable basis.

5. Suppose X, Y, and Z are topological spaces. Prove that if $f : X \to Y$ and $g : Y \to Z$ are continuous functions, then $g \circ f : X \to Z$ is continuous.

6. Suppose X and Y are topological spaces and $A \subset X$ is compact. Prove that if $f : X \to Y$ is continuous, then $f(U) \subset Y$ is compact.

7. Prove that if X is a compact topological space and $A \subset X$, then A is compact.

8. Let X and Y be topological spaces and suppose that $f: X \to Y$ is continuous. Verify that if $A \subset X$ is connected, then $f(A)$ is connected.

9. Prove Theorem 2.36.

10. Suppose X and Y are topological space, and $f: X \to Y$ is continuous. Prove that if Z is a topological space, $g: Z \to X$ is continuous, and $h = f \circ g$, then $h: Z \to Y$ is continuous.

2.2.4 Aspects of $\mathbb{R}^n$, Artificial Intelligence, Pattern Recognition, and Artificial Neural Networks

Euclidean space is one of the most utilized concept in pure and applied mathematics as well as in physics and engineering, and more recently in artificial intelligence (AI), pattern recognition (PR), and artificial neural networks (ANN). A few examples will suffice to verify this claim. It follows from Chapter 1 that $(\mathbb{R}^n, +)$ is an abelian group, $(\mathbb{R}, +, \times)$ is a field, and $\mathbb{R}^n$ is a vector space over the field $(\mathbb{R}, +, \times)$. Subsection 2.2.1 of this chapter is devoted to a multitude of different metrics on $\mathbb{R}^n$, while subsection 2.2.2 shows that $\mathbb{R}^n$ is also a topological space.

The most important topological spaces are *manifolds*. A topological space X is called an *n-dimensional manifold* if X is a Hausdorff space and each $x \in X$ has a neighborhood $N(x)$ that is homeomorphic to Euclidean n-space $\mathbb{R}^n$. Bernhard Riemann, a former student of Carl Gauss, was the first to coin the term "manifold" in his extensive studies of surfaces in higher dimensions. In modern physics, our universe is a manifold. Einstein's general relative theory relies on a four-dimensional spacetime manifold. The string theory model of our universe is based on a ten-dimensional manifold $R \times M$, where R denotes the four-dimensional spacetime manifold and M denotes the six-dimensional spacial Calabi-Yau manifold. Einstein's spacetime manifold is a generalized Riemannian manifold that uses the Ricci curvature tensor [32]. We do not expect the reader to be familiar with manifold theory and associated geometries. One reason for mentioning the subject of manifolds is that even Einstein's manifold, which tells us that that space and the gravitational field are one and the same, needs the locality of $\mathbb{R}^4$ at each point of the manifold.

Another reason is the usage of manifolds in disciplines outside the fields of pure mathematics and physics. The Riemannian manifold mentioned above is named after Riemann. A Riemannian manifold R belongs to the class of differentiable manifolds and as such has a tangent space $T_p \forall p \in R$. Riemannian manifolds and other differentiable manifolds have become the latest vogue in *artificial intelligence* and *machine learning* [161, 170, 174, 241, 43, 44].

Artificial intelligence is a field of study and experimentation based on mathematics. Current AI is task-dependent in that different tasks often need different approaches for successful solutions. These approaches consist of computer programs that process input data from the task environment and generate output data. At the present time, learning and *deep learning* approaches have been used for successfully solving a variety of tasks. On the downside, and in contrast to humans, these learning approaches cannot adapt after initial learning. The question whether machines can be built that are capable of *thinking* and performing various human tasks is a subject that has been fiercely debated ever since the publication in 1920 of Karel Capek's science fiction play R.U.R. (Rosu's Universal Robots) [39]. In 1950, Alan Turing published his influential paper on computing machines and intelligence, while the term "artificial intelligence" was first introduced by John McCarthy in 1956 [282, 186]. Intrinsically, AI is concerned with developing computer systems that have the ability to perform tasks that usually require human intelligence. These tasks are multifarious, ranging from autonomous surgery, self-driving vehicles, and translation between languages to robotic machines in manufacturing. Much has been achieved since the last century. Smart cellphones, automatic translations between various languages, decision-making, medical diagnostics, and machine learning are just a few examples from the substantial list of achievements. Despite these achievements, the question "can a machine think?" remains open. This question is important in the definition of AI since thinking is an activity of intelligence. Using intelligence in the definition of AI is a circular definition since there is no standard definition as to what constitutes intelligence. In this manual we interpret AI as having the ability to learn and solve problems by machine.

One common task in AI is the recognition of pattern in data. Recognizing patterns seems to be a basic attribute common to all members of biological kingdom Animalia. Animals of the phyla Chordate and Mollusca are capable of learning to recognize and remembering new patterns that do not necessarily exist in their natural environment [15, 113, 242, 280]. Humans, in particular, are very adept in recognizing and describing objects around them. They recognize faces of their friends and can often tell from their facial expression whether they are happy or sad. Most dog owners will agree that their dogs can do the same, and probably better. In fact, when a dog owner's close friend would put on a disguise, such as wearing a mask, wig, and different attire, the friend will most likely not be recognized by the dog owner. However, owing to its superior olfactory organ, the dog will not be fooled. On the other hand, a dog has no understanding of the content of a painting.

The field of pattern recognition, as related to AI, is concerned with the autonomous recognition and classification of patterns. Following general practice, we define a *pattern* as the description of an object. For instance, suppose you observe a speeding car hitting a dog and not stopping. Reporting the incident to the local police, you are being asked to describe the type of the speeding car you observed. In this case the object is a car and your answer "*a two-door blue Honda sedan*" is the description. In other words, you have seen an object and you were able to describe it to another person. The words *two-doors, blue, Honda*, and *sedan* are called the *features* of the pattern. The set of features describing a pattern is usually expressed in vector format and called a *feature vector*. In the above example, the feature vector would be $(two-door, blue, Honda, sedan)'$. The term *pattern vectors* is also often used when referring to feature vectors. Pattern recognition by machine refers to the ability of a computer to automatically recognize patterns from data. Data can be radio waves for analysis in radio astronomy, images of faces for face recognition, X-rays for the early detection of cancer cells, seismic waves in oil and gas exploration, hyperspectral imagery for monitoring water quality and tracking forest health are just a few examples taken from a multitude of data examples. A simple example for the generation of a feature vector from a pattern input is shown in Fig.2.3. In this scenario temperature serves as the data input to a sensor. The sensor measures the Celsius scale temperature and creates a continuous function $f(t) = c$, where t denotes time and c the temperature at time t. The next step is deciding the length of the time intervals t_i for digitization. This is followed by feature extraction which samples the function at the discrete points $t_1, t_2, \ldots, t_n$ and establishes the pattern vector $\mathbf{c} = (c_1, c_2, \ldots, c_n)'$, where $c_i = f(t_i)$ for $i = 1, 2, \ldots, n$. The feature vector $\mathbf{c}$ contains all the measured information of the pattern. For an autonomous pattern recognizer, the machine needs to make a decision about the pattern. These decisions depend on the problem to be solved.

For instance, suppose the autonomous system monitors the temperature of a nuclear reactor and α represents the heat index for operating the reactor. Specifically, we have three scenarios, namely to shut down the reactor and alert whenever $\frac{1}{n}\sum_i^n c_i \geq \alpha$, to alert the human operator in charge of operations whenever $\frac{1}{n}\sum_i^n c_i < \alpha$ but $c_i \geq \alpha$ for some $i \in \mathbb{N}_n$, or continue with the next input of length n from the sensor whenever $\frac{1}{n}\sum_i^n c_i < \alpha$ and $c_i < \alpha \; \forall \, i \in \mathbb{N}_n$.

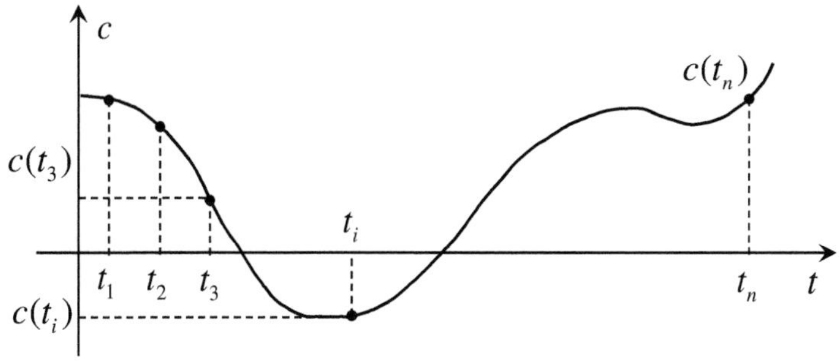

Figure 2.3 Generation of a pattern vector from a continuous temperature flow.

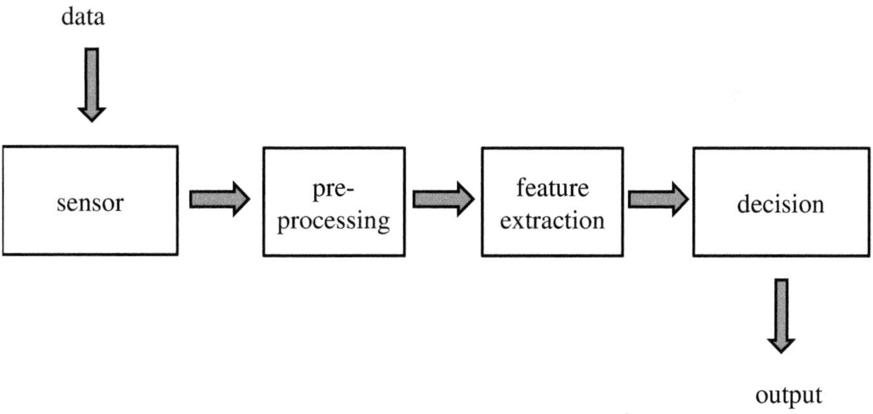

Figure 2.4 Basic flowchart for monitoring temperature in a nuclear reactor.

The basic component of the autonomous system can be summarized in terms of the flowchart shown in Figure 2.4

We must emphasize that, by and large, the tasks of preprocessing and feature extraction are far more complex than those in the preceding paragraph. In gas and oil exploration preprocessing is extremely important as well as extremely difficult. The raw data is 3-dimensional and represents location and arrival time of reflective seismic waves. These waves are created by sending shock waves into the ground (or water) and recording the reflected seismic waves on strategically located geophones. The initial seismic shock wave is created by a small group of vibrator trucks that are closely clustered together. The total combined energy the thumper trucks release at a their location can be anywhere between 100,000 to 200,000 pounds. The combined energy creates a seismic wave when the heavy blow from the trucks hits the ground. The seismic wave travels into the earth where it encounters the different layers of materials such as sand, clay and rocks. These layers will reflect part of the shock wave, and it is these reflected seismic waves that are recorded by the array of geophones. The geophones are arranged in parallel lines forming a digital rectangle or square whose center is the source seismic event. The spacing between the geophones depends on the terrain, decisions made by the individual companies and their geophysicists, time and costs, etc. Currently geophones can be several hundred feet apart. The closer the spacing, the more accurate the results. However, seismic surveys for oil explorations cover many square miles and are very expensive, costing more than $100,000$ per square mile. Tighter spacings of geophones increases the preprocessing time drastically, even on current advanced computers. Prohibitively high costs and increase in processing time make it currently impossible to reduce the geophone spacing problem.

The data recorded by the array of geophones is the *raw* data that needs to be processed. Preprocessing consist of three major tasks known as *stacking, filtering,* and *wave equation migration.* These processes are extremely complex, time consuming, very costly, and need specialized computers. Preprocessing is done by machine. Feature extraction may also be accomplished by machine with the aid of machine learning. Nonetheless, the resulting processed data must still be interpreted by a geophysicist or geologist before deciding to drill foe oil or gas.

Autonomous facial recognition is a much simpler process than autonomous seismic data preprocessing and feature extraction. Currently, autonomous face recognition has become a very successful business enterprise; a fact that can be easily ascertained by simply asking Google, "How does face recognition work?"

The data for facial recognition are faces, and the sensors are composed of one or more cameras. A single camera provides a 2-dimensional image while three or more cameras are usually employed for obtaining a 3-dimensional image. Because of business involvement and continued research advancement there exist a wide variety of different types of competing algorithms for face recognition. Different algorithms often require different preprocessing tasks. For instance, some tasks may involve finding faces in a crowd and tracking one or more of these faces. The detection of faces in images is a subset of computer vision (CV) algorithms that deal with object detection in imagery. Object detection is an extremely active area of research, owing to the fact of its applications in such areas as driverless vehicles, national defense, and medical diagnostics. If the aim is face recognition, then face detection is more often than not an essential first step.

Other tasks may involve image enhancement in case of different light

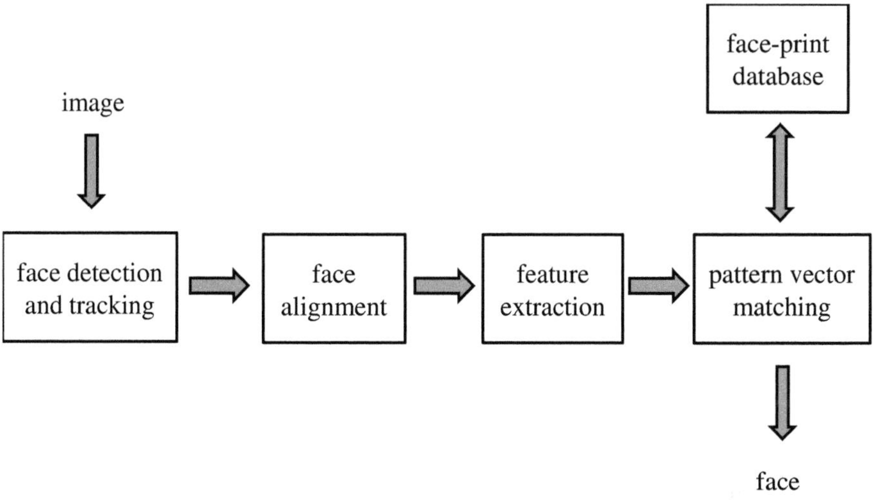

Figure 2.5 Flowchart of tasks for autonomous face recognition.

effects or/and isolating the face or/and background removal and centering of the face. The pros and cons of various tasks and approaches for face detection and face recognition can be found in the literature, e.g. [20, 311, 138, 64], as well as on the web. A general flowchart of the basic tasks for an autonomous face recognition system is given in Figure 2.5.

Feature extraction follows after completing the preliminary task of isolating and centering a face. Features are derived from the geometry of the face. These features are distances between *nodal points* on the face. Examples of nodal points are points around the nose that measure the length and width of the nose, two points that measures the distance between the top of the forehead to the top of the nose, and various others that measure the width and length of the chin, and so on. Figure 2.6 shows a few of these nodal point connections. The resulting pattern vector, also called a *face-print*, represents the geometry of the face taken by the camera.

For a given person, several images are usually taken in order to establish a robust autonomous facial recognizer systems. Some systems also use multi-view images by viewing the face's left, center, and right side. The resulting face-prints are stored in a database. Assuming that the dimension of the pattern vectors is $n \in \mathbb{N}$, and $k \in \mathbb{N}$ denotes the number of images taken of person $P(j)$, then the set $P(j) = \{\mathbf{p}^1, \ldots, \mathbf{p}^k\}$ forms a close cluster of points in $\mathbb{R}^n$.

Autonomous face recognition systems require databases. The databases are crucial in order to carry out facial recognition and verification. Verifica-

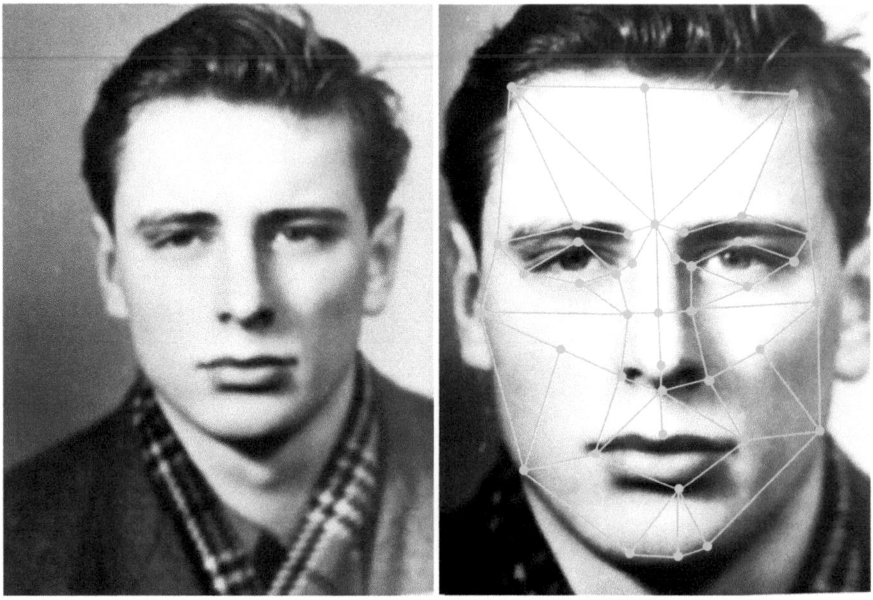

Figure 2.6 Left, input face image; right, centered face with nodal points and their connections.

tion means to verify whether or not the recognition was correct. Currently ANN approaches for solving the face detection, face recognition and verification problems have become very popular. The reason for the popularity is due to the surprising increase in the performance of face detection, recognition and verification. The success of ANNs in solving a wide variety of pattern recognition problems was known before their use in facial recognition. Thus, since faces are patterns, it makes sense to apply ANNs in order to solve the autonomous face recognition problem.

A common basic model of an artificial neural network is the *hidden layer* model shown in Figure 2.7. A variety of models have feedback loops used in training. If the network has two or more hidden layers, then it is called a *deep* network. Some deep networks are also be obtained by combining two or more different ANNs. Combining ANNs is used to increase performance, but it also increases the computational burden. Most AI and deep ANN algorithms are bandwidth hungry and rely on accelerators. An *accelerator* is a device that helps speed up the performance of an application, such as ANNs or computer vision.

An example of an accelerator is the Field Programmable Gate Array (FPGA). The FPGA is an integrated circuit of a collection of logic blocks,

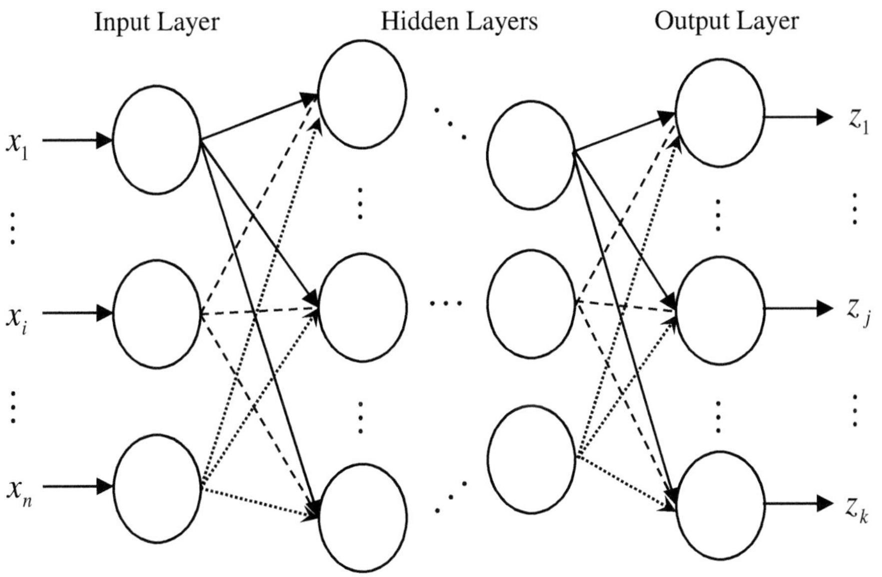

Figure 2.7 An artificial neural network with hidden layers of neurons.

where each logic block has the ability to do a simple logic operation such as AND, OR, and XOR. The real power of FPGAs is that the logic blocks can be easily reconfigured dynamically to perform massively parallel tasks in real time. Increased performance in real time pattern recognition has become an extraordinary achievement of the development of artificial neural networks.

Another important reason for the success of ANNs in pattern recognition is that the network can *learn* to recognize and verify a pattern in a data set that was never used as an input. What is important to note here is that learning algorithms are offline. As pointed out by Gary Marcus, most AI learning algorithms, particularly deep learning algorithms, are greedy, brittle, rigid and opaque [182]. According to Kjell Hole and Sabutai Ahmad, the algorithms are greedy because they demand big data sets to learn, brittle because they frequently fail when confronted with mildly different scenario than that in the training set, rigid because they cannot keep adapting after initial training, and opaque because the internal representations make it challenging to interpret their decisions [118]. To overcome the current shortcomings they recommend that the AI community needs to concentrate more on biomimetic approaches for the computational principles of the human brain in order to create a general AI with a performance close to humans at most cognitive tasks of interest. We realize that the differences between biological neurons and artificial neurons, and their associated networks, are different to the extreme. However, this knowledge should not prevent us from the challenge of exploring biomimetic approaches.

CHAPTER 3

Lattice Theory

3.1 HISTORICAL BACKGROUND

HISTORICALLY, lattice theory had its beginning 120 years ago when Richard Dedekind worked on a revised and enlarged edition of Peter Dirichlet's *Vorlesungen über Zahlentheorie* [71]. Dedekind discovered the basic properties of modular lattices (Dual Gruppen von Modultypus) and distributive lattices [70, 71]. Although research related to lattices continued throughout the early part of the 20th century, it was the 1930s that witnessed a rapid expansion of contributions directly related to the development of lattice theory as a field of mathematics in its own right. These contributions focused on applying lattice theory to the fields of set theory, algebra, geometry, and topological spaces [26, 27, 28, 29, 87, 127, 135, 136, 137, 175, 177, 210, 270, 271, 304]. A major milestone resulting from these efforts was Garrett Birkhoff's classic book on lattice theory [30]. Birkhoff's book presents a systematic development of lattice theory that generalizes and unifies various relationships between subsets of a set and the logic of set theory, between substructures of an algebraic structure such as groups and rings, between logic and quantum mechanics, and between concepts within geometric structures such as topological spaces.

Applications of lattices outside the domain of mathematics followed its establishment as a well-defined theory. Shannon (1953), Shimbel (1954), and Benzaken (1968) used lattices in electronic communication theory [22, 249, 256], Simon (1955) in economics [257], and Cuninghame-Green (1960) and Giffler (1960), followed by several other researchers in industrial systems engineering with focus on machine scheduling, optimization, and approximation problems [14, 61, 62, 63, 93, 208].

Application of a new theory generally involves the development of novel techniques, and results in further insights and expansion of the theory itself.

DOI: 10.1201/9781003154242-3

Cuninghame-Green's formulation of a matrix theory utilizing lattice operations turned out to be of vital importance in the development of lattice associative memories (LAMs) as well as in establishing important connections between lattice theory and the field of mathematical morphology as utilized in image processing [65, 66, 218, 219, 223]. A more recent development is the incorporation of dendritic and axonal structures into the morphology of LNN neurons in order to enhance the computational capabilities of LNNs [221, 222, 224].

3.2 PARTIAL ORDERS AND LATTICES

A function $f : X \to Y$ is a particular subset of $X \times Y$ that *relates* each element of X with some element of Y (see Definition 1.4). When $Y = X$, then f is an example of a *binary* relation. Binary relations were also discussed in some detail in subsection 1.2.1. The concept of an *order* relation of elements of a set X is fundamental in lattice theory and we begin our discussion by examining this concept more closely.

3.2.1 Order Relations on Sets

A binary relation $\mathfrak{R}$ on a set X is, intuitively, a proposition such that for each ordered pair (x, y) of elements of X, one can determine whether $x\mathfrak{R}y$ is or is not true. Here $x\mathfrak{R}y$ means that "x is related (by the relation $\mathfrak{R}$) to y." For example, if L is the set of all lines in a given plane, then "is parallel to" or "is perpendicular to" are binary relations on L.

The notion of a binary relation on a set can be rigorously defined by stating it formally in terms of the set concept.

Definition 3.1 A *binary relation* $\mathfrak{R}$ on a set X is a subset $\mathfrak{R} \subset X \times X$.

Thus, any subset $\mathfrak{R}$ of $X \times X$ is a binary relation on X and if such a subset is being used, then it is customary to write $x\mathfrak{R}y$ for $(x, y) \in \mathfrak{R}$.

Example 3.1

1. Set inclusion is a relation on any power set. In particular, if X is a set and

$$\mathfrak{R} = \{(A, B) : A \subset B,\ A,\ B \in 2^X\},$$

then $\mathfrak{R}$ is a binary relation of 2^X.

2. The relation of *less or equal*, denoted by $\leq$, between real numbers is the set $\{(x,y) : x \leq y\} \subset \mathbb{R} \times \mathbb{R}$.

3. For any set X, the *diagonal* $\Delta = \{(x,x) : x \in X\}$ is the relation of equality.

4. The *inverse relation* of $\mathfrak{R}$, denoted by $\mathfrak{R}^{-1}$, is the relation $\mathfrak{R}^{-1} = \{(y,x) : (x,y) \in \mathfrak{R}\}$. Thus, the inverse relation of $\leq$ in (2) above is the relation *greater or equal* $\geq$.

Note that in binary relations on X, a pair of elements of X need not be related. For instance, in Example 3.1(3.) above, if $x,y \in X$ and $x \neq y$, then neither (x,y) or (y,x) are in Δ.

Certain relations on a set allow elements of that set to be arranged in some order. For example, if a child arranges a set of sticks in the order from shortest to longest, it has an intuitive grasp of the relation "is shorter than." From this example we can see that there are at least two properties which a relation $\mathfrak{R}$ must have if it is to order a set. Specifically:

- $\mathfrak{R}$ must be antisymmetric. That is, given two sticks, one of them must be shorter than the other. Otherwise they could not be given a relative position in the order.

- $\mathfrak{R}$ must be transitive. That is, given three sticks x, y, and z, with x shorter than y and y shorter than z, then x must be shorter than z.

We collect these ideas in a definition.

Definition 3.2 A relation $\preccurlyeq$ on a set P is called a *partial order* on P if and only if for every x, y, $z \in P$ the following three conditions are satisfied:

1. $x \preccurlyeq x$ (reflexive)

2. $x \preccurlyeq y$ and $y \preccurlyeq x \Rightarrow x = y$ (antisymmetric)

3. $x \preccurlyeq y$ and $y \preccurlyeq z \Rightarrow x \preccurlyeq z$ (transitive)

The relations defined in Example 3.1 are all partial order relations. The relation of *less or equal* given in Example 3.1(2.) is also called the *natural order* on $\mathbb{R}$.

A set P together with a partial order $\preccurlyeq$, denoted by $(P, \preccurlyeq)$, is called a *partially ordered set* or simply a *poset*. If $x \preccurlyeq y$ in a partially ordered set, then we say that x *precedes* y or that x *is included in* y and that y *follows* x or that y *includes* x. If $(P, \preccurlyeq)$ is a poset, then we define the notation $x \prec y$, where $x, y \in P$, to mean that $x \preccurlyeq y$ and $x \neq y$. The next theorem is a trivial consequence of these definitions.

Theorem 3.1 *Suppose* $(P, \preccurlyeq)$ *is a poset.*

1. *If* $C \subset P$, *then* $(C, \preccurlyeq)$ *is also a poset.*
2. $\nexists x \in P \ni x \prec x$.
3. *If* $x \prec y$ *and* $y \prec z$, *then* $x \prec z$, *where* $x, y, z \in P$.

Note that Definition 3.2 does *not* imply that given $x, y, \in P$, then either $x \preccurlyeq y$ or $y \preccurlyeq x$. Consequently, in a partially ordered set not every pair of elements needs to be related. Partially ordered sets in which every pair of elements is related play a vital role in lattice theory. Specifically, we have

Definition 3.3 A poset $(P, \preccurlyeq)$ is called a *totally* (or *linearly*) *ordered* if and only if $\forall x, y \in P$, with $x \neq y$, either $x \prec y$ or $y \prec x$.

The set $\mathbb{R}$ together with the natural order of $\leq$ is an example of a totally ordered set while the relation *is a subset of* (Example 3.1.1) is a partial order which is not a total order. A poset of primary interest in this treatise is provided by the next example.

Example 3.2 Let the relation $\leq$ on $\mathbb{R}^n$ be defined by $\mathbf{x} \leq \mathbf{y} \Leftrightarrow x_i \leq y_i$ for $i = 1, 2, \ldots n$. It is easily verified that $\leq$ satisfies all three conditions of Definition 3.2. The partial order $\leq$ on $\mathbb{R}^n$ is said to be *induced* by the partial order of less or equal on $\mathbb{R}$. Note that for $n = 2$, $(2,3) \nleq (1,5) \nleq (2,3)$, but $(2,3) \leq (2,5)$ and $(1,5) \leq (2,5)$. Thus $(\mathbb{R}^n, \leq)$ is not totally ordered.

If $\mathfrak{R} = \preccurlyeq$ is a partial order on P, then it is easy to see that the inverse relation $\mathfrak{R}^{-1}$, denoted by $\succcurlyeq$, is also a partial order on P. The inverse partial order relation $\succcurlyeq$ of $\preccurlyeq$ is also called *the dual of* $\preccurlyeq$ and gives rise to the following definition:

Definition 3.4 The *dual* of a partially ordered set P is that partially ordered set P^* defined by the inverse partial order relation on the same elements.

Thus, if $(P, \preccurlyeq)$ is a poset, then $(P, \preccurlyeq)^* = (P^*, \succcurlyeq)$ is also a poset and since $(P^*)^* = P$, this terminology is legitimate. In the remainder of this section we suppose that $(R, \preccurlyeq)$, $(S, \preccurlyeq)$, and $(T, \succcurlyeq)$ are three posets.

Definition 3.5 A function $\psi : R \to S$ is called *order preserving* or *isotone* if

$x \preccurlyeq y \Rightarrow \psi(x) \preccurlyeq \psi(y)$ and *strictly isotone* if $x \prec y \Rightarrow \psi(x) \prec \psi(y)\ \forall\, x, y \in R$.

Similarly, a function $\varphi : S \to T$ is called *order reversing* or *antitone* if

$x \preccurlyeq y \Rightarrow \varphi(x) \succcurlyeq \varphi(y)$ and *strictly antitone* if $x \prec y \Rightarrow \varphi(x) \succ \varphi(y)\ \forall\, x, y \in S$.

If $\psi : R \to S$ is order preserving and has an order preserving inverse $\psi^{-1} : S \to R$, then ψ and ψ^{-1} are called *order preserving isomorphisms* and the posets R and S are said to be *isomorphic*. Again, the notation $R \approx S$ means that R and S are isomorphic. If $R = S$, then the isomorphism ψ is also called an *automorphism*. Similarly, if $\varphi : S \to T$ is order reversing and has an order reversing inverse $\varphi^{-1} : T \to S$, then φ and φ^{-1} are called *order reversing isomorphisms* and the posets S and T are also said to be *isomorphic* and again denoted by $S \approx T$. In case $T = S^*$, then the isomorphism φ is called a *dual* isomorphism.

Example 3.3 The posets $(\mathbb{R}, \leq)$ and $(\mathbb{R}^+, \geq)$ are isomorphic. An isomorphism is given by the function $\psi : \mathbb{R} \to \mathbb{R}^+$ defined by $\psi(x) = e^{-x}$. Since $x \leq y \Rightarrow e^{-x} \geq e^{-y}$, ψ is order reversing. The inverse of ψ is given by $\psi^{-1}(v) = -ln(v)$, the negative of the natural log of v, where $v \in \mathbb{R}^+$. Since $\psi^{-1}(\psi(x)) = \psi^{-1}(e^{-x}) = -ln(e^{-x}) = -(-x) = x$ and $v \geq w \Rightarrow \psi^{-1}(v) = -ln(v) \leq -ln(w) = \psi^{-1}(w)$, ψ^{-1} is an order reversing inverse. However, ψ is not a dual isomorphism since $\mathbb{R}^* \neq \mathbb{R}^+$.

Let P be a partially ordered set with partial order $\preccurlyeq$. An *upper bound* (if it exists) of a subset X of P is an element $y \in P$ such that $x \preccurlyeq y\ \forall x \in X$. If such an upper bound exists, then X is said to be *bounded from above*. The *least upper bound* of X, denoted by $lub(X)$ or $sup(X)$, is an upper bound $y_0 \in P$ of X such that $y_0 \preccurlyeq y$ for every upper bound y of X. By the antisymmetry property of partially ordered sets, $supX$, if it exists, is unique. In a likewise fashion we define a *lower bound* (if it exists) of X as an element $y \in P$ such that $y \preccurlyeq x \; \forall x \in X$. If such a lower bound exists, then X is said to be *bounded from below*. The set X is said to be *bounded* if and only if it is bounded from above and from below. The *greatest lower bound* of X, denoted by $glb(X)$ or $inf(X)$, is a lower bound $y_0 \in P$ for X such that $y \preccurlyeq y_0$ for every lower bound y of X. Again, by antisymmetry, $inf(X)$ is unique if it exists. If $inf(X) \in X$, then $inf(X)$ is also called the *least element* or *smallest element* or the *minimum* of X. Likewise, if $sup(X) \in X$, then $sup(X)$ is also called the *greatest element* or *largest element* or the *maximum* of X. This terminology is in agreement with the terminology introduced in subsection 2.1.1 and should not be confused with the terms of minimal or maximal elements of X.

Definition 3.6 If $X \subset P$ and $(P, \preccurlyeq)$ is a partially ordered set, then $y \in X$ is called a *maximal element of* $X \Leftrightarrow (y \preccurlyeq x \Rightarrow y = x) \forall x \in X$. Similarly, $y \in X$ is called a *minimal element of* $X \Leftrightarrow (x \preccurlyeq y \Rightarrow y = x) \forall x \in X$.

Obviously, a least element must be minimal and a greatest element must be maximal, but the converse is not true.

Example 3.4 Let $P_1 = \{a,b,c,d\}$ and $X = P$. Defining a partial order $\preccurlyeq$ on P_1 by setting $x \preccurlyeq x \; \forall x \in X$, $a \preccurlyeq b \preccurlyeq d$, and $a \preccurlyeq c \preccurlyeq d$, then obviously $d = lubX$ and $a = glbX$. Figure 3.1 (a) provides a graphical interpretation of this partially ordered set. Here the straight line segments between elements represent the ordering, with the *larger* element above the smaller element, and the *smallest* or *least* element on the bottom of the graph. On the other hand, setting $P_2 = \{a,b,c,d\}$ and defining the partial order $\preccurlyeq$ on P_2 by setting $x \preccurlyeq x \; \forall x \in P_2$, $a \preccurlyeq c$, $a \preccurlyeq d$, $b \preccurlyeq c$, and $b \preccurlyeq d$. For this partial order we have that if $x \in P_2$ and $c \preccurlyeq x$, then $x = c$. Thus c is a maximal element. Similarly, $d \preccurlyeq x \Rightarrow x = d$. Thus the poset P_2 has two distinct maximal elements. It is just as obvious that a and b are minimal elements of P_2. Consequently, neither maximal nor minimal elements need to be unique. Figure 3.1 provides a graphical interpretation of these two posets.

It is also easy to show that any finite nonempty subset X of a partially ordered set has minimal and maximal elements. These elements play an im-

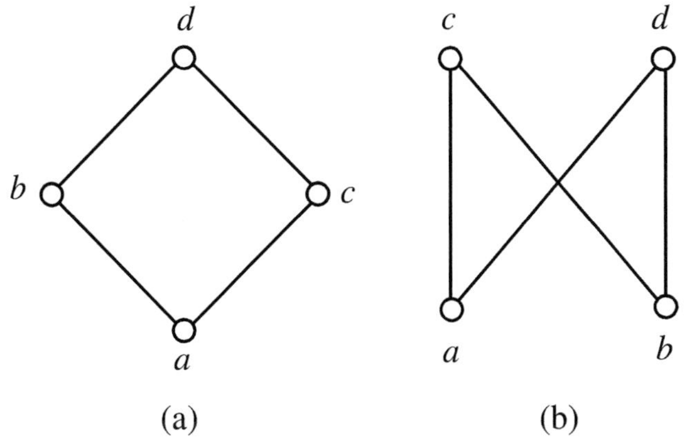

Figure 3.1 Graphical representation of finite posets: (a) the poset P_1 and (b) the poset P_2.

portant role in well-ordered bounded sets.

Definition 3.7 If $(P, \preccurlyeq)$ is a poset and $C \subset P$, then C is called a *chain* in P if and only if C is totally ordered. If C is a chain in P and $card(C) = n$, where $n \in \mathbb{Z}^+$, then the *length of* C is defined as $\ell(C) = n - 1$. The *length* of P is defined as $\ell(P) = lub\{\ell(C) : C \text{ is a chain in } P\}$ if the least upper bound exists. P is said to be of *finite length* if and only if $\ell(P)$ is finite.

Example 3.5 Let $P = \{a, b, c, d, e\}$ be partially ordered by $a \preccurlyeq e \preccurlyeq d$ and $a \preccurlyeq b \preccurlyeq c \preccurlyeq d$. Since the largest chain is $a \preccurlyeq b \preccurlyeq c \preccurlyeq d$, $\ell(P) = 3$ (see also Figure 3.1(a)). On the other hand, if $\leq$ denotes the natural order of the integers, then the poset $(\mathbb{Z}, \geq)$ has length $\ell(\mathbb{Z}) = \aleph_0$.

Applying the definition of a chain one can easily obtain the following result:

Theorem 3.2 *Any subset of a chain is a chain and every finite chain has a glb and lub.*

Definition 3.8 Let $(P, \preccurlyeq)$ be a poset and $a, b \in P$. We say that b *covers* a if and only if $a \prec b$ and $\nexists\, x \in P \ni a \prec x \prec b$.

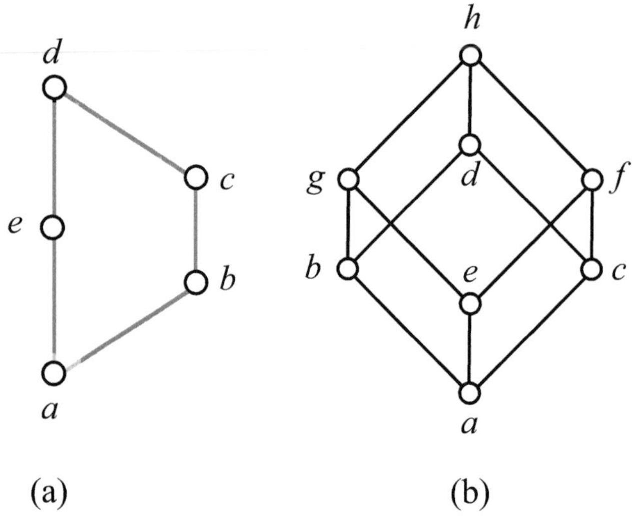

Figure 3.2 Hasse diagrams: (a) the poset of Example 3.5, (b) poset of Example 3.6.

For finite posets the covering relations can be used to obtain graphical representations of the sets and their order relations as was done in Fig. 3.2. In this representation, known as a *Hasse* diagram, each element of a poset is represented as a vertex in the plane and a line segment or curve that goes upward from x to y whenever y covers x. A segment may cross another segment but must not touch any vertices except at the segment's endpoints. In this text the vertices will be represented by small circles.

Example 3.6 Suppose $P = \{a,b,c,d,e,f,g,h\}$ with partial order defined by $a \prec b \prec g \prec h$, $a \prec d \prec f \prec h$, $a \prec e \prec f$, $b \prec c \prec h$, $e \prec c$, and $d \prec g$. The Hasse diagram representing the poset P is shown in Figure 3.2(b).

Definition 3.9 If P is a poset of finite length and $O = inf P$, then for $x \in P$ the *height of* x, denoted by $h(x)$, is defined as

$$h(x) = lub\{\ell(C) : C = \{x_0, x_1, \ldots, x_m\} \text{ with } O = x_0 < x_1 < \cdots < x_m = x\}.$$

As a direct consequence of this definition we have that $h(x) = 1 \Leftrightarrow x$ covers O, and if $I = supP$, then $h(I) = \ell(P)$. If $h(x) = 1$, then x is called an *atom*.

The height and length functions belong to a class of functions that provide specific measurements and information on lattice structures. These measures will be discussed in subsequent sections of this chapter.

Exercises 3.2.1

1. Prove Theorem 3.1.

2. How many different partially ordered sets can be obtained from $X = \{a,b,c\}$ and how many of these partially ordered sets are non-isomorphic?

3. Repeat Exercise 2 for the set $\{a,b,c,d\}$.

4. Suppose $\varnothing \neq X \subset P$ and $|X| = n$ for some $n \in \mathbb{N}$. Verify X has a minimal and a maximal element.

5. Prove Theorem 3.2.

3.2.2 Lattices

This section summarizes an assortment of relevant concepts from lattice theory, beginning with a short list of different types of lattices.

Definition 3.10

1. A *lattice* is a partially ordered set L such that for any two elements $x, y \in L$, $glb\{x,y\}$ and $lub\{x,y\}$ exist. If L is a lattice, then we denote $glb\{x,y\}$ by $x \wedge y$ and $lub\{x,y\}$ by $x \vee y$, respectively. The expression $x \wedge y$ is also referred to as the *meet* or *min* of x and y, while $x \vee y$ is referred to as the *join* or *max* of x and y.

2. A *sublattice* of a lattice L is a subset X of L such that for each pair $x, y \in X$, we have that $x \wedge y \in X$ and $x \vee y \in X$.

3. A lattice L is said to be *complete* if and only if for each of its subsets X, $inf(X)$ and $sup(X)$ exist. The symbols $\bigwedge X$ and $\bigvee X$ are also commonly used for $inf(X)$ and $sup(X)$, respectively.

4. A lattice L is said to be *conditionally complete* if and only if for every nonempty bounded subset X, $inf(X)$ and $sup(X)$ exist.

According to this definition, any nonempty complete lattice L contains a least element O and a greatest element I. These elements are commonly called the *universal bounds* of L. It is also obvious that the dual of a lattice is a lattice, and that the dual of any complete lattice is a complete lattice with *glb* and *lub* interchanged. For a finite subset $X = \{x_1, x_2, \ldots, x_n\}$ of a complete lattice the *glb* and *lub* are given by $\bigwedge X = \bigwedge_{i=1}^{n} x_i = x_1 \wedge x_2 \wedge \cdots \wedge x_n$ and $\bigvee X = \bigvee_{i=1}^{n} x_i = x_1 \vee x_2 \vee \cdots \vee x_n$, respectively.

Example 3.7

1. The poset P_1 in Example 3.4 is a lattice but the poset P_2 is not a lattice since neither the $glb\{a,b\}$ or the $lub\{c,d\}$ exist.

2. The real numbers $\mathbb{R}$ together with the relation of less or equal ($\leq$) constitute a totally ordered set. If $x \vee y = lub\{x,y\}$ and $x \wedge y = glb\{x,y\}$ $\forall x, y \in \mathbb{R}$, then $\mathbb{R}$ together with the operations $\vee$ and $\wedge$ is a lattice. But $(\mathbb{R}, \vee, \wedge)$ is not a complete lattice as there is no smallest or largest number. However, $\mathbb{R}$ is a conditionally complete lattice since every bounded nonempty subset has a *glb* and a *lub*.

3. Define $-\infty < x < \infty$ $\forall x \in \mathbb{R}$ and $-\infty \leq x \leq \infty$ for $x \in \{-\infty, \infty\}$. Then $(\mathbb{R}_{\pm\infty}, \vee, \wedge)$ is a complete lattice with largest element ∞ and smallest element $-\infty$. The dual of this lattice is obtained by replacing $\leq$ with the relation of *greater or equal*, $\geq$. Obviously, $(\mathbb{R}, \vee, \wedge)$ is a sublattice of $(\mathbb{R}_{\pm\infty}, \vee, \wedge)$.

4. For a given set X, the power set 2^X together with the relation of $\subset$ of set inclusion is a complete lattice with largest element X and smallest element $\varnothing$. For any family $\mathcal{A} = \{S_\lambda : \lambda \in \Lambda\}$ of subsets of X, $inf\mathcal{A} = \bigcap_{\lambda \in \Lambda} S_\lambda$ and $sup\mathcal{A} = \bigcup_{\lambda \in \Lambda} S_\lambda$.

5. The set $[0, \infty]$ with the relation $\leq$ is a complete lattice. Here 0 is the smallest element and ∞ is the largest element. $([0, \infty], \vee, \wedge)$ is a sublattice of $(\mathbb{R}_{\pm\infty}, \vee, \wedge)$ but not of $(\mathbb{R}, \vee, \wedge)$.

6. The unit interval $[0, 1] = \{x \in \mathbb{R} : 0 \leq x \leq 1\}$ together with the relation $\leq$ is a complete sublattice of $(\mathbb{R}, \vee, \wedge)$. The smallest and greatest elements are 0 and 1, respectively.

7. We partially order $\mathbb{R}^X$ by $f \leq g \Leftrightarrow f(x) \leq g(x)$ $\forall x \in X$ and define $h = f \vee g$ by

$$h(x) = \begin{cases} f(x) & \text{if } g(x) \leq f(x) \\ g(x) & \text{if } f(x) \leq g(x) \end{cases}$$

and $k = f \wedge g$ by

$$k(x) = \begin{cases} f(x) & \text{if } f(x) \le g(x) \\ g(x) & \text{if } g(x) \le f(x). \end{cases}$$

It follows that $(\mathbb{R}^X, \vee, \wedge)$ is a lattice but not a complete lattice. However, $(\mathbb{R}^X_{\pm\infty}, \vee, \wedge)$ is a complete lattice with smallest element the constant function O defined by $O(x) = -\infty \ \forall x \in X$ and largest element the constant function I defined by $I(x) = \infty \ \forall x \in X$.

The binary operations $\wedge$ and $\vee$ on lattices have several important properties, some of them being analogous to those of ordinary multiplication and addition. The following properties are easily verified [30].

Theorem 3.3 *If $(X, \preccurlyeq)$ is a partially ordered set, then the operations $\vee$ and $\wedge$ satisfy the following equations (whenever the expressions referred to exist):*

1. $x \wedge x = x$ *and* $x \vee x = x$ *(idempotent)*
2. $x \wedge y = y \wedge x$ *and* $x \vee y = y \vee x$ *(commutative)*
3. $x \wedge (y \wedge z) = (x \wedge y) \wedge z$ *and* $x \vee (y \vee z) = (x \vee y) \vee z$ *(associative)*
4. $x \wedge (x \vee y) = x \vee (x \wedge y) = x$ *(absorption)*
5. $x \preccurlyeq y \Leftrightarrow x \wedge y = x$ *and* $x \preccurlyeq y \Leftrightarrow x \vee y = y$ *(consistency)*
6. *If X has a least element O, then $O \wedge x = O$ and $O \vee x = x \ \forall x \in X$*
7. *If X has a largest element I, then $I \wedge x = x$ and $I \vee x = I \ \forall x \in X$*

Observe that in the above theorem we do not require that the partially ordered set X is a lattice. In fact, Dedekind [71] used properties *1* through *4* of Theorem 3.3 to define lattices, and Birkhoff [30] proved that these properties completely characterize lattices. By employing commutativity, associativity, and induction, the following associative laws become an easy consequence of Theorem 3.3.

Corollary 3.4 *If* $(L,\vee,\wedge)$ *is a lattice and* $x_{ij} \in L$ *for* $i = 1,\ldots,n$ *and* $j = 1,\ldots,m$, *then*

$$\bigvee_{i=1}^{n}\Big(\bigvee_{j=1}^{m} x_{ij}\Big) = \bigvee_{j=1}^{m}\Big(\bigvee_{i=1}^{n} x_{ij}\Big) \quad \text{and} \quad \bigwedge_{i=1}^{n}\Big(\bigwedge_{j=1}^{m} x_{ij}\Big) = \bigwedge_{j=1}^{m}\Big(\bigwedge_{i=1}^{n} x_{ij}\Big).$$

Additional algebraic relationships for lattices are given by the next theorem.

Theorem 3.5 *If* $(L,\vee,\wedge)$ *is a lattice with partial ordering* $\preccurlyeq$, *then*

1. $y \preccurlyeq z \Rightarrow x \wedge y \preccurlyeq x \wedge z$ *and* $x \vee y \preccurlyeq x \vee z$ *(isotone)*
2. $x \preccurlyeq z \Rightarrow x \vee (y \wedge z) \preccurlyeq (x \vee y) \wedge z$ *(modularity)*
3. $x \wedge (y \vee z) \preccurlyeq (x \wedge y) \vee (x \wedge z)$ ($\wedge$ *distributivity*)
4. $x \vee (y \wedge z) \preccurlyeq (x \vee y) \wedge (x \vee z)$ ($\vee$ *distributivity*)

Proof. We shall only prove property 2. The remaining properties are just as easy or follow directly from Theorem 3.3.
2. Obviously, $x \preccurlyeq x \vee y$ and, by hypothesis, $x \preccurlyeq z$. Hence, $x \preccurlyeq (x \vee y) \wedge z$. Also, $y \wedge z \preccurlyeq y \preccurlyeq x \vee y$. Thus, $y \wedge z \preccurlyeq (x \vee y) \wedge z$ and, therefore, $x \vee (y \wedge z) \preccurlyeq (x \vee y) \wedge z$.
□

It is often convenient and much simpler to deal with only one of the operations of $\vee$ or $\wedge$ and obtain equivalent relations for the other through duality. For this, the following notion is helpful.

Definition 3.11 A *semilattice* is a commutative semigroup $(S,\circ)$ which satisfies the idempotent law $s \circ s = s$.

It follows from Theorem 3.3 that every partially ordered set S for which the operation $x \vee y$ is defined for each pair $x, y \in S$ is a semilattice. Conversely, under the relation defined by

$$x \preccurlyeq y \quad \Leftrightarrow \quad x \circ y = y, \tag{3.1}$$

any semilattice with binary operation $\circ$ becomes a partially ordered set in which $x \circ y = lub\{x, y\}$. It should now be obvious that if S is a partially ordered set for which the operation $x \wedge y$ is defined for each pair $x, y \in S$, then S is also a semilattice. Whenever the operation $x \vee y$ or $x \wedge y$ are known or assumed, then the notation $(S, \vee)$ or $(S, \wedge)$ will be used to specify these semilattice in order to avoid any ambiguities. Some authors refer to the semilattices $(S, \vee)$ and $(S, \wedge)$ as the *join* and *meet* semilattices, respectively. In this text, however, the semilattice $(S, \vee)$ will be referred to as a *max*-semilattice and $(S, \wedge)$ as a *min*-semilattice.

Example 3.8 $(\mathbb{R}, \vee)$ is a semilattice with dual $(\mathbb{R}, \wedge)$. Obviously, $(\mathbb{R}_{-\infty}, \vee)$ is also a max-semilattice with dual the min-semilattice $(\mathbb{R}_{\infty}, \wedge)$. Note that the semilattice $(\mathbb{R}_{-\infty}, \vee)$ is a monoid with zero element $-\infty$ since $r \vee (-\infty) = (-\infty) \vee r = r \; \forall r \in \mathbb{R}_{-\infty}$. Similarly, $(\mathbb{R}_{\infty}, \wedge)$ is a monoid with zero element ∞.

As mentioned earlier, the operations of $\vee$ and $\wedge$ in a lattice are analogous to the arithmetic operations of addition $(+)$ and multiplication $(\times)$, respectively. This analogy is most striking in distributive lattices.

Definition 3.12 A lattice $(L, \vee, \wedge)$ is a *distributive lattice* if and only if the following equality holds in L:

$$x \wedge (y \vee z) = (x \wedge y) \vee (x \wedge z). \tag{3.2}$$

This equation expresses the similarity with the distributive law $x(y+z) = xy + xz$ of ordinary arithmetic. All lattices in Example 3.7 are distributive lattices. We also note that by duality we have that

$$x \vee (y \wedge z) = (x \vee y) \wedge (x \vee z) \tag{3.3}$$

in any distributive lattice. In fact, the following general laws hold for distributive lattices [30]:

$$a \wedge \left(\bigvee_{i=1}^{n} b_i \right) = \bigvee_{i=1}^{n} (a \wedge b_i), \quad a \vee \left(\bigwedge_{i=1}^{n} b_i \right) = \bigwedge_{i=1}^{n} (a \vee b_i), \tag{3.4}$$

and

$$\bigvee_{j=1}^{m} \left(\bigwedge_{i=1}^{n} x_{ij} \right) \leqslant \bigwedge_{i=1}^{n} \left(\bigvee_{j=1}^{m} x_{ij} \right). \tag{3.5}$$

The last inequality is known as the *minimax principle*.

Now if $x \preccurlyeq z$, then by Theorem 3.3 $z = x \vee z$. Thus, substituting z for $x \vee z$ in equation 3.2 we have that any distributive lattice satisfies the following law:

$$x \preccurlyeq z \Rightarrow x \vee (y \wedge z) = (x \vee y) \wedge z. \tag{3.6}$$

Any lattice satisfying equation 3.6 is called a *modular* lattice. Although every distributive lattice is modular, not every modular lattice is a distributive lattice.

Example 3.9 Let $L = \{O, a, b, c, I\}$ have partial order defined by $O \preccurlyeq a \preccurlyeq I$, $O \preccurlyeq b \preccurlyeq I$, and $O \preccurlyeq c \preccurlyeq I$. Then

$$a \wedge (b \vee c) = a \wedge I = a \neq O = O \vee O = (a \wedge b) \vee (a \wedge c).$$

This shows that L is not a distributive lattice. However, L is a modular lattice since equation (3.6) holds. For example,

$$b \vee (a \wedge I) = b \vee a = (b \vee a) \wedge I.$$

Changing the partial order on L to $O \preccurlyeq a \preccurlyeq b \preccurlyeq I$ and $O \preccurlyeq c \preccurlyeq I$ results in a lattice which is neither distributive nor modular since

$$a \preccurlyeq b \quad \text{and} \quad a \vee (c \wedge b) = a \vee O = a \neq b = I \wedge b = (a \vee c) \wedge b.$$

The next theorem demonstrates that despite the differences between distributive and modular lattices displayed in the example, the height function on these lattices exhibits the same important property whenever both lattices are of finite length.

Theorem 3.6 *If L is a modular lattice of finite length and h is the height function, then*

$$h(x) + h(y) = h(x \vee y) + h(x \wedge y) \quad \forall\, x, y \in L.$$

Suppose L is a lattice with smallest element O and largest element I, and $x \in L$. If there exists an element $x^c \in L$ such that $x^c \wedge x = O$ and $x^c \vee x = I$, then x^c is called the *complement* of x. Next suppose that L is a distributive lattice. If $a \wedge x = a \wedge y$ and $a \vee x = a \vee y$, then $x = x \wedge (a \vee x) = x \wedge (a \vee y) = (x \wedge a) \vee (x \wedge$

$y) = (a \wedge y) \vee (x \wedge y) = (a \vee x) \wedge y = (a \vee y) \wedge y = y$. Therefore, $x = y$. Thus, if x^c and x' are two complements so that $x \wedge x^c = O = x \wedge x'$ and $x \vee x^c = I = x \vee x'$, then by our analysis we have that $x^c = x'$. That is, complements are unique in distributive lattices. We say that a lattice is *complemented* if and only if all of its elements have a complement.

Modular and complemented lattices are of special interest for applications in such diverse fields as probability theory, including ergodic theory and multiplicative processes, linear algebra, computer science, and several engineering disciplines. For example, the set L of all subspaces of $\mathbb{R}^n$ is a complemented modular lattice. Here the orthogonal complement $S^\perp$ of any subspace S satisfies $S^\perp \cap S = \varnothing$ and $S^\perp \cup S = \mathbb{R}^n$. Also, by definition, a *Boolean* lattice is a complemented distributive lattice. The uses of Boolean algebras in computer science and engineering are manifold and range from the design of electrical networks to the theory of computing. Because of its importance in computer science and engineering we summarize the properties discussed above that define a Boolean lattice.

A *Boolean lattice* is a lattice L that is associative, commutative, distributive, and satisfies the following three properties $\forall x, y \in L$:

1. *absorption* $x \vee (x \wedge y) = x$ and $x \wedge (x \vee y) = x$

2. *identity* $x \vee O = x$ and $x \wedge I = x$

3. *complement* $x \vee x^c = I$ and $x \wedge x^c = O$

The next theorem is a consequence of this definition.

Theorem 3.7 *In any Boolean lattice, each element x has a unique complement x^c. Furthermore, complementation satisfies the following equations:*

1. $(x^c)^c = x$

2. $(x \wedge y)^c = x^c \vee y^c$ *and* $(x \vee y)^c = x^c \wedge y^c$.

Complementation should not be confused with conjugation or duality. For instance, the *dual* or *conjugate* of $r \in \mathbb{R}_{\pm\infty}$ is the unique element $r^* \in \mathbb{R}_{\pm\infty}$ defined as

$$r^* = \begin{cases} -r & \text{if } r \in \mathbb{R} \\ -\infty & \text{if } r = \infty \\ \infty & \text{if } r = -\infty \end{cases} \tag{3.7}$$

It follows that

1′. $(r^*)^* = r$

2′. $(r \wedge s)^* = r^* \vee s^*$ and $(r \vee s)^* = r^* \wedge s^*$.

Equations *1′* and *2′* have the appearance of *1* and *2* of Theorem 3.7. However, $r \wedge r^* = -\infty$ and $r \vee r^* = \infty$ if and only if $r = \infty$ or $r = -\infty$, respectively.

Since complements in a Boolean lattice L are unique, we can view complementation as a function $L \to L$ that maps $x \to x^c$ From this point of view, complementation is a unary operation. If L is a Boolean lattice, then the algebraic structure $(L, \vee, \wedge, ^c)$ determined by the two binary operations $\vee$ and $\wedge$, and the unary operation of complement on L is commonly called a *Boolean algebra.*

Example 3.10

1. It follows from the laws of set operations that $(2^X, \cup, \cap, ^c)$ is a Boolean algebra.

2. Let $f, g \in \{0,1\}^X$ and define $h = f \vee g$ by

$$h(x) = \begin{cases} f(x) & \text{if } g(x) \leq f(x) \\ g(x) & \text{if } f(x) < g(x) \end{cases}$$

and $k = f \wedge g$ by

$$k(x) = \begin{cases} f(x) & \text{if } f(x) \leq g(x) \\ g(x) & \text{if } g(x) < f(x). \end{cases}$$

Define the complement of f as the function $f^c : X \to \{0,1\}$ by

$$f^c(x) = \begin{cases} 1 & \text{if } f(x) = 0 \\ 0 & \text{if } f(x) = 1 \end{cases}$$

The least element is the zero function $O : X \to \{0,1\}$, defined as $O(x) = 0 \ \forall x \in X$, and the largest element is the constant function $I : X \to \{0,1\}$ defined by $I(x) = 1 \ \forall x \in X$. It is now easy to show that $(\{0,1\}^X, \vee, \wedge, ^c)$ is a Boolean algebra. This algebra provides a rigorous mathematical basis for the description of a wide range of Boolean image transformations.

3. The lattice $(\mathbb{R}^X_{\pm\infty}, \vee, \wedge)$ of Example 3.7(6) is a distributive lattice with a least and a largest elements Only the functions O and I have complements, namely $O^c = I$ and $I^c = O$.

The next theorem provides a tool for identifying which lattices are Boolean.

Theorem 3.8 *Let L be a lattice. If $\forall\, x \in L\ \exists x^c$ such that*

$$(x \wedge y)^c = x^c \vee y^c \quad \textit{and} \quad (x \vee y)^c = x^c \wedge y^c,$$

then L is a Boolean lattice.

Exercises 3.2.2

1. Show that $\mathbb{R}^n$ is a lattice.
2. Suppose C is a chain and $P \subset C$. Prove that P is a lattice.
3. Verify the seven equations of Theorem 3.3.
4. Prove Corollary 3.4.
5. Verify the equations 1,3, and 4 of Theorem 3.5.
6. Suppose A and B are lattices. Prove that $A \times B$ is a lattice.
7. Prove equation 3.3.
8. Prove equation 3.4.
9. Prove equation 3.5.
10. Construct two examples of lattices that are not modular.
11. Prove Theorem 3.6.
12. Suppose L is a distributive lattice and $x, y, z \in L$. Verify that if $z \wedge x = z \wedge y$ and $z \vee x = z \vee y$, then $x = y$.
13. Prove Theorem 3.7
14. Prove Theorem 3.8.
15. Suppose L is a Boolean lattice and $x, y \in L$. Prove that $(x \wedge y) \wedge (x^c \vee y) = O$.

3.3 RELATIONS WITH OTHER BRANCHES OF MATHEMATICS

This section presents a few examples of relations between lattice theory and other branches of mathematics that are relevant to the goals of this anthology. As discussed in Section 3.1, lattice theory, since its early beginnings, has been closely intertwined with other branches of mathematics. The number of these relationships and their applications is enormous and far beyond the objectives of this treatise. Here we only present a few examples of mathematical disciplines whose core axioms are based on lattice theory. For the interested researcher, a large number of examples are available in [30, 132].

3.3.1 Topology and Lattice Theory

An interesting example of the intertwinement between topology and lattice theory is given by the following theorem:

Theorem 3.9 *If (X,τ) is a topological space, then $(\tau,\leq)$ is a complete lattice.*

Proof. To show that τ is a lattice, simply define for any pair $U,V\in\tau$, $U\leq V \Leftrightarrow U\subset V$, $U\wedge V=U\cap V$, and $U\vee V=U\cup V$.

To prove completeness, we must show that $\forall\, t\subset\tau$, $\bigwedge t$ and $\bigvee t$ exist. Since $\bigvee t=\bigcup_{V\in t}\{V\subset\tau\}$, it follows from axiom 2 of τ that $\bigvee t$ exist.

On the other hand, the set $\bigwedge t=\bigcap_{V\in t}\{V\subset\tau\}$ may not be open, in which case $\bigwedge t$ is not an element of τ. In this case set $I=int\bigwedge t$. Since I is open, $I\in\tau$ and $I\subset\bigwedge t\subset V\ \forall V\in t$. Since we are assuming that $\bigwedge t$ is not open, $\bigwedge t\neq\varnothing\in\tau$ and $\varnothing\subset\bigwedge t$. Let $U\in\tau$ with $U\subset\bigwedge t$ and $U\subset V\ \forall V\in t$. Therefore I is the infimum of t □

Knowing that the topology τ is a complete lattice is an important tool in the study of various properties of different topologies for a given set X.

Another notion shared by topology and lattice theory is the interval. Given a lattice $L\neq\varnothing$, with $a\leq b$, then the *closed interval* $[a,b]$ is defined as $[a,b]=\{x\in L: a\leq x\leq b\}$. Since $[a,b]\subset L$, $[a,b]$ is a complete sublattice of L. Open and half-open intervals are defined, respectively, by $(a,b)=\{x\in L: a<x<b\}$, $(a,b]=\{x\in L: a<x\leq b\}$ and $[a,b)=\{x\in L: a\leq x<b\}$. If L is a complete lattice and a chain, then it becomes easy define a basis for a topology of X based on intervals. Explicitly, we define a basis for τ by defining the basic *open* sets in terms of the following sets: (1) X and $\varnothing$, (2) $(a,I]$ for any $a\in X$ and $\forall\, x>a$, $[O,a)\ \forall\, x<a$, and (a,b) for any $a<b$ in X. Since $\mathbb{R}_{\pm\infty}$ is a complete lattice chain, it is also a topological space whose topology

is generated by the basis described above. If L is a chain but not a complete lattice, then open intervals of type (a,b) are sufficient to form a basis for a topology.

Recall that if X is a topological space and $x \in X$, then any open set $U \in \tau$ with $x \in U$ is a neighborhood of x. If for every pair $x, y \in X$ with $x \neq y$ there exist two open sets U and V with $x \in U$, $y \in V$ and $U \cap V = \varnothing$, then (X, τ) is called a *Hausdorff space*. For example, the topological space (L, τ), where L is a chain and τ is generated by the intervals basis, is a Hausdorff space.

The Hausdorff condition of separation of points is one of the most frequently discussed and employed condition. All metric spaces are Hausdorff spaces and so are all manifolds. Most importantly, the separation condition implies the uniqueness of limits of sequences and provides a direct connection to Boolean algebra. For example, if X is a Hausdorff space, then according to Definition 2.15(2) $\varnothing$ and X are both open and closed sets; also known as *clopen* sets. Depending on the topology τ, X can have any number of clopen sets. For instance, if $X = \mathbb{Q}$ with the standard Euclidean topology, then $A = \{x \in \mathbb{Q} : x^2 > 2\}$ is a clopen set. Note that X is a totally disconnected space with an infinite number of clopen sets. If X is a disconnected Hausdorff space, then its topology contains at least three clopen sets. Now suppose that X is a totally disconnected compact Hausdorff space. Setting $\mathfrak{A} = \{A \subset X : A \text{ is a clopen set }\}$ and defining

$$A + B = (A \cap B') \cup (A' \cap B) \text{ and } A \cdot B = A \cap B,$$

turns $\mathfrak{A}$ into a Boolean ring. We complete this section with the following theorem:

The Stone Representation Theorem. If R is a Boolean ring, then there exists a totally disconnected Hausdorff space X such that R is isomorphic to the Boolean ring of all clopen subset of X.

Exercises 3.3.1

1. Prove that if L is a chain and τ is generated by the interval basis, then (L, τ) is a topological space. Prove this for the following cases: (1) $|L| = n$, (2) $|L| = \infty$ and $O \in L$, (3) $I \in L$, (4) $\{O, I\} \subset L$, and (5) $|L| = \infty$ and L having no maximal or minimal bounds.

2. Prove that the lattice $(\mathbb{R}^n, \vee, \wedge)$ is a topological space, assuming that the topology τ is generated by open intervals.

3. Prove that the topological spaces defined in Exercise 1 and Exercise 2 are Hausdorff spaces.

4. Prove that every metric space is a Hausdorff space.

5. Prove that $(\mathfrak{A}, +, \cdot)$ is indeed a Boolean ring.

3.3.2 Elements of Measure Theory

Some of the most useful computational tools for applications of various mathematical theories to real world problems are mappings to the real number system. Measure theory, probability theory, and fuzzy set theory are just three obvious examples that use such real-valued mappings. The intertwinement of these three theories and lattice theory is that the former theories are equipped with the operational structure $\{\varnothing, {}^c, \cup, \cap, \subset\}$ which, in many cases, can be replaced by $\{0, ', \vee, \wedge, \leq\}$.

Measure theory is the basis of modern integration theory and these two theories play a fundamental role in a wide range of mathematical endeavors. Measures for a set X depend on the set algebras of subsets of 2^X defined as follows:

Definition 3.13 If $\mathcal{M} \subset 2^X$, then $\mathcal{M}$ is called a *σ-algebra in X* if and only if the following three conditions are satisfied:

M_1. $X \in \mathcal{M}$.

M_2. $A \in \mathcal{M} \Rightarrow A' \in \mathcal{M}$.

M_3. $A = \bigcup_{i=1}^{\infty} A_i$ and $A_i \in \mathcal{M}\ \forall i \in \mathbb{N} \Rightarrow A \in \mathcal{M}$.

If $\mathcal{M}$ is a σ-algebra in X, then the pair $(X, \mathcal{M})$ is called a *measurable space* and each $A \in \mathcal{M}$ is called a *measurable set*. Again, in order to reduce notational overhead, it is common practice to call X a measurable space with the assumption that there exists an associated σ-algebra $\mathcal{M}$ in X. Some simple examples are $\mathcal{M} = \{\varnothing, X\}$, $\mathcal{M} = \{\varnothing, A, A', X\}$, and $\mathcal{M} = 2^X$. Note also that each example is a complete lattice. The next theorem is an immediate consequence of Definition 3.14.

Theorem 3.10 *A σ-algebra $\mathcal{M}$ in a set X has the following properties:*

1. $\varnothing \in \mathcal{M}$.

2. If $\{A_i\}_{i=1}^{\infty} \subset \mathcal{M}$, *then* $\bigcap_{i=1}^{\infty} A_i \in \mathcal{M}$.

3. If $A, B \in \mathcal{M}$, then $A \cup B \in \mathcal{M}$ and $A \setminus B \in \mathcal{M}$.

Proof. By conditions M_1 and M_2, $\varnothing = X' \in \mathcal{M}$. This proves property *1*. Next suppose $\{A_i\}_{i=1}^{\infty} \subset \mathcal{M}$. Then by conditions M_2 and M_3 we have $\bigcup_{i=1}^{\infty} A_i' \in \mathcal{M}$. It now follows from condition M_2 that

$$\bigcap_{i=1}^{\infty} A_i = (\bigcup_{i=1}^{\infty} A_i')' \in \mathcal{M}.$$

This proves property *2*.

To prove property *3*, simply set $A_1 = A$, $A_2 = B$, and for $i > 2$ set $A_i = \varnothing$. Then according to condition M_3, $A \cup B = \bigcup_{i=1}^{\infty} A_i \in \mathcal{M}$. A similar argument can be used to show that property *2* also holds for finite intersections. Finally, $A \setminus B = A \cap B' \in \mathcal{M}$. This proves property *3*. □

Considering Theorem 3.10 and the definition of a σ-algebra it is an easy exercise to show that $\mathcal{M}$ is lattice. Also, various useful measure spaces for a given set X depend on the selection of subsets of 2^X which may not be σ-algebras. In these cases the following theorem proves very helpful.

Theorem 3.11 *If X is a nonempty set and $\mathcal{S} \subset 2^X$, then there exists a smallest σ-algebra $\widetilde{\mathcal{M}}$ in X such that $\mathcal{S} \subset \widetilde{\mathcal{M}}$.*

Proof. Let $\Sigma = \{\mathcal{M} : \mathcal{S} \subset \mathcal{M}$ and $\mathcal{M}$ is a σ-algebra in $X\}$. Since $\mathcal{M} = 2^X \in \Sigma$, $\Sigma \neq \varnothing$. Define

$$\widetilde{\mathcal{M}} = \bigcap_{\mathcal{M} \in \Sigma} \mathcal{M}.$$

Since $X \in \mathcal{M}$ and $\mathcal{S} \subset \mathcal{M}\, \forall\, \mathcal{M} \in \Sigma$ we have that $X \in \widetilde{\mathcal{M}}$ and $\mathcal{S} \subset \widetilde{\mathcal{M}}$.

By definition of $\widetilde{\mathcal{M}}$, $\widetilde{\mathcal{M}} \subset \mathcal{M}\, \forall\, \mathcal{M} \in \Sigma$. Thus, if $A \in \widetilde{\mathcal{M}}$, then $A \in \mathcal{M}\, \forall\, \mathcal{M} \in \Sigma$. But then $A' \in \mathcal{M}\, \forall\, \mathcal{M} \in \Sigma$ so that $A' \in \widetilde{\mathcal{M}}$.

Finally, if $A_k \in \widetilde{\mathcal{M}}\, \forall\, k \in \mathbb{N}$, then $A_k \in \mathcal{M}\, \forall\, k \in \mathbb{N}$ and $\forall\, \mathcal{M} \in \Sigma$. But then $\bigcup_{k=1}^{\infty} A_k \in \mathcal{M}\ \forall\, \mathcal{M} \in \Sigma$. Therefore $\bigcup_{k=1}^{\infty} A_k \in \widetilde{\mathcal{M}}$. □

It is common practice to set $\widetilde{M} = \sigma(\mathcal{S})$ and call $\sigma(\mathcal{S})$ the σ-algebra *generated* by $\mathcal{S}$. For an example, suppose (X, τ) is a topological space and $\sigma(\tau)$ the σ-algebra generated by τ. In this specific case the elements of $\sigma(\tau)$ are called *Borel sets* and $(X, \sigma(\tau))$ a Borel measurable space, often denoted by $(X, \mathcal{B})$.

We defined the notion of a measurable space without defining a measure. The following definition will fill this apparent gap:

Definition 3.14 Suppose $\mathcal{M}$ is a σ-algebra in a set X. A set function $\mu : \mathcal{M} \to [0,\infty]$ is called a *measure* if and only if

$$\mu(\emptyset) = 0 \quad \text{and} \quad \mu(\bigcup_{i=1}^{\infty} A_i) = \sum_{i=1}^{\infty} \mu(A_i) \quad \text{whenever} \quad A_i \cap A_j = \emptyset \, \forall i, j.$$

If $\mathcal{M}$ is a σ-algebra in X and μ a measure on $\mathcal{M}$, then the triple $(X, \mathcal{M}, \mu)$ is called a *measure space* and the elements of $\mathcal{M}$ are said to be μ*-measurable*. Calling X a measure space, means that X refers to that triple. The symbol $\mathcal{M}(\mu)$ means that $\mathcal{M}$ is a σ-algebra and μ is a positive measure on $\mathcal{M}$ and the set X is assumed to be known from the discussion at hand. Conversely, when calling the set X a measure space, one assumes that $\mathcal{M}(\mu)$ is known. For convenience sake, the notation $(X, \mathcal{M}(\mu))$ is also often used to describe the triple $(X, \mathcal{M}, \mu)$. The next three theorems are foundational for applications of measure theory.

Theorem 3.12 *A measure space* $(X, \mathcal{M}, \mu)$ *has the following properties:*

1. *If* $A, B \in \mathcal{M}$ *and* $A \subset B$, *then* $\mu(A) \leq \mu(B)$.
2. *If* $A, B \in \mathcal{M}$, $A \subset B$ *and* $\mu(A) < \infty$, *then* $\mu(B \setminus A) = \mu(B) - \mu(A)$.
3. *If* $A_1 \subset A_2 \subset A_3 \subset \ldots \in \mathcal{M}$, *then* $\mu(\bigcup_{i=1}^{\infty} A_i) \leq \sum_{i=1}^{\infty} \mu(A_i)$.

Proof. If $A \subset B$. then $B = A \cup (B \setminus A)$ and by definition of the measure μ, $\mu(B) = \mu(A \cup (B \setminus A)) = \mu(A) + \mu(B \setminus A) \geq \mu(A)$. This proves property *1*.

The proofs of properties 2 and *3* are just as easy and left as an exercise.

□

Theorem 3.13 *If* $\{A_n\}_{n=1}^{\infty} \subset \mathcal{M}$ *and* $A_n \subset A_{n+1} \, \forall n \in \mathbb{N}$, *then*

$$\mu(\bigcup_{n=1}^{\infty} A_n) = \lim_{n\to\infty} \mu(A_n).$$

Proof. First let us assume that for some $j \in \mathbb{N}$, $\mu(A_j) = \infty$. Since $A_j \subset \bigcup_{n=1}^{\infty} A_n$, it follows from Theorem 3.12(*1*) and (*3*) that $\infty = \mu(A_j) \leq \mu(\bigcup_{n=1}^{\infty} A_n) \leq \sum_{i=1}^{\infty} \mu(A_i) = \infty$. Hence $\mu(\bigcup_{n=1}^{\infty} A_n) = \infty$. Furthermore, since $A_j \subset A_{j+k} \, \forall k \in \mathbb{N}$, we have that $\infty = \mu(A_j) \leq \mu(A_{j+k}) \forall k \geq 1$. Therefore $\lim_{n\to\infty} \mu(A_n) =$

$\lim_{k\to\infty}\mu(A_{j+k}) = \infty$ and $\mu(\bigcup_{n=1}^{\infty} A_n) = \lim_{n\to\infty}\mu(A_n)$. This proves the theorem in case the measure of some set A_j is infinite.

Next, suppose that $\mu(A_n) < \infty\,\forall\, n \in \mathbb{N}$. Note that

$$\bigcup_{n=1}^{\infty} A_n = A_1 \cup (A_2 \setminus A_1) \cup (A_3 \setminus A_2) \cup \ldots = A_1 \cup \bigcup_{n=2}^{\infty}(A_n \setminus A_{n-1})$$

Since the set $\{A_1, A_n \setminus A_{n-1} : n \in \mathbb{N}\ n \geq 2\}$ is a collection of mutually disjoint sets, we have that

$$\mu(\bigcup_{n=1}^{\infty} A_n) = \mu(A_1 \cup \bigcup_{n=2}^{\infty}(A_n \setminus A_{n-1})) = \mu(A_1) + \sum_{n=2}^{\infty}\mu(A_n \setminus A_{n-1}).$$

It now follows from Theorem 3.12(2) that

$$\mu(\bigcup_{n=1}^{\infty} A_n) = \mu(A_1) + \sum_{n=2}^{\infty}[\mu(A_n) - \mu(A_{n-1}] = \mu(A_1) + \lim_{n\to\infty}(\sum_{k=2}^{n}\mu(A_k) - \mu(A_{k-1}))$$
$$= \mu(A_1) + \lim_{n\to\infty}[\mu(A_n) - \mu(A_1)] = \lim_{n\to\infty}\mu(A_n). \quad \square$$

Theorem 3.14 *If* $\{A_n\}_{n=1}^{\infty} \subset \mathcal{M}$, $A_{n+1} \subset A_n\,\forall\, n \in \mathbb{N}$, *and* $\mu(A_1) < \infty$, *then*

$$\mu(\bigcap_{n=1}^{\infty} A_n) = \lim_{n\to\infty}\mu(A_n).$$

Proof. Mimicking the proof for Theorem 3.12, one obtains $\bigcap_{n=1}^{\infty} A_n = A_1 \cap \bigcap_{n=2}^{\infty}(A_n \setminus A_{n-1})$ and (1) $\mu(A_1 \setminus \bigcap_{n=1}^{\infty} A_n) = \mu(A_1) - \mu(\bigcap_{n=1}^{\infty} A_n)$. But we also have

$$A_1 \setminus \bigcap_{n=1}^{\infty} A_n = A_1 \cap (\bigcap_{n=1}^{\infty} A_n)' = A_1 \cap \bigcup_{n=1}^{\infty} A_n' = \bigcup_{n=1}^{\infty}(A_1 \cap A_n') = \bigcup_{n=1}^{\infty}(A_1 \setminus A_n)$$

Setting $B_n = A_1 \setminus A_n\,\forall\, n \in \mathbb{N}$, then $B_n \subset B_{n+1}$ and $B_n \cap B_{n+1} = \varnothing\ \forall\, n$. Thus,

$$\mu(A_1 \setminus \bigcap_{n=1}^{\infty} A_n) = \mu(A_1) - \mu(\bigcap_{n=1}^{\infty} A_n) = \mu(\bigcup_{n=1}^{\infty} B_n) = \lim_{n\to\infty}\mu(B_n),$$

where the last equality is due to Theorem 3.13. Therefore $\mu(A_1 \setminus \bigcap_{n=1}^{\infty} A_n) = \lim_{n\to\infty}\mu(B_n) = \lim_{n\to\infty}\mu(A_1 \setminus A_n) = \lim_{n\to\infty}[\mu(A_1) - \mu(A_n)] = \mu(A_1) - \lim_{n\to\infty}\mu(A_n) = \mu(A_1) - \mu(\bigcap_{n=1}^{\infty} A_n)$. where the last equality is due to equation (1) at the beginning of the proof. It now follows that $\mu(\bigcap_{n=1}^{\infty} A_n) =$

$\lim_{n\to\infty}\mu(A_n)$. □

By definition of the measure μ, $\mu(\varnothing) = 0$. It is natural to ask if there are sets $A \in \mathcal{M}$ with $A \neq \varnothing$ that are of measure zero. As an example, consider the measure space $(X, \sigma(\tau), \mu)$, where $X = \mathbb{R}$, the elements of τ are generated by open intervals, and $\mu([a,b]) = b - a$. Thus, if $a = b$, then $\mu([a,a]) = a - a = 0$. Furthermore, since $[a,b] = \{x \in \mathbb{R} : a \leq x \leq b\} = \{a\} \cup \{x \in \mathbb{R} : a < x < b\} \cup \{b\}$, it follows that $\mu([a,b]) = \mu([a,a] + \mu((a,b)) + \mu([b,b]) = \mu((a,b))$. A little more interesting is the fact that $\mu(\mathbb{Q}) = 0$. Here let $\mathbb{Q}$ be labeled by the natural integers so that $\mathbb{Q} = \{q_n \in \mathbb{Q} : n = 1, 2, 3, \ldots\}$ and set $A_n = \{q_n\}$. Since $A_n \cap A_m = \varnothing$ whenever $n \neq m$ we have that $\mu(\mathbb{Q}) = \mu(\bigcup_{n=1}^{\infty} A_n) = \sum_{n=1}^{\infty}\mu(A_n) = \sum_{n=1}^{\infty}\mu([q_n, q_n]) = 0$.

Since $\mu(\mathbb{Q}) = 0$, it follows that every countable set is a null set (see Exercise 3.3.2(2). Thus the question may arise as to whether or not there exist sets that are uncountable and are of measure zero. The next example provides the answer to this question.

Example 3.11 Recall from Example 2.3 that the Cantor set C was defined as $C = \bigcap_{n=1}^{\infty} C_n$, where $C_1 \supset C_2 \supset C_3 \supset \cdots \supset C_n \supset \cdots$ with each set C_n being defined as the union of 2^n closed disjoint intervals, namely $C_1 = [0, \frac{1}{3}] \cup [\frac{2}{3}, 1]$, $C_2 = [0, \frac{1}{9}] \cup [\frac{2}{9}, \frac{3}{9}] \cup [\frac{6}{9}, \frac{7}{9}] \cup [\frac{8}{9}, 1]$ and so on. Thus, $\mu(C_1) = \mu([0, \frac{1}{3}) + \mu([\frac{2}{3}, 1]) = \frac{1}{3} + \frac{1}{3} = \frac{2}{3}$, $\mu(C_2) = \frac{4}{9} = (\frac{2}{3})^2$ and, inductively, $\mu(C_n) = (\frac{2}{n})^n$. In view of Theorem 3.14, we now have

$$\mu(C) = \mu(\bigcap_{n=1}^{\infty} C_n) = \lim_{n\to\infty}\mu(C_n) = \lim_{n\to\infty} = \left(\frac{2}{n}\right)^n = 0.$$

According to Theorem 2.35, C is uncountable. This shows that non-countable sets can be of measure zero.

One connection of measure theory and other mathematical systems is via functions. For a particular example, suppose that X is a measurable space, (Y, τ) is a topological space, and $f : X \to Y$ is a function, then f is called a *measurable* function if and only if $f^{-1}(U)$ is a measurable set in X $\forall U \in \tau$. The proof of the following theorem is an easy consequence of the theorem described in Exercise 2.2.3(2).

Theorem 3.15 *Suppose X and Y are topological spaces and $f : X \to Y$ is continuous. If Z is a measurable space and $g : Z \to X$ is measurable, then $h = f \circ g : Z \to Y$ is measurable.*

Exercise 3.3.2(1) provides another connection between lattice theory and measure theory.

Exercises 3.3.2

1. Prove that $\mathcal{M}$ is a complete lattice.

2. Prove properties 2 and 3 of Theorem 3.12.

3. Prove that every countable set is a null set.

4. Prove Theorem 3.15.

3.3.3 Lattices and Probability

Connections between lattice theory and probability theory were already well-known in the mid-twentieth century [30, 238]. More recent advances can be found in [145, 146, 147].

The notion of measure theory is of fundamental importance in probability theory. Explicitly, suppose Ω is a nonempty set and $(\Omega, \mathcal{F}, P)$ is a measure space where the measure P satisfies the following conditions:

1. $P(A) \geq 0 \ \forall A \in \mathcal{F}$

2. $P(\Omega) = 1$

3. If $A_i \in \mathcal{F} \ \forall i \in \mathbb{N}$ and $A_i \cap A_j = \emptyset$ whenever $i \neq j$, then $P(\bigcup_{i=1}^{\infty} A_i) = \sum_{i=1}^{\infty} P(A_i)$.

If these conditions are satisfied, then $(\Omega, \mathcal{F}, P)$ is called a *probability space* with *sample space* Ω, *event space* $\mathcal{F}$, and *probability measure* P. The elements of $\mathcal{F}$ are called *events* and if A is an event, then $P(A)$ is referred to as he *probability of* A. If $a \in \Omega$ and $A = \{a\}$, then A is called an *elementary event* or *simple event.*

Since $\mathcal{F}$ is a σ-algebra, we have that if $A, B \in \mathcal{F}$ with $A \cap B = \emptyset \in \mathcal{F}$, then $P(A \cap B) = P(\emptyset) = 0$. It also follows from Theorem 3.12 that

P_1. If A, B are events with $A \subset B$, then $P(A) \leq P(B)$ and $P(B \backslash A) = P(B) - P(A)$.

The following properties are just as easy to verify.

P_2. If A is an event, then $0 \leq P(A) \leq 1$ and $P(A') = 1 - P(A)$.

P_3. If A and B are events, then $P(A \cup B) = P(A) + P(B) - P(A \cap B)$ and $P(A) = P(A \cap B) + P(A \cap B')$.

Since $\mathcal{F}$ is a lattice with least element $\varnothing$ and largest element Ω, P is simply a special "measure" function of the lattice $\mathcal{F}$ to the interval $[0, 1]$. However, in probability theory there is also the concept of a conditional probability, which refers to finding the probability of some event B while knowing that another event A has occurred. Examining this problem leads to the definition

$$P(B|A) = \frac{P(A \cap B)}{P(A)} \quad \text{or, equivalently,} \tag{3.8}$$

$$P(A \cap B) = P(A)P(B|A) \tag{3.9}$$

According to equation 3.9 the probability that both A and B occur is equal to the probability that A occurs times the probability that B occurs given that A has occurred. The expression $P(B|A)$ is called the *conditional probability* of B given that A occurred. If $P(B|A) = P(B)$, then A and B are called *independent events*. This is equivalent to

$$P(A \cap B) = P(A)P(B) \tag{3.10}$$

Example 3.12 Let $\Omega = \{1, 2, 3, 4, 5, 6\}$ correspond to the numbers on a die and the die is tossed twice. Suppose the problem is to find the probability of obtaining a 4, 5, or 6 on the first toss and a 1, 2, 3, or 4 on the second toss. Let A denote the event of the first toss and B the event of the second toss. Assuming a unbiased, the second toss is independent of the first toss so that $P(A \cap B) = P(A)P(B|A) = P(A)P(B) = \frac{3}{6} \times \frac{4}{6} = \frac{1}{3}$.

There are other methods to compute $P(A \cap B)$. For instance, property P_3 is equivalent to

$$P(A \cap B) = P(A) + P(B) - P(A \cup B) \tag{3.11}$$

so that $P(A \cap B) = \frac{3}{6} + \frac{4}{6} - \frac{5}{6} = \frac{1}{3}$.

Note that equation 3.11 avoids the multiplication in equation 3.10, making it more pliable for converting certain probability problems into lattice-based problems.

Exercises 3.3.3

1. Prove that $\mathcal{F}$ is a complete lattice.

2. Verify properties P_2 and P_3.

3. Suppose that for $i \in \mathbb{N}$, B_i is an event, $B_i \cap B_j = \varnothing \; \forall i \neq j$, and $\bigcup_{i=1}^{\infty} B_i = \Omega$. Prove that if $P(B_i) \neq 0$, then $P(A) = \sum_{i=1}^{\infty} P(A|B_i)P(B_i)$.

4. Suppose two cards are drawn from a well-shuffled deck of 52 cards. What is the probability that they are both kings if the first card is not replaced?

5. What is the probability in Exercise 3 if the first card is replaced?

6. Suppose that $A_1, A_2, \ldots, A_n$ are events, $A_i \cap A_j = \varnothing$, and $\bigcup_{i=1}^{n} A_i = \Omega$. Prove that if A is any event, then

$$P(A_i|A) = \frac{P(A_i)P(A|A_i)}{\sum_{j=1}^{n} P(A_j)P(A|A_j)}.$$

3.3.4 Fuzzy Lattices and Similarity Measures

Fuzzy lattices are a variant of fuzzy sets. For our purposes we use the definition provided by D. Klaua and L.A. Zadeh in 1965 [143, 326].

Definition 3.15 If X is a nonempty set, then a *fuzzy set* is an ordered pair

$$\widetilde{A} = \{(x, \mu_{\tilde{A}}(x)) : x \in X\},$$

where $\mu_{\tilde{A}}$ is a function $\mu_{\tilde{A}} : X \longrightarrow [0,1]$ is called the *membership* function and $\mu_{\tilde{A}}(x)$ represents the *degree of membership*.

Example 3.13 Let $X = \mathbb{R}$ and one wants the degree of membership of numbers close to zero. Then the membership function C may be satisfactory. Note that if $x = 0$, then $\mu_{\tilde{A}}(x) = 1$.

There exists a plethora of different approaches to fuzzy lattices [4, 130, 253, 254, 279, 328], and a sundry more. Many of these actually deal with lattice fuzzy sets often denoted L-fuzzy sets and not directly with fuzzy lattices. A particular appealing approach proposed by V.G. Kaburlasos [131] is based on the following simple definition.

Definition 3.16 A *fuzzy lattice* is a triple $(L, \leq, \mu)$, where $(L, \leq)$ is a complete lattice and $(L \times L, \mu)$ is a fuzzy set with the property that $\mu(x,y) = 1 \Leftrightarrow x \leq y$.

It follows from Definition 3.15 that $(L \times L, \mu) = \{((x,y), \mu(x,y)) : (x.y) \in L \times L\}$.

In our discussion, we drop the $\leq$ notation since it is understood that any lattice L is a partially ordered set. Thus we simply say that if L is a lattice, then the pair (L,μ), where is a lattice and $(L\times L,\mu)$ is a fuzzy set and $\mu(x,y) = 1 \Leftrightarrow x \leq y$.

The following theorem, proven by Kaburlasos, is an immediate consequence of this definition.

Theorem 3.16 *If L is a lattice and* $\sigma : L\times L \longrightarrow [0,1]$ *is an inclusion measure on L, then* (L,σ) *is a fuzzy lattice.*

The inclusion measure as defined by Kaburlasos is as follows:

Definition 3.17 Suppose L is a complete lattice with least element O and largest element I. An *inclusion measure* on L is a mapping $\sigma : L\times L \longrightarrow [0,1]$ satisfying the following conditions:

C_0. $\sigma(x,O) = 0,\ x \neq O$

C_1. $\sigma(x,x) = 1, \forall\, x \in L$

C_2. $u \leq w \ \Rightarrow\ \sigma(x,u) \leq \sigma(x,w)$

C_3. $x \wedge y < x \ \Rightarrow\ \sigma(x,y) < 1$

The usefulness of this definition is that it applies to any lattice. If the lattice is not complete, simply drop property C_0.

There exist many variants of Definition 3.17. The inclusion measure σ and its variants belong to a larger class called *similarity* measures. As defined in [60], distance measures such as metrics, measure numerically how unlike or different two data points are, while similarity measures measure numerically how alike two data points are. In short, a similarity measure is the antithesis of a distance measure since a *higher* value indicates a greater similarity, while for a distance measure a *lower* value indicates greater similarity. For instance, if d is a metric on X, then $m(x,y) = \frac{1}{1+d(x,y)}$ is a similarity measure on X.

The notion of an *interval* was well-known to mathematicians of ancient Greece, while multidimensional intervals made their entry with the rise of Cartesian geometry. Intervals are basic in mathematical analysis as well as in several chapters of this volume. Their generalization and applications in lattice theory are due to G. Birkhoff [30] who defines intervals in terms of a sublattices as follows.

Definition 3.18 If $\mathbb{L}$ is a lattice and $a, b \in \mathbb{L}$ with $a \leq b$, then the set $[a,b] \subset \mathbb{L}$ defined by $[a,b] = \{x \in \mathbb{L} : a \leq x \leq b\}$ is called a *(closed) interval*. A subset $S \subset \mathbb{L}$ is called a *convex sublattice* if $\forall\, a, b \in S \;\Leftrightarrow\; [a \wedge b, a \vee b] \subset S$.

The importance of this definition becomes obvious as Birkhoff devotes a whole section of his book to interval topology. Kaburlasos employed these generalized intervals in order to define *fuzzy interval numbers* [131]. A more in-depth discussion of fuzzy interval analysis can be found in [75]. For our objectives, we conclude this section with the following observation. Note that if $\mathbb{L}$ is a lattice and $[a,b] \subset \mathbb{L}$, then $[a,b] \in 2^{\mathbb{L}}$. Thus, if $I_{\mathbb{L}} = \{[a,b] : a, b \in \mathbb{L}\}$ denotes the set of all intervals of $\mathbb{L}$, then $I_{\mathbb{L}} \subset 2^{\mathbb{L}}$. By replacing the notation for subset, union, and intersection of the elements of $I_{\mathbb{L}}$, respectively by

$$[a,b] \preccurlyeq [c,d] := [a,b] \subset [c,d], \quad [a,b] \vee [c,d] := [a,b] \cup [c,d],$$
$$\text{and } [a,b] \wedge [c,d] := [a,b] \cap [c,d],$$

we see that the poset $(I_{\mathbb{L}}, \preccurlyeq)$ is a lattice. The lattice $(I_{\mathbb{L}}, \preccurlyeq)$ has the following relationship with $\mathbb{L}$.

Theorem 3.17 *If $\mathbb{L}$ is a complete lattice, then $I_{\mathbb{L}}$ is a complete lattice.*

Exercises 3.3.4

1. Prove Theorem 3.16.
2. Provide three distinct examples of similarity measures on a lattice.
3. Prove Theorem 3.17.

CHAPTER 4

Lattice Algebra

LATTICE algebra refers to the mathematical edifice obtained when combining lattice operations with operations of the abstract algebraic systems discussed in Section 1.2 as well as metrics and other measures. The lattice algebra obtained in this manner shares many similarities with linear algebra even though it is very different from it.

4.1 LATTICE SEMIGROUPS AND LATTICE GROUPS

In addition to being a distributive lattice, the set of real numbers is also a ring, and our early experience in elementary algebra has taught us the useful properties:

L_1. $x \leq y \Rightarrow x+z \leq y+z$

L_2. $0 \leq x$ and $0 \leq y \Rightarrow 0 \leq xy$

L_3. $0 \leq z \Rightarrow z(x \vee y) = zx \vee zy$ and $z(x \wedge y) = zx \wedge zy$, where $x, y, z \in \mathbb{R}$.

These properties exhibit the interplay between the lattice and the group and semigroup operations of addition and multiplication. The aim of this section is to explore such interactions in a more general setting by considering the algebraic properties of the structure $(\mathbb{F}, \preccurlyeq, \circ)$, where $(\mathbb{F}, \preccurlyeq)$ is a poset and $(\mathbb{F}, \circ)$ is a semigroup. While many properties can be couched in terms of lattice related concepts, the main purpose of approaching these structures from a general algebraic viewpoint is to develop an analogy to linear operator theory.

Suppose $\mathbb{F}$ is a set with binary operation $\circ$ and partial order $\preccurlyeq$, and that the system $(\mathbb{F}, \preccurlyeq, \circ)$ satisfies the following property:

L_4. $x \preccurlyeq y \Rightarrow a \circ x \circ b \preccurlyeq a \circ y \circ b \; \forall a, b \in \mathbb{F}$.

DOI: 10.1201/9781003154242-4

Assuming property L_4, we define the following algebraic structures:

1. If $(\mathbb{F}, \circ)$ is a semigroup, then $\mathbb{F}$ is a *partially ordered semigroup* or *po-semigroup*.

2. If $(\mathbb{F}, \circ)$ is a group, then $\mathbb{F}$ is a *partially ordered group* or *po-group*.

3. If $(\mathbb{F}, \circ)$ is a semigroup and $(\mathbb{F}, \vee)$ is a semilattice, then $\mathbb{F}$ is a *semilattice semigroup* or *sℓ-semigroup*.

4. If $(\mathbb{F}, \circ)$ is a group and $(\mathbb{F}, \vee)$ is a semilattice, then $\mathbb{F}$ is a *semilattice group* or *sℓ-group*.

5. If $(\mathbb{F}, \circ)$ is a semigroup and $(\mathbb{F}, \vee, \wedge)$ is a lattice, then $\mathbb{F}$ is a *lattice semigroup* or *ℓ-semigroup*.

6. If $(\mathbb{F}, \circ)$ is a group and $(\mathbb{F}, \vee, \wedge)$ is a lattice, then $\mathbb{F}$ is a *lattice group* or *ℓ-group*.

More generally, if $(\mathbb{F}, \circ)$ is an algebraic structure and $\mathbb{F}$ is a poset, semilattice, or lattice, then the structure will have the appropriate prefix. For example, if $(\mathbb{F}, \circ)$ is a monoid and $\mathbb{F}$ is a poset, semilattice, or lattice, then $\mathbb{F}$ is a *po-monoid*, *sℓ-monoid*, or *ℓ-monoid*, respectively.

It follows from property L_4 that in any *po*-semigroup, *po*-group, *sℓ*-semigroup, *sℓ*-group, *ℓ*-semigroup, and *ℓ*-group every semigroup and group translation is order preserving. Obviously, $(\mathbb{R}, \vee, \wedge, +)$ is an ℓ-group since $L_1 \Rightarrow L_4$. On the other hand, $(\mathbb{R}, \vee, \wedge, \times)$ is not an ℓ-semigroup since L_4 does not hold for multiplication. However, $(\mathbb{R}^+, \vee, \wedge, \times)$ is an ℓ-group. For ℓ-semigroups we have the following useful result.

Theorem 4.1 *If* $(\mathbb{F}, \vee, \wedge, \circ)$ *is a ℓ-semigroup and* $a, b, x, y \in \mathbb{F}$, *then*

$$a \circ (x \vee y) \circ b = (a \circ x \circ b) \vee (a \circ y \circ b) \quad \text{and} \quad a \circ (x \wedge y) \circ b = (a \circ x \circ b) \wedge (a \circ y \circ b).$$

Proof. If $x \preccurlyeq y$, then it follows from the definition of an ℓ-semigroup that $a \circ x \circ b \preccurlyeq a \circ y \circ b$. But this implies that

$$(a \circ x \circ b) \vee (a \circ y \circ b) = a \circ y \circ b = a \circ (x \vee y) \circ b$$

and

$$(a \circ x \circ b) \wedge (a \circ y \circ b) = a \circ x \circ b = a \circ (x \wedge y) \circ b.$$

If $y \leqslant x$, then $a \circ y \circ b \leqslant a \circ x \circ b$. In this case we have

$$(a \circ x \circ b) \vee (a \circ y \circ b) = a \circ x \circ b = a \circ (x \vee y) \circ b$$

and

$$(a \circ x \circ b) \wedge (a \circ y \circ b) = a \circ y \circ b = a \circ (x \wedge y) \circ b.$$

Therefore, in either of the two possible cases, the conclusion of the theorem is true. □

The next result is an immediate consequence of Theorem 4.1.

Corollary 4.2 *Suppose $a, b, x, y \in \mathbb{F}$. If $(\mathbb{F}, \vee, \circ)$ is an $s\ell$-semigroup, then*

$$a \circ (x \vee y) \circ b = (a \circ x \circ b) \vee (a \circ y \circ b).$$

If $(\mathbb{F}, \wedge, \circ)$ is an $s\ell$-semigroup, then

$$a \circ (x \wedge y) \circ b = (a \circ x \circ b) \wedge (a \circ y \circ b).$$

The distributive properties guaranteed by Theorem 4.1 easily generalize to the following equations for $s\ell$-semigroups:

$$a \circ (\bigvee_{i=1}^{n} x_i) \circ b = \bigvee_{i=1}^{n} (a \circ x_i \circ b) \quad \text{and} \quad a \circ (\bigwedge_{i=1}^{n} x_i) \circ b = \bigwedge_{i=1}^{n} (a \circ x_i \circ b). \quad (4.1)$$

Note that if $\mathbb{F}$ is an ℓ-monoid or $s\ell$-monoid with identity ϕ, then by setting $b = \phi$ in equation 4.1, we obtain

$$a \circ \bigvee_{i=1}^{n} x_i = \bigvee_{i=1}^{n} (a \circ x_i) \quad \text{and} \quad a \circ \bigwedge_{i=1}^{n} x_i = \bigwedge_{i=1}^{n} (a \circ x_i). \quad (4.2)$$

As a matter of fact, a more general relationship holds [30].

Theorem 4.3 *The following equalities hold in any ℓ-group:*

$$\bigwedge_i [\bigwedge_j (x_i \circ y_j)] = \bigwedge_i [x_i \circ \bigwedge_j y_j] \quad \textit{and} \quad \bigvee_i [\bigvee_j (x_i \circ y_j)] = \bigvee_i [x_i \circ \bigvee_j y_j].$$

If the $s\ell$-semigroup or $s\ell$-group $(\mathbb{F}, \vee, \circ)$ is also an $s\ell$-semigroup or an $s\ell$-group $(\mathbb{F}, \wedge, \circ^*)$ under a semilattice operation $\wedge$ and a semigroup or group operation $\circ^*$ and satisfies the equation

$$a \vee (b \wedge a) = a \wedge (b \vee a) = a \ \forall a, b \in \mathbb{F}, \tag{4.3}$$

then we say that $\mathbb{F}$ is an *$s\ell$-semigroup with duality* or an *$s\ell$-group with duality*, respectively. If the operations $\circ$ and $\circ^*$ coincide, then the operation $\circ$ is called *self-dual*. An $s\ell$-semigroup or an $s\ell$-group with duality and self-dual semigroup or group operation $\circ$ is simply an ℓ-semigroup or ℓ-group, respectively.

It has been shown (Birkhoff [30]) that a partially ordered group cannot have universal bounds except in the trivial case where $\mathbb{F} = \{0\}$. Thus, an ℓ-group cannot be a complete lattice (unless it is $\{0\}$); that is, each subset U of $\mathbb{F}$ cannot have a *lub* and *glb* in $\mathbb{F}$. In particular, the ℓ-group $(\mathbb{R}, \vee, \wedge, +)$, cannot be a complete lattice. Extending the ℓ-group $(\mathbb{R}, \vee, \wedge, +)$ in a mathematically well-defined manner to include the universal bounds ∞ and $-\infty$ will mean that the complete lattice $(\mathbb{R}_{\pm\infty}, \vee, \wedge)$, together with the extended addition, will degenerate into an ℓ-semigroup since neither element ∞ or $-\infty$ can have an inverse under addition. More specifically, extending the addition + by setting

$$a + \infty = \infty + a = \infty \qquad \forall a \in \mathbb{R}_{\infty} \tag{4.4}$$

$$a + -\infty = -\infty + a = -\infty \qquad \forall a \in \mathbb{R}_{\pm\infty} \tag{4.5}$$

results in the equation

$$-\infty + \infty = \infty + -\infty = -\infty. \tag{4.6}$$

These equations establish a well-defined additive operation on the lattice $\mathbb{R}_{\pm\infty}$ with the property that $a + 0 = 0 + a = a \ \forall a \in \mathbb{R}_{\pm\infty}$. However, we now have the problem that neither element ∞ or $-\infty$ has an additive inverse. Therefore, $\mathbb{R}_{\pm\infty}$ is an ℓ-monoid but not an ℓ-group.

The dual of the ℓ-monoid $(\mathbb{R}_{\pm\infty}, \vee, \wedge, +)$ is $(\mathbb{R}_{\pm\infty}, \wedge, \vee, +^*)$ with the dual operation $+^*$ defined by

$$a +^* -\infty = -\infty +^* a = -\infty \qquad \forall a \in \mathbb{R}_{-\infty} \tag{4.7}$$

$$a +^* \infty = \infty +^* a = \infty \qquad \forall a \in \mathbb{R}_{\pm\infty}. \tag{4.8}$$

Accordingly, we have

$$-\infty +^* \infty = \infty +^* -\infty = \infty \tag{4.9}$$

and the dual operations $+$ and $+^*$ are identical (self-dual) operations except on the subset $\{-\infty, \infty\}$ of $\mathbb{R}_{\pm\infty}$. Equations 4.6 and 4.9 show that $+$ and $+^*$ introduce an asymmetry between ∞ and $-\infty$.

The two $s\ell$-monoids $(\mathbb{R}_{\pm\infty}, \vee, +)$ and $(\mathbb{R}_{\pm\infty}, \wedge, +^*)$ are coupled by duality. The function $\varphi : \mathbb{R}_{\pm\infty} \to \mathbb{R}_{\pm\infty}$, defined by $\varphi(r) = r^*\ \forall r \in \mathbb{R}_{\pm\infty}$, provides for another way of looking at this dual relationship. Here r^* denotes the dual of r as defined in equation 3.7. It is easy to show that φ is a bijection, and that $\varphi(r+s) = \varphi(r) +^* \varphi(s)$ and, since $r \leq s \Leftrightarrow -r \geq -s$, $\varphi(r \vee s) = \varphi(r) \wedge \varphi(s)$. Thus, φ is an isomorphism and the two structures $(\mathbb{R}_{\pm\infty}, \vee, +)$ and $(\mathbb{R}_{\pm\infty}, \wedge, +^*)$ are algebraically equivalent. For these two reasons, it is easy and often convenient to combine these two structures into one coherent algebraic system $(\mathbb{R}_{\pm\infty}, \vee, \wedge, +, +^*)$. The combined algebraic structure is commonly known as a *bounded lattice ordered group* or simply a *blog*. The rationale for this name is that the underlying set $\mathbb{R}$ with the self-dual operation $+ = +^*$ is a group and $(\mathbb{R}_{\pm\infty}, \vee, \wedge)$ is a bounded lattice. Also, given any ℓ-group $(\mathbb{F}_{\pm\infty}, \vee, \wedge, \circ)$ and defining universal bounds ∞ and $-\infty$ by $-\infty < a < \infty\ \forall a \in \mathbb{F}$, one can extend $\mathbb{F}$ to a blog $(\mathbb{F}_{\pm\infty}, \vee, \wedge, \circ, \circ^*)$ by using equations analogous to equations 4.4 through 4.8 for extending the self-dual operation $\circ$.

Exercises 4.1

1. Prove that if $(\mathbb{F}, \circ)$ is a po-group and $a, b, c, d \in \mathbb{F}$, then $a \leq b$ and $c \leq d \Rightarrow a \circ c \leq b \circ d$.

2. Prove that if $\mathbb{F}$ is a lattice semigroup, then eqn. 4.1 holds.

3. Prove Theorem 4.3.

4.2 MINIMAX ALGEBRA

For the remainder of this chapter we discuss lattice-based algebras from a viewpoint that differs from the one presented in most current textbooks and research papers on the subject. Currently, the common approach is reflected in the preceding section where ℓ-groups, ℓ-semigroups, and $s\ell$-semigroups were defined in terms of groups or semigroups that also happen to be lattices or semilattices. In fact, the common definition of an ℓ-group, ℓ-ring, $s\ell$-ring, or a vector lattice assumes that the underlying algebraic structure is a group $(G, \circ)$, a ring $(R, +, \times)$, or vector space $\mathbb{V}$, respectively, that also happen to be lattices or semilattices. Thus, for example, $(\mathbb{R}, \vee, \wedge, +)$ is an ℓ-group and $(\mathbb{R}, \vee, +)$ an $s\ell$-group. The approach taken here is one of analogy of two algebraic structures having the *same* number of operations. This approach was used by Cunninghame-Green in his treatise on minimax algebra [63]. Central questions that arise using this approach are:

1. In what sense are the operations analogous or similar?
2. What laws, if any, are common to both structures?
3. What practical advantages can be gained by choosing one structure over another for solving a given problem?

Answers to these questions can be gleaned from Cunninghame-Green's contributions as well as this text. L-monoids serve as an excellent introductory example for illustrating the proposed viewpoint as well as for addressing some of the above questions. Any ℓ-monoid $(\mathbb{F}, \vee, \circ)$ with identity ϕ satisfies the following properties:

L_1'. $x \preccurlyeq y \Rightarrow x \vee z \preccurlyeq y \vee z$

L_2'. $\phi \preccurlyeq x$ and $\phi \preccurlyeq y \Rightarrow \phi \preccurlyeq x \circ y$

L_3'. $z \circ (x \vee y) = (z \circ x) \vee (z \circ y)$ and $z \circ (x \wedge y) = (z \circ x) \wedge (z \circ y)$.

With the exception of property L_3', these properties are exactly the same as properties L_1 and L_2 for the ring $(\mathbb{R}, +, \times)$ listed at the beginning of Section 4.1, with addition $(+)$ being replaced by the binary maximum $(\vee)$ operation, and multiplication $(\times)$ being replaced by $(\circ)$. Property L_3' is stronger than property L_3 since the requirement that $0 \leq z$ is not necessary. However, $(\mathbb{F}, \vee, \circ)$ is not a ring since axiom R_1 for rings is not satisfied. On the other hand, an $s\ell$-semigroup $(\mathbb{F}, \vee, \circ)$ is *almost* a ring since

R_1'. $(\mathbb{F}, \vee)$ is an abelian semigroup,

R_2'. $(\mathbb{F}, \circ)$ is a semigroup, and

R_3'. $\forall a, b, c \in \mathbb{F}, a \circ (b \vee c) = (a \circ b) \vee (a \circ c)$ and $(a \vee b) \circ c = (a \circ c) \vee (b \circ c)$.

Note that properties R_2' and R_3' are the same as ring axioms R_2 and R_3 with $\vee$ replacing $+$ and $\circ$ replacing $\times$. Only property R_1' is weaker than ring axiom R_1. But since $(\mathbb{F}, \vee)$ is an abelian semigroup, $(\mathbb{F}, \vee, \circ)$ is a semiring. We distinguish between the two semirings $(\mathbb{F}, \vee, \circ)$ and $(\mathbb{F}, \wedge, \circ)$ by using term *max-semiring* for the former and *min-semiring* for the latter. The combined dual structure $(\mathbb{F}, \vee, \wedge, \circ)$, where $\circ$ is selfdual, is called a *minimax-semiring*.

We need to sound a note of caution in regards to terminology. We already noted that lattice-based semirings are also known as *semilattice ordered semigroups* or *$s\ell$-semigroups* and *lattice ordered semigroups* or *ℓ-semigroups*.

Cunninghame-Green calls these semirings *belts* as the word "belt" is in certain situations synonymous with the word "ring" [63]. We prefer the name "semiring" since problems arising in certain lattice algebras will then assume the flavor of problems in linear algebra. Furthermore, the term *semiring* is accepted standard terminology in the mathematical community and, in particular, among algebraists. Thus, unless stated otherwise or it is clear from the discussion, we use the term *semiring* when we refer to any of the three lattice-based semirings defined above.

In analogy with our discussion on rings (Subsection 1.2.3), we note that the max-semiring $(\mathbb{R}_{-\infty}, \vee, +)$ (or its dual $(\mathbb{R}_{\infty}, \wedge, +^*)$) possesses several additional properties that are not necessarily properties enjoyed by max-semirings or rings in general. Among these properties are:

P'_1. The existence of an identity element for the operation $\circ = +$, called the *unit* element, which in this case is 0.

P'_2. The commutativity of the operation $\circ = +$.

P'_3. $(\mathbb{R}_{-\infty}, \vee)$ is a monoid with identity $-\infty$, called the *null* element of the semiring.

P'_4. For every non-null element $a \in \mathbb{R}_{-\infty}$ there is an inverse, namely $-a$ in $\mathbb{R}_{-\infty}$ such that $a + -a = -a + a = 0$.

If $a \neq -\infty$, then $\bigvee_{i=1}^{2} a = a \vee a = a \neq -\infty$. By associativity of the operation $\vee$, it follows that $\bigvee_{i=1}^{n} a = a \neq -\infty$ for any positive integer n. Thus, any max-semiring is a semiring with characteristic zero. Similarly, any min-semiring is a semiring with characteristic zero. Therefore, the special ring property P_3 is not a special semiring property for min- or max-semirings. For this reason it was not listed among the special semiring properties. The ring $(\mathbb{R}, +, \times)$ possesses properties P_1 through P_4. Note that for $i \neq 3$, properties P'_i and P_i are algebraically the same. This means that various theorems relying on these properties hold for both structures, $(\mathbb{R}, +, \times)$ and $(\mathbb{R}_{-\infty}, \vee, +)$ (or its dual $(\mathbb{R}_{\infty}, \wedge, +^*)$). Table 4.1 compares some basic properties common to the ring $(\mathbb{R}, +, \times)$ and the semiring $(\mathbb{R}_{-\infty}, \vee, +)$.

Thus, the unit for $(\mathbb{R}, +, \times)$ is 1, while the unit for $(\mathbb{R}_{-\infty}, \vee, +)$ is 0. The null element for $(\mathbb{R}, +, \times)$ is 0, while the null element for $(\mathbb{R}_{-\infty}, \vee, +)$ is $-\infty$. Although the unit and null elements are completely different in the two structures, they reflect the same properties common to these special rings and semirings. It is, therefore, not surprising that various other laws and theorems in linear algebra have analogies in minimax algebra.

As before, the special properties lead to some further definitions.

TABLE 4.1 Properties common to a ring and semiring.

$(\mathbb{R},+,\times)$	$(\mathbb{R}_{-\infty},\vee,+)$
$a\times 0=0\times a=0$	$a+-\infty=-\infty+a=-\infty$
$a+0=0+a=a$	$a\vee-\infty=-\infty\vee a=a$
$1\times a=a\times 1=a$	$0+a=a+0=a$
$a\times(b+c)=(a\times b)+(a\times c)$	$a+(b\vee c)=(a+b)\vee(a+c)$

Definition 4.1 A semiring satisfying P_1' is called a *semiring with unity*. If a semiring satisfies P_2', it is called a *commutative* or *abelian semiring*. If it satisfies P_3', then it is called a *semiring with a null element*. If it satisfies P_4', it is called a *division semiring*. If it is an abelian semiring as well as a division semiring, then it is called a *semifield*.

A *max-semifield* is a max-semiring which is also a semifield, and a min-semiring that is a semifield is called a *min-semifield*. If $(\mathbb{F}, \vee, \circ)$ is a max-semifield with dual $(\mathbb{F}, \wedge, \circ)$ a min-semifield, then the combined lattice structure $(\mathbb{F}, \vee, \wedge, \circ)$ is called a *minimax-semifield*. If $\mathbb{F}_{\pm\infty}$ is a bounded lattice with extended dual operations $\circ$ and $\circ^*$ as described in Section 4.1, where $\circ$ and $\circ^*$ are asymmetric on $\{\infty, -\infty\}$, then the combined structure $(\mathbb{F}_{\pm\infty}, \vee, \wedge, \circ, \circ^*)$ is a *bounded minimax-semifield* whenever $(\mathbb{F}_{-\infty}, \vee, \circ)$ and $(\mathbb{F}_{\infty}, \wedge, \circ^*)$ are semifields. Standard examples of max-semifields are $(\mathbb{R}, \vee, +)$ and $(\mathbb{R}_{-\infty}, \vee, +)$. Their duals $(\mathbb{R}, \wedge, +)$ and $(\mathbb{R}_{\infty}, \wedge, +)$ are min-semifields, while $(\mathbb{R}, \vee, \wedge, +)$ is a minimax-semifield and the blog $(\mathbb{R}_{\pm\infty}, \vee, \wedge, +, +^*)$ is a *bounded* minimax-semifield.

Examples of min-, max-, and minimax-semifields involving multiplication instead of addition are $((0,\infty], \wedge, \times)$, $([0,\infty), \vee, \times)$, and $(\mathbb{R}^+, \vee, \wedge, \times)$, respectively. The structures $((0,\infty], \wedge, \times)$ and $([0,\infty), \vee, \times)$ are duals and can be combined to form the bounded minimax-semifield $([0,\infty], \vee, \wedge, \times, \times^*)$, where the two multiplicative operations $\times$ and $\times^*$ differ only on the set $\{0,\infty\}$. Specifically, $a\times b = a\times^* b \quad \forall a, b \in [0,\infty)$, and $a\times\infty = a\times^*\infty = \infty \quad \forall a \in (0,\infty]$. Asymmetry occurs on the set $\{0,\infty\}$, where we define $0\times\infty=\infty\times 0=0$ and $0\times^*\infty=\infty\times^* 0=\infty$. The *dual* or *conjugate* of $r\in[0,\infty]$ is defined as

$$r^* = \begin{cases} 1/r & \text{if } r\in\mathbb{R}^+ \\ 0 & \text{if } r=\infty, \\ \infty & \text{if } r=0 \end{cases} \tag{4.10}$$

and the duality isomorphism is the function $\varphi : [0,\infty) \to (0,\infty]$ defined by $\varphi(r) = r^*$.

The function $\psi : [-\infty, \infty) \to [0, \infty)$, defined by

$$\psi(x) = \begin{cases} e^x & \text{if } x \neq -\infty \\ 0 & \text{if } x = -\infty, \end{cases} \tag{4.11}$$

has the property that $\psi(x+y) = \psi(x) \times \psi(y)$. For $x \neq -\infty \neq y$, this claim is a consequence of Example 1.11. In this case we also have $\psi(x \vee y) = e^{x \vee y} = e^x \vee e^y = \psi(x) \vee \psi(y)$. If one of x or y equals $-\infty$, say $x = -\infty \neq y$, then

$$\begin{aligned} \psi(x+y) = \psi(-\infty + y) = \psi(-\infty) = 0 = 0 \times \psi(y) = \psi(x) \times \psi(y) \text{ and} \\ \psi(x \vee y) = \psi(-\infty \vee y) = \psi(y) = e^y = 0 \vee e^y = \psi(x) \vee \psi(y). \end{aligned}$$

If $x = -\infty = y$, then

$$\begin{aligned} \psi(x+y) = \psi(-\infty + -\infty) = \psi(-\infty) = 0 = 0 \times 0 = \psi(x) \times \psi(y) \text{ and} \\ \psi(x \vee y) = \psi(-\infty \vee -\infty) = \psi(-\infty) = 0 = 0 \vee 0 = \psi(x) \vee \psi(y). \end{aligned}$$

Hence, ψ is a homomorphism. It easily follows from Example 1.11 that ψ is a bijection with inverse defined by

$$\psi^{-1}(x) = \begin{cases} \ln(x) & \text{if } x > 0 \\ -\infty & \text{if } x = 0. \end{cases} \tag{4.12}$$

Therefore, $([-\infty, \infty), \vee, +)$ and $([0, \infty), \vee, \times)$ are isomorphic. This proves the following isomorphism theorem.

Theorem 4.4 $(\mathbb{R}_{\infty}, \wedge, +) \approx (\mathbb{R}_{-\infty}, \vee, +) \approx ([0, \infty), \vee, \times) \approx ((0, \infty], \wedge, \times)$.

The first and last isomorphisms in the theorem follow from the two duality isomorphisms, each mapping $r \to r^*$, where the respective duals are given by equations 3.7 and 4.10.

Extending the function $\psi : \mathbb{R}_{-\infty} \to [0, \infty)$, defined in equation 4.11, to $\psi : \mathbb{R}_{\pm\infty} \to [0, \infty]$ by setting

$$\psi(x) = \begin{cases} e^x & \text{if } x \in \mathbb{R} \\ \infty & \text{if } x = \infty, \\ 0 & \text{if } x = -\infty \end{cases} \tag{4.13}$$

results again in an isomorphism. It is a simple bookkeeping task to check that

1. $\psi(x+y) = \psi(x) \times \psi(y)$ and $\psi(x +^* y) = \psi(x) \times^* \psi(y)$.

2. $\psi(x \vee y) = \psi(x) \vee \psi(y)$ and $\psi(x \wedge y) = \psi(x) \wedge \psi(y)$.

3. ψ is a bijection with inverse given by

$$\psi^{-1}(x) = \begin{cases} \ln(x) & \text{if } x \in \mathbb{R}^+ \\ \infty & \text{if } x = \infty. \\ -\infty & \text{if } x = 0 \end{cases} \tag{4.14}$$

This provides a proof for the next isomorphism theorem.

Theorem 4.5 $(\mathbb{R}_{\pm\infty}, \vee, \wedge, +, +^*) \approx ([0, \infty], \vee, \wedge, \times, \times^*)$.

The importance of Theorem 4.4 and Theorem 4.5 lies in the fact that if one does not want to deal with negative numbers one can choose the structure $([0, \infty), \vee, \times)$ (or $([0, \infty] \vee, \wedge, \times, \times^*)$) instead of $(\mathbb{R}_{-\infty}, \vee, +)$ (or $(\mathbb{R}_{\pm\infty}, \vee, \wedge, +, +^*)$) and obtain the same results since these two structures are algebraically the same. Computationally, however, addition is faster than multiplication. Thus, one may prefer to use $\vee$ and $+$ instead of $\vee$ and $\times$. To the neurobiologist, on the other hand, dealing with negative synaptic weights may be bothersome.

The ring $(\mathbb{R}^X, +, \times)$ provided an example of a non-numeric ring which was not a division ring and, hence, not a field (Example 1.11). Addition and multiplication on $\mathbb{R}^X$ were defined in terms of addition and multiplication of the field $\mathbb{R}$. By using a max-, min-, or minimax-semifield $\mathbb{F}$ instead, it is possible to define a lattice semiring structure on $\mathbb{F}^X$ as well.

Example 4.1 Consider the max-semifield $(\mathbb{R}, \vee, +)$ and let $f, g, h \in \mathbb{R}^X$ be arbitrary real valued functions on X. The function $(f \vee g)$ is defined by $(f \vee g)(x) = f(x) \vee g(x)\ \forall x \in X$ (see also Examples 1.7 and 3.7(7)). Using this definition, it is trivial to show that

$$f \vee g = g \vee f \quad \text{and} \quad f \vee (g \vee h) = (f \vee g) \vee h.$$

Therefore, $(\mathbb{R}^X, \vee)$ is an abelian semigroup. Also, from Example 1.11 we know that $(\mathbb{R}^X, +)$ is an abelian group. Commutativity of addition of real numbers and Corollary 4.2 applied to the max-semiring $(\mathbb{R}, \vee, +)$ imply that

$$\begin{aligned}[f + (g \vee h)](x) &= f(x) + (g \vee h)(x) = f(x) + (g(x) \vee h(x)) \\ &= (f(x) + g(x)) \vee (f(x) + h(x)) = (g(x) + f(x)) \vee (h(x) + f(x)) \\ &= (g(x) \vee h(x)) + f(x) = (g \vee h)(x) + f(x) = [(g \vee h) + f](x).\end{aligned}$$

This shows that $f + (g \vee h) = (f + g) \vee (f + h)$ and $(g \vee h) + f = (g + f) \vee (h + f)$. Thus, property R_3' is satisfied. It follows that $(\mathbb{R}^X, \vee, +)$ is a commutative division semiring and, hence, a max-semifield.

This example shows that a disadvantage of $(\mathbb{R}^X, \vee, +)$ versus $(\mathbb{R}^X, +, \times)$ is that $(\mathbb{R}^X, \vee)$ is an abelian semigroup while $(\mathbb{R}^X, +)$ is an abelian group. On the other hand, however, is the advantage that $(\mathbb{R}^X, \vee, +)$ satisfies property P_4 while $(\mathbb{R}^X, +, \times)$ does not.

It is an easy exercise to show that the function $\varphi : \mathbb{R}^X \to \mathbb{R}^X$ defined by $\varphi(f) = -f$ is an isomorphism. For instance, since $f(x) \leq g(x) \Leftrightarrow -f(x) \geq -g(x)$, we have $\varphi(f \vee g) = -(f \vee g) = -f \wedge -g = \varphi(x) \wedge \varphi(g)$. Thus, $(\mathbb{R}^X, \vee, +) \approx (\mathbb{R}^X, \wedge, +)$ and, hence, $(\mathbb{R}^X, \wedge, +)$ is also a commutative division semiring.

As can be ascertained from the deliberation of this section, minimax-semifields are closely linked to the concept of duality. It is the existence of inverses with respect to the operation $\circ$ that provide for duality isomorphisms. In order to realize such isomorphisms we need to establish a few simple facts related to inverses. In the following we let $-a$ denote the inverse of a with respect to the operation $\circ$.

Theorem 4.6 *Suppose* $(\mathbb{F}, \vee, \wedge, \circ)$ *is a minimax-semifield with unit* 0 *and partial order* $\preccurlyeq$. *The following relationships hold for all* $x, y \in \mathbb{F}$:

1. $0 \preccurlyeq x \Leftrightarrow -x \preccurlyeq 0$,

2. $x \preccurlyeq y \Leftrightarrow -y \preccurlyeq -x$, *and*

3. $x \preccurlyeq y \Rightarrow a \circ (-y) \circ b \preccurlyeq a \circ (-x) \circ b \quad \forall a, b \in \mathbb{F}$.

Proof. (*1*) Suppose $0 \preccurlyeq x$. Then $-x = 0 \circ -x \preccurlyeq x \circ -x = 0$. Conversely, if $-x \preccurlyeq 0$, then $0 = x \circ -x \preccurlyeq x \circ 0 = x$.

(2) Suppose $x \preccurlyeq y$. Then $0 = x \circ -x \preccurlyeq y \circ -x$. Thus, by part (*1*), $-y \circ x = -(y \circ -x) \preccurlyeq 0$ and, hence, $-y = -y \circ 0 = -y \circ (x \circ -x) = -(y \circ -x) \circ -x \preccurlyeq 0 \circ -x = -x$. The equation $-y \circ x = -(y \circ -x)$ follows from the uniqueness of inverses. To prove the converse, suppose that $-y \preccurlyeq -x$. Then $x \circ -y \preccurlyeq x \circ -x = 0$. Again, by part (*1.*), $0 \preccurlyeq -x \circ y$. Therefore, $x = x \circ 0 \preccurlyeq -x \circ (-x \circ y) = (x \circ -x) \circ y = 0 \circ y = y$.

Claim (*3*) is an easy consequence of (2) and left as an exercise. □

Consider the function $\varphi : \mathbb{F} \to \mathbb{F}$ defined by $\varphi(x) = -x$. If $x \preccurlyeq y$, then $\varphi(x \vee y) = -(x \vee y) = -y$ and $\varphi(x) \wedge \varphi(y) = -x \wedge -y = -y$ since by Theorem 4.6 we have $-y \preccurlyeq -x$. Likewise, if $y \preccurlyeq x$, then $\varphi(x \vee y) = -x$ and $\varphi(x) \wedge \varphi(y) =$

$-x$. Therefore, $\varphi(x \vee y) = \varphi(x) \wedge \varphi(y)$. Similarly, $\varphi(x \wedge y) = \varphi(x) \vee \varphi(y)$. This verifies the equations:

$$-(a \vee b) = -a \wedge -b \quad \text{and} \quad -(a \wedge b) = -a \vee -b \tag{4.15}$$

$$a \vee b = -(-a \wedge -b) \quad \text{and} \quad a \wedge b = -(-a \vee -b) \tag{4.16}$$

These equations signify that the function $\psi(x) = a \circ (-x) \circ b$ is a dual isomorphism for any fixed pair $a, b \in \mathbb{F}$. Thus, in any minimax-semifield the following identities hold:

$$a \circ -(x \vee y) \circ b = (a \circ -x \circ b) \wedge (a \circ -y \circ b) \tag{4.17}$$

$$a \circ -(x \wedge y) \circ b = (a \circ -x \circ b) \vee (a \circ -y \circ b). \tag{4.18}$$

In analogy with equations 4.1 and 4.2, this easily generalizes to

$$a \circ (-\bigvee_{i=1}^{n} x_i) \circ b = \bigwedge_{i=1}^{n} (a \circ -x_i \circ b) \; ; \; a \circ (-\bigwedge_{i=1}^{n} x_i) \circ b = \bigvee_{i=1}^{n} (a \circ -x_i \circ b), \tag{4.19}$$

and, hence,

$$a \circ (-\bigvee_{i=1}^{n} x_i) = \bigwedge_{i=1}^{n} (a \circ -x_i) \quad \text{and} \quad a \circ (-\bigwedge_{i=1}^{n} x_i) = \bigvee_{i=1}^{n} (a \circ -x_i). \tag{4.20}$$

Exercises 4.2

1. Prove claim 3 of Theorem 4.6.
2. Prove that equation 4.19 holds $\forall n \in \mathbb{N}$.
3. Verify that $(\mathbb{R}^X, \vee, +) \approx (\mathbb{R}^X, \wedge, +)$ by proving that the function $\varphi : \mathbb{R}^X \to \mathbb{R}^X$ defined by $\varphi(f) = -f$ is an isomorphism.

4.2.1 Valuations, Metrics, and Measures

There are various functions that map lattices to $\mathbb{R}$ or other lattice groups. Among these the most important one is probably the valuation function which is defined as follows.

Definition 4.2 Suppose $(\mathbb{F}, \circ)$ is a ℓ-Abelian group. A *valuation* on a lattice L is a function $v : L \to \mathbb{F}$ that satisfies

$$v(x) \circ v(y) = v(x \vee y) \circ v(x \wedge y) \quad \forall x, y \in L.$$

A valuation is said to be *isotone* if and only if $x \preccurlyeq y \Rightarrow v(x) \leq v(y)$, and *positive* if and only if $x \prec y \Rightarrow v(x) < v(y)$.

For our goal we assume that the Abelian ℓ-group is $(\mathbb{R},+)$ so that the defining valuation equation corresponds to

$$v(x)+v(y)=v(x\vee y)+v(x\wedge y) \quad \forall x,y\in L. \tag{4.21}$$

It is interesting to note that equation 4.21 is equivalent to $v(x\wedge y)=v(x)+v(y)-v(x\vee y)$ with the latter equation having the same form as equation 3.11.

Example 4.2

1. Let X be a nonempty finite set and $L=2^X$ be partially ordered by $x\preccurlyeq y \Leftrightarrow x\subset y$, and define $x\vee y := x\cup y$ and $x\wedge y := x\cap y$. Let $v: L\to\mathbb{R}$ be defined by $v(x)=card(x)$. Since $card(x\cup y)=card(x)+card(y)-card(x\cap y)$, it follows that $v(x)+v(y)=v(x\vee y)+v(x\wedge y)$. If x is a strict subset of y, then $card(x)<card(y)$. This implies that v is a positive and isotone valuation on L.

2. Consider the lattice $(\mathbb{R}^n, \vee, \wedge)$. The function $v:\mathbb{R}^n\to\mathbb{R}$ given by $v(\mathbf{x})=\sum_{i=1}^n c_i x_i$ is a valuation. This valuation is positive if and only if c_i is positive for $i=1,\ldots,n$.

3. Let $L=\{a, b, c, d, e\}$ be the lattice derived from the poset of Example 3.5. Define a valuation $v:L\to\mathbb{R}$ by $v(a)=0$, $v(d)=1$, and $v(x)=\frac{1}{2}$ for $x\in\{b.c, e\}$. Then v is an isotone valuation, but v is not positive.

4. If L is a modular lattice of finite length, then it follows from Theorem 3.6 that the height function $h(x)$ is a positive valuation.

Theorem 4.7 *If L is a lattice and v is an isotone valuation on L, then the function* $d: L\times L\to\mathbb{R}$ *defined by*

$$d(x,y)=v(x\vee y)-v(x\wedge y) \tag{4.22}$$

satisfies

1. $d(x,y)\geq 0$ *and* $d(x,x)=0$,

2. $d(x,y)=d(y,x)$,

3. $d(x,y)\leq d(x,z)+d(z,y)$,

4. $d(x,y)\geq d(a\vee x,a\vee y)+d(a\wedge x,a\wedge y)$, $\forall x,y,z,a\in L$.

The proof of this theorem is straightforward and left as an exercise.

The valuation v defined in Example 4.2(3) is an isotone valuation. However, $d(a,b) = v(a \vee b) - v(a \wedge b) = v(b) - v(a) = 0$. Since $a \neq b$, d is not a metric, but just a pseudo-metric. Note that in this case we have $a \vee b > a \wedge b$, but $v(a \vee b) \not> v(a \wedge b)$. In fact, the condition

$$x \vee y > x \wedge y \Leftrightarrow v(x \vee y) > v(x \wedge y) \tag{4.23}$$

is equivalent to $d(x,y) = 0 \Leftrightarrow x = y$. This equivalence yields the following corollary of Theorem 4.7:

Corollary 4.8 *Suppose L is a lattice and v is an isotone valuation on L. The function $d(x,y) = v(x \vee y) - v(x \wedge y)$ is a metric on L if and only if the valuation v is positive.*

Consequently, the function $d(x,y) = v(x \vee y) - v(x \wedge y)$ is a metric for the valuation given in Example 4.2(1) as well as for the valuation in Example 4.2(2) as long as $c_i > 0$.

The metric d defined on a lattice L in terms of an isotone positive valuation is called a *lattice metric* or simply an *ℓ-metric*, and the pair (L, d) is called a *metric lattice* or a *metric lattice space*. The importance of ℓ-metrics is due to the fact that they can be computed using only the operations of $\vee$, $\wedge$, and $+$. Because of this, they require far less computational time than any ℓ_p metric whenever $1 < p < \infty$. As will be shown in subsequent chapters, they are easily implemented and partially computed within the dendritic structure of an artificial neuron.

Just as different ℓ_p norms give rise to different ℓ_p metrics on $\mathbb{R}^n$, different positive valuations on a lattice will yield different ℓ-metrics. For instance, if $L = \mathbb{R}^n$, then the two positive valuations $v_1(\mathbf{x}) = \sum_{i=1}^n x_i$ and $v_\infty(\mathbf{x}) = \bigvee_{i=1}^n x_i$ define two different ℓ-metrics on L. In particular, we have

Theorem 4.9

1. $d_1(\mathbf{x},\mathbf{y}) = v_1(\mathbf{x} \vee \mathbf{y}) - v_1(\mathbf{x} \wedge \mathbf{y})$,

2. $d_\infty(\mathbf{x},\mathbf{y}) = v_\infty(\mathbf{x} \vee \mathbf{y}) - v_\infty(\mathbf{x} \wedge \mathbf{y})$.

Proof. Considering equations 4.15 through 4.18 establishes the following equalities:

$$
\begin{aligned}
v_1(\mathbf{x} \vee \mathbf{y}) - v_1(\mathbf{x} \wedge \mathbf{y}) &= \sum_{i=1}^{n}(x_i \vee y_i) - \sum_{i=1}^{n}(x_i \wedge y_i) = \sum_{i=1}^{n}[(x_i \vee y_i) - (x_i \wedge y_i)] \\
&= \sum_{i=1}^{n}[(x_i \vee y_i) - x_i] \vee [(x_i \vee y_i) - y_i] \\
&= \sum_{i=1}^{n}[(x_i - x_i) \vee (y_i - x_i)] \vee [(x_i - y_i) \vee (y_i - y_i)] \\
&= \sum_{i=1}^{n}(y_i - x_i) \vee (x_i - y_i) = \sum_{i=1}^{n}|x_i - y_i| = d_1(\mathbf{x}, \mathbf{y}).
\end{aligned}
$$

Replacing the sum $\sum$ by the maximum operation $\bigvee$ and using an analogous argument will prove part 2 of the theorem. □

The importance of Theorem 4.9 is that the ℓ_1 and ℓ_∞ metrics can be computed using only the operations of $\vee$, $\wedge$, and $+$. Because of this, they require far less computational time than any ℓ_p metric whenever $1 < p < \infty$. As will be shown in subsequent chapters, they can also be easily implemented and partially computed within the dendritic structure of an artificial neuron.

Another metric that is computable using only addition and the operations of *max* and *min* is the *Hausdorff* metric. The Hausdorff distance, named after Felix Hausdorff who introduced this distance in 1914 [110], measures the distance between two nonempty subsets of a metric space. More specifically, given a metric space (M, d) and two nonempty subsets X and Y of M, define

$$d(x,Y) = inf\{d(x,y) : y \in Y\} \quad \text{and} \quad d(X,Y) = sup\{d(x,Y) : x \in X\} \quad (4.24)$$

Equation 4.24 extends the metric d between points of M to a distance function between nonempty subsets of M. The *Hausdorff distance*, denoted by d_H, between nonempty subsets is now defined as

$$d_H(X,Y) = d(X,Y) \vee d(Y,X) \quad (4.25)$$

Since the existence of *inf* or *sup* is not guaranteed, equation 4.25 does not represent a metric. However, if $\mathfrak{K} \subset 2^M$ represents the set of all nonempty compact subsets of M, then d_H is a metric on $\mathfrak{K}$ and $(\mathfrak{K}, d_H)$ is a metric space.

By choosing $d = d_1$ or $d = d_\infty$ as a metric on $\mathbb{R}^n$, then d_H is a metric on $\mathfrak{K} \subset 2^{\mathbb{R}^n}$ whose computation involves only the operations $\vee$, $\wedge$, and $+$.

We conclude this section with a closer look at the lattice structure on which ℓ-metrics can be defined. Observe that the lattice L in Example 4.2(3) is not modular. The ℓ-metric determining a metric lattice space forces modularity on the underlying lattice structure. Specifically, we have the following

Theorem 4.10 *A metric lattice space is modular.*

Proof. Let (L, d) be a metric lattice space. We must show that if $x \preccurlyeq z$, then $x \vee (y \wedge z) = (x \vee y) \wedge z$, where $x, y, z, \in L$.

Assuming that $x \preccurlyeq z$,

$$\begin{aligned} 0 = v(y) - v(y) &= [v(y \vee x) + v(y \wedge x) - v(x)] - v(y) \\ &= [v(y \vee x) + v(y \wedge x) - v(x)] - [v(y \vee z) + v(y \wedge z) - v(z)] \\ &= [v(z) - v(z \vee y) - v(z \wedge y)] + [v(x \vee y) + v(x \wedge y) - v(x)] \\ &= v((x \vee y) \wedge z) - v(x \vee (y \wedge z)). \end{aligned}$$

Therefore,

$$v(x \vee (y \wedge z)) = v((x \vee y) \wedge z). \tag{4.26}$$

By Theorem 3.5(2) (*modularity*), $x \preccurlyeq z \Rightarrow x \vee (y \wedge z) \preccurlyeq (x \vee y) \wedge z$. That v is positive follows from Corollary 4.8. This makes the relation $x \vee (y \wedge z) \prec (x \vee y) \wedge z$ impossible because of equation 4.26. Therefore, $x \vee (y \wedge z) = (x \vee y) \wedge z$. □

Valuations and metrics derived from valuations give rise to some interesting as well as useful measures. For instance, if $v : L \to \mathbb{R}$, then the function $\sigma(x, u) = \frac{v(u)}{b(x \vee u)}$ is a similarity measure and also an inclusion measure since it satisfied the four conditions of Definition 3.17. For metric spaces it is often desirable as well as easy to define similarity measures that are partially dependent on the metric.

Definition 4.3 Suppose (X, d) is a metric space and $x, y, z \in X$. A mapping $S : X \times X \to [0, 1]$ satisfying the following conditions:

S_1. $S(x, x) = 1, \forall x \in X$

S_2. $S(x, z) < 1, \forall x \neq z$

S_3. $d(x, y) \leq d(z, y) \Rightarrow S(z, y) \leq S(x, y)$

is a similarity measure *dependent* on d.

Example 4.3 Let (X,d) be the metric lattice space $(\mathbb{R}^n, d_1)$. Then the map S defined by

$$S(\mathbf{x},\mathbf{y}) = \frac{1}{\|\mathbf{y}-\mathbf{x}\|_1 + 1} \tag{4.27}$$

is a similarity measure dependent on d_1.

We conclude this section with another similarity measure based on valuations.

Definition 4.4 Suppose L be a lattice with $inf L = O \in L$ and $x, y \in L$. Define a similarity measure $S : L \times L \longrightarrow [0,1]$ as a mapping satisfying the following conditions:

S_1. $S(x,O) = 0, \ \forall x \neq O$

S_2. $S(x,x) = 1, \forall x \in L$

S_3. $S(x,y) < 1, \forall x \neq y$

The map S defined by

$$S(x,y) = \frac{v(y)}{v(y \vee x)} \wedge \frac{v(y \wedge x)}{v(y)} \tag{4.28}$$

is a similarity measure satisfying Definition 4.4.

Exercises 4.2.1

1. Prove Theorem 4.7.
2. Prove that if L is a metric distributive lattice, then $d(x,y) + d(y,z) = d(x,z) \Leftrightarrow y \in \{x \wedge z, x \vee z\}$.
3. Suppose L is complete lattice and v is an isotone valuation on L. Define three distinct similarity measure $S : L \times L \rightarrow [0,1]$.

4.3 MINIMAX MATRIX THEORY

This section extends lattice-based semiring operations to matrices. These extensions are analogous to matrix operations based on a field $\mathbb{F}$ as given in Example 1.8 and Theorem 1.9.

Let $(\mathbb{F}, \vee, \wedge, \circ)$ be a minimax-semiring and $M_{mn}(\mathbb{F})$ the set of all $m \times n$ matrices with values in the set $\mathbb{F}$. Suppose $A = (a_{ij})$ and $B = (b_{ij})$ are elements of $M_{mn}(\mathbb{F})$. The *max*-sum of A and B, denoted by $A \vee B$, is the matrix $C = (c_{ij}) \in M_{mn}(\mathbb{F})$ defined by

$$A \vee B = (a_{ij}) \vee (b_{ij}) = (a_{ij} \vee b_{ij}) = (c_{ij}) = C. \tag{4.29}$$

That is, $c_{ij} = a_{ij} \vee b_{ij}$ for $i = 1, \ldots, m$ and $j = 1, \ldots, n$. Similarly, the *min*-sum of A and B, denoted by $A \wedge B$, is the matrix $C = (c_{ij})$ defined by

$$A \wedge B = (a_{ij}) \wedge (b_{ij}) = (a_{ij} \wedge b_{ij}) = (c_{ij}) = C. \tag{4.30}$$

The notion of multiplication of a real valued matrix by a scalar also applies to semirings. Thus, if $\alpha \in \mathbb{F}$ and $A \in M_{mn}(\mathbb{F})$, then

$$\alpha \circ A = \alpha \circ (a_{ij}) = (\alpha \circ a_{ij}). \tag{4.31}$$

Since vectors are one-dimensional matrices of form $\mathbf{v} \in M_{1\times n}(\mathbb{F})$ or $\mathbf{w} \in M_{n\times 1}(\mathbb{F})$, the operations defined by equations 4.21 through 4.23 apply to vectors as well. Specifically, if $\mathbf{v} = (v_1, \ldots, v_n)'$, $\mathbf{w} = (w_1, \ldots, w_n)'$, and $\alpha \in \mathbb{F}$, then

$$\mathbf{v} \vee \mathbf{w} = (v_i) \vee (w_i) = (v_i \vee w_i), \quad \mathbf{v} \wedge \mathbf{w} = (v_i) \wedge (w_i) = (v_i \wedge w_i)$$
$$\text{and} \quad \alpha \circ \mathbf{v} = \alpha \circ (v_i) = (\alpha \circ v_i), \tag{4.32}$$

where $(v_i) = (v_{1i}) = \mathbf{v}$. The following theorem provides some useful rules for doing matrix arithmetic based on the minimax-semiring $\mathbb{F}$:

Theorem 4.11 *The following statements are valid for any scalars $\alpha, \beta \in \mathbb{F}$ and matrices $A, B, C \in M_{mn}(\mathbb{F})$.*

1. $A \vee B = B \vee A$ *and* $A \wedge B = B \wedge A$.
2. $A \vee (B \vee C) = (A \vee B) \vee C$ *and* $A \wedge (B \wedge C) = (A \wedge B) \wedge (A \wedge C)$.
3. $(\alpha \circ \beta) \circ A = \alpha \circ (\beta \circ A)$.
4. $(\alpha \vee \beta) \circ A = (\alpha \circ A) \vee (\beta \circ A)$.
5. $(\alpha \wedge \beta) \circ A = (\alpha \circ A) \wedge (\beta \circ A)$.
6. $\alpha \circ (A \vee B) = (\alpha \circ A) \vee (\alpha \circ B)$.
7. $\alpha \circ (A \wedge B) = (\alpha \circ A) \wedge (\alpha \circ B)$.

If, in addition, $\mathbb{F}$ is a distributive lattice, then we also have

8. $A \vee (B \wedge C) = (A \vee B) \wedge (A \vee C)$ *and* $A \wedge (B \vee C) = (A \wedge B) \vee (A \wedge C)$.

Proof. We will prove only two of the rules and leave the remaining for the reader to verify.

1. Since $(\mathbb{F}, \vee)$ and $(\mathbb{F}, \wedge)$ are abelian semigroups, we have that

$$A \vee B = (a_{ij}) \vee (b_{ij}) = (a_{ij} \vee b_{ij}) = (b_{ij} \vee a_{ij}) = (b_{ij}) \vee (a_{ij}) = B \vee A.$$
$$A \wedge B = (a_{ij}) \wedge (b_{ij}) = (a_{ij} \wedge b_{ij}) = (b_{ij} \wedge a_{ij}) = (b_{ij}) \wedge (a_{ij}) = B \wedge A.$$

4. Here we need to use property P_3'.

$$\begin{aligned}(\alpha \vee \beta) \circ A &= (\alpha \vee \beta) \circ (a_{ij}) = ((\alpha \vee \beta) \circ a_{ij}) = ((\alpha \circ a_{ij}) \vee (\beta \circ a_{ij})) \\ &= (\alpha \circ a_{ij}) \vee (\beta \circ a_{ij}) = (\alpha \circ (a_{ij})) \vee (\beta \circ (a_{ij})) \\ &= (\alpha \circ A) \vee (\beta \circ A).\end{aligned}$$

The proof of rule (*5*) is identical with $\wedge$ replacing $\vee$. □

Rules (*1*), (*2*), and (*7*) can be summarized as follows:

Corollary 4.12 *If* $(\mathbb{F}, \vee, \wedge)$ *is a lattice, then* $(M_{mn}(\mathbb{F}), \vee, \wedge)$ *is also a lattice. Furthermore, if* $\mathbb{F}$ *is a distributive lattice, then so is* $M_{mn}(\mathbb{F})$.

Thus far we have not used the fact that $\mathbb{F}$ is a minimax-semiring. To obtain a matrix *product* we need to take the ring operation $\circ$ into account. If $A \in M_{mp}(\mathbb{F})$ and $B \in M_{pn}(\mathbb{F})$, then the *max-product* of A and B, denoted by $A \boxtimes B$, is the matrix $C \in M_{mn}(\mathbb{F})$ defined by

$$A \boxtimes B = (\bigvee_{k=1}^{p} (a_{ik} \circ b_{kj})) = (c_{ij}) = C. \tag{4.33}$$

That is, $c_{ij} = \bigvee_{k=1}^{p} (a_{ik} \circ b_{kj})$ for $i = 1, \ldots, m$ and $j = 1, \ldots, n$.

The *min-product* of A and B, denoted by $A \boxtimes B$, is the matrix C defined by

$$A \boxtimes B = (\bigwedge_{k=1}^{p} (a_{ik} \circ b_{kj})) = (c_{ij}) = C. \tag{4.34}$$

It is obvious that the max-product (or min-product) formulation mirrors the formulation of the usual matrix product $c_{ij} = \sum_{k=1}^{p} (a_{ik} \cdot b_{kj})$, with $\sum$ being replaced by $\bigvee$ (or $\bigwedge$) and multiplication by the operation $\circ$. The two matrix products $\boxtimes$ and $\boxtimes$ are collectively called *minimax matrix products*.

Example 4.4

1. Consider the minimax-semifield $(\mathbb{R}, \vee, \wedge, +)$ and suppose $A, B \in M_{2\times 2}(\mathbb{R})$ are given by

$$A = \begin{pmatrix} 1 & 2 \\ -2 & 3 \end{pmatrix} \quad \text{and} \quad B = \begin{pmatrix} 4 & -1 \\ 3 & 2 \end{pmatrix}.$$

$$\text{Then} \quad A \boxed{\vee} B = \begin{pmatrix} 5 & 4 \\ 6 & 5 \end{pmatrix}, \quad \text{while} \quad B \boxed{\vee} A = \begin{pmatrix} 5 & 6 \\ 4 & 5 \end{pmatrix}.$$

Similarly, we have

$$A \boxed{\wedge} B = \begin{pmatrix} 5 & 0 \\ 2 & -3 \end{pmatrix} \quad \text{and} \quad B \boxed{\wedge} A = \begin{pmatrix} -3 & 2 \\ 0 & 5 \end{pmatrix}.$$

2. Using the minimax-semiring $(\mathbb{R}^+, \vee, \wedge, \times)$ and the matrices

$$A = \begin{pmatrix} 1 & 2 \\ 2 & 3 \end{pmatrix} \quad \text{and} \quad B = \begin{pmatrix} 4 & 1 \\ 3 & 2 \end{pmatrix} \quad \text{instead}$$

$$\text{then} \quad A \boxed{\vee} B = \begin{pmatrix} 6 & 4 \\ 9 & 6 \end{pmatrix}, \quad \text{while} \quad B \boxed{\vee} A = \begin{pmatrix} 4 & 8 \\ 4 & 6 \end{pmatrix},$$

$$A \boxed{\wedge} B = \begin{pmatrix} 4 & 1 \\ 8 & 2 \end{pmatrix}, \quad \text{and} \quad B \boxed{\wedge} A = \begin{pmatrix} 2 & 3 \\ 3 & 6 \end{pmatrix}.$$

The example shows that neither product, $\boxed{\vee}$ or $\boxed{\wedge}$, is commutative. However, several other important properties of these products do hold.

Theorem 4.13 *If* $(\mathbb{F}, \vee, \wedge, \circ)$ *is a minimax-semifield,* $A, B, C \in M_{nn}(\mathbb{F})$, *and* $\alpha \in \mathbb{F}$, *then*

1. $A \boxed{\vee} (B \boxed{\vee} C) = (A \boxed{\vee} B) \boxed{\vee} C$,

2. $A \boxed{\vee} (B \vee C) = (A \boxed{\vee} B) \vee (A \boxed{\vee} C)$,

3. $(A \vee B) \boxed{\vee} C = (A \boxed{\vee} C) \vee (B \boxed{\vee} C)$,

4. $A \boxed{\wedge} (B \boxed{\wedge} C) = (A \boxed{\wedge} B) \boxed{\wedge} C$,

5. $A \boxed{\wedge} (B \wedge C) = (A \boxed{\wedge} B) \wedge (A \boxed{\wedge} C)$,

6. $(A \wedge B) \boxed{\wedge} C = (A \boxed{\wedge} C) \wedge (B \boxed{\wedge} C)$,

7. $\alpha \circ (A \boxed{\vee} B) = (\alpha \circ A) \boxed{\vee} B = A \boxed{\vee} (\alpha \circ B)$, *and*

8. $\alpha \circ (A \boxed{\wedge} B) = (\alpha \circ A) \boxed{\wedge} B = A \boxed{\wedge} (\alpha \circ B)$.

Proof. We only prove equations (*1*), (2), and (*7*), and leave the proofs of the remaining equalities as exercises since they are similar to the proofs presented.

1. Let $M = B \boxed{\vee} C$, $D = A \boxed{\vee} M$, $N = A \boxed{\vee} B$, $E = N \boxed{\vee} C$, and $r, s \in J_n$. Then

$$\begin{aligned} d_{rs} &= \bigvee_{k=1}^{n} (a_{rk} \circ m_{ks}) = \bigvee_{k=1}^{n} ((a_{rk} \circ \bigvee_{j=1}^{n} (b_{kj} \circ c_{js})) \\ &= \bigvee_{k=1}^{n} \Big(\bigvee_{j=1}^{n} (a_{rk} \circ (b_{kj} \circ c_{js})) \Big) = \bigvee_{k=1}^{n} \Big(\bigvee_{j=1}^{n} ((a_{rk} \circ b_{kj}) \circ c_{js}) \Big) \\ &= \bigvee_{j=1}^{n} \Big(\bigvee_{k=1}^{n} ((a_{rk} \circ b_{kj}) \circ c_{js}) \Big) = \bigvee_{j=1}^{n} (n_{rj} \circ c_{js}) = e_{rs}, \end{aligned}$$

where the first two equalities follow from the symbol and product definitions, the third from Theorem 4.3, the fourth equality from associativity of the operation $\circ$, the fifth follows from Corollary 3.4, and the sixth and seventh equalities from the symbol and product definition. This proves that $D = E$ and, hence, equation (*1.*).

2. Let $M = B \vee C$, $D = A \boxed{\vee} M$, $N = A \boxed{\vee} B$, $G = A \boxed{\vee} C$, and $E = N \vee G$. Thus, we must show that $D = E$.

$$\begin{aligned} d_{rs} &= \bigvee_{k=1}^{n} (a_{rk} \circ m_{ks}) = \bigvee_{k=1}^{n} ((a_{rk} \circ (b_{ks} \vee c_{ks})) \\ &= \bigvee_{k=1}^{n} ((a_{rk} \circ b_{ks}) \vee (a_{rk} \circ c_{ks})) = \Big(\bigvee_{k=1}^{n} (a_{rk} \circ b_{ks}) \Big) \vee \Big(\bigvee_{k=1}^{n} (a_{rk} \circ c_{ks}) \Big) \\ &= n_{rs} \vee g_{rs} = e_{rs}. \end{aligned}$$

Note that in order to establish the third equality in the proof of (2.) we need the distributive semiring law R_3'.

7. Let $C = A \boxed{\vee} B$. Then

$$\alpha \circ C = (\alpha \circ c_{ij}) = (\alpha \circ \bigvee_{k=1}^{n} (a_{ik} \circ b_{kj})) = (\bigvee_{k=1}^{n} (\alpha \circ (a_{ik} \circ b_{kj})))$$
$$= (\bigvee_{k=1}^{n} ((\alpha \circ a_{ik}) \circ b_{kj})) = (\alpha \circ A) \boxed{\vee} B.$$

Since $\mathbb{F}$ is an abelian semiring,

$$(\bigvee_{k=1}^{n} ((\alpha \circ a_{ik}) \circ b_{kj})) = (\bigvee_{k=1}^{n} (a_{ik} \circ (\alpha \circ b_{kj}))) = A \boxed{\vee} (\alpha \circ B). \quad \square$$

The following is a direct consequence of Theorem 4.11 and Theorem 4.13:

Corollary 4.14 *If $\mathbb{F}$ is a max-semifield, then $(M_{nn}(\mathbb{F}), \vee, \boxed{\vee})$ is a max-semiring. If $\mathbb{F}$ is a min-semifield, then $(M_{nn}(\mathbb{F}), \wedge, \boxed{\wedge})$ is a min-semiring.*

Properties *1.* through *8.* of Theorem 4.13 are stated in terms of square matrices. This requirement is not necessary as the theorem easily extends to non-square matrices as long the dimensions satisfy equations 4.29 through 4.34. However, since minimax matrix products are not commutative binary operations, neither the max-semiring $M_{nn}(\mathbb{F})$ nor the min-semiring $M_{nn}(\mathbb{F})$ can be semifields. Nevertheless, if $\mathbb{F}$ is a minimax-semifield with duality, then it follows from Theorems 4.11 and 4.13 and their corollaries that $(M_{nn}(\mathbb{F}), \vee, \boxed{\vee})$ and $(M_{nn}(\mathbb{F}), \wedge, \boxed{\wedge})$ can be combined into the minimax-semiring $(M_{nn}(\mathbb{F}), \vee, \wedge, \boxed{\vee}, \boxed{\wedge})$. In this setting the properties of mixed matrix multiplications using $\boxed{\vee}$ and $\boxed{\wedge}$ become an object of interest. In this connection, the following theorem is of fundamental importance.

Theorem 4.15 *If $\mathbb{F}$ is a minimax-semifield and A, B, C are matrices of dimension $m \times p$, $p \times n$, and $n \times q$, respectively, with values in $\mathbb{F}$, then the*

$$A \boxed{\vee} (B \boxed{\wedge} C) \preccurlyeq (A \boxed{\vee} B) \boxed{\wedge} C \quad \textit{and} \quad (A \boxed{\wedge} B) \boxed{\vee} C \preccurlyeq A \boxed{\wedge} (B \boxed{\vee} C)$$

Proof. Let $E = (e_{ij}) = B \boxed{\wedge} C$ and $D = (d_{ij}) = A \boxed{\vee} B$. We shall show that $A \boxed{\vee} E \preccurlyeq D \boxed{\wedge} C$ which will prove the first inequality.

$$
\begin{aligned}
A \boxed{\vee} E &= (\bigvee_{k=1}^{p} (a_{ik} \circ e_{kj})) = (\bigvee_{k=1}^{p} (a_{ik} \circ \bigwedge_{\ell=1}^{n} (b_{k\ell} \circ c_{\ell j}))) = (\bigvee_{k=1}^{p} \bigwedge_{\ell=1}^{n} (a_{ik} \circ (b_{k\ell} \circ c_{\ell j}))) \\
&= (\bigvee_{k=1}^{p} \bigwedge_{\ell=1}^{n} (a_{ik} \circ b_{k\ell} \circ c_{\ell j})) \preccurlyeq (\bigwedge_{\ell=1}^{n} \bigvee_{k=1}^{p} (a_{ik} \circ b_{k\ell} \circ c_{\ell j})) \\
&= (\bigwedge_{\ell=1}^{n} (\bigvee_{k=1}^{p} (a_{ik} \circ b_{k\ell})) \circ c_{\ell j}) = (\bigwedge_{\ell=1}^{n} (d_{i\ell} \circ c_{\ell j})) = D \boxed{\wedge} C.
\end{aligned}
$$

The inequality follows from equation 3.5 by setting $x_{k\ell} = a_{ik} \circ b_{k\ell} \circ c_{\ell j}$. The remaining equalities are a consequence of the definitions of the max and min matrix products and equations 4.1 and 4.2. The proof of the second inequality relationship is analogous. □

The notion of conjugacy in dual lattices extends to matrices if $\mathbb{F}$ is a semiring with duality. In this case, the *conjugate* of a matrix $A = (a_{ij})$ is the matrix $A^* = (a^*_{ij})$, where a^*_{ij} is the conjugate of a_{ji} defined by $a^*_{ij} = -a_{ji}$ if the additive notation is used for the binary operation $\circ$ and $a^*_{ij} = a^{-1}_{ji}$ if the multiplicative notation is used for $\circ$. The respective null elements (if they exist) are defined as in equations 3.7 and 4.10.

The *transpose* of a matrix $A = (a_{ij}) \in M_{mn}(\mathbb{F})$, denoted by A', is the same as defined in linear algebra, namely $A' = (a'_{ij})$, where $a'_{ij} = a_{ji}$. It follows directly from the definition of A' that $A^* = -A'$. Other fundamental properties of the conjugate and transpose are given by the next two theorems.

Theorem 4.16 *Let $A, B \in M_{mn}(\mathbb{F})$ and $\alpha \in \mathbb{F}$.*

1. $(A^*)^* = A$ *and* $(A')' = A$
2. $(\alpha \circ A)^* = \alpha \circ A^*$ *and* $(\alpha \circ A)' = \alpha \circ A'$
3. $(A \vee B)^* = A^* \wedge B^*$ *and* $(A \vee B)' = A' \vee B'$
4. $(A \wedge B)^* = A^* \vee B^*$ *and* $(A \wedge B)' = A' \vee B'$

Proof. The proofs of *1* and *2* are trivial. To prove *3*, let $C = A \vee B$. Then

$$\begin{aligned}(A \vee B)^* &= (a_{ij} \vee b_{ij})^* = (c_{ij})^* = (c^*_{ij}) = -(c_{ji}) = -(a_{ji} \vee b_{ji}) \\ &= (-a_{ji} \wedge -b_{ji}) = (a^*_{ij} \wedge b^*_{ij}) = (a_{ij})^* \wedge (b^*_{ij}) = A^* \wedge B^*.\end{aligned}$$

The proof of *4* is analogous. □

If $A = (a_{ij}) \in M_{nn}(\mathbb{F})$, then the *diagonal of* A is denoted by $diag(A)$ and defined as $diag(A) = (a_{11}, a_{22}, \ldots, a_{nn})$. Supposing that $\mathbb{F}$ is a semiring with unit 0, A an $m \times n$ matrix with entries in $\mathbb{F}$, $B = A \boxed{\vee} A^*$ and $C = A \boxed{\wedge} A^*$, then $diag(B) = \mathbf{0} = diag(C)$, where $\mathbf{0} = (0, 0, \ldots, 0)$. This is an immediate consequence of the conjugation process since

$$b_{ii} = \bigvee_{k=1}^{n} (a_{ik} \circ a^*_{ki}) = \bigvee_{k=1}^{n} (a_{ik} - a_{ik}) = 0 = \bigwedge_{k=1}^{n} (a_{ik} - a_{ik}) = \bigwedge_{k=1}^{n} (a_{ik} \circ a^*_{ki}) = c_{ii}, \tag{4.35}$$

for $i = 1, \ldots, n$. Obviously, the same is true for the diagonals of $A^* \boxed{\vee} A$ and $A^* \boxed{\wedge} A$.

Theorem 4.17 *Suppose* $\mathbb{F}$ *is a semifield with duality. If A and B are matrices with entries from* $\mathbb{F}$ *and dimension* $m \times p$ *and* $p \times n$*, respectively, then the following properties hold:*

1. $A \boxed{\wedge} B = (B^* \boxed{\vee} A^*)^*$ *and* $A \boxed{\vee} B = (B^* \boxed{\wedge} A^*)^*$
2. $(A \boxed{\vee} A^*) \boxed{\wedge} B \preccurlyeq B \preccurlyeq (A^* \boxed{\wedge} A) \boxed{\vee} B$ *and* $B \boxed{\wedge} (A \boxed{\vee} A^*) \preccurlyeq B \preccurlyeq B \boxed{\vee} (A^* \boxed{\wedge} A)$
3. $(A \boxed{\vee} A^*) \boxed{\wedge} A = A = (A \boxed{\wedge} A^*) \boxed{\vee} A$ *and* $A \boxed{\wedge} (A^* \boxed{\vee} A) = A = A \boxed{\vee} (A^* \boxed{\wedge} A)$

Proof. *1*. Let $C = B^* \boxed{\vee} A^*$. Then $c_{ij} = \bigvee_{k=1}^{p} (b^*_{ik} \circ a^*_{kj})$ and

$$\begin{aligned}(B^* \boxed{\vee} A^*)^* = C^* = (c^*_{ij}) = (-c_{ji}) &= (-\bigvee_{k=1}^{p} (b^*_{jk} \circ a^*_{ki})) = (-\bigvee_{k=1}^{p} (-b_{kj} \circ (-a_{ik}))) \\ &= (-\bigvee_{k=1}^{p} -(a_{ik} \circ b_{kj})) = (\bigwedge_{k=1}^{p} (a_{ik} \circ b_{kj})) = A \boxed{\wedge} B.\end{aligned}$$

The proof of the second property of (*1*) is analogous.
2. Let $D = A \boxed{\vee} A^*$. Using the fact that $d_{ii} = 0$ we now have

$$D \boxed{\wedge} B = (\bigwedge_{k=1}^{p} (d_{ik} \circ b_{kj})) \preccurlyeq (d_{ii} \circ b_{ij}) = B.$$

Thus, $(A \boxed{\vee} A^*) \boxed{\wedge} B \preccurlyeq B$. Similarly, for $C = A^* \boxed{\wedge} A$ we have

$$B = (b_{ij}) = (c_{ii} \circ b_{ij}) \preccurlyeq (\bigvee_{k=1}^{p} (c_{ik} \circ b_{kj})) = C \boxed{\vee} B.$$

This proves property *2.*

3. Observe that

$$A \preccurlyeq A \boxed{\vee} (A^* \boxed{\wedge} A) \preccurlyeq (A \boxed{\vee} A^*) \boxed{\wedge} A \preccurlyeq A,$$

where the first inequality is a consequence of the second equation of (*2*) by setting $A = B$, the second inequality follows from Theorem 4.15, while the third inequality is a consequence of the first equation of (*2*). Since $A = A$, the three inequalities become equalities. The second equation of (*3*) can be derived in a likewise fashion. □

The two relations $(c_{ii} \circ b_{ij}) \preccurlyeq (\bigvee_{k=1}^{p} (c_{ik} \circ b_{kj}))$ and $(\bigwedge_{k=1}^{p} (d_{ik} \circ b_{kj})) \preccurlyeq (d_{ii} \circ b_{ij})$ appearing in the proof of Theorem 4.17(*2.*) are closely related to the concepts of diagonally max dominant and diagonally min dominant square matrices.

Definition 4.5 If $A \in M_{nn}(\mathbb{F})$ where $\mathbb{F}$ is a semifield with duality, then A is said to be *diagonally max dominant* if and only if

$$\bigvee_{k=1}^{n} (a_{ik} \circ a^*_{kj}) \preccurlyeq (a_{ii} \circ a^*_{ij}) \; \forall i, j \in \{1, \ldots, n\}.$$

Similarly, A is said to be *diagonally min dominant* if and only if

$$(a_{ii} \circ a^*_{ij}) \preccurlyeq \bigwedge_{k=1}^{n} (a_{ik} \circ a^*_{kj}) \; \forall i, j \in \{1, \ldots, n\}.$$

Thus, if A is diagonally max dominant and $\mathbb{F} = \mathbb{R}$ with $\circ = +$, then $a_{ii} - a_{ji} = \bigvee_{k=1}^{p} (a_{ik} - a_{jk}) \; \forall i, j \in \{1, \ldots, n\}$. If A is diagonally min dominant, then $a_{ii} - a_{ji} = \bigwedge_{k=1}^{p} (a_{ik} - a_{jk}) \; \forall i, j \in \{1, \ldots, n\}$. The importance of these diagonally dominant matrices will become apparent in lattice vector space theory.

If $\mathbb{F}$ is a max-semiring with unit ϕ and null element $-\infty$, then there exists an identity (unit) matrix $I \in M_{nn}(\mathbb{F})$ and a null matrix $O \in M_{nn}(\mathbb{F})$ given by

$$I = \begin{pmatrix} \phi & -\infty & \cdots & -\infty & -\infty \\ -\infty & \phi & \cdots & -\infty & -\infty \\ \vdots & \vdots & \ddots & \vdots & \vdots \\ -\infty & -\infty & \cdots & \phi & -\infty \\ -\infty & -\infty & \cdots & -\infty & \phi \end{pmatrix} \text{ and } O = \begin{pmatrix} -\infty & \cdots & -\infty \\ \vdots & \ddots & \vdots \\ -\infty & \cdots & -\infty \end{pmatrix}. \quad (4.36)$$

The following equations are trivial consequences of the definition of I and O:

$$\begin{aligned} I \boxed{\vee} A &= A \boxed{\vee} I = A, \\ A \vee O &= O \vee A = A, \\ A \boxed{\vee} O &= O \boxed{\vee} A = O. \end{aligned} \tag{4.37}$$

These equations verify the next theorem.

Theorem 4.18 *If $\mathbb{F}$ is a max-semiring with unity and null element, then so is $M_{nn}(\mathbb{F})$. Similarly, if $\mathbb{F}$ is a min-semiring with unity and null element, then so is $M_{nn}(\mathbb{F})$.*

In particular, if ∞ denotes the unit and ϕ the null element of $(\mathbb{F}, \wedge, \circ)$, then simply replace all $-\infty$ entries in equation 4.36 with ∞ in order to obtain the identity and null matrices for $(M_{nn}(\mathbb{F}), \wedge, \boxed{\wedge})$. Next, replace $\boxed{\vee}$ and $\vee$ in equation 4.37 with $\boxed{\wedge}$ and $\wedge$, respectively. This proves the second claim of Theorem 4.18. If $\mathbb{F}$ is a semifield with dual structure $\mathbb{F}^*$, then they become isomorphic structures. For example, if $\mathbb{F} = \mathbb{R}_{-\infty}$ and $\mathbb{F}^* = \mathbb{R}_{\infty}$, then the following theorem can be easily verified:

Theorem 4.19 $(M_{nn}(\mathbb{R}_{-\infty}), \vee, \boxed{\vee}) \approx (M_{nn}(\mathbb{R}_{\infty}), \wedge, \boxed{\wedge})$.

This again allows for combining min- and max-semiring structures into an algebraically well-defined lattice-based matrix algebra $(M_{nn}(\mathbb{R}_{\pm\infty}), \vee, \wedge, \boxed{\vee}, \boxed{\wedge},)$.

Exercises 4.3

1. Prove the remaining six properties of Theorem 4.11.
2. Prove equations 3 through 6 and equation 8 of Theorem 4.13.
3. Prove property 4 of Theorem 4.20.

4.3.1 Lattice Vector Spaces

The theory of ℓ-groups, $s\ell$-groups, $s\ell$-semigroups, and lattice-based vector spaces provide an extremely rich setting in which many concepts from linear algebra and abstract algebra can be transferred to the lattice domain via analogies. Lattice-based vector spaces as defined in this section provide an excellent example of such an analogy. In the following discussion $(\mathbb{V}, \diamond)$ will denote a semilattice.

Definition 4.6 An *$s\ell$-vector space* $\mathbb{V}$ *over a semifield* $(\mathbb{F}, \diamond, \circ)$, denoted by $\mathbb{V}(\mathbb{F})$, is a semilattice $(\mathbb{V}, \diamond)$ together with an operation called *scalar addition* of each element of $\mathbb{V}$ by an element of $\mathbb{F}$ on the left, such that $\forall \alpha, \beta \in \mathbb{F}$ and $\mathbf{v}, \mathbf{w} \in \mathbb{V}$ the following five conditions are satisfied:

V_1. $\alpha \circ \mathbf{v} \in \mathbb{V}$

V_2. $\alpha \circ (\beta \circ \mathbf{v}) = (\alpha \circ \beta) \circ \mathbf{v}$

V_3. $(\alpha \diamond \beta) \circ \mathbf{v} = (\alpha \circ \mathbf{v}) \diamond (\beta \circ \mathbf{v})$

V_4. $\alpha \circ (\mathbf{v} \diamond \mathbf{w}) = (\alpha \circ \mathbf{v}) \diamond (\alpha \circ \mathbf{w})$

V_5. $\phi \circ \mathbf{v} = \mathbf{v}$

The elements of $\mathbb{V}$ are called *vectors* and the elements of $\mathbb{F}$ are called *scalars*.

Unless otherwise stated or elucidated, in the remainder of this text we shall use the column vector notation for vectors consisting of multiple components (i.e., vector coordinates). The reason for this is that in many engineering applications systems of equations are often expressed in the form $A\mathbf{x} = \mathbf{y}$, where A is an $m \times n$ matrix and $\mathbf{x}$ represents an $n \times 1$ column vector. Thus, the column vector notation aids one of the goals of this treatise, which is to seek similarities between linear algebra and lattice semigroup algebras.

If $\mathbb{F}$ is a min-semifield $(\mathbb{F}, \wedge, \circ)$ and $\mathbb{V}$ a min-semilattice $(\mathbb{V}, \wedge)$, then $\mathbb{V}$ is also called a *min-vector space*. Similarly, if $\diamond = \vee$, then the $s\ell$-vector space $\mathbb{V}$ is also called a *max-vector space*.

The definition of an $s\ell$-vector space $\mathbb{V}$ over a semifield $\mathbb{F}$ mimics that of a linear vector space $\mathbb{V}$ over a field $\mathbb{F}$ (Definition 1.14). In fact, the five postulates of the two definitions look identical except for the symbols denoting the two binary operators. For *lattice vector spaces*, also called ℓ-vector spaces, the semilattice $\mathbb{V}$ is replaced by a lattice and $\mathbb{F}$ is a minimax semifield. More precisely, we have the following definition:

Definition 4.7 Suppose $(\mathbb{V}, \vee, \wedge)$ is a lattice and $(\mathbb{F}, \vee, \wedge, \circ)$ is a minimax semifield. Then $\mathbb{V}(\mathbb{F})$ is called an *ℓ-vector space* if and only if conditions V_1 through V_5 are satisfied for the two binary operations $\{\circ, \vee\}$ as well as for the pair $\{\circ, \wedge\}$.

It is important to note that the lattice vector space definition given here is different from that of *vector lattices* as postulated by Birkhoff and others [29, 30, 87, 134, 127]. In these references vector lattices are simply partially ordered vector spaces satisfying the isotone property. The ℓ-vector space definition given above is basically the same as Cuninghame-Green's definition of band spaces [62]. A major difference between ℓ-vector spaces and vector spaces associated with linear algebra is the lack of inverses of elements of a lattice with respect to the binary operations $\vee$ and $\wedge$. Nevertheless, $s\ell$-vector spaces and ℓ-vector spaces share many important properties and concepts found in standard vector space theory. These include the analogous notions of linear independence and linear dependence, eigenvalues and eigenvectors, and linear transformations.

The lattice vector spaces $(\mathbb{R}^n, \vee, \wedge)$ over the semifield $(\mathbb{R}, \vee, \wedge, +)$ and $(\mathbb{R}^n_{\pm\infty}, \vee, \wedge)$ over the bounded semifield $(\mathbb{R}_{\pm\infty}, \vee, \wedge, +, +^*)$ provide the standard models of ℓ-vector space and are the primary focus of this treatise. However, there are many other $s\ell$-vector spaces and ℓ-vector spaces that have proven useful in a wide variety of application domains.

Example 4.5

1. Let $\mathbb{V} \subset \mathbb{R}^2_{-\infty}$ be defined by $\mathbb{V} = \{\mathbf{v} \in \mathbb{R}^2_{-\infty} : v_2 \geq v_1 - 1\}$. Given an arbitrary pair of vectors $\mathbf{v}, \mathbf{w} \in \mathbb{V}$ and setting $\mathbf{u} = \mathbf{v} \vee \mathbf{w}$, then

$$u_2 = v_2 \vee w_2 \geq (v_1 - 1) \vee (w_1 - 1) = (v_1 \vee w_1) - 1 = u_1 - 1.$$

Thus, $\mathbf{u} \in \mathbb{V}$ and $(\mathbb{V}, \vee)$ is a semilattice. The semilattice $\mathbb{V}$ is a max-vector space over the semifield $(\mathbb{R}_{-\infty}, \vee, +)$ since $\mathbb{V}(\mathbb{R}_{-\infty})$ satisfies properties V_1 through V_5. For instance, for $\mathbf{v} \in \mathbb{V}$ and $\alpha \in \mathbb{R}_{-\infty}$, set $\mathbf{w} = \alpha + \mathbf{v}$. then $\mathbf{w} = (w_1, w_2)' = (\alpha + v_1, \alpha + v_2)'$, where $(\cdot)'$ denotes the transpose and $w_2 = \alpha + v_2 \geq \alpha + (v_1 - 1) = (\alpha + v_1) - 1 = w_1 - 1$. This proves that V_1 is satisfied. Showing that properties V_2 and V_5 are satisfied is just as easy. To prove V_3 observe that

$$\begin{aligned}(\alpha \vee \beta) + \mathbf{v} &= ((\alpha \vee \beta) + v_1, (\alpha \vee \beta) + v_2) \\ &= ((\alpha + v_1) \vee (\beta + v_1), (\alpha + v_2) \vee (\beta + v_2)) \\ &= (\alpha + v_1, \alpha + v_2) \vee (\beta + v_1, \beta + v_2) = (\alpha + \mathbf{v}) \vee (\beta + \mathbf{v}).\end{aligned}$$

Here the first equality follows from the definition of scalar addition (4.32) and the second from property R_3' of a semiring. The third equality follows from the pointwise maximum of two vectors (4.32) and the fourth from the pointwise addition of a scalar and a vector. The proof of property V_4 is left as an exercise.

2. Let $\mathbb{C} = \{a+bi : a,b \in \mathbb{R}$ and $i = \sqrt{-1}\}$ and define $(a+bi) \vee (c+di) = (a \vee c)+(b+d)i$. Then $(\mathbb{C}, \vee)$ is a semilattice and $\mathbb{C}(\mathbb{R})$ is a max-vector space over the semifield $(\mathbb{R}, \vee, +)$ if scalar addition is defined as $\alpha+(a+bi) = (\alpha+a)+(\alpha+b)i$ for $\alpha \in \mathbb{R}$. Property V_1 follows from the definition of scalar addition. The remaining properties can also be easily verified. For example, for $\alpha,\beta \in \mathbb{R}$,

$$
\begin{aligned}
(\alpha \vee \beta)+(a+bi) &= ((\alpha \vee \beta)+a)+((\alpha \vee \beta)+b)i = ((\alpha+a) \vee (\beta+b))i \\
&= ((\alpha+a)+(\alpha+b)i) \vee ((\beta+a)+(\beta+b)i) \\
&= (\alpha+(a+bi)) \vee (\beta+(a+bi)).
\end{aligned}
$$

This verifies property V_3

3. According to Corollaries 4.12 and 4.14, $(M_{mn}(\mathbb{F}), \vee, \wedge)$ is a lattice whenever $(\mathbb{F}, \vee, \wedge)$ is a lattice. If in addition $\mathbb{F}$ is a minimax semifield, then it follows from Theorem 4.11 that the required ℓ-vector space axioms V_1 through V_5 are all satisfied. Thus, $(M_{mn}(\mathbb{F}), \vee, \wedge)$ is an ℓ-vector space over the minimax semifield $\mathbb{F}$. Observe how well this example mimics Examples 1.8 and 1.9.

Whenever $\mathbb{V}(\mathbb{F}) = (M_{mn}(\mathbb{F}))$ is an $s\ell$ or ℓ-vector space over the respective $s\ell$ or ℓ-semifield $\mathbb{F}$ and $m = 1$, then we shall also use the notation $\mathbb{V}_n(\mathbb{F})$ to specify this vector space. In case $\mathbb{F}^n = \mathbb{V}_n$, the symbol $\mathbb{F}^n$ is commonly used for this $s\ell$ or ℓ-vector space.

Given an $s\ell$-vector space or an ℓ-vector space $\mathbb{V}$, it is often possible to form another $s\ell$-vector space or an ℓ-vector space by taking a subset $\mathbb{W}$ of $\mathbb{V}$ and employing the operations of $\mathbb{V}$. For $\mathbb{W}$ to be a subspace of $\mathbb{V}$ it must satisfy the five properties listed in Definition 4.6. However, in analogy with Theorem 1.10, it suffices to show that in addition to axiom V_1, all one needs to prove is that $\mathbb{W}$ is closed under the lattice operations of $\mathbb{V}$.

Theorem 4.20 *If $\mathbb{W}$ is a nonempty subset of an $s\ell$ or ℓ-vector space $\mathbb{V}$ over the semifield $\mathbb{F}$, then $\mathbb{W}$ is a subspace of $\mathbb{V}$ $\Leftrightarrow$ the following conditions hold:*

1. *if $\mathbf{v}, \mathbf{w} \in \mathbb{W}$, then $\mathbf{v} \diamond \mathbf{w} \in \mathbb{W}$, and*

2. *if $\alpha \in \mathbb{F}$ and $\mathbf{w} \in \mathbb{W}$, then $\alpha \circ \mathbf{w} \in \mathbb{W}$.*

As before, the operation $\diamond$ represents one of the generic lattice operation $\vee$ or $\wedge$ so that condition (*1.*) has to be satisfied for each operation $\vee$ and $\wedge$ whenever $\mathbb{V}$ is an ℓ-vector space. In light of Theorem 4.20, it becomes easy to show that the $s\ell$-vector space $\mathbb{V}$ in Example 4.5(1.) is a subspace of the $s\ell$-vector space $(\mathbb{R}^2, \vee)$ over $(\mathbb{R}, \vee, +)$ and the line $\{(x,y) : y = x - 1\}$ is a one-dimensional subspace of both max-vector spaces $(\mathbb{R}^2, \vee)$ and $(\mathbb{V}, \vee)$.

Exercises 4.3.1

1. Show that the set $\mathbb{V} = \{\mathbf{v} \in \mathbb{R}^2_{-\infty} : v_2 \geq v_1 - 1\}$ defined in Example 4.5(1.) satisfies property V_4.
2. Prove that the max-vector space $\mathbb{C}(\mathbb{R})$ defined in Exercise 4.5(2.) satisfies properties V_1, V_2, V_4, and V_5.
3. Prove Theorem 4.20.

4.3.2 Lattice Independence

In conventional linear algebra, linear independence is often defined in different ways. However, these definitions are all logically equivalent. In contrast, in minimax algebra and general lattice theory there exist several competing but logically nonequivalent definitions of independence with all of them having some common similarities with linear independence [30, 62, 63, 223]. The basic problem arises with noncomplete (unbounded) lattices such as $(\mathbb{R}, \vee, \wedge)$. The absence of zero elements prevents the existence of equations of form $\bigwedge_{i=1}^n(\alpha_i \circ x_i) = O$ and $\bigvee_{i=1}^n(\alpha_i \circ x_i) = O^*$ and thus forestalls the formulation of a coherent definition of a basis for $s\ell$ and ℓ-vector spaces. Nevertheless, it does not prevent the definition of $s\ell$ and ℓ-independence.

For a finite subset $X = \{\mathbf{x}^1, \ldots .\mathbf{x}^k\}$ of an ℓ-vector space and a nonempty subset Ξ of $\{1, \ldots, k\}$, equations of form $\mathbf{v} = \bigvee_{\xi\in\Xi}(\alpha_\xi \circ \mathbf{x}^\xi)$ and $\mathbf{w} = \bigwedge_{\xi\in\Xi}(\alpha_\xi \circ \mathbf{x}^\xi)$ are respectively called *linear max sums* and *linear min sums* of vectors from X. These linear max and min sums, are often referred to as *max* and *min sums*, are well-defined with $\mathbf{v}$ and $\mathbf{w}$ representing the respective *lub* and *glb* of the set $\{\alpha_\xi \circ \mathbf{x}^\xi : \xi \in \Xi\}$. Linear min and max sums are the basic building blocks of linear minimax sums.

Definition 4.8 Suppose $\mathbb{V}$ is an ℓ-vector space over the semifield $(\mathbb{F}, \vee, \wedge, \circ)$, $X = \{\mathbf{x}^1, \ldots, \mathbf{x}^k\} \subset \mathbb{V}$, and $\Xi \subset \{1, \ldots, k\}$. A vector $\mathbf{x} \in \mathbb{V}$ is said to be a *linear minimax combination* of vectors from the set X if $\mathbf{x}$ is of form

$$\mathbf{x} = \bigvee_{j\in J} \bigwedge_{\xi\in\Xi} (a_{\xi,j(\xi)} \circ \mathbf{x}^\xi), \tag{4.38}$$

where $\mathbf{x}^\xi \in X$ and J is a finite set of indices with $j(\xi)$ denoting the fact that the index depends on the value $\xi \in \Xi$. The expression on the right of the equation is also called the *standard form* of a linear minimax sum of vectors from X.

It follows that any finite expression involving the symbols $\vee, \wedge$, and vectors of form $a \circ \mathbf{x}^\xi$, where $a \in \mathbb{F}$ and $\mathbf{x}^\xi \in X$, can be written in the form of equation 4.30. For instance, if $\mathbf{x} = a \circ \mathbf{x}^\lambda$, then setting $J = \{1\}$, $\Xi = \{\lambda\}$, and $a_{\lambda j(\lambda)} = a$ results in $\mathbf{x} = \bigvee_{j\in J} \bigwedge_{\xi\in\Xi}(a_{\xi,j(\xi)} \circ \mathbf{x}^\xi)$. Other expressions that involve the meet and join operations are just as simple to change into the standard form but may require a little more bookkeeping.

Example 4.6 Let $\mathbb{V}$ be an ℓ-vector space over the semifield $(\mathbb{F}, \vee, \wedge, \circ)$ with duality $(\cdot)^*$, $X = \{\mathbf{x}^1, \ldots .\mathbf{x}^k\} \subset \mathbb{V}$, and $\mathbf{x} = (a \circ \mathbf{x}^\lambda) \vee (b \circ \mathbf{x}^\gamma)$ with $\lambda, \gamma \in \{1, \ldots, k\}$. If the vectors $\mathbf{v} \in \mathbb{V}$ are of form $\mathbf{v} = (v_1, \ldots, v_n)'$ with $n \geq 1$, then In order to express $\mathbf{x}$ in the standard minimax sum format, let $\Xi = \{\lambda, \gamma\}$ and define $\alpha_{\lambda i} = \bigvee_{\xi\in\Xi} x_i^\lambda \circ (x_i^\xi)^*$. Thus, for any $i \in \{1, \ldots, n\}$,

$$\bigvee_{\xi\in\Xi} x_i^\lambda \circ (x_i^\xi)^* \geq x_i^\lambda \circ (x_i^\gamma)^* \Rightarrow [\bigvee_{\xi\in\Xi} x_i^\lambda \circ (x_i^\xi)^*] \circ x_i^\gamma \geq x_i^\lambda \circ [(x_i^\gamma)^* \circ x_i^\gamma] = x_i^\lambda,$$

since $[(x_i^\gamma)^* \circ x_i^\gamma]$ equals the unit element of the group $(\mathbb{F}, \circ)$. Hence, $\alpha_{\lambda i} \circ x_i^\gamma \geq x_i^\lambda\ \forall\, i \in \{1, \ldots, n\}$. It now follows that for $\alpha_\lambda = a \circ \bigvee_{i=1}^n \alpha_{\lambda i}$, $\alpha_\lambda \circ \mathbf{x}^\gamma \geq a \circ \mathbf{x}^\lambda$ and, hence, $a \circ \mathbf{x}^\lambda = (a \circ \mathbf{x}^\lambda) \wedge (\alpha_\lambda \circ \mathbf{x}^\gamma)$. In a likewise fashion one can construct β_γ such that $b \circ \mathbf{x}^\gamma = (b \circ \mathbf{x}^\gamma) \wedge (\beta_\gamma \circ \mathbf{x}^\lambda)$. Setting $J = \{1,2\}$ and defining

$$a_{\xi 1} = \begin{cases} a & \text{if } \xi = \lambda \\ \alpha_\lambda & \text{if } \xi \neq \lambda \end{cases} \quad \text{and} \quad a_{\xi 2} = \begin{cases} b & \text{if } \xi = \gamma \\ \beta_\gamma & \text{if } \xi \neq \gamma \end{cases}$$

results in

$$(a \circ \mathbf{x}^\lambda) \vee (b \circ \mathbf{x}^\gamma) = \bigvee_{j\in J} \bigwedge_{\xi\in\Xi} (a_{\xi j} \circ \mathbf{x}^\xi).$$

The methodology used in this example reveals how any vector $\mathbf{x}$ of form $\mathbf{x} = (a \circ \mathbf{x}^\lambda) \vee (b \circ \mathbf{x}^\gamma)$ can just as easily be expressed in terms of *all* the elements of X by simply setting $\Xi = \{1, \ldots, k\}$, $J = \{1, \ldots, n\}$, where n corresponds to the number of vector coordinates, and defining $\alpha_{\lambda i}$, α_λ, $\beta_{\gamma i}$, and β_γ as in the above example. Extending the definition of $a_{\xi j}$ to all indices by setting

$$a_{\xi j} = \begin{cases} a & \text{if } j \text{ is odd and } \xi = \lambda \\ \alpha_\lambda & \text{if } j \text{ is odd and } \xi \neq \lambda \end{cases} \quad \text{and} \quad a_{\xi j} = \begin{cases} b & \text{if } j \text{ is even and } \xi = \gamma \\ \beta_\gamma & \text{if } j \text{ is even and } \xi \neq \gamma \end{cases}$$

one obtains the formulation

$$\mathbf{x} = \bigvee_{j=1}^{n} \bigwedge_{\xi=1}^{k} (a_{\xi j} \circ \mathbf{x}^{\xi}). \tag{4.39}$$

The following theorem generalizes the results of Example 4.6.

Theorem 4.21 *Suppose* $\mathbb{V}$ *is an* ℓ*-vector space over the semifield* $(\mathbb{F}, \vee, \wedge, \circ)$ *with duality* $(\cdot)^*$, $X = \{\mathbf{x}^1, \ldots, \mathbf{x}^k\} \subset \mathbb{V}$, *and* A, B *are nonempty subsets of* $\Xi = \{1, \ldots, k\}$. *If*

$$\mathbf{x} = \bigvee_{\gamma \in A} (a_\gamma \circ \mathbf{x}^\gamma) \quad \textit{and} \quad \mathbf{y} = \bigwedge_{\lambda \in B} (b_\lambda \circ \mathbf{x}^\lambda),$$

then there exist scalars $a_{\xi\gamma}, b_{\xi\lambda} \in \mathbb{F}$ *such that*

$$\mathbf{x} = \bigvee_{\gamma \in A} \bigwedge_{\xi \in \Xi} (a_{\xi\gamma} \circ \mathbf{x}^\xi) \quad \textit{and} \quad \mathbf{y} = \bigvee_{\lambda \in B} \bigwedge_{\xi \in \Xi} (b_{\xi\lambda} \circ \mathbf{x}^\xi).$$

Proof. For each pair $(\xi, \gamma) \in \Xi \times A$, define $a_{\xi\gamma} = a_\gamma \circ \bigvee_{j=1}^{n} [x_j^\gamma \circ (x_j^\xi)^*]$. A consequence of this definition is that $a_{\xi\gamma} \geq a_\gamma \circ [x_i^\gamma \circ (x_i^\xi)^*]$ and hence $a_{\xi\gamma} \circ x_i^\xi \geq a_\gamma \circ [x_i^\gamma \circ (x_i^\xi)^*] \circ x_i^\xi = a_\gamma \circ x_i^\gamma \; \forall i \in \{1, \ldots, n\}$. But this implies $a_{\xi\gamma} \circ \mathbf{x}^\xi \geq a_\gamma \circ \mathbf{x}^\gamma$ and, therefore, $\bigwedge_{\xi \in \Xi} (a_{\xi\gamma} \circ \mathbf{x}^\xi) = a_\gamma \circ \mathbf{x}^\gamma$ for every $\gamma \in A$. Thus $\bigvee_{\gamma \in A} \bigwedge_{\xi \in \Xi} (a_{\xi\gamma} \circ \mathbf{x}^\xi) = \bigvee_{\gamma \in A} (a_\gamma \circ \mathbf{x}^\gamma) = \mathbf{x}$.

The proof that $\mathbf{y}$ can be expressed as $\bigvee_{\lambda \in B} \bigwedge_{\xi \in \Xi} (b_{\xi\lambda} \circ \mathbf{x}^\xi)$ is left as an exercise. □

A note of caution is necessary when manipulating minimax sums: Due to the minimax principle (equation 3.5) some care must be taken when interchanging the operators $\bigvee$ and $\bigwedge$. The indexing set J in equation 4.30 plays an important role in turning the inequality of equation 3.5 into an equality. More precisely, for each $\xi \in \Xi$ define a nonempty subset of indices Γ_ξ and let J denote the set of functions j with domain Ξ satisfying the property that for each $\xi \in \Xi$, $j(\xi) \in \Gamma_\xi$. If $\mathbb{V}$ is a distributive lattice, then

$$\bigwedge_{\xi \in \Xi} \bigvee_{\ell \in \Gamma_\xi} (a_{\xi,\ell} \circ \mathbf{x}^\xi) = \bigvee_{j \in J} \bigwedge_{\xi \in \Xi} (a_{\xi,j(\xi)} \circ \mathbf{x}^\xi). \tag{4.40}$$

We shall not prove this equality as its verification can be found in [30]. A consequence of equation 4.40 is that the left side expression of the equation could just as well have been utilized in defining linear minimax sums. Although linear minimax sums are commonly referred to as minimax sums, we used the term *linear* in order to emphasize the similarity between traditional linear sums and $s\ell$ sums; i.e., between $\sum a_\xi \cdot \mathbf{x}^\xi$ and $\bigvee a_\xi \circ \mathbf{x}^\xi$ or $\bigwedge a_\xi \circ \mathbf{x}^\xi$. The terminology also accentuates the differences between linear lattice-based sums and lattice-based polynomials. Within the confines of minimax algebra, a *max*-polynomial in x is of form $p(x) = \bigvee_{i=0}^{n}(a_i + r_i x)$, where $a_i \in \mathbb{R}_{-\infty}$, $r_i \in \mathbb{Z}$, and $r_0 \leq r_1 \leq \cdots \leq r_n$ (see also [166, 167]).

The value $sup\{i : i \in \mathbf{N}_{[0,\infty)} \ni a_i \neq -\infty\}$ is called the *degree* of $p(x)$. Thus, if $sup\{i : i \in \mathbf{N}_{[0,\infty)} \ni a_i \neq -\infty\} = n$, then $p(x) = \bigvee_{i=0}^{n}(a_i + ix)$ and $p(x) = \bigvee_{i=0}^{n}(i + x)$ are both max-polynomials. The *min*-polynomial in x is defined analogously by simply replacing the operation $\bigvee$ with $\bigwedge$ and the lower bound $-\infty$ with ∞. Combining the two dual notions of min and max-polynomials will result in the notion of a minimax polynomial.

Example 4.7 1. If $p(x) = a_0 \vee (a_1 + x) \vee \cdots \vee (a_m + mx)$ and $q(x) = b_0 \vee (b_1 + x) \vee \cdots \vee (b_n + nx)$ are max-polynomials, and $m \leq n$, then

$$p(x) \vee q(x) = c_0 \vee (c_1 + x) \vee \cdots \vee (c_n + nx),$$

where $c_i = a_i \vee b_i$ for $0 \leq i \leq n$ and $a_i = -\infty$ for $m < i \leq n$. This shows that $p(x) \vee q(x)$ is a max-polynomial.

2. Let $X = \{\mathbf{x}^1, \mathbf{x}^2, \mathbf{x}^3\}$, where X is a subset of the ℓ-vector space $(\mathbb{R}^2, \vee, \wedge)$ over the ℓ-semifield $(\mathbb{R}^2, \vee, \wedge, +)$ and $\mathbf{x}^1 = \binom{10}{0.5}$, $\mathbf{x}^2 = \binom{6.5}{5.5}$, and $\mathbf{x}^3 = \binom{4}{-1.5}$. If $\mathbf{x} = \binom{10}{3}$, then $\mathbf{x}$ can be expressed as a max sum of $\mathbf{x}^1$ and $\mathbf{x}^3$, namely $\mathbf{x} = \mathbf{x}^1 \vee (4.5 + \mathbf{x}^3)$. To express $\mathbf{x}$ as a minimax sum of $\{\mathbf{x}^1, \mathbf{x}^3\}$, let $\Xi = \{1, 3\} \subset \{1, 2, 3\}$, $J = \{1, 2\}$, and set $a_{1,1} = 0$, $a_{1,2} = 2.5$, $a_{3,1} = 6$, and $a_{3,2} = 4.5$. Then

$$\begin{aligned}\bigvee_{j\in\{1,2\}} \bigwedge_{\xi\in\Xi}(a_{\xi,j} + \mathbf{x}^\xi) &= \bigvee_{j\in\{1,2\}} [(a_{1,j} + \mathbf{x}^1) \wedge (a_{3,j} + \mathbf{x}^3)] \\ &= [(a_{1,1} + \mathbf{x}^1) \wedge (a_{3,1} + \mathbf{x}^3)] \vee [(a_{1,2} + \mathbf{x}^1) \wedge (a_{3,2} + \mathbf{x}^3)] \\ &= \left\{\begin{pmatrix}10\\0.5\end{pmatrix} \wedge \begin{pmatrix}10\\4.5\end{pmatrix}\right\} \vee \left\{\begin{pmatrix}12.5\\3\end{pmatrix} \wedge \begin{pmatrix}8.5\\3\end{pmatrix}\right\} = \begin{pmatrix}10\\3\end{pmatrix}.\end{aligned}$$

In order to interchange the operators $\bigvee$ and $\bigwedge$, we need to define the two indexing sets Γ_1 and Γ_3. We already know that the set J consists of two indices with each depending on $\xi \in \{1, 3\}$. Thus we need to define two functions j_1 and j_2 with domain Ξ and values $j_1(1) = 1$, $j_1(3) = 1$,

$j_2(1) = 2$, and $j_2(3) = 2$. Hence, $\Gamma_1 = \{1,2\} = \Gamma_3$. Therefore,

$$\bigwedge_{\xi\in\Xi}\bigvee_{\ell\in\Gamma_\xi}(a_{\xi,j}+\mathbf{x}^\xi) = \Big\{\bigvee_{\ell\in\Gamma_1}[(a_{1,\ell}+\mathbf{x}^1)\Big\}\wedge\Big\{\bigvee_{\ell\in\Gamma_3}(a_{3,\ell}+\mathbf{x}^3)\Big\}$$
$$= [(a_{1,1}+\mathbf{x}^1)\vee(a_{1,2}+\mathbf{x}^1)]\wedge[(a_{3,1}+\mathbf{x}^3)\vee(a_{3,2}+\mathbf{x}^3)]$$
$$= \left\{\begin{pmatrix}10\\0.5\end{pmatrix}\vee\begin{pmatrix}12.5\\3\end{pmatrix}\right\}\wedge\left\{\begin{pmatrix}10\\4.5\end{pmatrix}\vee\begin{pmatrix}8.5\\3\end{pmatrix}\right\} = \begin{pmatrix}10\\3\end{pmatrix}.$$

The notion of independence provides another example of the similarities of various concepts found in both linear algebra and minimax algebra.

Definition 4.9 Let $X = \{\mathbf{x}^1, \ldots .\mathbf{x}^k\} \subset \mathbb{V}$.

1. If $\mathbb{V}$ is an $s\ell$-vector space, then a vector $\mathbf{x} \in \mathbb{V}$ is said to be *$s\ell$-dependent* on $X \Leftrightarrow \mathbf{x}$ can be expressed as a linear $s\ell$-sum of vectors from X. If $\mathbf{x}$ is not $s\ell$-dependent on X, then $\mathbf{x}$ is said to be *$s\ell$-independent* of X.

2. The set X is said to be *$s\ell$-independent* $\Leftrightarrow \forall\xi \in \{1, \ldots, k\}$, $\mathbf{x}^\xi$ is $s\ell$-independent of $X \setminus \{\mathbf{x}^\xi\}$.

3. If $\mathbb{V}$ is an ℓ-vector space, then a vector $\mathbf{x} \in \mathbb{V}$ is said to be *ℓ-dependent* on $X \Leftrightarrow \mathbf{x}$ can be expressed as a linear minimax sum of vectors from X. If $\mathbf{x}$ is not ℓ-dependent on X, then $\mathbf{x}$ is said to be *ℓ-independent* of X.

4. The set X is said to be *ℓ-independent* $\Leftrightarrow \forall\xi \in \{1, \ldots, k\}$, $\mathbf{x}^\xi$ is ℓ-independent of $X \setminus \{\mathbf{x}^\xi\}$.

An obvious consequence of Definitions 4.8 and 4.9 is the fact that if $X = \{\mathbf{x}^1, \mathbf{x}^2\} \subset \mathbb{R}^n$, where $\mathbb{R}^n$ is either an $s\ell$ or ℓ-vector space, then X is $s\ell$ or ℓ-independent if and only if $L(\mathbf{x}^1) \neq L(\mathbf{x}^2)$.

If $\mathbb{V}$ is a max-vector space or a min-vector space, then the terms *max independence* or *min independence* are also used instead of $s\ell$-independence. In a similar vein, whenever it is convenient we shall also use the term *lattice independence* for ℓ-independence.

Example 4.8 Consider the max-vector space $(\mathbb{R}^2, \vee)$ over the semifield $(\mathbb{R}, \vee, +)$ and let $X = \{\mathbf{x}^1, \mathbf{x}^2, \mathbf{x}^3\} \subset \mathbb{R}^2$, where $\mathbf{x}^1 = \binom{-1}{0}$, $\mathbf{x}^2 = \binom{1}{0}$, and $\mathbf{x}^3 = \binom{1}{1}$. The shaded region R in Figure 4.1 is bounded by the two lines $f(\mathbf{x}) = x_1 - x_2 + 1 = 0$ and $g(\mathbf{x}) = x_1 - x_2 - 1 = 0$. Thus, if $\mathbf{x} \in R$, then $x_2 - 1 \leq x_1 \leq x_2 + 1$ Setting $a = x_2$ and $b = x_1 - 1$ results in $\mathbf{x} = (a + \mathbf{x}^1) \vee (b + \mathbf{x}^2)$. This shows

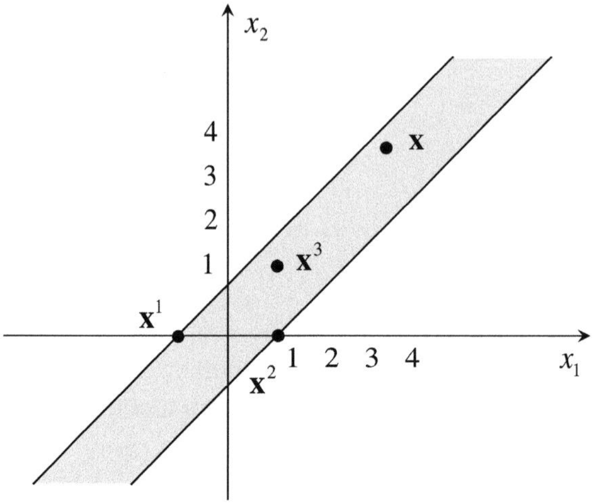

Figure 4.1 Every point in the shaded region is max dependent on X.

that any point $\mathbf{x} \in R$ is max dependent on X. However, the set X is not max independent as $\mathbf{x}^3 = (1 + \mathbf{x}^1) \vee \mathbf{x}^2$. It is also easy to show that any nonempty strict subset of X is max independent. Replacing $(\mathbb{R}^2, \vee)$ by the ℓ-vector space $(\mathbb{R}^2, \vee, \wedge)$ over the semifield $(\mathbb{R}, \vee, \wedge, +)$ the same arguments show that $\mathbf{x}$ is ℓ-dependent on X.

Since the concept of dependence in minimax algebra resembles the concept of dependence in linear algebra, it is not surprising that the *span* of a set of vectors in minimax algebra is a mirror image of this concept in linear algebra.

Definition 4.10 Let X denote a finite nonempty subset of $\mathbb{V}$. If $\mathbb{V}$ is an $s\ell$-vector space, then the *$s\ell$-span of* X, denoted by $S(X)$, is defined as $S(X) = \{\mathbf{x} \in \mathbb{V} : \mathbf{x}\ is\ s\ell\text{-dependent on}\ X\}$.

Similarly, if $\mathbb{V}$ is an ℓ-vector space, then the *ℓ-span of* X is defined as $S(X) = \{\mathbf{x} \in \mathbb{V} : \mathbf{x}\ is\ \ell\text{-dependent on}\ X\}$.

According to this definition, if $X = \{\mathbf{x}^1, \ldots . \mathbf{x}^k\} \subset \mathbb{V}$ and $(\mathbb{V}, \vee)$ is a max-vector space over the semifield $\mathbb{F}$, then $\mathbf{x} \in S(X)$ is of form $\mathbf{x} = \bigvee_{j=1}^{k}(\alpha_j \circ \mathbf{x}^j)$. Hence for $\alpha \in \mathbb{F}$, $\alpha \circ \mathbf{x} = \bigvee_{j=1}^{k}(\beta_j \circ \mathbf{x}^j) \in S(X)$, where $\beta_j = \alpha \circ \alpha_j$. It is just as easy to show that if $\mathbf{x}, \mathbf{y} \in S(X)$, then $\mathbf{x} \vee \mathbf{y} \in S(X)$. In view of Theorem 4.20, this shows that $S(X)$ is a max-vector subspace of $\mathbb{V}$. Analogous reasoning proves that if X is a finite subset of a min-vector space (or an ℓ-vector space), then $S(X)$ is a min-vector subspace (or an ℓ-vector subspace). These results are summarized in the following theorem:

Theorem 4.22 *If $\mathbb{V}$ is an $s\ell$-vector space (or an ℓ-vector space) over the semifield $\mathbb{F}$ and X is a finite subset of $\mathbb{V}$, then $S(X)$ is an $s\ell$-vector subspace (or an ℓ-vector subspace) of $\mathbb{V}$.*

Although the concepts of $s\ell$ and ℓ-dependence and independence resemble the concepts of linear dependence and independence, direct comparison is generally impossible.

Example 4.9 Let $\mathbb{V}$ be the $s\ell$-vector subspace of $(\mathbb{R}^2_{-\infty}, \vee)$ defined by $\mathbb{V} = \{\mathbf{v} \in \mathbb{R}^2_{-\infty} : v_2 \geq v_1 - 1\}$ and let $X = \{\binom{-\infty}{0}, \binom{0}{-1}\}$. Copying the argument given in Example 4.5(1), shows that $\mathbb{V}$ is indeed a max-vector space. Now, if $\mathbf{x} = \binom{x_1}{x_2} \in \mathbb{V}$, then

$$\mathbf{x} = \begin{pmatrix} -\infty \\ x_2 \end{pmatrix} \vee \begin{pmatrix} x_1 \\ x_1 - 1 \end{pmatrix} = \left[x_2 + \begin{pmatrix} -\infty \\ 0 \end{pmatrix}\right] \vee \left[x_1 + \begin{pmatrix} 0 \\ -1 \end{pmatrix}\right] \in S(X),$$

which shows that $S(X) = \mathbb{V}$. Furthermore, $2 = |X| = dim\mathbb{V} = dimS(X)$. Obviously, X is also $s\ell$-independent. However, since most vector spaces discussed in linear algebra are vector spaces over the field $(\mathbb{R}, +, \times)$ and $-\infty \notin \mathbb{R}$, one cannot test X for linear dependence or independence.

This example illustrates that for complete lattice vector spaces that contain the non-numeric elements ∞ or $-\infty$, testing a set of vectors for linear or affine independence may not be possible. Nonetheless, when considering the $s\ell$ or ℓ-vector space $\mathbb{R}^n$ over the $s\ell$ or ℓ-semifield $\mathbb{R}$, then any subset of $\mathbb{R}^n$ can be tested directly for affine or linear independence. For example, if $X \subset \mathbb{R}^2$ is as in Example 4.8, then $S(X)$ contains the shaded region including the two boundary lines $L(\mathbf{x}^1)$ and $L(\mathbf{x}^2)$ shown in Figurer 4.1. As mentioned in the example, the set X is not ℓ-independent. But since the three points of X correspond to the vertices of a 2-simplex, X is affine independent (for the definition of affine independence see subsection 4.4.1). The two vectors $\mathbf{x}^1$ and $\mathbf{x}^2$ are max-independent as well as minimax or ℓ-independent. However, they are not linearly independent. On the other hand, they are affine independent. Additionally, the two points $\mathbf{x}^2$ and $\mathbf{v} = \binom{3}{2}$ are linearly independent, but not ℓ-independent since $3 + \mathbf{x}^2 = \mathbf{v}$. This shows that there are no algebraic relationships between linear independence and ℓ-independence.

Another divergence from linear algebra concerns the dimensionality of an ℓ-subspace generated by a set of k lattice independent vectors of an n-dimensional ℓ-space.

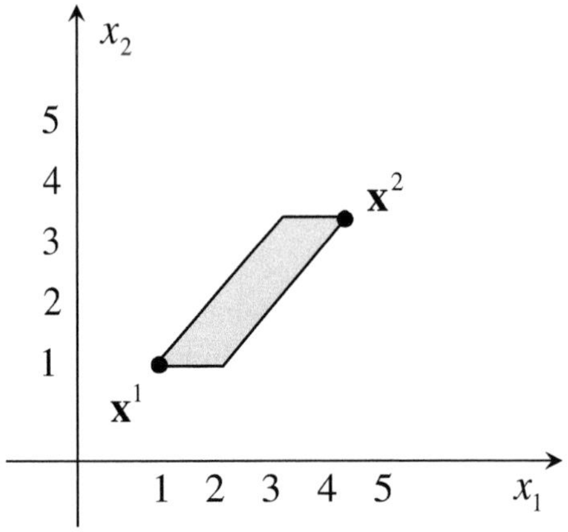

Figure 4.2 The shaded region represents a subset of the ℓ-span of X, where X is as in Example 4.10.

Example 4.10 Suppose $X = \{\mathbf{x}^1, \mathbf{x}^2\} \subset \mathbb{R}^3$, where $\mathbb{R}^3$ represents the ℓ-vector space $(\mathbb{R}^3, \vee, \wedge)$ over the minimax field $\mathbb{R}$, $\mathbf{x}^1 = (1,1,0)'$, and $\mathbf{x}^2 = (4,3,0)'$. The shaded region R shown in Figure 4.2 is a subset of $S(X)$. To verify this fact, note that R is bounded by the lines $x_2 = 1$, $x_2 = 3$ and the lines $f(\mathbf{x}) = x_1 - x_2 = 0$ and $g(\mathbf{x}) = x_1 - x_2 - 1 = 0$. Thus, if $\mathbf{x} = (x_1, x_2, 0)' \in R$, then $1 \leq x_1 \leq 4$, $1 \leq x_2, \leq 3$, and $0 \leq x_1 - 1 \leq x_2 \leq x_1$. Setting $a = x_1 - 4$ and $b = x_2 - 1$, one obtains $\mathbf{x} = [(a+\mathbf{x}^1) \vee (b+\mathbf{x}^2)] \wedge \mathbf{x}^2 \in S(X)$. Furthermore, since $\mathbb{R}$ is a minimax semifield, it follows that for any $\mathbf{v} \in L(\mathbf{x})$, $\mathbf{v} = c + \mathbf{x} = c + [(a+\mathbf{x}^1) \vee (b+\mathbf{x}^2)] \wedge \mathbf{x}^2 = [(c+a)+\mathbf{x}^1] \vee [(c+b)+\mathbf{x}^2] \wedge (c+\mathbf{x}^2) \in S(X)$. Thus, $L(\mathbf{x}) \subset \bigcup_{\mathbf{y} \in R} L(\mathbf{y}) \subset S(X)$. Since $\bigcup_{\mathbf{y} \in R} L(\mathbf{y})$ is 3-dimensional, $dimS(X) = 3$.

The example demonstrates that in contrast to vector spaces encountered in linear algebra, the dimension of an ℓ-subspace generated by a set X of lattice independent vectors can be larger than the cardinality of X.

As we observed earlier, any two points $\mathbf{x}, \mathbf{y} \in \mathbb{R}^2$ with the property that $L(\mathbf{x}) \neq L(\mathbf{y})$ are ℓ-independent. However, no three points in $\mathbb{R}^2$ are ℓ-independent. Since any two distinct points in the plane $\mathbb{R}^2$ are affinely independent it follows that ℓ-independence implies affine independence in $\mathbb{R}^2$. The converse does not hold, affine independence does not imply ℓ-independence. Thus, the question arises as to whether or not ℓ-independence implies affine independence in $\mathbb{R}^n$ for $n \geq 3$. The next example answers this question in the negative.

Example 4.11 Let $X = \{\mathbf{x}^1, \mathbf{x}^2, \mathbf{x}^3, \mathbf{x}^4\} \subset \mathbb{R}^3$, where $\mathbf{x}^1 = (10,3,0)'$, $\mathbf{x}^2 =$

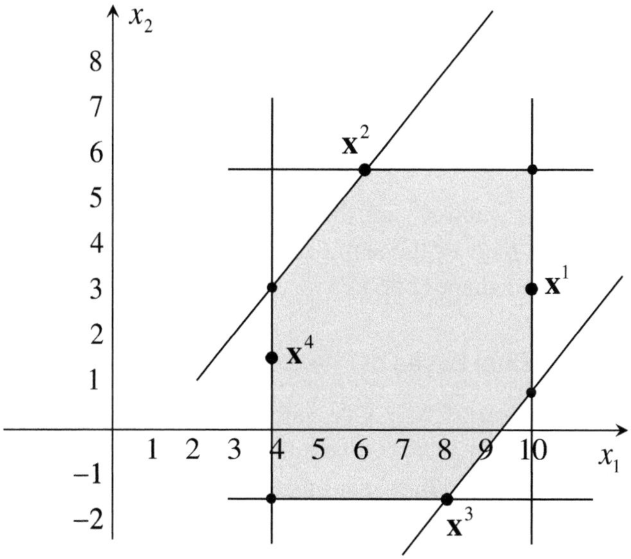

Figure 4.3 The set $X = \{\mathbf{x}^1, \mathbf{x}^2, \mathbf{x}^3, \mathbf{x}^4\} \subset \mathbb{R}^3$ with the shaded region representing the planar subset of $S(X)$ in the plane $x_3 = 0$.

$(6.5, 5.5, 0)'$, $\mathbf{x}^3 = (8, -1.5, 0)'$, and $\mathbf{x}^4 = (4, 1.5, 0)'$. Since all four vectors of X lie in the plane $x_3 = 0$, X is not affine independent. However, X is ℓ-independent.

The shaded region R shown in Figure 4.3 is bounded by the lines $x_1 = 4$, $x_1 = 10$ $x_2 = -1.5$, $x_2 = 5.5$, and the lines given by $f(\mathbf{x}) = x_1 - x_2 - 1 = 0$ and $g(\mathbf{x}) = x_1 - x_2 - 9.5 = 0$. Consequently, any point $\mathbf{x} \in R$ must satisfy the inequalities $4 \leq x_1 \leq 10$, $-1.5 \leq x_2 \leq 5.5$, and $x_2 \leq x_1 - 1 \leq x_2 + 8.5$. Given this information it is easy to show (see Example 4.10) that any $\mathbf{x} \in R$ can be expressed as a linear minimax sum of elements of X. Again reasoning as in Example 4.10 proves that $dimS(X) = 3$.

This example not only demonstrates that ℓ-independence does not imply affine independence, but also shows that the cardinality of a set X of ℓ-independent vectors can exceed the dimension of $S(X)$.

Exercises 4.3.2

1. Complete the proof of Theorem 4.21 by proving that
   $\mathbf{y} = \bigvee_{\lambda \in B} \bigwedge_{\xi \in \Xi} (b_{\xi\lambda} \circ \mathbf{x}^{\xi})$.

2. Prove Theorem 4.22.

3. Prove equation 4.40.

4. Suppose $p(x)$ and $q(x)$ are the two max-polynomials defined in Example 4.7(1). Show that $p(x) + q(x) = c_0 \vee (c_1 + x) \vee \cdots \vee (c_{m+n} + (m+n)x)$ by defining c_i in terms of a_i and b_j.

5. Suppose $\mathbf{x}^1$, $\mathbf{x}^2$, and $\mathbf{x}^3$ are three distinct points in the lattice vector spaces $(\mathbb{R}^3, \vee, \wedge)$ over the semifield $(\mathbb{R}, \vee, \wedge, +)$. What are the possible dimensions and shapes of $S(X)$ for $X = \{\mathbf{x}^1, \mathbf{x}^2\}$ and for $X = \{\mathbf{x}^1, \mathbf{x}^2, \mathbf{x}^3\}$?

4.3.3 Bases and Dual Bases of ℓ-Vector Spaces

According to subsection 1.2.4, if X is a subset of linearly independent vectors of a vector space $\mathbb{V}$, then $S(X)$ is a vector subspace of $\mathbb{V}$ and $|X| = dimS(X)$. The examples of the preceding section demonstrate that this equation may not necessarily be true for ℓ-independent (or $s\ell$-independent) subsets of ℓ-vector spaces (or $s\ell$-vector spaces). Since a goal of minimax algebra is to capture similarities common to vector spaces and ℓ-vector spaces, we consider the fundamental building blocks of the most common vector space $\mathbb{R}^n$. These building blocks are the vectors of the standard basis $\mathfrak{E} = \{\mathbf{e}^1, \ldots, \mathbf{e}^n\}$ (see Example 1.10(1)). Observe that the coordinates of the basis vectors are defined in terms of the null and unit elements of the field $(\mathbb{R}, +, \times)$. Since the null and unit elements of the $s\ell$-semifield $(\mathbb{R}_{-\infty}, \vee, +)$ are $-\infty$ and 0, respectively, it is reasonable to define the *standard $s\ell$-basis* for the $s\ell$-vector space $(\mathbb{R}^n_{-\infty}, \vee)$ as the set $\mathfrak{E}_{-\infty} = \{\mathfrak{e}^1, \ldots, \mathfrak{e}^n\}$, where

$$\mathfrak{e}^j_i = \begin{cases} 0 & \text{if } i = j \\ -\infty & \text{if } i \neq j \end{cases} \tag{4.41}$$

and $i, j \in \{1, \ldots, n\}$.

The fact that $S(\mathfrak{E}_{-\infty}) = \mathbb{R}^n_{-\infty}$ follows from the observation that if $\mathbf{x} = (x_1, \ldots, x_n)' \in \mathbb{R}^n_{-\infty}$, then $\mathbf{x} = \bigvee_{j=1}^n (x_j + \mathfrak{e}^j)$. Also, $n = |\mathfrak{E}_{-\infty}| = dimS(\mathfrak{E}_{-\infty}) = dim(\mathbb{R}^n_{-\infty})$. This provides one strong algebraic similarity between the standard basis $\mathfrak{E}$ and the standard $s\ell$-basis $\mathfrak{E}_{-\infty}$. Note also that coordinates of the summation vector $\mathbf{e} = \sum_{j=1}^n \mathbf{e}^j$ all have unit value $1 = e_i$ for $i \in \{1,2,\ldots,n\}$ and the coordinates of the maximum vector $\mathfrak{e} = \bigvee_{j=1}^n \mathfrak{e}^j$ all have unit value $0 = \mathfrak{e}_i$ for $i \in \{1,2,\ldots,n\}$. In a similar vein, observe that the columns of the identity matrix

$$I = (\mathrm{e}^1, \ldots, \mathrm{e}^n) = \begin{pmatrix} 0 & -\infty & \cdots & -\infty & -\infty \\ -\infty & 0 & \cdots & -\infty & -\infty \\ \vdots & \vdots & \ddots & \vdots & \vdots \\ -\infty & -\infty & \cdots & 0 & -\infty \\ -\infty & -\infty & \cdots & -\infty & 0 \end{pmatrix}, \tag{4.42}$$

for vectors from the $s\ell$-vector space $\mathbb{R}^n_{-\infty}$ mirror those of the identity matrix for vectors from the vector space $\mathbb{R}^n$, where

$$I = (\mathbf{e}^1, \ldots, \mathbf{e}^n) = \begin{pmatrix} 1 & 0 & \cdots & 0 & 0 \\ 0 & 1 & \cdots & 0 & 0 \\ \vdots & \vdots & \ddots & \vdots & \vdots \\ 0 & 0 & \cdots & 1 & 0 \\ 0 & 0 & \cdots & 0 & 1 \end{pmatrix}. \tag{4.43}$$

In both cases, the columns of the identity matrices correspond to the basis vectors of their respective spaces. The standard basis for the $s\ell$-vector space $(\mathbb{R}^n_{\infty}, \wedge)$ can now be defined in terms of the duality operation $(\cdot)^*$ with the basis vectors given by the columns of the matrix

$$I^* = (\mathrm{e}^1, \ldots, \mathrm{e}^n)^* = \begin{pmatrix} 0 & \infty & \cdots & \infty & \infty \\ \infty & 0 & \cdots & \infty & \infty \\ \vdots & \vdots & \ddots & \vdots & \vdots \\ \infty & \infty & \cdots & 0 & \infty \\ \infty & \infty & \cdots & \infty & 0 \end{pmatrix}. \tag{4.44}$$

Consequently, the ℓ-vector space $(\mathbb{R}^n_{\pm\infty}, \vee, \wedge)$ over the bounded minimax semifield $(\mathbb{R}_{\pm\infty}, \vee, \wedge, +, +^*)$ has a dual basis system $\mathfrak{E}_{\pm\infty} = \{\mathfrak{E}_{-\infty}, \mathfrak{E}_{\infty}\}$, where $\mathfrak{E}_{\infty} = (\mathfrak{E}_{-\infty})^*$ consists of the column vectors of the matrix I^* of equation 4.36. Due to the duality property, any vector of $\mathbb{R}^n_{\pm\infty}$ can be expressed as a linear minimax sum using either basis.

Example 4.12 Let $\mathbf{x} = \binom{x_1}{x_2} \in \mathbb{R}^2_{\pm\infty}$. Then $\mathbf{x} = (x_1 + \mathrm{e}^1) \vee (x_2 + \mathrm{e}^2)$. For instance, if $\mathbf{x} = \binom{\infty}{-\infty}$, then

$$\begin{aligned} \mathbf{x} &= \begin{pmatrix} \infty \\ -\infty \end{pmatrix} \vee \begin{pmatrix} -\infty \\ -\infty \end{pmatrix} = \left[\infty + \begin{pmatrix} 0 \\ -\infty \end{pmatrix}\right] \vee \left[-\infty + \begin{pmatrix} -\infty \\ 0 \end{pmatrix}\right] \\ &= \left[\infty +^* \begin{pmatrix} 0 \\ \infty \end{pmatrix}\right] \wedge \left[-\infty +^* \begin{pmatrix} \infty \\ 0 \end{pmatrix}\right] = \begin{pmatrix} \infty \\ \infty \end{pmatrix} \wedge \begin{pmatrix} \infty \\ -\infty \end{pmatrix} = \mathbf{x}. \end{aligned}$$

The reason for the close resemblance of either basis $\mathfrak{E}_{-\infty}$ or $\mathfrak{E}_{\infty}$ with the standard basis $\mathfrak{E}$ is due to the fact that conventional vector spaces are defined over fields, while ℓ-vector spaces are defined over ℓ-semifields which are *almost* fields. This definition of lattice bases is very different from the concept of a lattice basis as defined by G. Birkhoff in [30]. In his deliberation Birkhoff points out that the basis concept does not extend to lattices that are not complete. His argument also applies to the concept of a basis of an ℓ-vector space where the lattice is not complete. In particular, it is impossible to define a basis for the ℓ-vector space $\mathbb{R}^n$ over the ℓ-semifield $\mathbb{R}$. However, the fact that $\mathbb{R}^n$ and $\mathbb{R}$ are conditionally complete allows for a basis of any ℓ-subspace of $\mathbb{R}^n$ generated by the span of a compact subset X of $\mathbb{R}^n$. More precisely, for $i \in \mathbb{N}_n$, let $p_i : X \to \mathbb{R}$ denote the ith projection defined by $p_i(\mathbf{x}) = x_i$. Since $X \subset \mathbb{R}^n$ is compact and p_i is continuous on X, the numbers $v_i = inf\{p_i(\mathbf{x}) : \mathbf{x} \in X\}$ and $u_i = sup\{p_i(\mathbf{x}) : \mathbf{x} \in X\}$ exist for every $i \in \mathbb{N}_n$. By setting $\mathbf{v} = (v_1, \ldots, v_n)'$ and $\mathbf{u} = (u_1, \ldots, u_n)'$ one obtains $X \subset [\mathbf{v}, \mathbf{u}]$. It follows that

$$v_i - u_j \leq x_i - x_j \leq u_i - v_j \quad \text{for} \quad i, j \in \{1, \ldots, n\}. \tag{4.45}$$

Thus, the set $\{d_{ij} : d_{ij} = x_i - x_j, \mathbf{x} \in X\}$ is a bounded subset of $\mathbb{R}$ and, therefore, $\bigwedge_{\mathbf{x}\in X}(x_i - x_j)$ and $\bigvee_{\mathbf{x}\in X}(x_i - x_j)$ exist. Given this fact, we define two collections of vectors $\mathfrak{W} = \{\mathfrak{w}^1, \ldots, \mathfrak{w}^n\}$ and $\mathfrak{M} = \{\mathfrak{m}^1, \ldots, \mathfrak{m}^n\}$ by setting

$$\mathfrak{w}_i^j = \bigwedge_{\mathbf{x}\in X}(x_i - x_j) \quad \text{and} \quad \mathfrak{m}_i^j = \bigvee_{\mathbf{x}\in X}(x_i - x_j). \tag{4.46}$$

Observe that for $j = i$, $\mathfrak{w}_i^i = 0 = \mathfrak{m}_i^i$. The next theorem is a consequence of equation 4.46.

Theorem 4.23 *Given the column matrices* $I_X = (\mathfrak{w}^1, \ldots, \mathfrak{w}^n)$ *and* $J_X = (\mathfrak{m}^1, \ldots, \mathfrak{m}^n)$, *then* $I_X^* = J_X$.

Proof. By definition of the dual $(\cdot)^*$ of a matrix, we have that $\forall\, i, j \in \{1, \ldots, n\}$

$$\mathfrak{m}_i^j = \bigvee_{\mathbf{x}\in X}(x_i - x_j) = -\bigwedge_{\mathbf{x}\in X} -(x_i - x_j) = -\bigwedge_{\mathbf{x}\in X}(x_j - x_i) = -\mathfrak{w}_j^i \quad \square \tag{4.47}$$

In view of this theorem we shall denote the matrix J_X by I_X^*. The respective columns of the matrices I_X and I_X^* form the elements of the sets $\mathfrak{W}$ and $\mathfrak{M}$ that contain the minimax basis vectors for the ℓ-vector space $S(X)$.

Definition 4.11 Suppose X is a compact subset of $\mathbb{R}^n$. A *standard basis* for the ℓ-vector space $S(X)$ over the ℓ-semifield $\mathbb{R}^n$ is any ℓ-independent subset of $\mathfrak{W}$ of largest cardinality. If $\mathfrak{W}$ is ℓ-independent, then $\mathfrak{W}$ is said to be a *maximal standard basis* for $S(X)$.

Similarly, a *standard dual basis* for the ℓ-vector space $S(X)$ is any ℓ-independent subset of $\mathfrak{M}$ of largest cardinality. It follows from Theorem 4.23 that if $\mathfrak{W}$ is ℓ-independent, then $\mathfrak{M}$ is also ℓ-independent and called the *maximal dual basis* for $S(X)$.

Example 4.13

1. Let $X \subset \mathbb{R}^3$ be the set specified in Example 4.10. Then

$$\mathfrak{w}^1 = \begin{pmatrix} 0 \\ \mathfrak{w}_2^1 \\ \mathfrak{w}_3^1 \end{pmatrix} = \begin{bmatrix} 0 \\ \bigwedge_{\xi=1}^{2}(x_2^\xi - x_1^\xi) \\ \bigwedge_{\xi=1}^{2}(x_3^\xi - x_1^\xi) \end{bmatrix} = \begin{pmatrix} 0 \\ -1 \\ -4 \end{pmatrix}. \tag{4.48}$$

Similarly, $\mathfrak{w}^2 = (0,0,-3)'$ and $\mathfrak{w}^3 = (1,1,0)'$. Since $\mathfrak{w}^2 = (1+\mathfrak{w}^1) \wedge (-1+\mathfrak{w}^3)$, $\mathfrak{W} = \{\mathfrak{w}^1, \mathfrak{w}^2, \mathfrak{w}^3\}$ is not ℓ-independent and, hence, not a standard basis for $S(X)$. The standard basis for $S(X)$ is given by the set $\{\mathfrak{w}^1, \mathfrak{w}^3\}$. Thus, $S(X)$ does not have a maximal standard basis. It is also just as easy to show that $\mathfrak{m}^1 \vee (-3+\mathfrak{m}^3) = \mathfrak{m}^2$, which means that $\mathfrak{M}$ is not ℓ-independent. Therefore the dual standard basis of $S(X)$ is given by the ℓ-independent subset $\{\mathfrak{m}^1, \mathfrak{m}^3\}$.

2. The ℓ-vector space generated by the set $X \subset \mathbb{R}^3$ of Example 4.11 has a maximal standard basis $\mathfrak{W} = \{\mathfrak{w}^1, \mathfrak{w}^2, \mathfrak{w}^3\}$ and a maximal dual basis $\mathfrak{M} = \{\mathfrak{m}^1, \mathfrak{m}^2, \mathfrak{m}^3\}$, where $\mathfrak{w}^1 = (0,-9.5,-10)'$, $\mathfrak{w}^2 = (1,0,-5.5)'$, and $\mathfrak{w}^3 = (4,-1.5,0)'$. The dual basis can be computed with ease by using equation 4.47.

As discussed earlier, the columns of the identity matrices defined by equations 4.42, 4.43, and 4.44 contain the vectors that provide the canonical bases of their respective spaces. This also holds for the matrices I_X and I_X^*. In order to verify this claim, it is necessary to show that I_X and I_X^* are indeed identity matrices for the ℓ-space $S(X)$. Since we are dealing with matrices we use the notation $\mathfrak{w}_{ij} = \mathfrak{w}_i^j$ and $\mathfrak{m}_{ij} = \mathfrak{m}_i^j$ to represent the value of the (i,j)th coordinate of the matrix I_X and I_X^*, respectively. In the following lemma and theorem we let $\Xi = \{1, \ldots, k\}$ and $J = \{1, \ldots, n\}$.

Lemma 4.24 *If $X = \{\mathbf{x}^1, \ldots, \mathbf{x}^k\} \subset \mathbb{R}^n$, then for any $\xi \in \Xi$ and $i \in J$*

$$\bigwedge_{j=1}^{n} [(x_i^\xi - x_j^\xi) - \mathfrak{w}_{ij}] = 0 = \bigvee_{j=1}^{n} [(x_i^\xi - x_j^\xi) - \mathfrak{m}_{ij}].$$

Proof. For any $\xi \in \Xi$ and $\forall\, i, j \in J$, $x_i^\xi - x_j^\xi \geq \bigwedge_{\lambda \in \Xi}(x_i^\lambda - x_j^\lambda) = \mathfrak{w}_{ij}$. Hence, $[(x_i^\xi - x_j^\xi) - \mathfrak{w}_{ij}] \geq 0\ \forall\, \xi \in \Xi$ and $\forall\, i, j \in J$. It follows that $\bigwedge_{j \in J}[(x_i^\xi - x_j^\xi) - \mathfrak{w}_{ij}] \geq 0$ for any $\xi \in \Xi$ and $\forall\, i \in J$. But $\bigwedge_{j \in J}[(x_i^\xi - x_j^\xi) - \mathfrak{w}_{ij}] \leq [(x_i^\xi - x_i^\xi) - \mathfrak{w}_{ii}] = 0$. Therefore, $\bigwedge_{j \in J}[(x_i^\xi - x_j^\xi) - \mathfrak{w}_{ij}] = 0$. The proof that $\bigvee_{j \in J}[(x_i^\xi - x_j^\xi) - \mathfrak{m}_{ij}] = 0$ is analogous. □

Theorem 4.25 $I_X \boxed{\vee} \mathbf{x}^\xi = \mathbf{x}^\xi = I_X^* \boxed{\wedge} \mathbf{x}^\xi \quad \forall\, \mathbf{x}^\xi \in X.$

Proof. Arbitrarily select a $\xi \in \{1, \ldots, k\}$ and let $[I_X \boxed{\vee} \mathbf{x}^\xi]_i$ denote the ith coordinate of the vector $I_X \boxed{\vee} \mathbf{x}^\xi$. We shall show that $[I_X \boxed{\vee} \mathbf{x}^\xi]_i = x_i^\xi\ \forall i \in \{1, \ldots, n\}$. Applying Lemma 4.24 we have

$$\bigwedge_{j \in J} [(x_i^\xi - x_j^\xi) - \mathfrak{w}_{ij}] = 0 \Leftrightarrow x_i^\xi + \bigwedge_{j \in J} (-\mathfrak{w}_{ij} - x_j^\xi) = 0$$

$$\Leftrightarrow\ x_i^\xi - \bigvee_{j=1}^{n} -(-\mathfrak{w}_{ij} - x_j^\xi) = 0 \Leftrightarrow x_i^\xi - \bigvee_{j=1}^{n} (\mathfrak{w}_{ij} + x_j^\xi) = 0$$

$$\Leftrightarrow\ \bigvee_{j=1}^{n} (\mathfrak{w}_{ij} + x_j^\xi) = x_i^\xi \Leftrightarrow [I_X \boxed{\vee} \mathbf{x}^\xi]_i = x_i^\xi.$$

This proves that $I_X \boxed{\vee} \mathbf{x}^\xi = \mathbf{x}^\xi$. The proof that $\mathbf{x}^\xi = I_X^* \boxed{\wedge} \mathbf{x}^\xi$ is similar. □

Unless stated otherwise, for the remainder of this section we shall assume that $X = \{\mathbf{x}^1, \ldots, \mathbf{x}^k\} \subset \mathbb{R}^n$ even though most theorems that follow can be extended to compact subsets of $\mathbb{R}^n$. The matrix I_X and its dual may also be viewed as functions $\mathfrak{W}_X, \mathfrak{W}_X^* : \mathbb{R}^n \longrightarrow \mathbb{R}^n$ defined by $\mathfrak{W}_X(\mathbf{x}) = (\mathfrak{w}^1, \ldots, \mathfrak{w}^n) \boxed{\vee} \mathbf{x}$ and $\mathfrak{W}_X^*(\mathbf{x}) = (\mathfrak{m}^1, \ldots, \mathfrak{m}^n) \boxed{\wedge} \mathbf{x}$ so that the restricted functions $\mathfrak{W}_X|_X = I_X$ and $\mathfrak{W}_X^*|_X = I_X^*$. By definition of these functions it will be more convenient to simply set $\mathfrak{W}_X = (\mathfrak{w}^1, \ldots, \mathfrak{w}^n) = I_X$ and think of $\mathfrak{W}_X$ as either the *extension* of I_X to $\mathbb{R}^n$ or I_X as the restriction of $\mathfrak{W}_X$ to X. The same comment applies to the function $\mathfrak{W}_X^* = \mathfrak{M}_X$ and I_X^*. The utility of the extended functions $\mathfrak{W}_X$ and $\mathfrak{W}_X^*$ will be discussed in Chapter 5. One property of these extended functions is that $\mathfrak{W}_X^*(\mathbf{x}) \leq \mathbf{x} \leq \mathfrak{W}_X(\mathbf{x})\ \forall\, \mathbf{x} \in \mathbb{R}^n$. This property is due to the next theorem.

Theorem 4.26 $I_X^* \boxtimes \mathbf{x} \leq \mathbf{x} \leq I_X \boxtimes \mathbf{x} \ \forall \mathbf{x} \in \mathbb{R}^n$.

Proof. For every $i \in \{1, \ldots, n\}$, we have

$$[I_X^* \boxtimes \mathbf{X}]_i = \bigwedge_{j=1}^{n} (\mathfrak{m}_{ij} + x_j) \leq \mathfrak{m}_{ii} + x_i = x_i = \mathfrak{w}_{ii} + x_i \leq \bigvee_{j=1}^{n} (\mathfrak{w}_{ij} + x_j) = [I_X \boxtimes \mathbf{x}]_i. \quad \square$$

The set of fixed points of I_X and I_X^*, when viewed as functions of $\mathbb{R}^n \longrightarrow \mathbb{R}^n$, will be denoted by $F(I_X) = \{\mathbf{x} \in \mathbb{R}^n : I_X \boxtimes \mathbf{x} = \mathbf{x}\}$ and $F(I_X^*) = \{\mathbf{x} \in \mathbb{R}^n : I_X^* \boxtimes \mathbf{x} = \mathbf{x}\}$, respectively. According to Theorem 4.25, $X \subset F(I_X) \cap F(I_X^*)$. The next theorem provides a stronger result.

Theorem 4.27 $F(I_X) = F(I_X^*)$.

Proof. Suppose $I_X \boxtimes \mathbf{x} = \mathbf{x}$. It follows that for $i = 1, \ldots, n$, $x_i = (I_X \boxtimes \mathbf{x})_i = \bigvee_{j=1}^{n} (\mathfrak{w}_{ij} + x_j)$. Hence, $x_i \geq \mathfrak{w}_{ij} + x_j$ for $j = 1, \ldots, n$. Since this holds for all $i = 1, \ldots, n$, we have that

$$x_i \geq \mathfrak{w}_{ij} + x_j \quad \forall i, j \in \{1, \ldots, n\}. \tag{4.49}$$

By Theorem 4.26, $I_X^* \boxtimes \mathbf{x} \leq \mathbf{x}$. If equality does not hold, then $\exists j \in \{1, \ldots, n\}$ such that $(I_X^* \boxtimes \mathbf{x})_j < x_j$. Thus, $\bigwedge_{k=1}^{n} (\mathfrak{m}_{jk} + x_k) < x_j$. Also, for some $i \in \{1, \ldots, n\}$ we must have $\mathfrak{m}_{ji} + x_i = \bigwedge_{k=1}^{n} (\mathfrak{m}_{jk} + x_k)$ and, therefore, $\mathfrak{m}_{ji} + x_i < x_j$. Using duality we now obtain

$$x_i - x_j < -\mathfrak{m}_{ji} = -\bigvee_{\xi=1}^{k} (x_j^\xi - x_i^\xi) = -\bigvee_{\xi=1}^{k} -(x_i^\xi - x_j^\xi) = \bigwedge_{\xi=1}^{k} (x_i^\xi - x_j^\xi) = \mathfrak{w}_{ij}.$$

Hence, $x_i < \mathfrak{w}_{ij} + x_j$, which contradicts the inequality (4.49). Therefore, the equality $I_X^* \boxtimes \mathbf{x} = \mathbf{x}$ must hold.

The case $I_X^* \boxtimes \mathbf{x} = \mathbf{x} \Rightarrow I_X \boxtimes \mathbf{x} = \mathbf{x}$ is proven in an analogous fashion. $\square$

Since the two fixed point sets are equal, we let $F(X)$ represent the common set of fixed points. The next two theorems establish the fact that I_X and I_X^* with the respective operations of $\boxtimes$ and $\boxtimes$ are identity matrices on the ℓ-vector space $S(X)$.

Theorem 4.28 *If* $\mathbf{x}, \mathbf{y} \in F(X)$, *then* $(a + \mathbf{x}) \in F(X)$, $(a + \mathbf{x}) \vee (b + \mathbf{y}) \in F(X)$, *and* $(c + \mathbf{x}) \wedge (d + \mathbf{y}) \in F(X) \ \forall a, b, c, d \in \mathbb{R}$.

Proof. Suppose $\mathbf{x} \in F(X)$ and $a \in \mathbb{R}$. Then for $i = 1,\ldots,n$,

$$\begin{aligned}
[I_X \boxtimes (a+\mathbf{x})]_i &= \bigvee_{j=1}^{n} [\mathfrak{w}_{ij} + (a + x_j)] = \bigvee_{j=1}^{n} [a + (\mathfrak{w}_{ij} + x_j)] \\
&= a + \bigvee_{j=1}^{n} (\mathfrak{w}_{ij} + x_j) \quad \text{by equations 4.1 and 4.2} \\
&= a + [I_X \boxtimes \mathbf{x}]_i \\
&= a + x_i \quad \text{since } \mathbf{x} \in F(X).
\end{aligned}$$

Therefore, $a + \mathbf{x} \in F(X)$.

Next let $\mathbf{z} = (a+\mathbf{x}) \vee (b+\mathbf{y})$ and $i \in \{1,2,\ldots,n\}$. Using Theorem 4.1 and equation 4.2 we obtain

$$\begin{aligned}
(I_X \boxtimes \mathbf{z})_i &= \bigvee_{j=1}^{n} (\mathfrak{w}_{ij} + z_j) = \bigvee_{j=1}^{n} [\mathfrak{w}_{ij} + (a+x_j) \vee (b+y_j)] \\
&= \bigvee_{j=1}^{n} [(\mathfrak{w}_{ij} + (a+x_j)) \vee (\mathfrak{w}_{ij} + (b+y_j))] \\
&= (\bigvee_{j=1}^{n} [\mathfrak{w}_{ij} + (a+x_j)]) \vee (\bigvee_{j=1}^{n} [\mathfrak{w}_{ij} + (b+y_j)]) \\
&= [a + \bigvee_{j=1}^{n} (\mathfrak{w}_{ij} + x_j)] \vee [b + \bigvee_{j=1}^{n} (\mathfrak{w}_{ij} + y_j)] \quad \text{by equation 4.2} \\
&= [a + (I_X \boxtimes \mathbf{x})_i] \vee [b + (I_X \boxtimes \mathbf{x})_i] \\
&= (a + x_i) \vee (b + y_i) = z_i.
\end{aligned}$$

Since i was arbitrary, $(I_X \boxtimes \mathbf{z})_i = z_i \; \forall\, i = 1,2,\ldots,n$. Hence, $I_X \boxtimes \mathbf{z} = \mathbf{z}$. An analogous argument shows that $(c+\mathbf{x}) \wedge (d+\mathbf{y})$ is also a fixed point. □

Setting a, b, c, and d equal to 0 in Theorem 4.28 shows that $F(X)$ satisfies condition (*1*) of Theorem 4.19 while the remaining conclusions of Theorem 4.28 satisfy condition (2) of Theorem 4.19. Thus, $F(X)$ is an ℓ-vector space over the ℓ-semifield $(\mathbb{R}, \vee, \wedge, +)$.

Theorem 4.29 $F(X) = S(X)$.

Proof. To prove the equality, we show that $S(X) \subset F(X)$ and $F(X) \subset S(X)$.

If $\mathbf{x} \in S(X)$, then $\mathbf{x} = \bigvee_{j \in J} \bigwedge_{\xi=1}^{k} (a_{\xi j} + \mathbf{x}^{\xi})$. But by Theorem 4.25 each $\mathbf{x}^{\xi}$

is a fixed point of I_X. Hence, by Theorem 4.28 every finite linear minimax combination is also a fixed point of I_X. Therefore, $\mathbf{x} \in F(X)$, which implies that $S(X) \subset F(X)$.

Conversely, suppose that $\mathbf{x} \in F(X)$, where $\mathbf{x} = (x_1, \ldots, x_n)'$. We will show that $\mathbf{x}$ is a linear minimax sum of elements of X. Specifically, for each $j = 1, \ldots, n$ and $\xi = 1, \ldots, k$, define $a_{\xi j} = (x_j - x_j^\xi)$. Then

$$
\begin{aligned}
\bigvee_{j=1}^{n} \bigwedge_{\xi=1}^{k} (a_{\xi j} + \mathbf{x}^\xi) &= \bigvee_{j=1}^{n} \bigwedge_{\xi=1}^{k} [(x_j - x_j^\xi) + \mathbf{x}^\xi] \\
&= \bigvee_{j=1}^{n} [x_j + \bigwedge_{\xi=1}^{k} (-x_j^\xi + \mathbf{x}^\xi)] \\
&= [x_1 + \bigwedge_{\xi=1}^{k} (-x_1^\xi + \mathbf{x}^\xi)] \vee \cdots \vee [x_n + \bigwedge_{\xi=1}^{k} (-x_n^\xi + \mathbf{x}^\xi)] \\
&= [x_1 + \bigwedge_{\xi=1}^{k} \begin{pmatrix} x_1^\xi - x_1^\xi \\ x_2^\xi - x_1^\xi \\ \vdots \\ x_n^\xi - x_1^\xi \end{pmatrix}] \vee \cdots \vee [x_n + \bigwedge_{\xi=1}^{k} \begin{pmatrix} x_1^\xi - x_n^\xi \\ x_2^\xi - x_n^\xi \\ \vdots \\ x_n^\xi - x_n^\xi \end{pmatrix}] \\
&= \begin{pmatrix} \mathfrak{w}_{11} + x_1 \\ \mathfrak{w}_{21} + x_1 \\ \vdots \\ \mathfrak{w}_{n1} + x_1 \end{pmatrix} \vee \begin{pmatrix} \mathfrak{w}_{12} + x_2 \\ \mathfrak{w}_{22} + x_2 \\ \vdots \\ \mathfrak{w}_{n2} + x_2 \end{pmatrix} \vee \cdots \vee \begin{pmatrix} \mathfrak{w}_{1n} + x_n \\ \mathfrak{w}_{2n} + x_n \\ \vdots \\ \mathfrak{w}_{nn} + x_n \end{pmatrix} \\
&= I_X \boxed{\vee} \mathbf{x} = \mathbf{x}.
\end{aligned}
$$

Thus, $\mathbf{x} \in S(X)$, which implies that $F(X) \subset S(X)$ □

A consequence of Theorem 4.23 is that I_X and I_X^* are identity matrices for the ℓ-vector space $S(X)$.

Exercises 4.3.3

1. Complete the proof of Lemma 4.24 by proving that $\bigvee_{j=1}^{n}[(x_i^\xi - x_j^\xi) - \mathfrak{m}_{ij}] = 0$.

2. Complete the proof of Theorem 4.25 by proving that $\mathbf{x}^\xi = I_X^* \boxed{\wedge} \mathbf{x}^\xi$.

3. Complete the proof of Theorem 4.27 by proving that $I_X^* \boxed{\wedge} \mathbf{x} = \mathbf{x} \Rightarrow I_X \boxed{\vee} \mathbf{x} = \mathbf{x}$.

4. Complete the proof of Theorem 4.28 by proving that $(c+\mathbf{x}) \wedge (d+\mathbf{y}) \in F(X)$.

4.4 THE GEOMETRY OF $S(X)$

Theorem 4.29 shows that the ℓ-span of a compact set $X \subset \mathbb{R}^n$ is identical to the set of fixed points of I_X. However, this knowledge does not provide any direct information concerning the geometry of $S(X)$. Specifically, the equation $S(X) = F(X)$ does not furnish explicit geometric properties such as the shape of $S(X)$. The aim of this section is to provide a concise geometric description of the ℓ-span $S(X)$ and some associated properties.

4.4.1 Affine Structures in $\mathbb{R}^n$

In this section we review some well-known linear structures in Euclidean n-space $\mathbb{R}^n$. The review covers only concepts and facts concerning these structures that proved pertinent to the goals of the next two chapters.

Viewing $\mathbb{R}^n$ as a vector space over the reals, a set $E \subset \mathbb{R}^n$ is called an *affine subspace* or a *linear subspace* of $\mathbb{R}^n$ if and only if for $\mathbf{w} \in E$, the set $\overrightarrow{E} = \{\mathbf{x} - \mathbf{w} : \mathbf{x} \in E\}$ is a vector subspace of $\mathbb{R}^n$. The major difference between an affine (or linear) space E and $\overrightarrow{E}$ is that $\mathbf{0} \in \overrightarrow{E}$, while the origin may not be an element of E. The *dimension* of an affine space E is defined as $dim(E) = dim(\overrightarrow{E})$. Through much of this treatise we will use the symbol L^k to denote a k-dimensional affine subspace of $\mathbb{R}^n$.

The concept of linear independence in vector spaces gives rise to the notion of affine independence. Specifically, a set $S = \{\mathbf{s}^0, \mathbf{s}^1, \ldots, \mathbf{s}^m\} \subset \mathbb{R}^n$ is said to be *affine independent* if and only if the set $\{\mathbf{s}^j - \mathbf{s}^0 : j = 1, 2, \ldots, m\}$ is linearly independent. Thus linear independence, by definition, implies affine independence. Consequently for $n \geq 2$, any two distinct points in $\mathbb{R}^n$ are affine independent but not necessarily linearly independent. Any three non-collinear points are affine independent, any four non-coplanar points are affine independent, and in general for $m \leq n$, any $m+1$ points in $\mathbb{R}^n$ are affine independent if and only if they are not points of a common $(m-1)$-dimensional affine subspace of $\mathbb{R}^n$. The fact that the determinant of a square matrix does not equal zero if and only if columns of the matrix are linearly independent is a fundamental theorem of linear algebra. Hence in case $m = n$, one common method of testing as to whether or not the set S is affine independent is to compute the determinant of the $n \times n$ matrix $(\mathbf{s}^1 - \mathbf{s}^0, \mathbf{s}^2 - \mathbf{s}^0, \ldots, \mathbf{s}^m - \mathbf{s}^0)$. A consequence of the definition of an affine subspace of $\mathbb{R}^n$ is that if L^k is an affine subspace of $\mathbb{R}^n$ with $2 \leq k \leq n$, then for any two distinct points $\mathbf{x}, \mathbf{y} \in L^k$

the set

$$L(\mathbf{x},\mathbf{y}) = \{\mathbf{z} \in \mathbb{R}^n : \mathbf{z} = \lambda\mathbf{x} + (1-\lambda)\mathbf{y},\ \lambda \in \mathbb{R}\} \subset L^k, \tag{4.50}$$

and $L(\mathbf{x},\mathbf{y}) = L^1$.

Equation 4.50 represents the unique line determined by the two points $\mathbf{x}$ and $\mathbf{y}$. The set $\langle\mathbf{x},\mathbf{y}\rangle = \{\mathbf{z} \in \mathbb{R}^n : \mathbf{z} = \lambda\mathbf{x} + (1-\lambda)\mathbf{y}, 0 \leq \lambda \leq 1\} \subset L(\mathbf{x},\mathbf{y})$ is called a *line segment*. Line segments are basic to the concept of convexity. A set $X \subset \mathbb{R}^n$ is said to be *convex* if and only if for any two points $\mathbf{x}, \mathbf{y} \in X$ the line segment $\langle\mathbf{x},\mathbf{y}\rangle \subset X$. It follows that every affine subspace of $\mathbb{R}^n$ is convex. If m denotes the maximal number of affine independent points in a convex set $X \subset \mathbf{R}^n$, then the dimension of X is given by $dim(X) = m - 1$. Another well-known fact is that the intersection of convex subsets of $\mathbb{R}^n$ is either empty or convex (see for example [81, 292]). Thus, the intersection of affine subspaces of $\mathbb{R}^n$ is also convex.

Hyperplanes, intersections of hyperplanes, and lines are affine subspaces of $\mathbb{R}^n$ that play a vital role in describing the shape of $S(X)$. For points $\mathbf{x} \in \mathbb{R}^n$, the lines defined by the equation

$$L(\mathbf{x}) = \{\mathbf{y} \in \mathbb{R}^n : \mathbf{y} = \lambda + \mathbf{x},\ \lambda \in \mathbb{R}\}, \tag{4.51}$$

are of particular interest in studying the geometry of $S(X)$. The connection between $L(\mathbf{x})$ and $S(X)$ is a due to Theorem 4.28, which guarantees that if $\mathbf{x} \in S(X)$, then $\lambda + \mathbf{x} \in S(X)\ \forall \lambda \in \mathbb{R}$ and, hence, $L(\mathbf{x}) \subset S(X)$.

Another set of affine subspaces associated with $S(X)$ consists of specific types of hyperplanes. A *hyperplane* E in $\mathbb{R}^n$ is defined as the set of all points $\mathbf{x} \in \mathbb{R}^n$ satisfying the equation

$$a_1x_1 + a_2x_2 + \cdots + a_nx_n = b, \tag{4.52}$$

where the a_i's and b are constants and not all the a_i's are zero. It follows from equation 4.52 that E is an $(n-1)$-dimensional affine subspace of $\mathbb{R}^n$.

Any hyperplane $E \subset \mathbb{R}^n$ separates $\mathbb{R}^n$ into two open half-spaces E^+ and E^- whose common boundary is E. If E is given by equation 4.52, then E can also be expressed in terms of the polynomial function

$$f(\mathbf{x}) = a_1x_1 + a_2x_2 + \cdots + a_nx_n - b = 0. \tag{4.53}$$

We follow the convention of identifying the half-spaces E^+ and E^- with the half-spaces $\{\mathbf{x} \in \mathbb{R}^n : f(\mathbf{x}) > 0\}$ and $\{\mathbf{x} \in \mathbb{R}^n : f(\mathbf{x}) < 0\}$, respectively. The *closure* of E^+ is the convex set $\overline{E^+} = \{\mathbf{x} \in \mathbb{R}^n : f(\mathbf{x}) \geq 0\}$. Similarly,

$\overline{E^-} = \{\mathbf{x} \in \mathbb{R}^n : f(\mathbf{x}) \leq 0\}$. Therefore $\overline{E^+} \cap \overline{E^-} = E$.

Suppose X is a subset of $\mathbb{R}^n$ and $E \subset \mathbb{R}^n$ is a hyperplane, then E is called a *support hyperplane* of X if and only if the following two conditions are satisfied:

1. Either $X \subset \overline{E^+}$ or $X \subset \overline{E^-}$, and
2. $\exists \mathbf{x} \in E$ such that $\mathbf{x} \in \partial X$.

A geometric structure closely associated with the notion of support hyperplanes and pertinent to our discussion is the convex polyhedron. A nonempty subset $K \subset \mathbb{R}^n$ is called a *polyhedron* if it is the intersection of a finite number of closed half-spaces. Thus, if $E_1, E_2, \ldots, E_k$ are the hyperplanes of K whose half-spaces determine K, then $K = \bigcap_{i=1}^k \overline{E_i^\pm}$. In this equation the symbol $\pm$ corresponds for each i to either $\pm = +$ if $K \subset \overline{E_i^+}$ or $\pm = -$ if $K \subset \overline{E_i^-}$. It follows from the definition of a polyhedron that a polyhedron need not be a bounded set. If K is bounded, then K is also called a *polytope*. Thus, polytopes are compact since they are closed and bounded sets.

A special case of a polytope is the simplex. Given an affine independent set $S = \{\mathbf{s}^0, ,\mathbf{s}^1, \ldots, \mathbf{s}^k\} \subset \mathbb{R}^n$, where $0 \leq k \leq n$, then S determines a k-*dimensional simplex* denoted by $\langle \mathbf{s}^0, \mathbf{s}^1, \ldots, \mathbf{s}^k \rangle$, or simply by σ^k, and defined by

$$\langle \mathbf{s}^0, \mathbf{s}^1, \ldots, \mathbf{s}^k \rangle = \{\mathbf{x} \in \mathbb{R}^n : \mathbf{x} = \sum_{i=0}^{k} \lambda_i \mathbf{s}^i, \sum_{i=0}^{k} \lambda_i = 1, \lambda_i \geq 0 \; \forall i \in \mathbb{Z}_{k+1}\}. \quad (4.54)$$

Note that σ^k is a subset of the of the k-dimensional affine space

$$L^k = \{\mathbf{x} \in \mathbb{R}^n : \mathbf{x} = \sum_{i=0}^{k} \lambda_i \mathbf{s}^i, \sum_{i=0}^{k} \lambda_i = 1, \lambda_i \in \mathbb{R}\},$$

and for $k = 0$, $k = 1$, $k = 2$, and $k = 3$, σ^k is a point, a line segment, a triangle, and a tetrahedron, respectively.

It will be convenient to associate an *orientation* with a given hyperplane and a set of parallel *directions* for an affine subspace of $\mathbb{R}^n$. Directions and orientations will be specified in terms of unit vectors emanating from the origin $\mathbf{0}$ with endpoints lying on the $(n-1)$-dimensional unit sphere $S^{n-1} = \{\mathbf{x} \in \mathbb{R}^n : \sum_{i=1}^n x_i^2 = 1\}$ centered at the origin. Thus, a direction $\mathbf{d}$ is uniquely determined by a system of directional cosines that locates the coordinates of **d** on the unit sphere. This system is defined by

$$\cos\theta_i = \mathbf{e}^i \cdot \mathbf{d} = d_i, \text{ where } \sum_{i=1}^{n} \cos^2\theta_i = 1, \tag{4.55}$$

$\mathbf{e}^i \cdot \mathbf{d}$ denotes the inner product of the vectors $\mathbf{e}^i$ and $\mathbf{d}$, and $\{\mathbf{e}^1, \ldots, \mathbf{e}^n\}$ is the standard basis of $\mathbb{R}^n$. Two directional vectors pertinent to our discussion are the vectors $\mathbf{e}$ and $\mathbf{d}_{ij}$ defined by

$$\mathbf{e} = \frac{\mathbf{e}^1 + \cdots + \mathbf{e}^n}{\|\mathbf{e}^1 + \cdots + \mathbf{e}^n\|} \text{ and } \mathbf{d}_{ij} = \frac{\mathbf{e}^i - \mathbf{e}^j}{\|\mathbf{e}^i - \mathbf{e}^j\|}, \tag{4.56}$$

where $i < j$, $1 \leq i < n$, and $1 < j \leq n$.

It follows from equations 4.55 and 4.56 that the inner product $\mathbf{e} \cdot \mathbf{d}_{ij} = 0$. For example, if $n = 3$, then $\mathbf{e} = (\frac{1}{\sqrt{3}}, \frac{1}{\sqrt{3}}, \frac{1}{\sqrt{3}})'$ and $\mathbf{d}_{12} = (\frac{1}{\sqrt{2}}, \frac{-1}{\sqrt{2}}, 0)'$, then $\mathbf{e} \cdot \mathbf{d}_{12} = 0$, where the two vectors are located on the 2-sphere S^2 as shown in Figure 4.4

A directional vector **d** is a *parallel direction*, or simply a *direction* for a k-dimensional affine subspace $L^k \subset \mathbb{R}^n$ with $0 < k < n$ if and only if for every point $\mathbf{x} \in L^k$, $\mathbf{x} + \mathbf{d} \in L^k$. For example, the vector $\mathbf{e}$ defined in equation 4.56 is a direction for the 1-dimensional affine subspace $L(\mathbf{x}) \subset \mathbb{R}^n$ defined in eqn. 4.51. If $D(L^k) = \{\mathbf{d} : \mathbf{d} \text{ is a parallel direction for } L^k\}$, then $D(L^k)$ is called the *set of parallel directions affiliated with* L^k. Thus, $D(L(\mathbf{x})) = \{\mathbf{e}, -\mathbf{e}\}$.

An *oriented* hyperplane E with *orientation* **d**, denoted by $E(\mathbf{d})$, is a hyperplane with an associated directional unit vector **d** that is normal (perpendicular) to E. Given two oriented hyperplanes $E_1(\mathbf{d})$ and $E_2(\mathbf{v})$, then E_1 and E_2 are said to be *parallel*, denoted by $E_1 \parallel E_2$, whenever $\mathbf{d} = \pm\mathbf{v}$ or, equivalently, when $\mathbf{d} \cdot \mathbf{v} = \pm 1$. If $\mathbf{d} \cdot \mathbf{v} \neq \pm 1$, then $E_1 \cap E_2$ is an affine subspace of dimension $n - 2$. A special case occurs when $\mathbf{d} \cdot \mathbf{v} = 0$. In this case E_1 and E_2 are said to be *perpendicular*, which will be denoted by $E_1 \perp E_2$. If $E(\mathbf{d})$ is an oriented hyperplane and $\mathbf{y} \in E(\mathbf{d})$, then the pair $\{\mathbf{y}, \mathbf{d}\}$ completely specifies the hyperplane E. We shall use the notation $E_{\mathbf{y}}(\mathbf{d})$ to denote the assumption or known fact that $\mathbf{y} \in E$ and E has orientation **d**.

Oriented hyperplanes of type $E(\mathbf{d}_{ij})$ are of particular importance in the study of partially bounded sublattices of $\mathbb{R}^n$. For any compact subsets X of $\mathbb{R}^n$ they form the boundaries of the sublattice $S(X)$. For a simple example, consider the set $X = \{\mathbf{x}^1, \mathbf{x}^2, \mathbf{x}^3\}$ defined in Example 4.8. As noted in the example, the boundaries of the set of points of $\mathbb{R}^2$ that are ℓ-dependent on X are

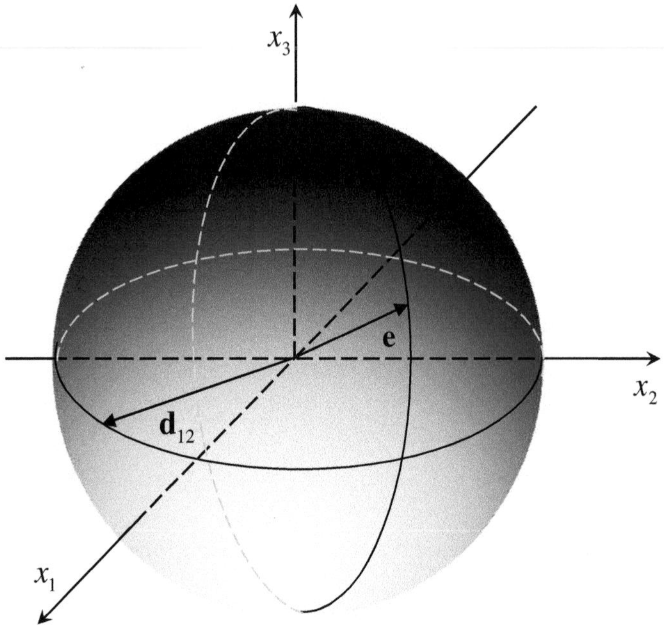

Figure 4.4 The directional vectors $\mathbf{d}_{12}$ and $\mathbf{e}$ in the vector space $\mathbb{R}^3$. The tip of the vectors are points on the unit sphere $S^2 \subset \mathbb{R}^3$. In this case the vector $\mathbf{d}_{12}$ is located on the unit sub-sphere $S^1 = \{\mathbf{x} \in S^2 : x_3 = 0\}$ while the tip of the vector $\mathbf{e}$ is located on the 1-dimensional sub-sphere $S = E_{\mathbf{0}}(\mathbf{d}_{12}) \cap S^2$.

given by the lines

$$E_{\mathbf{x}^1}(\mathbf{d}_{12}) = \{\mathbf{x} \in \mathbb{R}^2 : x_1 - x_2 = x_1^1 - x_2^1\} = L(\mathbf{x}^1) \quad \text{and} \tag{4.57}$$

$$E_{\mathbf{x}^2}(\mathbf{d}_{12}) = \{\mathbf{x} \in \mathbb{R}^2 : x_1 - x_2 = x_1^2 - x_2^2\} = L(\mathbf{x}^2), \tag{4.58}$$

where $\mathbf{x}^1 = \binom{-1}{0}$, $\mathbf{x}^2 = \binom{1}{0}$, and $\mathbf{d}_{12} = (\frac{1}{\sqrt{2}}, \frac{-1}{\sqrt{2}})'$. The shaded region R corresponding to the set $S(X)$ in Figure 4.1 is identical to the region given by $\overline{E^+_{\mathbf{x}^1}}(\mathbf{d}_{12}) \cap \overline{E^-_{\mathbf{x}^2}}(\mathbf{d}_{12})$. Therefore,

$$S(X) = \overline{E^+_{\mathbf{x}^1}}(\mathbf{d}_{12}) \cap \overline{E^-_{\mathbf{x}^2}}(\mathbf{d}_{12}), \tag{4.59}$$

with $E_{\mathbf{x}^1}(\mathbf{d}_{12}) \parallel E_{\mathbf{x}^2}(\mathbf{d}_{12})$.

Computing the standard basis for $S(X)$, where X is as in Example 4.8, one obtains $\mathfrak{w}^1 = \binom{0}{-1}$, $\mathfrak{w}^2 = \binom{-1}{0} = \mathbf{x}^1$ and the dual basis $\mathfrak{m}^1 = \binom{0}{1}$, and $\mathfrak{m}^2 =$

$\binom{1}{0} = \mathbf{x}^2$. It follows that

$$E_{\mathbf{x}^1}(\mathbf{d}_{12}) = E_{\mathfrak{w}^2}(\mathbf{d}_{12}) = E_{\mathfrak{m}^1}(\mathbf{d}_{12}) \quad \text{and}$$
$$E_{\mathbf{x}^2}(\mathbf{d}_{12}) = E_{\mathfrak{m}^2}(\mathbf{d}_{12}) = E_{\mathfrak{w}^1}(\mathbf{d}_{12}). \tag{4.60}$$

Consequently, there are multiple ways of rewriting equation 4.50 in terms of the mini-max bases of $S(X)$. For instance,

$$\begin{aligned} S(X) &= \overline{E^+_{\mathfrak{m}^1}}(\mathbf{d}_{12}) \cap \overline{E^-_{\mathfrak{m}^2}}(\mathbf{d}_{12}) \\ &= \overline{E^-_{\mathfrak{w}^1}}(\mathbf{d}_{12}) \cap \overline{E^+_{\mathfrak{w}^2}}(\mathbf{d}_{12}) \\ &= \overline{E^+_{\mathfrak{m}^1}}(\mathbf{d}_{12}) \cap \overline{E^-_{\mathfrak{w}^1}}(\mathbf{d}_{12}). \end{aligned} \tag{4.61}$$

In order to generalize equations 4.59 and 4.61, it will be instructive to illuminate the geometric properties of hyperplanes of type $E(\mathbf{d}_{ij})$. To begin with, note that the unit vector $\mathbf{d}_{ij} = (d_{ij,1}, \ldots, d_{ij,\ell}, \ldots, d_{ij,n})'$ has the property that for $\ell = 1, 2, \ldots, n$,

$$d_{ij,\ell} = \begin{cases} \frac{1}{\sqrt{2}} & \text{if } \ell = i \\ -\frac{1}{\sqrt{2}} & \text{if } \ell = j \\ 0 & \text{if } i \neq \ell \neq j \end{cases} \tag{4.62}$$

and, hence,

$$\mathbf{d}_{ij} \cdot \mathbf{d}_{ij} = 1 \quad \text{and} \quad \mathbf{d}_{ij} \cdot \mathbf{d}_{is} = \frac{1}{2} \text{ if } j \neq s, \tag{4.63}$$

$$\mathbf{d}_{ij} \cdot \mathbf{d}_{js} = -\frac{1}{2} \text{ since } j < s \quad \text{and} \quad \mathbf{d}_{ij} \cdot \mathbf{d}_{rs} = 0 \text{ if } \{i, j\} \cap \{r, s\} = \varnothing. \tag{4.64}$$

A consequence of the definition of the vector $\mathbf{d}_{ij}$ is that for a given point $\mathbf{y} \in \mathbb{R}^n$, there exist exactly $\sum_{i=1}^{n-1} i = [n(n-1)]/2$ distinct hyperplanes with orientation $\mathbf{d}_{ij}$, where $1 \leq i < j \leq n$, and containing the point $\mathbf{y}$. For example, for $\mathbf{y} \in \mathbb{R}^3$ there exist exactly three unique planes of type $E_{\mathbf{y}}(\mathbf{d}_{ij})$ containing $\mathbf{y}$, namely $E_{\mathbf{y}}(\mathbf{d}_{12})$, $E_{\mathbf{y}}(\mathbf{d}_{13})$, and $E_{\mathbf{y}}(\mathbf{d}_{23})$ as shown in Figure 4.5. Note that the intersection of these three planes is the line $L(\mathbf{y})$.

Since

$$E_{\mathbf{y}}(\mathbf{d}_{ij}) = \{\mathbf{x} \in \mathbb{R}^n : x_i - x_j = y_i - y_j\}, \tag{4.65}$$

it is easy to verify that each of the hyperplanes contains the line $L(\mathbf{y})$ and that the equality $\bigcap_{i<j} E_{\mathbf{y}}(\mathbf{d}_{ij}) = L(\mathbf{y})$ holds in general. Further properties of intersection of hyperplanes of type $E_{\mathbf{y}}(\mathbf{d}_{ij})$ are given by the following theorem:

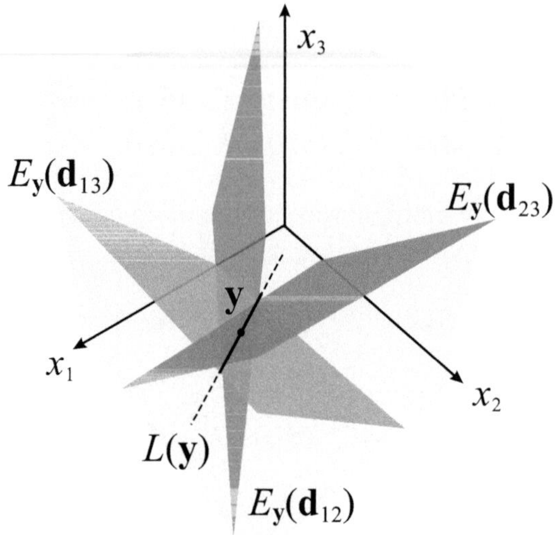

Figure 4.5 The three planes $E_\mathbf{y}(\mathbf{d}_{12})$, $E_\mathbf{y}(\mathbf{d}_{13})$, and $E_\mathbf{y}(\mathbf{d}_{23})$ containing the point $\mathbf{y} \in \mathbb{R}^3$ and hence the line $L(\mathbf{y})$.

Theorem 4.30 *Suppose $E_1 = E_\mathbf{y}(\mathbf{d}_{ij})$ and $E_2 = E_\mathbf{y}(\mathbf{d}_{rs})$ are two oriented hyperplanes in $\mathbb{R}^n$. If $(i, j) \neq (r, s)$, then $E_1 \cap E_2$ is an $(n-2)$-dimensional affine subspace of $\mathbb{R}^n$. Furthermore, if $\{i, j\} \cap \{r, s\} = \varnothing$, then E_1 and E_2 are perpendicular.*

For instance, suppose

$$E_{\mathfrak{w}^1}(\mathbf{d}_{12}) = \{\mathbf{x} \in \mathbb{R}^3 : x_1 - x_2 = \mathfrak{w}_1^1 - \mathfrak{w}_2^1\} \text{ and } E_{\mathfrak{w}^1}(\mathbf{d}_{13}) = \{\mathbf{x} \in \mathbb{R}^3 : x_1 - x_3 = \mathfrak{w}_1^1 - \mathfrak{w}_3^1\}.$$

Since $\mathfrak{w}_1^1 = 0$, the two equalities can be rewritten as

$$E_{\mathfrak{w}^1}(\mathbf{d}_{12}) = \{\mathbf{x} \in \mathbb{R}^3 : x_2 = x_1 + \mathfrak{w}_2^1\} \text{ and } E_{\mathfrak{w}^1}(\mathbf{d}_{13}) = \{\mathbf{x} \in \mathbb{R}^3 : x_3 = x_1 + \mathfrak{w}_3^1\},$$

which results in the one-dimensional line

$$E_{\mathfrak{w}^1}(\mathbf{d}_{12}) \cap E_{\mathfrak{w}^1}(\mathbf{d}_{13}) = \{\mathbf{x} \in \mathbb{R}^3 : \mathbf{x} = a + \mathfrak{w}^1 \text{ and } a = x_1\} = L(\mathfrak{w}^1)$$

The proof of Theorem 4.30 follows from equations 4.63, 4.64, and 4.65 and is left as an exercise.

Exercises 4.4.1

1. Suppose $X \subset \mathbb{R}^n$ is convex. Show that either $intX = \varnothing$ or convex.

2. Prove that if $X \subset \mathbb{R}^n$ is convex and $\mathbf{y} \notin X$, then the set $\langle \mathbf{y}, X \rangle = \bigcup_{\mathbf{x} \in X} \langle \mathbf{y}, \mathbf{x} \rangle$ is convex.

3. Let $X = \{\mathbf{x}^1, \mathbf{x}^2\}$ denote the set defined in Example 4.10. Use eqn. 4.65 in order to find the four planes $E_{\mathbf{y}}(\mathbf{d}_{ij})$ that determine the boundary of $S(X)$ and prove that $S(X)$ is the intersection of closed half spaces of these four planes.

4. Prove Theorem 4.30.

4.4.2 The Shape of $S(X)$

Suppose $X \subset \mathbb{R}^n$ is compact and $\mathfrak{W} = \{\mathfrak{w}^1, \ldots, \mathfrak{w}^n\}$ and $\mathfrak{W}^* = \{\mathfrak{m}^1, \ldots, \mathfrak{m}^n\}$ represent the sets of vectors obtained from the columns of I_X and I_X^*, respectively. Now consider the parallel hyperplanes

$$\begin{aligned} E_{\mathfrak{m}^i}(\mathbf{d}_{ij}) &= \{\mathbf{x} \in \mathbb{R}^n : x_i - x_j = \mathfrak{m}_i^i - \mathfrak{m}_j^i\} = \{\mathbf{x} \in \mathbb{R}^n : x_i - x_j + \mathfrak{m}_j^i = 0\} \\ E_{\mathfrak{w}^i}(\mathbf{d}_{ij}) &= \{\mathbf{x} \in \mathbb{R}^n : x_i - x_j = \mathfrak{w}_i^i - \mathfrak{w}_j^i\} = \{\mathbf{x} \in \mathbb{R}^n : x_i - x_j + \mathfrak{w}_j^i = 0\}, \\ &\text{where } \mathfrak{w}^i \leq \mathfrak{m}^i \text{ (see equation 4.46).} \end{aligned} \tag{4.66}$$

If $\mathbf{x} \in \mathbb{R}^n$ has the property that $x_i - x_j + \mathfrak{m}_j^i \geq 0$, then $\mathbf{x} \in \overline{E_{\mathfrak{m}^i}^+}(\mathbf{d}_{ij})$. Similarly, if $\mathbf{x}$ has the property that $x_i - x_j + \mathfrak{w}_j^i \leq 0$, then $\mathbf{x} \in \overline{E_{\mathfrak{w}^i}^-}(\mathbf{d}_{ij})$. If both properties are satisfied by $\mathbf{x}$, then $\mathbf{x} \in \overline{E_{\mathfrak{m}^i}^+}(\mathbf{d}_{ij}) \cap \overline{E_{\mathfrak{w}^i}^-}(\mathbf{d}_{ij})$, $L(\mathbf{x}) \subset \overline{E_{\mathfrak{m}^i}^+}(\mathbf{d}_{ij}) \cap \overline{E_{\mathfrak{w}^i}^-}(\mathbf{d}_{ij})$, and $\mathfrak{w}_j^i \leq x_j - x_i \leq \mathfrak{m}_j^i$ for $j = 1, \ldots, n$. Consequently, $\mathfrak{w}^i \leq -x_i + \mathbf{x} \leq \mathfrak{m}^i$ and since $\mathfrak{w}_i^i = 0 = \mathfrak{m}_i^i$, $\{\mathfrak{w}^i, \mathfrak{m}^i\} \subset E_0(\mathbf{e}^i)$ so that $x_i + \mathbf{x} \in \langle \mathfrak{w}^i, \mathfrak{m}^i \rangle \subset E_0(\mathbf{e}^i)$.

Another useful property of the hyperplanes with orientation $\mathbf{d}_{ij}$ and containing elements of the sets $\mathfrak{W}$ and $\mathfrak{M}$ is given by the following theorem.

Theorem 4.31 $E_{\mathfrak{m}^i}(\mathbf{d}_{ij}) = E_{\mathfrak{w}^j}(\mathbf{d}_{ij})$ *and* $E_{\mathfrak{w}^i}(\mathbf{d}_{ij}) = E_{\mathfrak{m}^j}(\mathbf{d}_{ij})$.

Proof.

$$\begin{aligned} E_{\mathfrak{m}^i}(\mathbf{d}_{ij}) &= \{\mathbf{x} \in \mathbb{R}^n : x_i - x_j + \mathfrak{m}_j^i = 0\} = \{\mathbf{x} \in \mathbb{R}^n : x_i - x_j - \mathfrak{w}_i^j = 0\} \\ &= \{\mathbf{x} \in \mathbb{R}^n : x_i - x_j = \mathfrak{w}_i^j - \mathfrak{w}_j^j\} = E_{\mathfrak{w}^j}(\mathbf{d}_{ij}) \end{aligned}$$

An analogous argument shows that $E_{\mathfrak{w}^i}(\mathbf{d}_{ij}) = E_{\mathfrak{m}^j}(\mathbf{d}_{ij})$ □

The following observation is pertinent in proving the next theorem. Note that if $\mathbf{y} \in \mathbb{R}^n$, then the half-open interval $[\mathbf{y}, \infty) = \Pi_{i=1}^n [y_i, \infty)$ can also be expressed in terms of half-spaces derived from hyperplanes of type $E_{\mathbf{y}}(\mathbf{e}^i)$. Specifically, we have

$$[\mathbf{y}, \infty) = \{\mathbf{x} \in \mathbb{R}^n : \mathbf{x} \geq \mathbf{y}\} = \bigcap_{i=1}^{n} \overline{E_{\mathbf{y}}^{+}}(\mathbf{e}^i), \tag{4.67}$$

and $\{\mathbf{y}\} = \bigcap_{i=1}^{n} E_{\mathbf{y}}(\mathbf{e}^i)$ since $E_{\mathbf{y}}(\mathbf{e}^i) \perp E_{\mathbf{y}}(\mathbf{e}^j)$ for $i \neq j$. Similarly, we have

$$[\mathbf{y}, -\infty) = \{\mathbf{x} \in \mathbb{R}^n : \mathbf{y} \geq \mathbf{x}\} = \bigcap_{i=1}^{n} \overline{E_{\mathbf{y}}^{-}}(\mathbf{e}^i). \tag{4.68}$$

Theorem 4.32 *If $X \subset \mathbb{R}^n$ is compact, then*

$$S(X) = \bigcap_{i=1}^{n-1} \bigcap_{j=i+1}^{n} [\overline{E_{\mathfrak{m}^i}^{+}}(\mathbf{d}_{ij}) \cap \overline{E_{\mathfrak{w}^i}^{-}}(\mathbf{d}_{ij})]$$

Before proving this theorem we simplify the mathematical symbology by defining $B(i, j) = \overline{E_{\mathfrak{m}^i}^{+}}(\mathbf{d}_{ij}) \cap \overline{E_{\mathfrak{w}^i}^{-}}(\mathbf{d}_{ij})$ and $P(X) = \bigcap_{i=1}^{n-1} \bigcap_{j=i+1}^{n} B(i, j)$.

Proof. We shall prove that $S(X) = P(X)$ by showing that $P(X) \subset S(X)$ and $S(X) \subset P(X)$. Suppose $\mathbf{y} \in P(X)$. Then $\mathbf{y} = inf(P(X) \cap [\mathbf{y}, \infty))$ and since $\mathbf{y} \in B(i, j)$ $\forall$ pairs (i, j) with $1 \leq i < j \leq n$, $L(\mathbf{y}) \subset P(X)$. Hence $-y_i + \mathbf{y} \in P(X)$ with $\mathfrak{w}^i \leq -y_i + \mathbf{y} \leq \mathfrak{m}^i$ or, equivalently, $y_i + \mathfrak{w}^i \leq \mathbf{y} \leq y_i + \mathfrak{m}^i$. Therefore $\langle y_i + \mathfrak{w}^i, y_i + \mathfrak{m}^i \rangle \subset E_{\mathbf{y}}(\mathbf{e}^i)$ and $\langle \mathbf{y}, y_i + \mathfrak{m}^i \rangle \subset P(X) \cap [\mathbf{y}, \infty)$ while $\langle y_i + \mathfrak{w}^i, \mathbf{y} \rangle \subset P(X) \cap [\mathbf{y}, -\infty)$. It now follows that $\bigvee_{i=1}^{n}(y_i + \mathfrak{w}^i) = \mathbf{y} = \bigwedge_{i=1}^{n}(y_i + \mathfrak{m}^i)$. Since $\mathbf{y}$ is equal to a min or max combination of the bases vectors $\mathfrak{w}^i$ or $\mathfrak{m}^i$, respectively, $\mathbf{y} \in S(X)$. This proves that $P(X) \subset S(X)$.

To prove the converse, suppose that $\mathbf{x} \in S(X)$. Since $F(X) = S(X)$, we have that

$$\mathfrak{w}^i + x_i \leq I_X \boxtimes \mathbf{x} = \mathbf{x} = I_X^* \boxtriangle \mathbf{x} \leq \mathfrak{m}^i + x_i \quad \forall\, i = 1, \ldots, n,$$

so that $\mathfrak{w}^i_j + x_i \leq x_j \leq \mathfrak{m}^i_j + x_i$ for $j = 1, \ldots, n$. But this implies that $(x_i - x_j) + \mathfrak{w}^i_j \leq 0$ and $(x_i - x_j) + \mathfrak{m}^i_j \geq 0$, which proves that $\mathbf{x} \in \overline{E_{\mathfrak{m}^i}^{+}}(\mathbf{d}_{ij}) \cap \overline{E_{\mathfrak{w}^i}^{-}}(\mathbf{d}_{ij})$ $\forall$ pairs $i, j \in \{1, \ldots, n\}$ with $1 \leq i < j \leq n$. Therefore $S(X) \subset P(X)$. □

Note that the expression $\bigcap_{i=1}^{n-1} \bigcap_{j=i+1}^{n} [\overline{E_{\mathfrak{m}^i}^{+}}(\mathbf{d}_{ij}) \cap \overline{E_{\mathfrak{w}^i}^{-}}(\mathbf{d}_{ij})]$ seems to involve only the vectors $\mathfrak{w}^i$ and $\mathfrak{m}^i$ for $i = 1, \ldots, n-1$. However, the relationships established in Theorem 4.31 make it clear that the expression also involves the vectors $\mathfrak{w}^n$ and $\mathfrak{m}^n$. Specifically, we have that $\overline{E_{\mathfrak{m}^i}^{+}}(\mathbf{d}_{in}) \cap \overline{E_{\mathfrak{w}^i}^{-}}(\mathbf{d}_{in}) =$

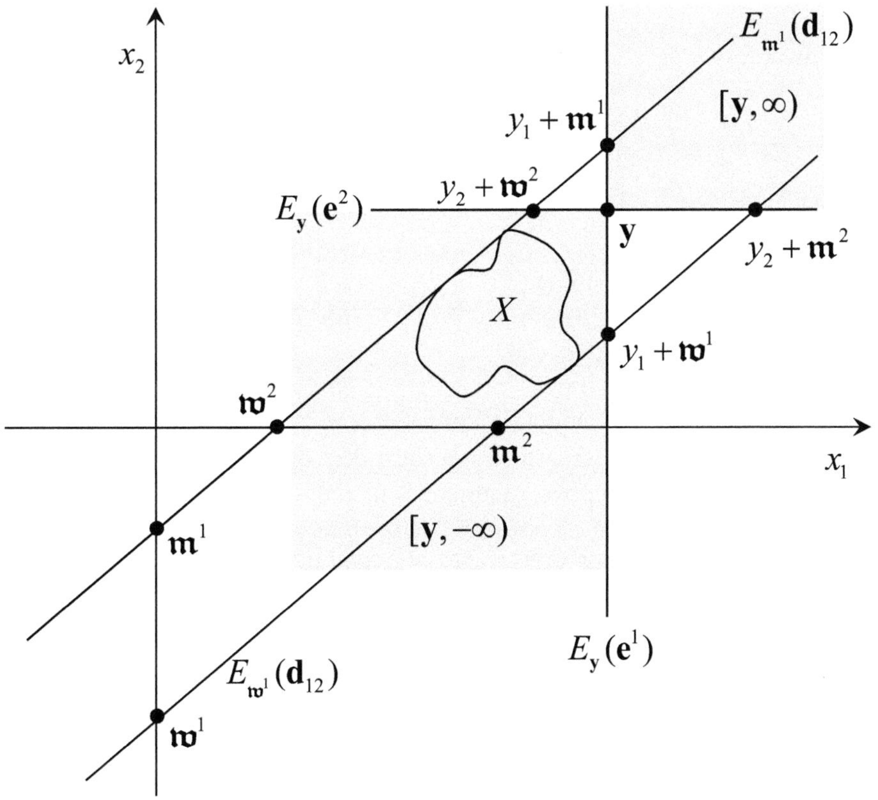

Figure 4.6 The key elements used in the proof of Theorem 4.32 illustrate how any point $\mathbf{y} \in \overline{E^+_{\mathfrak{m}^1}}(\mathbf{d}_{12}) \cap \overline{E^-_{\mathfrak{w}^1}}(\mathbf{d}_{12}) \subset \mathbb{R}^2$ can be expressed as a min or max combination of the bases element $\{\mathfrak{m}^1, \mathfrak{m}^2\}$ or $\{\mathfrak{w}^1, \mathfrak{w}^2\}$, respectively.

$\overline{E^+_{\mathfrak{w}^n}}(\mathbf{d}_{in}) \cap \overline{E^-_{\mathfrak{m}^n}}(\mathbf{d}_{in})$ for $i = 1, \ldots, n-1$. Figure 4.6 provides a graphical interpretation of all key elements associated with the proof of the theorem in case $n = 2$.

Another observation concerns the boundary of $S(X)$. Since the boundary of $B(i, j)$ is given by $\partial B(i, j) = E_{\mathfrak{m}^i}(\mathbf{d}_{ij}) \cup E_{\mathfrak{w}^i}(\mathbf{d}_{ij})$, it follows from Theorem 4.32, that

$$\partial S(X) = [\bigcup_{i=1}^{n-1} \bigcup_{j=i+1}^{n} \partial B(i, j)] \cap P(X) = [\bigcup_{i=1}^{n-1} \bigcup_{j=i+1}^{n} E_{\mathfrak{m}^i}(\mathbf{d}_{ij}) \cup E_{\mathfrak{w}^i}(\mathbf{d}_{ij})] \cap S(X). \tag{4.69}$$

For instance, the points $\mathfrak{w}^\ell$ and $\mathfrak{m}^\ell$ are boundary points since they are points on the lines

$$L(\mathfrak{w}^\ell) = \begin{cases} \bigcap_{j=2}^{n} E_{\mathfrak{w}^1}(\mathbf{d}_{1j}) & \text{if } \ell = 1 \\ (\bigcap_{i=1}^{\ell-1} E_{\mathfrak{w}^\ell}(\mathbf{d}_{i\ell})) \cap (\bigcap_{j=\ell+1}^{n} E_{\mathfrak{w}^\ell}(\mathbf{d}_{\ell j})) & \text{if } 1 < \ell < n \\ \bigcap_{i=1}^{n-1} E_{\mathfrak{w}^n}(\mathbf{d}_{in}) & \text{if } \ell = n \end{cases} \tag{4.70}$$

and

$$L(\mathfrak{m}^\ell) = \begin{cases} \bigcap_{j=2}^{n} E_{\mathfrak{m}^1}(\mathbf{d}_{1j}) & \text{if } \ell = 1 \\ (\bigcap_{i=1}^{\ell-1} E_{\mathfrak{m}^\ell}(\mathbf{d}_{i\ell})) \cap (\bigcap_{j=\ell+1}^{n} E_{\mathfrak{m}^\ell}(\mathbf{d}_{\ell j})) & \text{if } 1 < \ell < n \\ \bigcap_{i=1}^{n-1} E_{\mathfrak{m}^n}(\mathbf{d}_{in}) & \text{if } \ell = n \end{cases} \tag{4.71}$$

Although equations 4.70 and 4.71 look identical, it is important to remember that generally $E_{\mathfrak{w}^\ell}(\mathbf{d}_{i\ell}) \cap E_{\mathfrak{m}^\ell}(\mathbf{d}_{i\ell}) = \varnothing$ since they are parallel hyperplanes. A direct consequence of the two equations is that $\mathfrak{w}^\ell \in L(\mathfrak{w}^\ell) \subset \partial S(X)$ and $\mathfrak{m}^\ell \in L(\mathfrak{m}^\ell) \subset \partial S(X)$. Since these correspond to the intersections of the hyperplanes that form the boundary of $S(X)$, they constitute the edges of the unbounded polyhedron $S(X)$. More generally, if L is a line in a polyhedron K, then L is called an *edge* of K if and only if for any $\mathbf{x} \in L$, $\nexists \{\mathbf{y}, \mathbf{z}\} \subset K \setminus L \ni \mathbf{x} \in \langle \mathbf{y}, \mathbf{z} \rangle$.

As half-spaces determined by hyperplanes are convex and the intersection of convex sets is again convex, an immediate consequence of Theorem 4.32 is that $S(X)$ is a convex lattice with a boundary consisting of pairwise parallel hyperplanes. Although $S(X)$ has a boundary (the set of boundary points), it is not a bounded set since it is an infinite beam. The dimension of $S(X)$ depends on the set X and is equal to n if and only if all parallel pairs $E_{\mathfrak{w}^j}(\mathbf{d}_{ij}) \parallel E_{\mathfrak{w}^i}(\mathbf{d}_{ij})$ are disjoint. If for some pair $i < j$, $E_{\mathfrak{w}^j}(\mathbf{d}_{ij}) = E_{\mathfrak{w}^i}(\mathbf{d}_{ij})$ but all remaining pairs of parallel hyperplanes remain disjoint, then the dimension of $S(X)$ is the dimension of the hyperplane $E_{\mathfrak{w}^i}(\mathbf{d}_{ij})$, namely $n-1$. The same comments hold in case $E_{\mathfrak{m}^j}(\mathbf{d}_{ij}) = E_{\mathfrak{m}^i}(\mathbf{d}_{ij})$. As the equation of Theorem 4.32 indicates, every time a pair of parallel hyperplanes collapses into a single hyperplane, the dimension of $S(X)$ reduces by one. Thus, if m pairs of parallel support hyperplanes are equal, then the dimension of $S(X)$ is $n-m$, where $1 \leq m \leq n-1$.

The geometric shape of $S(X)$, as formulated in Theorem 4.32, represents a special type of a polyhedron, also referred to as a *prismatic beam* or a *polytopic beam*. Note that for any two elements $\mathbf{x}$ and $\mathbf{y}$ of $\mathbb{R}^n$, the hyperplanes $E_{\mathbf{x}}(\mathbf{e})$ and $E_{\mathbf{y}}(\mathbf{d}_{ij})$ are perpendicular since $\mathbf{e} \cdot \mathbf{d}_{ij} = 0 \; \forall$ pairs $\{i, j\} \subset \mathbb{N}_n$ with $i < j$. It follows that the intersection of $S(X) \cap E_{\mathbf{x}}(\mathbf{e})$ is a convex polytope of exactly the same type and shape as $S(X) \cap E_{\mathbf{y}}(\mathbf{e})$ for $\mathbf{y} \neq \mathbf{x}$; hence the names prismatic or polytopic beam.

The dimension of $S(X) \cap E_{\mathbf{x}}(\mathbf{e})$ is $(n-m)-1$ whenever $dimS(X) = n-m$.

If for instance $n = 3$ and $dimS(X) = 3$, then the polytope $S(X) \cap E_{\mathbf{x}}(\mathbf{e})$ is a polygon. For example, if $X \subset \mathbb{R}^3$ denotes the set defined in Example 4.10, then $S(X) \cap E_{\mathbf{x}}(\mathbf{e})$ is a parallelogram. If, on the other hand, $X \subset \mathbb{R}^n$ corresponds to the set defined in Example 4.11, then $S(X) \cap E_{\mathbf{x}}(\mathbf{e})$ is a hexagon and $S(X)$ is a hexagonal beam. In terms of the number of sides of the polyhedron $S(X) \cap E_{\mathbf{x}}(\mathbf{e})$, the hexagon is maximal for any compact set $X \subset \mathbb{R}^3$. On the other extreme there are various compact sets X for which $dimS(X) = 1$. For example, if $X = \{\mathbf{x} \in \mathbb{R}^3 : \mathbf{x} = \lambda\mathbf{p} + (1-\lambda)\mathbf{q},\ 0 \leq \lambda \leq 1\}$, where $\mathbf{p} = (5,2,9)'$ and $\mathbf{q} = (18, 15, 22)'$, then $dimS(X) = 1$ and $S(X) \cap E_{\mathbf{x}}(\mathbf{e})$ is a single point.

For a 2-dimensional example consider the set $X = \{\mathbf{x} \in \mathbb{R}^3 : \mathbf{x} = \lambda\mathbf{p} + (1-\lambda)\mathbf{q},\ 0 \leq \lambda \leq 1\}$ with $\mathbf{p} = (10,10,6)'$ and $\mathbf{q} = (10,8,6)'$. In this case $S(X)$ is a 2-dimensional strip of infinite length, bounded by the lines $L(\mathbf{p})$ and $L(\mathbf{q})$ as illustrated in results in Figure 4.7. Selecting the plane $E_{\mathbf{p}}(\mathbf{e})$, then $S(X) \cap E_{\mathbf{p}}(\mathbf{e}) = \langle \mathbf{p}, a + (4,2,0)' \rangle$, where $a = \frac{20}{3}$. In this particular example the following equalities hold: $E_{\mathfrak{w}^1}(\mathbf{d}_{13}) = E_{\mathfrak{w}^3}(\mathbf{d}_{13})$, $E_{\mathfrak{m}^1}(\mathbf{d}_{13}) = E_{\mathfrak{m}^3}(\mathbf{d}_{13})$, and $E_{\mathfrak{w}^2}(\mathbf{d}_{23}) \cap E_{\mathfrak{w}^3}(\mathbf{d}_{23}) = \varnothing$. This is in agreement with the comments made above.

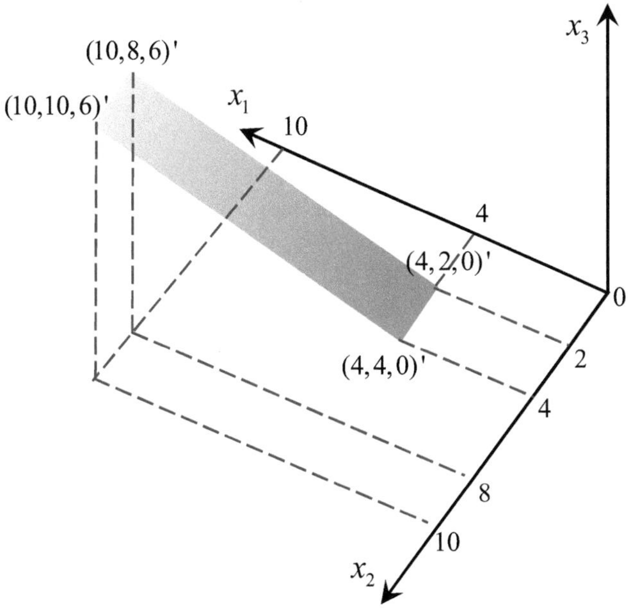

Figure 4.7 The infinite strip $S(X)$, where $X = \langle \mathbf{p}, \mathbf{q} \rangle$ with $\mathbf{p} = (10,10,6)'$ and $\mathbf{q} = (10,8,6)'$.

Exercises 4.4.2

1. Prove that $E_{\mathfrak{w}^i}(\mathbf{d}_{ij}) = E_{\mathfrak{m}^j}(\mathbf{d}_{ij})$.

2. Let $X \subset \mathbb{R}^3$ be the set defined in Example 4.11. Express $S(X)$ in terms of the geometric shape given in Theorem 4.32.

3. Provide a pictorial representation of $S(X)$, where X is defined in Example 4.11.

CHAPTER 5

Matrix-Based Lattice Associative Memories

ASSOCIATIVE memories are a special branch of pattern recognition. In this anthology a *pattern* is simply a vector in $\mathbb{R}^n$ with the vector components representing specific features of the pattern. The lattice-based approaches to pattern recognition and artificial neural networks presented in this and subsequent chapters had their roots in mathematical morphology. Mathematical morphology had proven to be an important tool in image processing and computer vision and it was realized that morphological operations were based on the Minkowski algebra of sets in $\mathbb{R}^n$ [65, 66, 112, 220]. Consequently these operations can be easily expressed in terms of the operations of the ℓ-vector space $(\mathbb{R}^n_{\pm\infty}, \vee, \wedge)$ over the bounded semifield $(\mathbb{R}_{\pm\infty}, \vee, \wedge, +, +^*)$. Since minimax matrix products were used to implement the morphological operations of *erosion* and *dilation* in various computer vision algorithms, minimax matrix-based lattice associative memories were initially called *Morphological Associative Memories* and this older terminology is still being used by some authors.

5.1 HISTORICAL BACKGROUND

The concept of an associative memory is a fairly intuitive one: Associative memory seems to be one of the primary functions of the brain. The ability of human beings to retrieve information on the basis of associated cues continues to elicit great interest among researchers. For example, a few pictures from a movie clip can evoke memory of the entire story of the movie; a glimpse of a partially occluded face in a crowd can be sufficient for recognizing an old friend. Investigations of how the brain is capable of making

DOI: 10.1201/9781003154242-5

such associations from partial information have led to a variety of theoretical neural network models that act as associative memories. The basic goal of these artificial associative memories is the retrieval of complete stored patterns from noisy or incomplete data. An associative memory is said to be *robust in the presence of noise* if presented with a corrupted version of a prototype input pattern it is still capable of retrieving the correct association.

Advances in mathematical theory related to associative memories played an important historical role in the revitalization of artificial neural networks (ANNs) research. Early research concerned with ANNs came to a virtual standstill during the 1970s. A widely publicized book by M. Minsky and S. Papert showed the limitations of the highly touted neural network model known as a perceptron [192]. Probably as much as any other single factor, the efforts of J.J. Hopfield during the early 1980s brought about a profound change in the perception of ANNs within the scientific community. As a well-known physicist at the California Institute of Technology, Hopfield's scientific credentials lent renewed credibility to the field of ANNs which had been badly tarnished by the hype of the mid-1960s. Several applications of Hopfield's early papers include associative or content-addressable memories [121, 122, 123]. It is however important to note that some significant work on associative memories did occur during the 1970s. In 1972, T. Kohonen proposed a correlation matrix model for associative memories. The model was trained—using the outer vector product rule (also known as the Hebb rule)—to learn an association between input and output patterns [151, 152]. James Anderson published a closely related paper at the same time, even though he and Kohonen worked independently [7].

Although all the above named early researchers in artificial neural networks have, justifiably, received accolades from their peers, credit must be given to Karl Steinbuch, the German pioneer of ANNs. Steinbuch introduced the first associative memory, called the "Lernmatrix" (learn matrix) in 1961 [264]. This was followed by the world's first monograph on artificial neural networks [265], which was revised and expanded three times [266, 267, 268]. Because Steinbuch's publications were in German, his work did not become widely known outside of Germany. He tried to remedy this situation with an English publication [269]. Deplorably, he was never afforded adequate attention by the international ANN research community. Nonetheless he is considered by many German researchers as the forgotten pioneer of artificial neural networks.

5.1.1 The Classical ANN Model

Artificial neural network models are specified by the network topology, node characteristics, and training or learning rules. Here the term "topology" does not appertain to the meaning of topology in the mathematical sense, but refers to the graphical layout of nodes, called *neurons*, and line segments or edges, called *axons*, where the axons represent the various connections between the nodes. Given a set $A = \{a_1, a_2, \ldots, a_n\}$ of neurons and $a_i \in A$, then the two basic equations governing the theory of computation in the standard classical neural network model are:

$$\tau_i(t) = \sum_{j=1}^{n} (w_{ij}(t) \cdot a_j(t)) \tag{5.1}$$

$$\text{and} \quad a_i(t+1) = f(\tau_i(t) - \theta_i), \tag{5.2}$$

where $a_j(t)$ denotes the value of the jth neuron at time t, $w_{ij}(t)$ represents the synaptic strength or connectivity value between the neuron a_i and the jth neuron at time t, with $w_{ij} = 0$ if there is no connection between a_i and a_j, $\tau_i(t)$ denotes the total input effect on neuron a_i at time t, and θ_i denotes a threshold, and f represents the next state function—also known as the *activation function* for a_i—which usually introduces a nonlinearity into the network. Generally, the activation function also plays an important role in the *learning* rules that determine the changes of the weight values $w_{ij}(t+1)$ at time $t+1$. Although not all current (artificial) network models can be precisely described by these two equations, they nevertheless can be viewed as variation of these two fundamental equations.

The computation represented by equations 5.1 and 5.2 are based on the operation of the field $(\mathbb{R}, +, \times)$. Replacing the operations in these equation with operations from the minimax-semifields $(\mathbb{R}, \vee, \wedge, +)$ or $(\mathbb{R}_{\pm\infty}, \vee, \wedge, +, +^*)$ results in the basic equations underlying the theory of computation in lattice-based neural networks. Specifically, we have

$$\tau_i(t) = \bigvee_{j=1}^{n} (w_{ij} + a_j(t)) \tag{5.3}$$

$$\text{and} \quad a_i(t+1) = f(\tau_i(t) - \theta_i), \tag{5.4}$$

or

$$\tau(t) = \bigwedge_{j=1}^{n}(w_{ij} + a_j(t)) \tag{5.5}$$

$$\text{and} \quad a_i(t+1) = f(\tau_i(t) - \theta_i). \tag{5.6}$$

Observe that equations 5.2, 5.4, and 5.6 are identical. Thus, the difference between the classical model and the early models of lattice-based neural networks is the computation of the next total input effect on the ith neuron which is given by equation 5.1 versus the computations given by equations 5.3 and 5.5. In the classical model equation 5.1 represents a linear process, making the need for a nonlinear activation function a necessity. In the lattice model equations 5.3 and 5.5 represent a nonlinear process so that in many cases the activation function f can be a linear function such as the identity function.

5.2 LATTICE ASSOCIATIVE MEMORIES

Associative memories based on lattice algebra are called *Lattice Associative Memories* or simply LAMs. The lattice associative memories discussed in this chapter mirror the structures of the matrix-based memories that resulted from the work of Steinbuch, Hopfield, and Kohonen. However, since the matrix operations are lattice-based, the properties and behavior of these memories are drastically different from those based on equation 5.1. Equations 5.3 and 5.5, which are basic equations for lattice neural networks, also represent the respective operations of *dilation* and *erosion* that form the foundation of mathematical morphology [246]. Since minimax matrix products were used to implement the morphological operations of erosion and dilation in various computer vision algorithms, minimax matrix-based LAMs were initially called *Morphological Associative Memories*. This older terminology is still being used by a few authors.

As in correlation encoding or the Hopfield net, the lattice associative memory provides a simple method for adding new associations. A weakness in correlation encoding is the requirement of orthogonality of the key vectors in order to exhibit perfect recall of the fundamental associations.

We also need to emphasize that the meaning of "associative" in LAMs refers to the interrelationship between data and has little to do with the storage mechanism of data within a computer. This does not mean, however, that these associations cannot be implemented in distributed physical systems.

5.2.1 Basic Properties of Matrix-Based LAMs

In the field of pattern recognition, patterns are usually viewed as column vectors in Euclidean space. Each component of a pattern vector $\mathbf{x} = (x_1, x_2, \ldots, x_n)' \in \mathbb{R}^n$ correspond to one of the pattern's features. The numerical value x_i of a pattern feature can represent a variety of objects or physical features such as signal strength, curvature, a probability value, mean mass, and so on. One goal in the theory of associative memories is for the memory to recall a stored pattern $\mathbf{y} \in \mathbb{R}^m$ when presented a pattern $\mathbf{x} \in \mathbb{R}^n$, where the pattern association expresses some desired pattern correlation. More precisely, suppose $X = \{\mathbf{x}^1, \ldots, \mathbf{x}^k\} \subset \mathbb{R}^n$ and $Y = \{\mathbf{y}^1, \ldots, \mathbf{y}^k\} \subset \mathbb{R}^m$ are two sets of pattern vectors with desired association given by the diagonal $\{(\mathbf{x}^\xi, \mathbf{y}^\xi) : \xi = 1, \ldots, k\}$ of $X \times Y$. The goal is to store these pattern pairs in some memory M such that for $\xi = 1, \ldots, k$, M recalls $\mathbf{y}^\xi$ when presented with the pattern $\mathbf{x}^\xi$. If such a memory M exists, then we shall express this association symbolically by $\mathbf{x}^\xi \rightarrow M \rightarrow \mathbf{y}^\xi$. Additionally, it is generally desirable for M to be able to recall $\mathbf{y}^\xi$ even when presented with a somewhat corrupted version of $\mathbf{x}^\xi$. If $X = Y$, then M is called an *auto-associative* memory. In this case we also have that $m = n$. The acronym LAAM will refer to a Lattice-based Auto-Associative Memory. If $X \neq Y$, then M may also be referred to as a *hetero-associative* memory.

If M represents a simple classical neural network such as the linear perceptron, then M consists of an input layer $\mathbf{a}$ of neurons which connects to an output layer $\mathbf{b}$ of neurons. The number of neurons in layers $\mathbf{a}$ and $\mathbf{b}$ are specified by the data under consideration. For instance, if X and Y are as above, then $\mathbf{a} = \{a_1, a_2, \ldots, a_n\}$ and $\mathbf{b} = \{b_1, b_2, \ldots, b_m\}$ with each a_j connected to every neuron in layer $\mathbf{b}$. In the simple perceptron case, if the pattern $\mathbf{x}^\xi$ is fed into the network, then simply set $a_j(t) = x_j^\xi$ for $j = 1, \ldots, n$ for all t used in the learning cycle. Viewing $\mathbf{a}$ and $\mathbf{b}$ as the column vectors $\mathbf{a} = (a_1, a_2, \ldots, a_n)'$ and $\mathbf{b} = (b_1, b_2, \ldots, b_m)'$, we can express $\mathbf{a}(t)$ as $\mathbf{a}(t) = \mathbf{x}^\xi$ during the learning cycle for the associated pair $(\mathbf{x}^\xi, \mathbf{y}^\xi)$. The weights $w_{ij}(t)$ get updated by a particular learning rule until $b_j(t) = y_j^\xi$ for all $j \in \mathbb{N}_m$ or t exceeds some preset time constraint and another pattern input is used in the training algorithm. The goal—which may not be achieved—is to obtain a set of weights $\{w_{ij} : i \in \mathbb{N}_m \text{ and } j \in \mathbb{N}_n\}$ such that for some time step t_s, $\mathbf{b}(t_s) = \mathbf{y}^\xi$ for every input $\mathbf{x}^\xi \in X$.

According to equation 5.1 the total network computation for a time step t is given by

$$T(t) = W(t) \cdot \mathbf{a}(t), \tag{5.7}$$

where $\mathbf{a}(t) = (a_1(t), \cdots, a_n(t))'$, $T(t) = (\tau_1(t), \cdots, \tau_m(t))'$, and W denotes the $m \times n$ synaptic weight matrix whose i, jth entry is $w_{ij}(t)$.

Analogous to equation 5.7, the total computational effort resulting from equations 5.3 and 5.5 are given by

$$T(t) = W(t) \boxed{\vee} \mathbf{a}(t) \quad \text{and} \quad T(t) = W(t) \boxed{\wedge} \mathbf{a}(t), \tag{5.8}$$

respectively.

The matrix correlation memories resulting from the work of Kohonen and Hopfield that were mentioned in the Introduction were the earliest ANN approaches for solving this particular problem. In these approaches, one starts out with an $m \times n$ matrix M defined in terms of the sum of outer products of the associated pattern vectors, namely

$$M = \sum_{\xi=1}^{k} \mathbf{y}^{\xi} \cdot \left(\mathbf{x}^{\xi}\right)'. \tag{5.9}$$

It follows that the $(i, j)th$ entry of M is given by $m_{ij} = \sum_{\xi=1}^{k} y_i^{\xi} x_j^{\xi}$. Furthermore, if the input patterns $\mathbf{x}^1, \ldots, \mathbf{x}^k$ are orthonormal, that is if

$$(\mathbf{x}^{\gamma})' \cdot \mathbf{x}^{\xi} = \begin{cases} 1 \text{ if } \xi = \gamma \\ 0 \text{ if } \xi \neq \gamma \end{cases}, \tag{5.10}$$

then $M \cdot \mathbf{x}^{\xi} = \mathbf{y}^{\xi}\left((\mathbf{x}^{\xi})' \cdot \mathbf{x}^{\xi}\right) + \sum_{\gamma \neq \xi} \mathbf{y}^{\gamma}\left((\mathbf{x}^{\gamma})' \cdot \mathbf{x}^{\xi}\right) = \mathbf{y}^{\xi}$.

Thus, we have *perfect recall* of the output patterns $\mathbf{y}^1, \ldots, \mathbf{y}^k$. If $\mathbf{x}^1, \ldots, \mathbf{x}^k$ are not orthonormal (as in most realistic cases), then the term

$$N = \sum_{\xi \neq \gamma} \mathbf{y}^{\gamma} \cdot ((\mathbf{x}^{\gamma})' \cdot \mathbf{x}^{\xi}) \neq \mathbf{0} \tag{5.11}$$

is called the *noise term*. Filtering processes using activation functions become necessary for changing the weights in order to retrieve the desired output pattern. The Hopfield net, the Lernmatrix approach, the Kohonen content addressable memory, and the various modifications of these three fundamental matrix-based memories are prime examples for the need of learning algorithms in order to achieve recall improvements.

The matrix-based LAMs are similar to the correlation-based approach when replacing the field operation of $(\mathbb{R}, +, \times)$ by the minimax-semifield operations of $(\mathbb{R}_{\pm\infty}, \vee, \wedge, +)$. For example, the correlation product in equation 5.9 is a $m \times n$ matrix, and the matrix M is the sum of these matrices. In the minimax domain we have a similar setup.

Definition 5.1 If $\mathbf{x} \in \mathbb{R}^n$ and $\mathbf{y} \in \mathbb{R}^m$ then the *lattice outer product* of $\mathbf{y}$ by $\mathbf{x}$ is defined as

$$\mathbf{y} \boxed{\vee} \mathbf{x}^* = \begin{pmatrix} y_1 - x_1 & \cdots & y_1 - x_n \\ \vdots & \ddots & \vdots \\ y_m - x_1 & \cdots & y_m - x_n \end{pmatrix} \tag{5.12}$$

Since $\mathbf{y} \boxed{\vee} \mathbf{x}^* = \mathbf{y} \boxed{\wedge} \mathbf{x}^*$, the lattice outer product of vectors is not dependent on the type of minimax matrix product being used and will henceforth be denoted by $\mathbf{y} \otimes \mathbf{x}$.

If $M = \mathbf{y} \otimes \mathbf{x}$, then the jth coordinate of the vector $W \boxed{\vee} \mathbf{x}$ is given by

$$(M \boxed{\vee} \mathbf{x})_j = \bigvee_{i=1}^{n} (y_j - x_i + x_i) = y_j, \tag{5.13}$$

so that $M \boxed{\vee} \mathbf{x} = \mathbf{y}$. Equivalently, we have

$$\mathbf{y} = \mathbf{y} \boxed{\vee} (\mathbf{x}^* \boxed{\vee} \mathbf{x}) = (\mathbf{y} \otimes \mathbf{x}) \boxed{\vee} \mathbf{x}. \tag{5.14}$$

Henceforth, we let $(\mathbf{x}^1, \mathbf{y}^1), \ldots, (\mathbf{x}^k, \mathbf{y}^k)$ denote k vector pairs, where $\mathbf{x}^\xi = (x_1^\xi, \ldots, x_n^\xi)' \in \mathbb{R}^n$ and $\mathbf{y}^\xi = (y_1^\xi, \ldots, y_n^\xi)' \in \mathbb{R}^m$, and $\xi = 1, 2, \ldots, k$. For a given pattern association $\{(\mathbf{x}^\xi, \mathbf{y}^\xi) : \xi = 1, 2, \ldots, k\}$, we define a pair of associated pattern matrices (X, Y), where $X = (\mathbf{x}^1, \ldots, \mathbf{x}^k)$ is of dimension $n \times k$ and $Y = (\mathbf{y}^1, \ldots, \mathbf{y}^k)$ is of dimension $m \times k$. Each pair (X, Y) of associated pattern matrices gives rise to two *canonical* lattice associative memories denoted by W_{XY} and M_{XY} which are defined by

$$M_{XY} = \bigvee_{\xi=1}^{k} \mathbf{y}^\xi \otimes \mathbf{x}^\xi \quad \text{and} \quad W_{XY} = \bigwedge_{\xi=1}^{k} \mathbf{y}^\xi \otimes \mathbf{x}^\xi. \tag{5.15}$$

Here the subscript XY serves as a reminder that M and W are associative memories tasked with mapping elements (column vectors) of X to elements of Y. Note that both W_{XY} and M_{XY} are of dimension $m \times n$. It follows from equation 5.15 that the i, jth element of the matrices M_{XY} and W_{XY} are given by

$$m_{ij} = \bigvee_{\xi=1}^{k} (y_i^\xi - x_j^\xi) \quad \text{and} \quad w_{ij} = \bigwedge_{\xi=1}^{k} (y_i^\xi - x_j^\xi), \tag{5.16}$$

respectively.

Given two real-valued matrices $A = (a_{ij})_{m \times n}$ and $B = (b_{ij})_{m \times n}$, then we say that *A is less or equal to B*, denoted by $A \leq B$, if and only if $a_{ij} \leq b_{ij}$ for

$i = 1, \ldots, m$ and $j = 1, \ldots, n$. If $A \le B$ and $\mathbf{x} \in \mathbb{R}^n$, then since $\bigvee_{j=1}^{n}(a_{ij} + x_j) \le \bigvee_{j=1}^{n}(b_{ij} + x_j)$ and $\bigwedge_{j=1}^{n}(a_{ij} + x_j) \le \bigwedge_{j=1}^{n}(b_{ij} + x_j)$ we obtain the following relationships:

$$A \le B \quad \text{and} \quad \mathbf{x} \in \mathbb{R}^n \Rightarrow A \boxtimes \mathbf{x} \le B \boxtimes \mathbf{x} \quad \text{and} \quad A \boxtimes \mathbf{x} \le B \boxtimes \mathbf{x}. \tag{5.17}$$

As an immediate consequence of the definition of the canonical memories we have the following result:

Theorem 5.1 $W_{XY} \boxtimes X \le Y \le M_{XY} \boxtimes X$.

Proof. By definition of W_{XY} and M_{XY},

$$W_{XY} \le \mathbf{y}^\xi \otimes \mathbf{x}^\xi \le M_{XY} \; \forall \xi = 1, 2, \ldots, k.$$

In view of equation 5.14, this means that

$$W_{XY} \boxtimes \mathbf{x}^\xi \le (\mathbf{y}^\xi \otimes \mathbf{x}^\xi) \boxtimes \mathbf{x}^\xi = \mathbf{y}^\xi = (\mathbf{y}^\xi \otimes \mathbf{x}^\xi) \boxtimes \mathbf{x}^\xi \le M_{XY} \boxtimes \mathbf{x}^\xi$$

$\forall \xi = 1, 2, \ldots, k$. □

Example 5.1

1. Consider the association $(\mathbf{x}^1, \mathbf{y}^1)$, $(\mathbf{x}^2, \mathbf{y}^2)$, and $(\mathbf{x}^3, \mathbf{y}^3)$, where

$$\mathbf{x}^1 = \begin{pmatrix} 0 \\ 0 \end{pmatrix}, \; \mathbf{x}^2 = \begin{pmatrix} 0 \\ -2 \end{pmatrix}, \; \mathbf{x}^3 = \begin{pmatrix} 0 \\ -3 \end{pmatrix}, \; \text{and}$$

$$\mathbf{y}^1 = \begin{pmatrix} 0 \\ 1 \\ 0 \end{pmatrix}, \; \mathbf{y}^2 = \begin{pmatrix} -1 \\ -1 \\ 0 \end{pmatrix}, \; \mathbf{y}^3 = \begin{pmatrix} -1 \\ -2 \\ 0 \end{pmatrix}.$$

Then

$$W_{XY} = \bigwedge_{\xi=1}^{3} \mathbf{y}^\xi \otimes \mathbf{x}^\xi = \begin{pmatrix} 0 & 0 \\ 1 & 1 \\ 0 & 0 \end{pmatrix} \wedge \begin{pmatrix} -1 & 1 \\ -1 & 1 \\ 0 & 0 \end{pmatrix} \wedge \begin{pmatrix} -1 & 2 \\ -2 & 1 \\ 0 & 3 \end{pmatrix} = \begin{pmatrix} -1 & 0 \\ -2 & 1 \\ 0 & 0 \end{pmatrix}$$

and $W_{XY} \boxtimes \mathbf{x}^\xi = \mathbf{y}^\xi$ for $\xi = 1, 2, 3$.

Similarly, $M_{XY} = \bigvee_{\xi=1}^{3} \mathbf{y}^\xi \otimes \mathbf{x}^\xi = \begin{pmatrix} 0 & 2 \\ 1 & 1 \\ 0 & 3 \end{pmatrix}$ and $M_{XY} \boxtimes \mathbf{x}^\xi = \mathbf{y}^\xi$ for $\xi = 1$ and $\xi = 3$. For $\xi = 2$ we only obtain the inequality $M_{XY} \boxtimes \mathbf{x}^2 = \begin{pmatrix} 0 \\ -1 \\ 0 \end{pmatrix} \ge \mathbf{y}^2$, which was guaranteed by Theorem 5.1.

2. Let

$$\mathbf{x}^1 = \begin{pmatrix} 2 \\ 0 \\ 2 \end{pmatrix}, \mathbf{x}^2 = \begin{pmatrix} 0 \\ 1 \\ 1 \end{pmatrix}, \mathbf{x}^3 = \begin{pmatrix} 1 \\ 2 \\ -1 \end{pmatrix}, \text{ and}$$

$$\mathbf{y}^1 = \begin{pmatrix} 1 \\ 1 \end{pmatrix}, \mathbf{y}^2 = \begin{pmatrix} 0 \\ 0 \end{pmatrix}, \mathbf{y}^3 = \begin{pmatrix} 0 \\ 1 \end{pmatrix}.$$

Then

$$W_{XY} = \begin{pmatrix} -1 & -2 & -1 \\ -1 & -1 & -1 \end{pmatrix} \quad \text{and} \quad M_{XY} = \begin{pmatrix} 0 & 1 & 1 \\ 0 & 1 & 2 \end{pmatrix}.$$

It is easily verified that $W_{XY} \boxed{\vee} \mathbf{x}^\xi = \mathbf{y}^\xi = M_{XY} \boxed{\wedge} \mathbf{x}^\xi$ holds for $\xi = 1, 2, 3$. However, using either one of the slightly distorted versions $\mathbf{x} = (2, 0, 1)'$ or $\mathbf{x} = (1, 0, 2)'$ of $\mathbf{x}^1$, one obtains $W_{XY} \boxed{\vee} \mathbf{x} = \mathbf{y}^1 = M_{XY} \boxed{\wedge} \mathbf{x}$. However, when using the vector $\mathbf{x} = (1, 0, 1)'$, which could represent a distorted version of either $\mathbf{x}^1$ or $\mathbf{x}^2$, one obtains $W_{XY} \boxed{\vee} \mathbf{x} = \mathbf{y}^2$ and $M_{XY} \boxed{\wedge} \mathbf{x} = \mathbf{y}^1$. Thus, $W_{XY} \boxed{\vee} \mathbf{x} \leq M_{XY} \boxed{\wedge} \mathbf{x}$ even though $\mathbf{x} \notin X$.

The above example raises three obvious questions, the first of which concerns the last inequality in Example 5.1(2). Specifically, does the inequality $W_{XY} \boxed{\vee} \mathbf{x} \leq M_{XY} \boxed{\wedge} \mathbf{x}$ hold for any vector $\mathbf{x}$? Here the answer is a resounding no. For example, if $\mathbf{x} = (5, 4, 1)'$, then

$$M_{XY} \boxed{\wedge} \mathbf{x} = \begin{pmatrix} 2 \\ 3 \end{pmatrix} < \begin{pmatrix} 4 \\ 4 \end{pmatrix} = W_{XY} \boxed{\vee} \mathbf{x}. \tag{5.18}$$

The second question concerns the existence of perfect recall memories. More precisely, for what sets of vector associations $(X, Y) = \{(\mathbf{x}^\xi, \mathbf{y}^\xi) : \xi = 1, 2, \ldots, k\}$ will W_{XY} or M_{XY} provide perfect recall? Once this question has been answered, the next logical question is to inquire as to the amount of distortion or noise the memories W_{XY} or M_{XY} can tolerate for perfect recall; that is, if $\tilde{\mathbf{x}}^\xi$ denotes a distorted version of $\mathbf{X}^\xi$, what are the conditions or bounds on $\tilde{\mathbf{x}}^\xi$ that guarantee that $W_{XY} \boxed{\vee} \tilde{\mathbf{x}}^\xi = \mathbf{y}^\xi$ or $M_{XY} \boxed{\wedge} \tilde{\mathbf{x}}^\xi = \mathbf{y}^\xi$? The next two theorems, established by Ritter *et al.* in [218], will address the second question.

Theorem 5.2 (*Optimality*) *Let* $(X, Y) = \{(\mathbf{x}^\xi, \mathbf{y}^\xi) : \xi = 1, 2, \ldots, k\}$ *denote the set of associate pattern vector pairs with* $X \subset \mathbb{R}^n$ *and* $Y \subset \mathbb{R}^m$. *Whenever there exist perfect recall memories A and B such that* $A \boxed{\vee} \mathbf{x}^\xi = \mathbf{y}^\xi$ *and* $B \boxed{\wedge} \mathbf{x}^\xi = \mathbf{y}^\xi$ *for* $\xi = 1, \ldots, k$, *then*

$$A \leq W_{XY} \leq M_{XY} \leq B \quad \textit{and} \quad W_{XY} \boxed{\vee} \mathbf{x}^\xi = \mathbf{y}^\xi = M_{XY} \boxed{\wedge} \mathbf{x}^\xi.$$

Proof. For $i = 1, \ldots, m$ and $j = 1, \ldots, n$, let a_{ij} denoted the i, jth element of the matrix A. Since A is a perfect recall memory for (X, Y), the ith coordinate of the vector $A \boxed{\vee} \mathbf{x}^\xi = \mathbf{y}^\xi$ is given by $(A \boxed{\vee} \mathbf{x}^\xi)_i = y_i^\xi$ for all $\xi = 1, 2, \ldots, k$ and $i = 1, \ldots, m$. Equivalently, we have

$$\bigvee_{j=1}^{n} (a_{ij} + x_j^\xi) = y_i^\xi \quad \forall \xi = 1, 2, \ldots, k \text{ and } i = 1, \ldots, m.$$

Thus, if $j \in \mathbb{N}_n$, then

$$\begin{aligned} & a_{ij} + x_j^\xi \leq y_i^\xi \quad \forall \xi = 1, \ldots, k \\ & \Leftrightarrow a_{ij} \leq y_i^\xi - x_j^\xi \quad \forall \xi = 1, \ldots, k \\ & \Leftrightarrow a_{ij} \leq \bigwedge_{\xi=1}^{k} (y_i^\xi - x_j^\xi) = w_{ij}. \end{aligned}$$

This proves that $A \leq W_{XY}$. In view of equation 5.17 and Theorem 5.1 we now have $\mathbf{y}^\xi = A \boxed{\vee} \mathbf{x}^\xi \leq W_{XY} \boxed{\vee} \mathbf{x}^\xi \leq \mathbf{y}^\xi \ \forall \xi = 1, \ldots, k$ and, therefore, $W_{XY} \boxed{\vee} \mathbf{x}^\xi = \mathbf{y}^\xi$ $\forall \xi = 1, \ldots, k$. An analogous argument proves that if $B \boxed{\wedge} \mathbf{x}^\xi = \mathbf{y}^\xi \ \forall \xi$, then $M_{XY} \leq B$ and $M_{XY} \boxed{\wedge} \mathbf{x}^\xi = \mathbf{y}^\xi \ \forall \xi = 1, \ldots, k$. □

The matrix memories A and B in the hypothesis of Theorem 5.2 are also, respectively, referred to as $\boxed{\vee}$-*perfect* and $\boxed{\wedge}$-*perfect* for the pattern association (X, Y). The next theorem answers the existence question of perfect recall memories. We assume that the set (X, Y) of associate pattern pairs are as in Theorem 5.2.

Theorem 5.3 *W_{XY} is $\boxed{\vee}$-perfect for (X, Y) $\Leftrightarrow$ for each $\xi = 1, 2, \ldots, k$, each row of the matrix $(\mathbf{y}^\xi \otimes \mathbf{x}^\xi) - W_{XY}$ contains a zero entry. Similarly, M_{XY} is $\boxed{\wedge}$-perfect for (X, Y) $\Leftrightarrow$ for each $\xi = 1, 2, \ldots, k$, each row of the matrix $M_{XY} - (\mathbf{y}^\xi \otimes \mathbf{x}^\xi)$ contains a zero entry.*

We only prove the first part of the theorem as the proof of the second part is analogous by simply replacing minimums with maximums, or vice versa, or by the use of some of the duality relationships discussed in Chapter 4.

Proof. W_{XY} is $\boxed{\vee}$-perfect for $(X,Y) \Leftrightarrow \forall \xi = 1,2,\ldots,k$ and $\forall i = 1,2,\ldots,m$

$$(W_{XY} \boxed{\vee} \mathbf{x}^\xi)_i = y_i^\xi \Leftrightarrow y_i^\xi - (W_{XY} \boxed{\vee} \mathbf{x}^\xi)_i = 0 \Leftrightarrow y_i^\xi - \bigvee_{j=1}^n (w_{ij} + x_j^\xi) = 0$$
$$\Leftrightarrow y_i^\xi + \bigwedge_{j=1}^n (-w_{ij} - x_j^\xi) = 0 \Leftrightarrow \bigwedge_{j=1}^n (y_i^\xi - x_j^\xi - w_{ij}) = 0$$
$$\Leftrightarrow \bigvee_{j=1}^n ([\mathbf{y}^\xi \otimes \mathbf{x}^\xi] - W_{XY})_{ij} = 0.$$

Since the last equality holds for all $i \in \mathbb{N}_m$ and all $\xi \in \mathbb{N}_k$, each column entry of the ith row of $(\mathbf{y}^\xi \otimes \mathbf{x}^\xi) - W_{XY}$ contains at least one zero entry. □

Example 5.2 The memory W_{XY} of Example 5.1(1) satisfies the condition of Theorem 5.3. For instance,

$$(\mathbf{y}^3 \otimes \mathbf{x}^3) - W_{XY} = \begin{pmatrix} -1 & 2 \\ -2 & 1 \\ 0 & 3 \end{pmatrix} - \begin{pmatrix} -1 & 0 \\ -2 & 1 \\ 0 & 0 \end{pmatrix} = \begin{pmatrix} 0 & 2 \\ 0 & 0 \\ 0 & 3 \end{pmatrix}.$$

On the other hand we have

$$M_{XY} - (\mathbf{y}^2 \otimes \mathbf{x}^2) = \begin{pmatrix} 0 & 2 \\ 1 & 1 \\ 0 & 3 \end{pmatrix} - \begin{pmatrix} -1 & 1 \\ -1 & 1 \\ 0 & 2 \end{pmatrix} = \begin{pmatrix} 1 & 1 \\ 2 & 0 \\ 0 & 1 \end{pmatrix},$$

and since the first row contains no zero we have that $M_{XY} \boxed{\vee} \mathbf{x}^2 \neq \mathbf{y}^2$.

The following is an easy consequence of Theorem 5.3.

Corollary 5.4

1. *$W_{XY} \boxed{\vee} X = Y$ if and only if for each row index $i \in \mathbb{N}_m$ and each $\gamma \in \mathbb{N}_k$, there exists a column index $j \in \mathbb{N}_n$—depending on i and γ—such that*

$$x_j^\gamma - y_i^\gamma = \bigvee_{\xi=1}^k (x_j^\xi - y_i^\xi).$$

2. *$M_{XY} \boxed{\wedge} X = Y$ if and only if for each row index $i \in \mathbb{N}_m$ and each $\gamma \in \mathbb{N}_k$, there exists a column index $j \in \mathbb{N}_n$—depending on i and γ—such that*

$$x_j^\gamma - y_i^\gamma = \bigwedge_{\xi=1}^k (x_j^\xi - y_i^\xi).$$

Proof. Let $i \in \mathbb{N}_m$ and $\gamma \in \mathbb{N}_k$. If $V^\gamma = [\mathbf{y}^\gamma \otimes \mathbf{x}^\gamma] - W_{XY}$, then $v_{ij}^\gamma = (y_i^\gamma - x_j^\gamma) - w_{ij}$. According to Theorem 5.3, $\exists\, j \ni v_{ij}^\gamma = (y_i^\gamma - x_j^\gamma) - w_{ij} = 0$ and, hence, $y_i^\gamma - x_j^\gamma = w_{ij} = \bigwedge_{\xi=1}^k (y_i^\xi - x_j^\xi)$. But this is equivalent to $x_j^\gamma - y_i^\gamma = -w_{ij} = \bigvee_{\xi=1}^k (x_j^\xi - y_i^\xi)$.

The proof of the second part of the corollary is analogous and left as an exercise. □

The conditions for $\boxed{\vee}$-perfect or $\boxed{\wedge}$-perfect recall imposed by Theorem 5.3 or its corollary are quite restrictive and—as Example 5.2 demonstrates—easily violated. As the next theorem corroborates, adding noise to input patterns makes direct recall in most cases impossible.

Theorem 5.5 *Let $\tilde{\mathbf{x}}^\gamma$ denote a distorted version of the pattern $\mathbf{x}^\gamma$. Then*
1. $W_{XY} \boxed{\vee} \tilde{\mathbf{x}}^\gamma = \mathbf{y}^\gamma \Leftrightarrow \forall\, j \in \mathbb{N}_n$ and $i \in \mathbb{N}_m$ the following two conditions are satisfied:

$$\tilde{x}_j^\gamma \le x_j^\gamma \vee \bigwedge_{i=1}^{m} [\bigvee_{\xi \ne \lambda} (y_i^\lambda - y_i^\xi + x_j^\xi)] \textit{ and } \exists\, j_i \in \mathbb{N}_n \ni \tilde{x}_{j_i}^\gamma = x_{j_i}^\gamma \vee [\bigvee_{\xi \ne \lambda} (y_i^\lambda - y_i^\xi + x_{j_i}^\xi)].$$

2. $M_{XY} \boxed{\vee} \tilde{\mathbf{x}}^\gamma = \mathbf{y}^\gamma \Leftrightarrow \forall\, j \in \mathbb{N}_n$ and $i \in \mathbb{N}_m$ the following two conditions are satisfied:

$$\tilde{x}_j^\gamma \ge x_j^\gamma \wedge \bigvee_{i=1}^{m} [\bigwedge_{\xi \ne \lambda} (y_i^\lambda - y_i^\xi + x_j^\xi)] \textit{ and } \exists\, j_i \in \mathbb{N}_n \ni \tilde{x}_{j_i}^\gamma = x_{j_i}^\gamma \wedge [\bigwedge_{\xi \ne \lambda} (y_i^\lambda - y_i^\xi + x_{j_i}^\xi)].$$

The notation j_i in the theorem stresses the fact that this specific index j depends on the row index i. The theorem emphasizes the ineffectiveness of employing the matrices W_{XY} and M_{XY} directly in order to solve pattern recognition problems. However, as we shall discover in the next sections, these matrices do play a vital role in constructing more robust matrix-based memories. We shall not prove Theorem 5.5 as the proof is simple but somewhat lengthy and can be found in [218].

Since generally $\tilde{\mathbf{x}}^\xi \notin X$ and $M_{XY} \boxed{\wedge} \mathbf{x}^\xi \ne \mathbf{y}^\xi$ as well as $W_{XY} \boxed{\vee} \mathbf{x}^\xi \ne \mathbf{y}^\xi$ in most cases, we will view the matrices M_{XY} and W_{XY} as functions mapping $\mathbb{R}^n \to \mathbb{R}^m$ defined by $\mathbf{x} \to M_{XY} \boxed{\wedge} \mathbf{x}$ and $\mathbf{x} \to W_{XY} \boxed{\vee} \mathbf{x}$ $\forall \mathbf{x} \in \mathbb{R}^n$, respectively.

Exercises 5.2.1

1. Complete the proof of Theorem 5.2 that if $B \boxed{\wedge} \mathbf{x}^\xi = \mathbf{y}^\xi \ \forall \xi$, then $M_{XY} \le B$ and $M_{XY} \boxed{\wedge} \mathbf{x}^\xi = \mathbf{y}^\xi \ \forall \xi = 1, \ldots, k$.

2. Complete the proof of Theorem 5.3 by showing that M_{XY} is ⊠-perfect for $(X,Y) \Leftrightarrow$ for each $\xi = 1,2,\ldots,k$, each row of the matrix $M_{XY} - (\mathbf{y}^\xi \otimes \mathbf{x}^\xi)$ contains a zero entry.

3. Prove part 2 of Corollary 5.4.

4. Prove Theorem 5.5.

5.2.2 Lattice Auto-Associative Memories

Among the various auto-associative networks, the Hopfield is the most widely known today [121, 122, 123]. A large number of researchers have exhaustively studied this network, its variations, and generalization [1, 6, 52, 72, 94, 139, 168, 193]. Hardware implementation issues of various associative memories have also been extensively studied [9, 159, 185, 190]. Unlike the various Hopfield type networks, the lattice matrix-based model provides the final result in one pass through the network without any significant amount of training.

The structure, but not the behavior, of the lattice-based auto-associative M_{XX} and W_{XX} are analogous to the Hopfield net. The weights (matrix elements) of the Hopfield net are defined by

$$w_{ij} = \begin{cases} \sum_{\xi=1}^{k}(x_i^\xi \cdot x_j^\xi) & \text{if } i \neq j \\ 0 & \text{if } i = j. \end{cases} \tag{5.19}$$

For a given input pattern $\mathbf{x} = (x_1, \ldots, x_n)'$ the Hopfield net algorithm proceeds by initializing $x_j(0) = x_j$ at time $t = 0$ and then iterates the recursive formula $x_j(t+1) = f(\sum_{i=1}^{n}(w_{ij} \cdot x_i(t)))$ until convergence. Generally, the function f is a hard-limiting nonlinearity.

The lattice-based analogues of the Hopfield memory are based on equation 5.16 with $X = Y$. Specifically, the respective i, jth elements of the matrices M_{XX} and W_{XX} are given by

$$m_{ij} = \bigvee_{\xi=1}^{k}(x_i^\xi - x_j^\xi) \quad \text{and} \quad w_{ij} = \bigwedge_{\xi=1}^{k}(x_i^\xi - x_j^\xi). \tag{5.20}$$

But this equation is identical to equation 4.46, which means that Theorem 4.23 applies so that

$$W_{XX} = I_X \text{ and } M_{XX} = J_X = I_X^* = W_{XX}^*. \tag{5.21}$$

Consequently Theorem 4.25 also applies and thus provides the following trivial corollary:

Corollary 5.6 *W_{XX} is $\boxed{\vee}$-perfect and M_{XX} is $\boxed{\wedge}$-perfect.*

Since the matrix transforms W_{XX} and M_{XX} correspond to the respective identity matrices I_X and J_X when restricted to $S(X) \subset \mathbb{R}^n$, it will be convenient to use the notation,

$$\mathfrak{W}_X = \begin{pmatrix} \mathfrak{w}_{11} & \cdots & \mathfrak{w}_{1n} \\ \vdots & \ddots & \vdots \\ \mathfrak{w}_{n1} & \cdots & \mathfrak{w}_{nn} \end{pmatrix} \quad \text{and} \quad \mathfrak{M}_X = \begin{pmatrix} \mathfrak{m}_{11} & \cdots & \mathfrak{m}_{1n} \\ \vdots & \ddots & \vdots \\ \mathfrak{m}_{n1} & \cdots & \mathfrak{m}_{nn} \end{pmatrix}, \tag{5.22}$$

where the elements $\mathfrak{w}_{ij}$ and $\mathfrak{m}_{ij}$ are as defined in equation 4.46. Since the memory transforms $\mathfrak{W}_X$ and $\mathfrak{M}_X$ correspond to the identity matrices I_X and J_X, it follows that these memories have unlimited storage capacity in the sense that they can store any finite number of patterns and recall them correctly. More precisely, if the dimension of the pattern vectors is n and k denotes the number of distinct exemplar vectors to be stored, then k is allowed to be any positive integer, no matter how large. In fact, it follows from equation 4.46 and Theorem 4.23 that the memories $\mathfrak{W}_X$ and $\mathfrak{M}_X$ can be constructed for any bounded infinite set $X \subset \mathbb{R}^n$. Of course, in case of strictly binary numbers, the limit is $k = 2^n$ as this is the maximum number of distinct binary patterns of length n. In comparison, McEliece et al. showed that for the Hopfield memory, the asymptotic limit capacity of k for the exact recovery of the exemplar pattern with high probability is $n/(2\log n)$ [179]. Additionally, the Hamming distance between a distorted version and its associated exemplar pattern $\mathbf{x}^\xi$ must be less than $n/2$. In brief, Hopfield memories have severe limitations in the number of patterns that can be stored and correctly recalled as well as patterns that share too many bits [171]. Likewise, the information storage capacity (the number of bits which can be stored and recalled associatively) of matrix-based lattice auto-associative also exceeds the respective number of certain linear matrix associative memories as calculated by Palm in [203, 204].

We will be using images in order to *visualize* various problems occurring in pattern recognition. Here we are not interested in image processing or image analysis techniques, but just the recognition of patterns expressed in vector format. For this reason it will be important to remember that patterns as discussed in this text are vectors, where the vector components represent numeric features such as probability values, telephone numbers, and other values mentioned in the introduction of subsection 5.2.1. Thus, we are not interested in recognizing an image after it has been rotated or undergone some other affine image transformation. Image recognition of images having undergone such transformations generally involves the extraction of various

Figure 5.1 The five patterns in the top row were used in constructing the Hopfield net and the two lattice auto-associative memories $\mathfrak{W}_X$ and $\mathfrak{M}_X$. The output for either $\mathfrak{W}_X$ or $\mathfrak{M}_X$ is identical to the input patterns. The bottom row represents the output patterns of the Hopfield net when presented with the respective patterns from the top row [218].

key image features associated with the image's content. Image content and feature extraction are usually goal-dependent and a multitude of image analysis tools are available for these tasks. These features or content descriptors can then be expressed in vector format for further pattern recognition decisions.

Example 5.3

Consider the five images $\mathbf{p}^1, \ldots, \mathbf{p}^5$ shown in the top row of Figure 5.1. Each $\mathbf{p}^\xi$ is an 18×18 Boolean image. Using the standard row-scan method, each image $\mathbf{p}^\xi$ can be converted into a vector format $\mathbf{x}^\xi = (x_1^\xi, \ldots, x_{324}^\xi)'$ by defining

$$x_{18(i-1)+j}^\xi = \begin{cases} 1 \text{ if } \mathbf{p}^\xi(i,j) = 1 & \text{(black pixel)} \\ 0 \text{ if } \mathbf{p}^\xi(i,j) = 0 & \text{(white pixel)} \end{cases} \tag{5.23}$$

Corollary 5.6 guarantees the perfect recall $\mathfrak{W}_X \boxed{\vee} \mathbf{x}^\xi = \mathbf{x}^\xi = \mathfrak{M}_X \boxed{\wedge} \mathbf{x}^\xi$ for $\xi = 1, \ldots, 5$. However, when employing the Hopfield memory for these patterns, we obtain the output displayed in the bottom row of Figure 5.1. Perfect recall is only achieved for the two patterns "A" and "X" while input patterns "B," "C," and "E" all converged to the same configuration not represented by any of the five input patterns. The reason for this is that these three patterns have too many bits in common and Hopfield type memories have difficulties in memorizing patterns that share too many bits.

Some improvements in the Hopfield approach can be obtained by using various techniques such as orthogonalization [263, 303]. Nevertheless, the limitation of the number of patterns that can be stored and successfully recalled remains severe. Another major advantage of the memories W_{XX} and $\mathfrak{M}_X$ is the one step convergence $\mathfrak{W}_X \boxed{\vee} \mathbf{x}^\xi = \mathbf{x}^\xi = \mathfrak{M}_X \boxed{\wedge} \mathbf{x}^\xi \; \forall \; \mathbf{x}^\xi \in X$.

Figure 5.2 The ten patterns used in constructing the Hopfield memory and the two lattice auto-associative memories $\mathfrak{W}_X$ and $\mathfrak{M}_X$. The output for either $\mathfrak{W}_X$ or $\mathfrak{M}_X$ is identical to the input patterns [218].

Figure 5.3 The output of the Hopfield memory. The output remains the same no matter which one of the patterns shown in Figure 5.2 is presented to the memory [218].

Example 5.4 Doubling the numbers of patterns defined in Example 5.3 by adding the lower case letters shown in Figure 5.2 results in complete recall failure when using the Hopfield network. Figure 5.3 shows the output from the Hopfield network when presented with any of the exemplar patterns shown in Figures 5.1 or 5.2. In contrast to the Hopfield memory, in the absence of noise the lattice auto-associative memories $\mathfrak{W}_X$ and $\mathfrak{M}_X$ will always provide perfect recall for noiseless input vectors from X.

In addition to being capable of storing and correctly recalling any finite number of pattern vectors $X \subset \mathbb{R}^n$, the memories $\mathfrak{W}_X$ and $\mathfrak{M}_X$ converge in one step for any input pattern.

Theorem 5.7 *Suppose $X \subset \mathbb{R}^n$ is a finite nonempty set and $\mathbf{x} \in \mathbb{R}^n$. If $\mathfrak{W}_X \boxvee \mathbf{x} = \mathbf{u}$ and $\mathfrak{M}_X \boxwedge \mathbf{x} = \mathbf{v}$, then $\mathfrak{W}_X \boxvee \mathbf{u} = \mathbf{u}$, $\mathfrak{M}_X \boxwedge \mathbf{v} = \mathbf{v}$, and $\mathbf{u} \geq \mathbf{v}$.*

Proof. Let $\mathfrak{W}_X \boxvee \mathbf{x} = \mathbf{u}$. Since $\mathfrak{w}_{ii} = 0 \; \forall\, i \in \mathbb{N}_n$, we obtain

$$(\mathfrak{W}_X \boxvee \mathbf{u})_i = \bigvee_{j=1}^{n} (\mathfrak{w}_{ij} + u_j) \geq \mathfrak{w}_{ii} + u_i = u_i \; \forall\, i \in \mathbb{N}_n.$$

Therefore

$$\mathbf{u} \leq \mathfrak{W}_X \boxvee \mathbf{u}. \tag{5.24}$$

Let $i, j, \ell \in \mathbb{N}_n$. Note that

$$\mathfrak{w}_{i\ell} = \bigwedge_{\xi=1}^{k}(x_i^\xi - x_\ell^\xi) \leq x_i^\gamma - x_\ell^\gamma \quad \text{and} \quad \mathfrak{w}_{\ell j} = \bigwedge_{\xi=1}^{k}(x_\ell^\xi - x_j^\xi) \leq x_\ell^\gamma - x_j^\gamma \text{ for any } \gamma \in \mathbb{N}_k.$$

Thus, $\mathfrak{w}_{i\ell} + \mathfrak{w}_{\ell j} \leq (x_i^\gamma - x_\ell^\gamma) + (x_\ell^\gamma - x_j^\gamma)\ \forall \gamma \in \mathbb{N}_k$ and, therefore,

$$\mathfrak{w}_{i\ell} + \mathfrak{w}_{\ell j} \leq \bigwedge_{\xi=1}^{k}(x_i^\xi - x_j^\xi) = \mathfrak{w}_{ij}. \tag{5.25}$$

According to the inequality 5.25 we now have $u_i = \bigvee_{j=1}^{n}(\mathfrak{w}_{ij} + x_j) \geq \bigvee_{j=1}^{n}(\mathfrak{w}_{i\ell} + \mathfrak{w}_{\ell j} + x_j)\ \forall \ell \in \mathbb{N}_n$. Therefore, if $i \in \mathbb{N}_n$, then

$$u_i \geq \bigvee_{\ell=1}^{n}\bigvee_{j=1}^{n}(\mathfrak{w}_{i\ell} + \mathfrak{w}_{\ell j} + x_j) = \bigvee_{\ell=1}^{n}[\mathfrak{w}_{i\ell} + (\bigvee_{j=1}^{n}(\mathfrak{w}_{\ell j} + x_j))] = \bigvee_{\ell=1}^{n}(\mathfrak{w}_{i\ell} + u_\ell) = (\mathfrak{W}_X \boxdot \mathbf{u})_i.$$

This shows that

$$\mathbf{u} \geq \mathfrak{W}_X \boxdot \mathbf{u}. \tag{5.26}$$

Equations 5.24 and 5.26 imply that $\mathfrak{W}_X \boxdot \mathbf{u} = \mathbf{u}$.

The proof of the claim $\mathfrak{M} \boxtimes \mathbf{v} = \mathbf{v}$ is similar and left as an exercise. To prove that $\mathbf{u} \geq \mathbf{v}$, let $i \in \mathbb{N}_n$ be arbitrarily chosen. Then

$$\begin{aligned}\mathbf{u}_i = (\mathfrak{W}_X \boxdot \mathbf{x})_i &= \bigvee_{j=1}^{n}(\mathfrak{w}_{ij} + x_j) \geq \mathfrak{w}_{ii} + x_i = x_i = \mathfrak{m}_{ii} + x_i \geq \bigwedge_{j=1}^{n}(\mathfrak{m}_{ij} + x_j) \\ &= (\mathfrak{M} \boxtimes \mathbf{x})_i = \mathbf{v}_i.\end{aligned}$$

Therefore $\mathbf{u} \geq \mathbf{v}$. □

The following theorem is a direct consequence of the proof of Theorem 5.7:

Theorem 5.8 *If X is a bounded, nonempty subset of $\mathbb{R}^n$ and $i, j, \ell \in \mathbb{N}_n$, then*

$$\mathfrak{w}_{i\ell} + \mathfrak{w}_{\ell j} \leq \mathfrak{w}_{ij} \quad \text{and} \quad \mathfrak{m}_{ij} \leq \mathfrak{m}_{i\ell} + \mathfrak{m}_{\ell j}.$$

The first inequality corresponds to equation 5.25 and the second inequality is left as an easy exercise.

Exercises 5.2.2

1. Prove the claim that $\mathfrak{M}_X \boxtimes \mathbf{v} = \mathbf{v}$ in the conclusion of Theorem 5.7.

2. Create a data set $X \subset \mathbb{R}^{10}$, with $|X| = 5$ and compute the auto-associative memories $\mathfrak{W}_X$ and $\mathfrak{M}_X$. Repeat this exercise for $|X| = 10$.

3. For readers familiar with image processing, use or create 10 Boolean images and–using equation 5.23–convert them into 10 vectors. Compute the auto-associative memories $\mathfrak{W}_X$ and $\mathfrak{M}_X$, where X denotes the set of the 10 vectors.

5.2.3 Pattern Recall in the Presence of Noise

Since the Hopfield net is incapable of recalling a large number of perfect exemplar patterns, the net's performance will be further degraded when these patterns are corrupted by noise or suffer from missing data. The memories $\mathfrak{W}_X$ and $\mathfrak{M}_X$ also exhibit poor performance when encountering noisy versions of exemplar patterns. However, they are extremely robust in the presence of certain types of noise and occlusions.

Definition 5.2 Suppose $X = \{\mathbf{x}^\xi : \xi \in \mathbb{N}_k\} \subset \mathbb{R}^n$ is a set of exemplar patterns and $\tilde{\mathbf{x}}^\lambda \in \mathbb{R}^n$ denotes a distorted version of the pattern $\mathbf{x}^\lambda \in X$. We say that the pattern $\tilde{\mathbf{x}}^\lambda$ is an *eroded version* of $\mathbf{x}^\lambda$ whenever $\tilde{\mathbf{x}}^\lambda \leq \mathbf{x}^\lambda$ and the pattern $\tilde{\mathbf{x}}^\lambda$ is a *dilated version* of $\mathbf{x}^\lambda$ whenever $\tilde{\mathbf{x}}^\lambda \geq \mathbf{x}^\lambda$.

Considering the five Boolean exemplar patterns shown in the top row of Figure 5.1, then a change in pattern values from $\mathbf{p}(i,j) = 1$ to $\mathbf{p}(i,j) = 0$ represents an erosive change, while a change from $\mathbf{p}(i,j) = 0$ to $\mathbf{p}(i,j) = 1$ represents a dilative change. The memory $\mathfrak{W}_X$ is extremely robust in recalling patterns that are distorted due to erosive changes. These changes can be random such as systems noise, partial pattern occlusion, etc., or nonrandom due to processing effects such as image filtering, image skeletonisation, and morphological operations involving erosions. Similarly, the memory $\mathfrak{M}_X$ is very robust in the presence of dilative noise. The reason for the robustness of $\mathfrak{W}_X$ in the presence of erosive noise and $\mathfrak{M}_X$ in the presence of dilative noise is a consequence of Theorem 5.5(*1*) and 5.5(*2*), respectively. The theorem provides strict limits on the type of allowable erosions and dilations for guaranteed perfect recall. These limits become very obvious when viewed in 2-dimensional space.

Example 5.5 Suppose $X = \{\mathbf{x}^1, \mathbf{x}^2, \mathbf{x}^3\} \subset \mathbb{R}^2$ is the set shown in Figure 5.4 Then any point $\mathbf{z} \in \mathbb{R}^2$ on the solid horizontal half-line $\{\mathbf{z} \in \mathbb{R}^2 : z_1 \leq x_1^2, z_2 =$

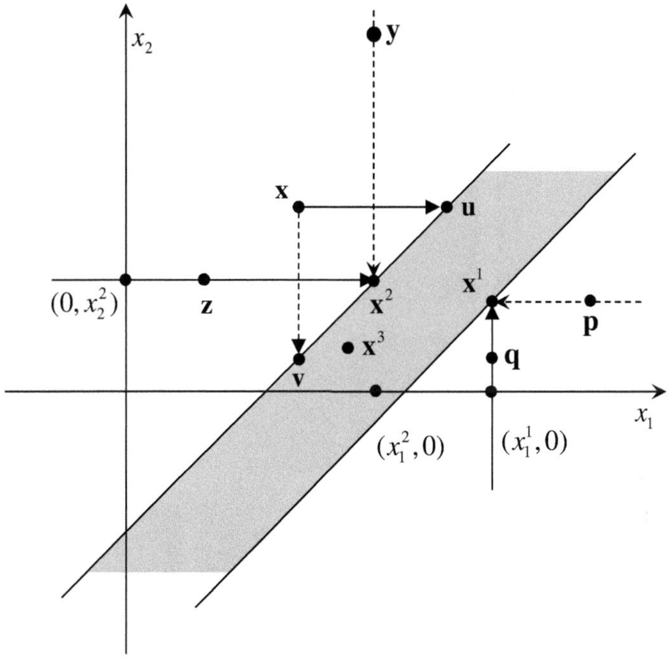

Figure 5.4 The orbits of various planar points when mapped by the matrices $\mathfrak{W}_X$ and $\mathfrak{M}_X$. Here any point on a solid half line will be mapped by $\mathfrak{W}_X$ to the half line's endpoint, as indicated by an arrowhead. Similarly, any point on a dashed half line maps to the endpoint of the half line under the action of $\mathfrak{M}_X$. The shaded area indicates part of the common fixed point set of the two matrices [218].

$x_2^2\}$ satisfies the conditions set by Theorem 5.5. (*1*) so that $\mathfrak{W}_X \boxtimes \mathbf{z} = \mathbf{x}^2$. Similarly, any point $\mathbf{y} \in \mathbb{R}^2$ on the vertical dashed half-line $\{\mathbf{y} \in \mathbb{R}^2 : y_1 = x_1^2, y_2 \geq x_2^2\}$ satisfies the conditions set by Theorem 5.5(2) and resulting in $\mathfrak{M}_X \boxtimes \mathbf{y} = \mathbf{x}^2$. The point $\mathbf{x}$ in the figure is spatially closer to the point $\mathbf{x}^2$ than either of the indicated points $\mathbf{y}$ or $\mathbf{z}$. However, $\mathbf{x}$ does not satisfy any of the conditions set by Theorem 5.5, and results in the inequality $\mathfrak{M}_X \boxtimes \mathbf{x} = \mathbf{v} < \mathbf{x}^2 < \mathbf{u} = \mathfrak{W}_X \boxtimes \mathbf{x}$. Since the pattern point $\mathbf{x}^3 \in F(X)$, no distorted input version $\tilde{\mathbf{x}}^3$ of $\mathbf{x}^3$ will be associated with $\mathbf{x}^3$ by either memory $\mathfrak{W}_X$ or $\mathfrak{M}_X$.

According to Theorem 5.7 the inequality $\mathfrak{M}_X \boxtimes \mathbf{x} \leq \mathfrak{W}_X \boxtimes \mathbf{x}$ holds for any $\mathbf{x} \in \mathbb{R}^n$. This inequality provides for a particularly attractive visual interpretation when $\mathbf{x}$ represents an image since a higher pixel values in an image correspond to brighter displays while lower values correspond to darker dis-

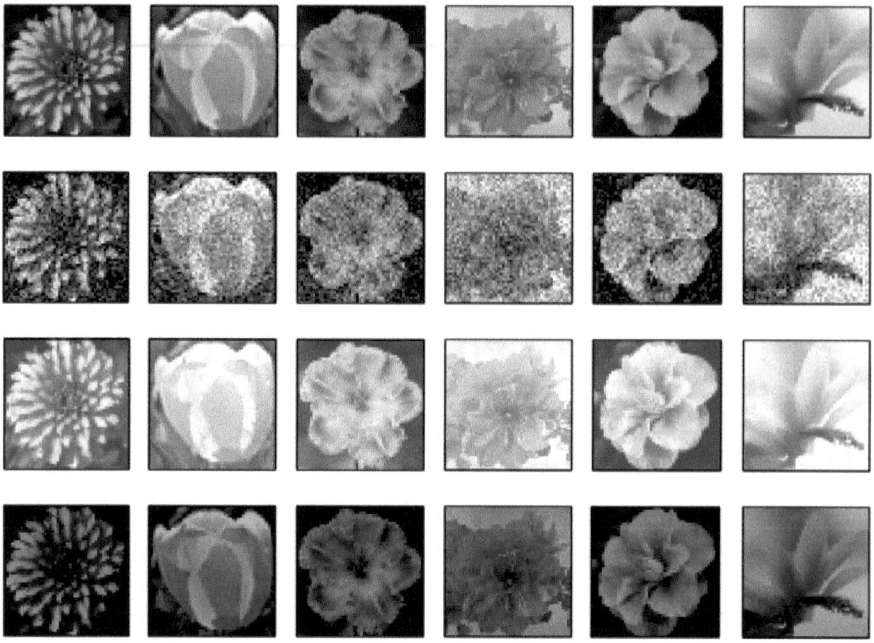

Figure 5.5 Top row: The patterns used in constructing the two lattice auto-associative memories $\mathfrak{W}_X$ and $\mathfrak{M}_X$. Second row: The corrupted input patterns. Third row: The output response of the memory $\mathfrak{W}_X$ when presented with the corrupted input above it. Fourth row: The output of $\mathfrak{M}_X$ when presented with the corrupted input [218].

plays. Figure 5.5 illustrates this phenomenon. In this figure the top row shows the exemplar patterns used in the construction of the memories $\mathfrak{W}_X$ and $\mathfrak{M}_X$, while the second row shows the input images to the two memories. These input image patterns were obtained by corrupting the exemplar patterns with 30% of randomly generated erosive and dilative noise. The third and fourth row show the respective outputs of the memories $\mathfrak{W}_X$ and $\mathfrak{M}_X$ for each corrupted input. The order of input-output patterns is as shown in the vertical columns.

It is noteworthy to observe that in the above example the memories $\mathfrak{W}_X$ and $\mathfrak{M}_X$ reassemble the basic structure of the image content. However, they are not capable in returning the original vector component values and may fail catastrophically when confronted with patterns corrupted by random noise. Nonetheless, their performance in the presence of only erosive or dilative noise can be very impressive.

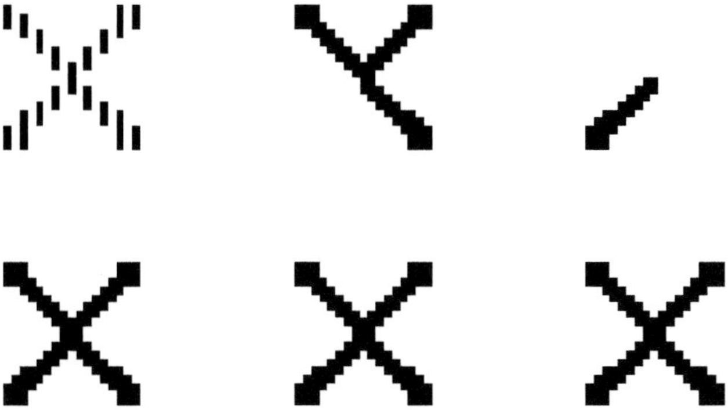

Figure 5.6 Top row: The three patterns corrupted by erosion used as input to the memory $\mathfrak{W}_X$. Bottom row: The output of $\mathfrak{W}_X$ for the corresponding input above it [218].

Example 5.6 Let X be the set of Boolean patterns shown in the top row of Figure 5.2. The pattern representing the letter **X** was artificially eroded in different ways as shown in the top row of Figure 5.6. Using these eroded patterns as input to the memory $\mathfrak{W}_X$ resulted in the perfect pattern outputs shown in the bottom row. Adding dilative noise to the pattern **X** by replacing background (white) pixels with black pixels as shown in the top row of Figure 5.7 and using these dilated patterns as input to the memory $\mathfrak{M}_X$ resulted in the perfect output shown in the bottom row of Figure 5.7. This is in stark contrast with the Hopfield net which fails to recall the exemplar pattern when any of the eroded or dilated versions of the **X** pattern is used as input.

Of course, complete failure also occurs when the corrupted versions of **X** shown in the top row of Figure 5.6 are used as input to the memory $\mathfrak{M}_X$ and the same happens when the memory $\mathfrak{W}_X$ is faced with inputs from the top row of Figure 5.7. The reason for this failure is a direct consequence of Theorem 5.5. In the next section we discuss approaches of constructing matrix-based memories that are more robust in the presence of random noise.

Exercises 5.2.3

1. Construct a set X containing ten Boolean vectors $\mathbf{x}^1, \mathbf{x}^2, \ldots, \mathbf{x}^{10}$ with $\mathbf{x}^i \in \mathbb{R}^{10}$ for $i \in \mathbb{N}_{10}$, and compute the auto-associative memories $\mathfrak{W}_X$ and $\mathfrak{M}_X$. For each vector $\mathbf{x}^i \in X$ randomly choose two coordinates x^i_j and x^i_ℓ having value 1, and replace the value with zero. Let Y denote the set of vectors obtained this way and compute the values $\mathfrak{W}_X \boxed{\vee} \mathbf{y}^\xi$ and $\mathfrak{M}_X \boxed{\wedge} \mathbf{y}^\xi$ for $\xi = 1, \ldots, 10$. Discuss the reasons for the output obtained.

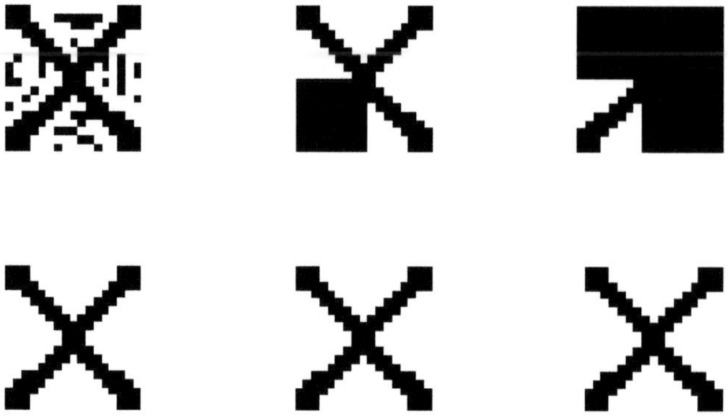

Figure 5.7 Top row: The three patterns corrupted by dilation used as input to the memory $\mathfrak{M}_X$. Bottom row: The output of $\mathfrak{M}_X$ for the corresponding input above it.

2. Let X, $\mathfrak{W}_X$, and $\mathfrak{M}_X$ denote the objects constructed in Exercise 1. Repeat Exercise 1 by randomly choosing two coordinates x^i_j and x^i_ℓ having value 0 and replacing them with the value with 1.

3. (For readers familiar with image processing) Let X, $\mathfrak{W}_X$, and $\mathfrak{M}_X$ denote the object constructed in Exercise **5.2.2**(3). Corrupt the elements of X with 30% of dilation noise and check the performance of $\mathfrak{W}_X$ and $\mathfrak{M}_X$.

4. Let X be as in Exercise 3, but corrupt the elements of X with 30% of erosive noise. Check the performance of $\mathfrak{W}_X$ and $\mathfrak{M}_X$.

5.2.4 Kernels and Random Noise

Since $\mathfrak{W}_X$ is well-suited for recognizing patterns corrupted by erosive noise and $\mathfrak{M}_X$ is equally well suited for recognizing patterns corrupted by dilative noise, it may seem suitable to process a noisy version $\tilde{\mathbf{x}}^\gamma$ of $\mathbf{x}^\gamma$ by a combination of $\mathfrak{W}_X$ and $\mathfrak{M}_X$. However, it is clear from Example 5.5—as well as Theorem 5.7—that such an approach will fail. In fact, P. Sussner proved that passing the output of $\mathfrak{M}_X \boxtimes \tilde{\mathbf{x}}^\gamma$ through the memory $\mathfrak{W}_X$ or, dually, $\mathfrak{W}_X \boxtimes \tilde{\mathbf{x}}^\gamma$ through the memory $\mathfrak{M}_X$, will generally not result in $\mathbf{x}^\gamma$ [272]. Nevertheless, a successful approach based on the intuitive idea of using the memories $\mathfrak{W}_X$ and $\mathfrak{M}_X$ in sequence was presented in [218, 223]. This method became known as the *kernel* method. The vocable "kernels"—as used in this treatise— refers to the *seed* or *core* components of a pattern vector $\mathbf{x}^\gamma$ that allow for the complete reconstruction of $\mathbf{x}^\gamma$. Unfortunately, the word "kernel" plays also a major role in machine learning where kernel methods refer to a class of algorithms for pattern recognition. These algorithms employ special types of similarity functions called kernel functions. Thus, the kernel methods used in machine learning are completely different from the kernel methods discussed in this chapter. Both of these distinct kernel methods also appear in the published literature where the content under discussion eliminates any possible confusion between the two concepts.

Suppose $X = \{\mathbf{x}^1, \ldots, \mathbf{x}^k\} \subset \mathbb{R}^n$ and $Y = \{\mathbf{y}^1, \ldots, \mathbf{y}^k\} \subset \mathbb{R}^m$ are two sets of pattern vectors and the goal is to construct a lattice-based matrix memory A such that $\mathbf{x}^\xi \to A \to \mathbf{y}^\xi \ \forall \xi \in \mathbb{N}_k$ and A is robust in the presence of random noise. The underlying idea in the kernel method is to define two associative memories M and W, where M associates with each input pattern $\mathbf{x}^\gamma$ an intermediate pattern $\mathbf{z}^\gamma$ and W associates each pattern $\mathbf{z}^\gamma$ the desired output $\mathbf{y}^\gamma$. In terms of the minimax matrix product, the desired equation is of form $W \boxvee (M \boxwedge \mathbf{x}^\gamma) = \mathbf{y}^\gamma$. If the $n \times k$ matrix $Z = (\mathbf{z}^1, \ldots, \mathbf{z}^k)$ of intermediate patterns satisfies certain conditions, then the matrices $\mathfrak{M}_Z$ and W_{ZY} can serve as M and W, respectively. Furthermore, if Z is properly chosen, then $W_{ZY} \boxvee (\mathfrak{M}_Z \boxwedge \tilde{\mathbf{x}}^\gamma) = \mathbf{y}^\gamma$ for most corrupted versions $\tilde{\mathbf{x}}^\gamma$ of $\mathbf{x}^\gamma$. If Z satisfies these basic properties, then Z is called a *kernel* for the associative pair (X, Y). The following formal definition of a kernel was proposed in [223]:

Definition 5.3 (*Restricted version*) Suppose $X = \{\mathbf{x}^1, \ldots, \mathbf{x}^k\} \subset \mathbb{R}^n$, $Y = \{\mathbf{y}^1, \ldots, \mathbf{y}^k\} \subset \mathbb{R}^m$, and $Z = \{\mathbf{z}^1, \ldots, \mathbf{z}^k\}$ a subset of $\mathbb{R}^n$. The set Z is said to be a *kernel* for (X, Y) if and only if the following three conditions are satisfied:

1. $Z \neq X$
2. $\mathfrak{M}_Z \boxwedge \mathbf{x}^\xi = \mathbf{z}^\xi \quad \forall \mathbf{x}^\xi \in X$
3. $W_{ZY} \boxvee \mathbf{z}^\xi = \mathbf{y}^\xi \quad \forall \mathbf{z}^\xi \in Z$.

The elements of Z are called *kernel vectors*. If $Y = X$, then Z is said to be a *kernel* for X.

It follows that if Z is a kernel for (X, Y), then

$$W_{ZY} \boxvee (\mathfrak{M}_Z \boxwedge \mathbf{x}^\gamma) = W_{ZY} \boxvee \mathbf{z}^\gamma = \mathbf{y}^\gamma \quad \forall \gamma \in \mathbb{N}_k \tag{5.27}$$

and W_{ZY} is $\boxvee$-perfect for the pattern association (Z, Y). Furthermore, a consequence of the second condition is that for every $\mathbf{z}^\xi \in Z$, $\mathbf{z}^\xi \leq \mathbf{x}^\xi$ since

$$z_i^\xi = [\mathfrak{M}_Z \boxwedge \mathbf{x}^\xi]_i = \bigwedge_{j=1}^{n} (\mathfrak{m}_{ij} + x_j^\xi) \leq \mathfrak{m}_{ii} + x_i^\xi = x_i^\xi \quad \forall i \in \mathbb{N}_n. \tag{5.28}$$

Therefore every $\mathbf{z}^\xi \in Z$ represents an eroded version of $\mathbf{x}^\xi$

For kernels to be effective in recognizing patterns that are severely corrupted by random noise, they need not only represent eroded subsets of X but should also be extremely sparse; i.e., for each γ, the corresponding kernel vector $\mathbf{z}^\gamma$ should consist mostly of zero entries when dealing with non-negative pattern vectors. The reason for the extreme sparseness is based on the following observation. If Z is sparse, then the corrupted version $\tilde{\mathbf{x}}^\gamma$ of $\mathbf{x}^\gamma$ will generally be able to afford a high degree of erosive noise and still satisfy the inequality $\mathbf{z}^\gamma \leq \tilde{\mathbf{x}}^\gamma$. Since $\mathfrak{M}_Z$ is robust in the presence of dilative noise, $\tilde{\mathbf{x}}^\gamma$ will be conceived as a dilated version of $\mathbf{z}^\gamma$ by the memory $\mathfrak{M}_Z$. On the other hand, if $\mathbf{z}^\lambda$ is not sparse and $\tilde{\mathbf{x}}^\gamma$ contains large amounts of erosive noise, then it is far more likely that $\mathbf{z}^\gamma \nleq \tilde{\mathbf{x}}^\gamma$ and $\mathfrak{M}_Z$ will have difficulty recognizing $\tilde{\mathbf{x}}^\gamma$. Ideally we would like that for each γ, the equality $z_j^\gamma = x_j^\gamma$ holds for exactly one $j \in \{1, \ldots, n\}$ and $z_i^\gamma = 0 \; \forall i \neq j$. If Z results in a kernel under these conditions, then we are guaranteed the recovery of $\mathbf{x}^\gamma$ from $\tilde{\mathbf{x}}^\gamma$ as long as $\mathbf{z}^\gamma \leq \mathfrak{M}_Z \boxtimes \tilde{\mathbf{x}}^\gamma \leq \mathbf{x}^\gamma$. These loose concepts lead to the definition of minimal representations of a pattern set X.

Definition 5.4 A set of patterns $Z \leq X$ is said to be a *minimal representation* of X if and only if for every $\gamma \in \mathbb{N}_k$ the following three conditions are satisfied:

1. $\mathbf{z}^\gamma \wedge \mathbf{z}^\xi = \mathbf{0} \; \forall \xi \neq \gamma$.

2. $\mathbf{z}^\gamma$ contains one and only one non-zero coordinate.

3. $W_{ZX} \boxtimes \mathbf{z}^\gamma = \mathbf{x}^\gamma$.

Such minimal representations of X are, in general, easily obtainable if the pattern vectors are Boolean and $|X| = k$ is much smaller than the dimension n. If these properties hold, then Z will be a kernel for X and the auto-associative memory

$$input \rightarrow \mathfrak{M}_Z \rightarrow W_{ZX} \rightarrow output \tag{5.29}$$

will be fairly robust in the presence of random Boolean noise and Z.

Example 5.7 Consider the set X and Z with elements

$$\mathbf{x}^1 = \begin{pmatrix} 1 \\ 1 \\ 0 \\ 0 \end{pmatrix}, \mathbf{x}^2 = \begin{pmatrix} 1 \\ 0 \\ 0 \\ 1 \end{pmatrix}, \mathbf{x}^3 = \begin{pmatrix} 1 \\ 0 \\ 1 \\ 0 \end{pmatrix}, \text{ and } \mathbf{z}^1 = \begin{pmatrix} 0 \\ 1 \\ 0 \\ 0 \end{pmatrix}, \mathbf{z}^2 = \begin{pmatrix} 0 \\ 0 \\ 0 \\ 1 \end{pmatrix}, \mathbf{z}^3 = \begin{pmatrix} 0 \\ 0 \\ 1 \\ 0 \end{pmatrix},$$

respectively. Then Z is a minimal representation of X. However, dilating any pattern $\mathbf{x}^\xi$ by replacing a single zero coordinate with the number 1 has the effect that $\exists \mathbf{z}^\lambda \in Z \setminus \{\mathbf{z}^\xi\} \ni \mathbf{z}^\lambda \leq \mathbf{x}^\lambda \leq \tilde{\mathbf{x}}^\xi$, thus making it impossible to recover $\mathbf{x}^\xi$.

The following connection between kernels and minimal representations for Boolean patterns was established by P. Sussner [272]:

Theorem 5.9 *Let X, Y and Z be sets of binary patterns with $Z \leq X$. If*

$$\forall \xi \neq \gamma,\ \mathbf{z}^\gamma \wedge \mathbf{z}^\xi = \mathbf{0} \text{ and } \mathbf{z}^\gamma \not\leq \mathbf{z}^\xi, \tag{5.30}$$

then Z is a kernel for (X, Y).

Thus, in order to construct a kernel Z for a Boolean pair (X, Y) of vector association one simply has to satisfy the two conditions of equation 5.30.

Example 5.8 The robustness in the presence of random noise of the memory $\{input \to \mathfrak{M}_Z \to W_{ZX} \to output\}$ can be best illustrated by converting Boolean images into binary pattern vectors. The five Boolean images used in this example are shown in the top row Figure 5.8. Each image is of size 32×32 pixels. The border framing each image serves as a viewing delimiter but is not part of the image. Using the row scanning approach (equation 5.23), each image was converted into a 1024-dimensional binary pattern vector $\mathbf{x}^\xi$, with $\mathbf{x}^1$ corresponding to the letter "A", $\mathbf{x}^2$ to the letter "E", and so on. The corresponding kernel vectors $\mathbf{z}^1, \ldots, \mathbf{z}^5$ were constructed by selecting for each $\xi = 1, \ldots, 5$ a black (value 1) pixel x_i^ξ of $\mathbf{x}^\xi$ such that $x_i^\xi \neq x_i^\lambda$ $\forall \lambda \in \{1, \ldots, 5\} \setminus \{\xi\}$ and setting

$$z_j^\xi = \begin{cases} 1 \text{ if } j = i \\ 0 \text{ if } j \neq i \end{cases} \tag{5.31}$$

Reverting these kernel vectors into Boolean images yields the images shown in the bottom row of Figure 5.8, and a visual inspection of these images shows that equation 5.30 in the hypothesis of Theorem 5.9 is satisfied.

Corrupting the exemplar patterns with 10% of random noise results in the patterns $\tilde{\mathbf{x}}^1, \ldots, \tilde{\mathbf{x}}^5$ corresponding to the images shown in the top row of Figure 5.9. Using these corrupted patterns as input to the memory $\{input \to \mathfrak{M}_Z \to W_{ZX} \to output\}$ results in complete recovery of the exemplar patterns as shown in the bottom row of Figure 5.9.

It is clear from Figure 5.8, that for any $\xi \in \{1, \ldots, 5\}$, $\mathbf{z}^\xi \not\leq \bigvee_{\lambda \neq \xi} \mathbf{z}^\lambda$. This

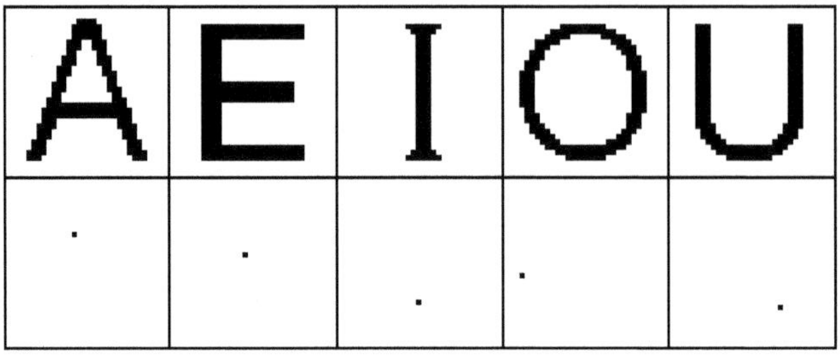

Figure 5.8 Top row: The set of exemplar patterns $X = \{\mathbf{x}^1, \ldots, \mathbf{x}^5\}$ used for selecting the elements of the kernel Z for X. Bottom row: The five corresponding kernel vectors selected for constructing the memories $\mathfrak{M}_Z$ and W_{ZX}.

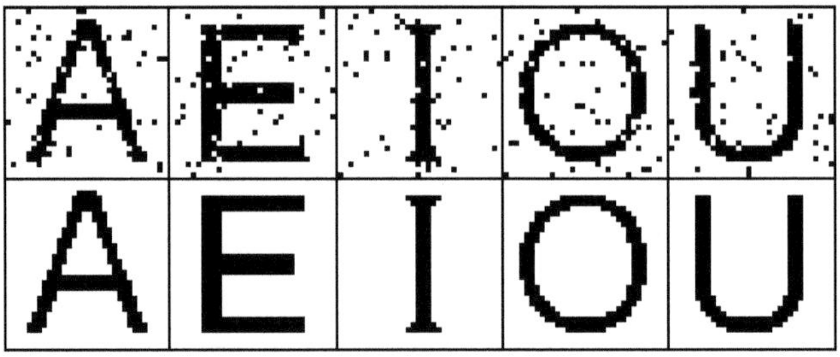

Figure 5.9 Top row: The five exemplar patterns corrupted by x% of random noise used as input to the memory $\mathfrak{M}_Z$. Bottom row: The output vector $W_{ZX} \boxed{\vee} [\mathfrak{M}_Z \boxed{\wedge} \tilde{\mathbf{x}}^{\xi}]$ converted to images for the corresponding input above it.

condition, known as *morphological independence*, is key in selecting a set Z of kernel vectors from X for a Boolean pattern association (X, Y). Although selecting kernel vectors in the Boolean case is easy, the selected kernels for (X, Y) are generally not unique.

Various attempts at generalizing Theorem 5.9 to the non-Boolean case have been unsuccessful. This failure is due to the fact that in the Boolean case the condition specified by equation 5.30 in the hypothesis of the theorem is implied by the notion of morphological independence and results in a kernel. As it turns out, the same is not true in the non-Boolean case. In order to illuminate this problem we begin by formally defining the notion of morphological independence.

Definition 5.5 Suppose $I \subset \mathbb{N}$. A set of pattern vectors $X = \{\mathbf{x}^\xi \in \mathbb{R}^n : \xi \in I\}$ is said to be *morphologically independent* if and only if for every $\lambda \in I$

$$\mathbf{x}^\lambda \nleq \bigvee_{\xi \in I \setminus \{\lambda\}} \mathbf{x}^\xi. \tag{5.32}$$

A consequence of the definition is that if X is morphologically independent, then it follows from equation 5.32 that there exists an index $j_\lambda \in \{1, \ldots, n\}$ such that $x^\xi_{j_\lambda} < x^\lambda_{j_\lambda}$ $\forall \xi \neq \lambda$. Hence $\mathbf{x}^\gamma \nleq \mathbf{x}^\xi$ $\forall \xi \neq \lambda$, which satisfies part of the hypothesis of Theorem 5.9. The converse, however, does not hold true.

Example 5.9 Let $X = \{\mathbf{x}^1, \mathbf{x}^2, \mathbf{x}^3\}$, where

$$\mathbf{x}^1 = \begin{pmatrix} 4 \\ 2 \\ 5 \\ 5 \end{pmatrix}, \mathbf{x}^2 = \begin{pmatrix} 2 \\ 6 \\ 8 \\ 4 \end{pmatrix}, \text{ and } \mathbf{x}^3 = \begin{pmatrix} 2 \\ 5 \\ 1 \\ 5 \end{pmatrix}.$$

$$\text{Then } \mathbf{x}^\lambda \nleq \mathbf{x}^\xi \ \forall \xi \neq \lambda, \text{ but } \mathbf{x}^3 \leq \bigvee_{\xi=1}^{2} \mathbf{x}^\xi = \begin{pmatrix} 4 \\ 6 \\ 8 \\ 5 \end{pmatrix}.$$

If the set X of pattern vectors is morphologically independent and the vector components are non-negative, then it is easy to define a set of pattern vectors $Z \leq X$ that satisfies equation 5.30 of Theorem 5.9. For each $\lambda \in \{1, \ldots, k\}$, we simply pick the index $j_\lambda \in \{1, \ldots, n\}$ for which $x^\xi_{j_\lambda} < x^\lambda_{j_\lambda}$ $\forall \xi \neq \lambda$ and define $\mathbf{z}^\lambda$ by setting

$$z^\lambda_i = \begin{cases} x^\lambda_{j_\lambda} & \text{if } i = j_\lambda \\ 0 & \text{if } i \neq j_\lambda \end{cases} \tag{5.33}$$

for $i = 1, \ldots, n$. It follows that equation 5.30 is satisfied. However, in this construction of Z we did not assume that X is Boolean. Hence it makes sense to ask whether or not Z is also a kernel in the non-Boolean case. The next example provides the answer this question.

Example 5.10 Slightly altering the set X defined in Example 5.9 by setting

$$\mathbf{x}^1 = \begin{pmatrix} 4 \\ 2 \\ 5 \\ 6 \end{pmatrix}, \mathbf{x}^2 = \begin{pmatrix} 2 \\ 3 \\ 8 \\ 4 \end{pmatrix}, \text{ and } \mathbf{x}^3 = \begin{pmatrix} 2 \\ 5 \\ 1 \\ 4 \end{pmatrix},$$

turns X into a morphologically independent set of vectors. Employing equation 5.33 results in the set $Z \leq X$ consisting of elements

$$\mathbf{z}^1 = \begin{pmatrix} 0 \\ 0 \\ 0 \\ 6 \end{pmatrix}, \mathbf{z}^2 = \begin{pmatrix} 0 \\ 0 \\ 8 \\ 0 \end{pmatrix}, \text{ and } \mathbf{z}^3 = \begin{pmatrix} 0 \\ 5 \\ 0 \\ 0 \end{pmatrix},$$

which satisfy equation 5.30. However, Z is not a kernel since

$$\mathfrak{M}_Z \boxed{\wedge} \mathbf{x}^1 = \begin{pmatrix} 0 & 0 & 0 & 0 \\ 5 & 0 & 5 & 5 \\ 8 & 8 & 0 & 8 \\ 6 & 6 & 6 & 0 \end{pmatrix} \boxed{\wedge} \begin{pmatrix} 4 \\ 2 \\ 5 \\ 6 \end{pmatrix} = \begin{pmatrix} 2 \\ 2 \\ 5 \\ 6 \end{pmatrix} \neq \mathbf{z}^1.$$

$$\text{Nevertheless, } \mathfrak{W}_X \boxed{\vee} (\mathfrak{M}_Z \boxed{\wedge} \mathbf{x}^1) = \begin{pmatrix} 0 & -3 & -6 & -2 \\ -2 & 0 & -5 & -4 \\ -1 & -4 & 0 & -3 \\ 2 & -1 & -4 & 0 \end{pmatrix} \boxed{\vee} \begin{pmatrix} 2 \\ 2 \\ 5 \\ 6 \end{pmatrix} = \mathbf{x}^1$$

and, similarly, $\mathfrak{W}_X \boxed{\vee} (\mathfrak{M}_Z \boxed{\wedge} \mathbf{x}^\xi) = \mathbf{x}^\xi$ for $\xi = 2$ and 3.

Observe that the vector $\mathfrak{M}_Z \boxed{\vee} \mathbf{x}^1$ in Example 5.10 is an eroded version of $\mathbf{x}^1$ and a greatly dilated version of $\mathbf{z}^1$. Furthermore, the auto-associative memory

$$input \to \mathfrak{M}_Z \to \mathfrak{W}_X \to output \tag{5.34}$$

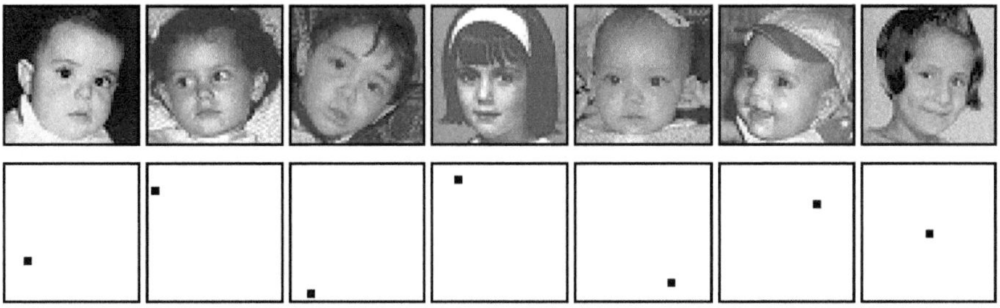

Figure 5.10 Top row: The seven morphologically independent exemplar patterns. Bottom row: The corresponding non-zero entries–enlarged for better visibility–of the seven elements of Z [223].

is a perfect recall memory for the set X. In view of this example it is natural to ask if the auto-associative memory defined by equation 5.34 is a perfect recall memory whenever X is a morphologically independent set of vectors containing no negative components and $Z \leq X$ is obtained via equation 5.33. In order to answer this question, consider the seven pattern images $\mathbf{p}^1, \ldots, \mathbf{p}^7$ displayed in the top row of Figure 5.10. Each $\mathbf{p}^\xi$ is a 50×50 256-grayscale image. Using the standard row-scan method each pattern image $\mathbf{p}^\xi$ can be converted into a pattern vector $\mathbf{x}^\xi = (x_1^\xi, \ldots, x_{2500}^\xi)$ by defining

$$x_{50(r-1)+c}^\xi = p^\xi(r,c) \text{ for } r, c = 1, \ldots, 50. \tag{5.35}$$

Using a computational tool such as Matlab, it can be easily verified that this set of seven vectors is morphologically independent.

The tabular construction displayed in Table 5.1 gives the row indices j_λ for $\lambda = 1, \ldots, 7$, satisfying the strict inequality $x_{j_\lambda}^\lambda > x_{j_\lambda}^\xi \ \forall \xi \neq \lambda$. Using these values of j_λ, the set Z can then be generated by employing equation 5.33. The bottom row of Figure 5.10 shows the pixel position corresponding to each pattern row index $j = j_\lambda$. For better rendering, each of these non-zero pixels was enlarged to a 3×3 block with the same value; also, a reference frame was drawn around each of the images.

However, Z is not a kernel since it is easily shown (using simple computational tools such as Matlab) that $\mathfrak{M}_Z \boxed{\triangle} X \neq Z$. Figure 5.11 also demonstrates that $\mathfrak{W}_X \boxed{\triangledown} (\mathfrak{M}_Z \boxed{\triangle} X) \neq X$. Partial reconstruction and the effect of crosstalk noise between patterns for perfect input is shown in the bottom row of the figure. Only pattern $\mathbf{x}^4$ is perfectly recalled. As an aside, direct computation of $W_{ZX} \boxed{\triangledown} Z$ also fails to recall X perfectly; visually the output is similar to the bottom row of Figure 5.11.

The concepts of *max dominance* and *min dominance* play a key role in

TABLE 5.1 Row index j, pixel position (r,c), and value (underscored) x_j^γ of each pattern γ used to build matrix Z [223].

γ	j	(r,c)	1	2	3	4	5	6	7
1	1759	(36,9)	<u>255</u>	141	165	94	142	159	196
2	453	(10,3)	25	<u>255</u>	71	164	136	101	184
3	2358	(48,8)	233	205	<u>237</u>	163	192	116	107
4	260	(6,10)	20	56	62	<u>255</u>	134	105	44
5	2186	(44,36)	175	112	102	89	<u>255</u>	173	200
6	737	(15,37)	195	70	96	116	136	<u>255</u>	164
7	1276	(26,26)	208	46	199	153	159	176	<u>251</u>

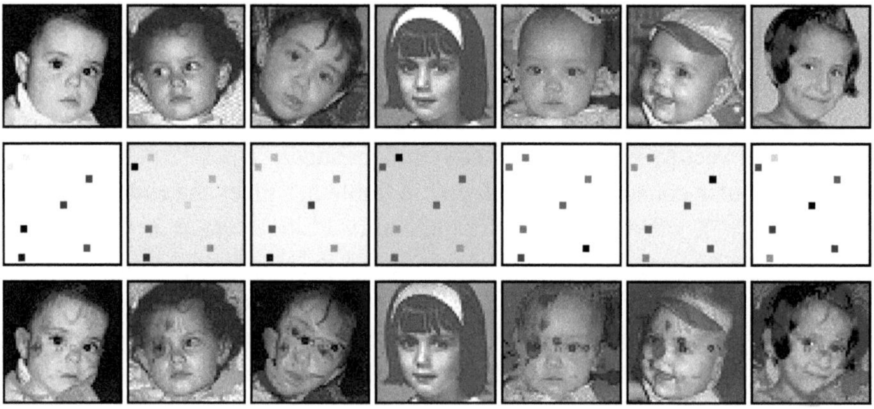

Figure 5.11 Top row: The input patterns for the memory $\mathfrak{M}_Z \to \mathfrak{W}_X$. Middle row: The output of the memory $\mathfrak{M}_Z$ when presented with the morphologically independent set in the top row. Bottom row: The output patterns of the combined memory $\mathfrak{M}_Z \to \mathfrak{W}_X$ [223].

refining the properties of lattice independence as well as morphological independence.

Definition 5.6 A set of vectors $X = \{\mathbf{x}^1, \ldots, \mathbf{x}^k\} \subset \mathbb{R}^n$ is said to be *max dominant* if and only if for every $\lambda \in \mathbb{N}_k$ there exists an index $j_\lambda \in \{1, \ldots, n\}$ such that

$$x^\lambda_{j_\lambda} - x^\lambda_i = \bigvee_{\xi=1}^{k} (x^\xi_{j_\lambda} - x^\xi_i) \quad \forall i \in \{1, \ldots, n\}. \tag{5.36}$$

Similarly, X is said to be *min dominant* if and only if for every $\lambda \in \mathbb{N}_k$ there exists an index $j_\lambda \in \{1, \ldots, n\}$ such that

$$x^\lambda_{j_\lambda} - x^\lambda_i = \bigwedge_{\xi=1}^{k} (x^\xi_{j_\lambda} - x^\xi_i) \quad \forall i \in \{1, \ldots, n\}. \tag{5.37}$$

Note that equations 5.36 and 5.37 are, respectively, equivalent to

$$x^\gamma_{j_\gamma} - x^\gamma_i = \mathfrak{m}_{j_\gamma i} = -\mathfrak{w}_{i j_\gamma} \text{ and } x^\gamma_{j_\gamma} - x^\gamma_i = \mathfrak{w}_{j_\gamma i} = -\mathfrak{m}_{i j_\gamma}. \tag{5.38}$$

For our current discussion it is more pertinent to observe that equations 5.36 and 5.37 can also be formulated by saying that for every $\lambda \in \mathbb{N}_k$ there exists an index $j_\lambda \in \{1, \ldots, n\}$ such that, respectively,

$$x^\lambda_{j_\lambda} - x^\lambda_i \leq x^\xi_{j_\lambda} - x^\xi_i \text{ and } x^\lambda_{j_\lambda} - x^\lambda_i \geq x^\xi_{j_\lambda} - x^\xi_i \tag{5.39}$$

$\forall i \in \{1, \ldots, n\}$ and $\forall \xi \in \{1, \ldots, k\}$.

Definition 5.7 A set of vectors $X = \{\mathbf{x}^1, \ldots, \mathbf{x}^k\} \subset \mathbb{R}^n$ is said to be *morphologically strongly independent* if and only if X is max dominant and for every $\lambda \in \mathbb{N}_k$, $\mathbf{x}^\lambda \not\leq \mathbf{x}^\xi \ \forall \xi \in \mathbb{N}_k \setminus \{\lambda\}$.

According to Example 5.9, if X is morphologically independent, then X satisfies the condition $\mathbf{x}^\lambda \not\leq \mathbf{x}^\xi \ \forall \xi \in \mathbb{N}_k \setminus \{\lambda\}$ of Definition 5.7. However, as the next example shows, the notions of morphological independence and strong independence are generally not equivalent.

Example 5.11 Let $X = (\mathbf{x}^1, \mathbf{x}^2, \mathbf{x}^3)$ be the morphologically independent set defined in Example 5.10. Straightforward computation shows that for $\gamma = 1$ there does not exist a $j \in \{1, 2, 3, 4\}$ such that

$$x^\xi_j - x^\xi_i \leq x^1_j - x^1_i$$

$\forall i = 1,2,3,4$ and $\forall \xi \neq 1$. Therefore, morphological independence does not imply strong morphological independence. To provide an example of a morphologically strongly independent set of patterns, let

$$\mathbf{x}^1 = \begin{pmatrix} 4 \\ 2 \\ 5 \\ 10 \end{pmatrix}, \mathbf{x}^2 = \begin{pmatrix} 2 \\ 3 \\ 8 \\ 4 \end{pmatrix}, \text{ and } \mathbf{x}^3 = \begin{pmatrix} 2 \\ 5 \\ 1 \\ 4 \end{pmatrix}.$$

Then for $\gamma = 1,2,3$ there exists $j = 4,3,2$, respectively, such that $x_j^\xi - x_i^\xi \leq x_j^\gamma - x_i^\gamma$ for $i = 1,2,3,4$. Note that X is essentially the same as in Example 5.10 except for $x_4^1 = 10$.

Since morphological strong independence requires that equation 5.36 is satisfied, the question may arise if max dominance implies the inequality $\mathbf{x}^\lambda \nleq \mathbf{x}^\xi \ \forall \xi \in \mathbb{N}_k \setminus \{\lambda\}$ or if $x_{j_\gamma}^\gamma \geq x_i^\gamma$ for $i = 1,\ldots,n$. Here again the answer is negative in both cases. By letting X consist of the vectors

$$\mathbf{x}^1 = \begin{pmatrix} 9 \\ 4 \end{pmatrix} \text{ and } \mathbf{x}^2 = \begin{pmatrix} 10 \\ 5 \end{pmatrix},$$

we have $x_2^\xi - x_i^\xi \leq x_2^1 - x_i^1$ and $x_1^\xi - x_i^\xi \leq x_1^2 - x_i^2$ for $\xi = 1,2$ and $i = 1,2$. Thus, X is max dominant. But since $\mathbf{x}^1 \leq \mathbf{x}^2$, X does not satisfy the second requirement $\mathbf{x}^\lambda \nleq \mathbf{x}^\xi$ for $\xi \neq \lambda$ of strong morphological independence.

In the discussion immediately following equation 5.32 it was shown that if X is morphologically independent, then $\forall \gamma = 1,\ldots,k$ there exists an index $j_\gamma \in \{1,\ldots,n\}$ such that $x_{j_\gamma}^\xi < x_{j_\gamma}^\gamma \ \forall \xi \neq \gamma$. As it turns out, the same property also holds for morphologically strongly independent sets.

Lemma 5.10 *If X is morphologically strongly independent, then $\forall \gamma = 1,\ldots,k$, there exists an index $j_\gamma \in \{1,\ldots,n\}$ such that $x_{j_\gamma}^\xi < x_{j_\gamma}^\gamma \ \forall \xi \neq \gamma$.*

Proof. Suppose that there exists an index $\lambda \neq \gamma$ such that $x_{j_\gamma}^\gamma \leq x_{j_\gamma}^\lambda$, where j_γ is as in equation 5.36. Then

$$\begin{aligned} x_i^\gamma &= x_{j_\gamma}^\gamma - (x_{j_\gamma}^\gamma - x_i^\gamma) = x_{j_\gamma}^\gamma - \bigvee_{\xi=1}^{k} \left(x_{j_\gamma}^\xi - x_i^\xi\right) \\ &= x_{j_\gamma}^\gamma + \bigwedge_{\xi=1}^{k} \left(x_i^\xi - x_{j_\gamma}^\xi\right) \leq x_{j_\gamma}^\lambda + \bigwedge_{\xi=1}^{k} \left(x_i^\xi - x_{j_\gamma}^\xi\right) \\ &\leq x_{j_\gamma}^\lambda + \left(x_i^\lambda - x_{j_\gamma}^\lambda\right) = x_i^\lambda \ \forall i = 1,\ldots,n. \end{aligned}$$

But $x_i^\gamma \leq x_i^\lambda \; \forall i = 1,\ldots,n$ implies that $\mathbf{x}^\gamma \leq \mathbf{x}^\lambda$, which contradicts our hypothesis of strong morphological independence. □

It is important to note that in the proof of Lemma 5.10 we made use of both conditions of the definition of strong independence. Also, since $x_{j_\gamma}^\xi < x_{j_\gamma}^\gamma \; \forall \xi \neq \gamma$ we have that

$$x_{j_\gamma}^\gamma > \bigvee_{\xi \neq \gamma} x_{j_\gamma}^\xi, \text{ hence } \mathbf{x}^\gamma \nleq \bigvee_{\xi \neq \gamma} \mathbf{x}^\xi.$$

This proves the next theorem:

Theorem 5.11 *Morphological strong independence implies morphological independence.*

As mentioned previously, in the Boolean case morphological independence and strong independence are equivalent notions.

Theorem 5.12 *Suppose X is Boolean. Then X is morphologically independent if and only if X is morphologically strongly independent.*

Proof. In view of Theorem 5.11, all we need to show is that morphological independence implies strong independence. Furthermore, assuming X is morphologically independent, then X satisfies condition 1 of strong independence. Therefore, it only remains to be shown that X also satisfies condition 2 of strong independence. Let $\gamma \in \{1,\ldots,k\}$; since X is morphologically independent, there exists an index $j_\gamma \in \{1,\ldots,n\}$ such that $x_{j_\gamma}^\xi < x_{j_\gamma}^\gamma \; \forall \xi \neq \gamma$. Since X is Boolean, this implies that $x_{j_\gamma}^\gamma = 1$ and $x_{j_\gamma}^\xi = 0 \; \forall \xi \neq \gamma$. Hence,

$$x_{j_\gamma}^\xi - x_i^\xi = -x_i^\xi \leq 0 \leq 1 - x_i^\gamma = x_{j_\gamma}^\gamma - x_i^\gamma \tag{5.40}$$

for $i = 1,\ldots,n$ and $\xi = 1,\ldots,k$. □

As indicated by Example 5.10, equation 5.30 can be satisfied by any set of non-negative vectors X of morphologically independent patterns. Furthermore, if X is Boolean, then the resulting set Z represents a set of kernel vectors but Z need not be a kernel if X is not Boolean. However, the set X in Example 5.10 is not strongly independent (see Example 5.11). Therefore, it seems natural to ask if strongly independent sets yield kernels when using the method described by equation 5.33. We examine this question in the remainder of this section and provide a modification of the definition of a kernel in order to circumvent the problems encountered by the kernel method used thus far.

A roadblock in obtaining meaningful kernels in the non-Boolean case is the overly restrictive requirement that $\mathfrak{M}_Z \boxtimes \mathbf{x}^\gamma = \mathbf{z}^\gamma$. However, if we simply require that there exists a memory W such that

$$W \boxtimes (\mathfrak{M}_Z \boxtimes \mathbf{x}^\gamma) = \mathbf{y}^\gamma, \tag{5.41}$$

which agrees with the intuitive idea expressed in the introduction of this section, namely that $\mathfrak{M}_Z \boxtimes \mathbf{x}^\gamma$ need only be some intermediate pattern, then several results that mirror Sussner's theorem can be obtained for non-Boolean patterns. These observations provide the rationale for replacing Definition 5.3 with the following less restrictive definition of a kernel:

Definition 5.8 (*Relaxed version*) Let $Z = (\mathbf{z}^1, \ldots, \mathbf{z}^k)$ be an $n \times k$ matrix. The set Z is called a *kernel* for (X, Y) if and only if $Z \leq X$ with $Z \neq X$ and there exists a memory W such that

$$W \boxtimes (\mathfrak{M}_Z \boxtimes \mathbf{x}^\gamma) = \mathbf{y}^\gamma. \tag{5.42}$$

If $Y = X$, then Z is a said to be a *kernel* for X.

Example 5.12 Consider the two subsets X and Y of $\mathbb{R}^3$ with elements

$$\mathbf{x}^1 = \begin{pmatrix} 0 \\ 0 \\ 0 \end{pmatrix}, \mathbf{x}^2 = \begin{pmatrix} 0 \\ -2 \\ -4 \end{pmatrix}, \mathbf{x}^3 = \begin{pmatrix} 0 \\ -3 \\ 0 \end{pmatrix}, \text{ and}$$

$$\mathbf{y}^1 = \begin{pmatrix} 0 \\ 1 \\ 0 \end{pmatrix}, \mathbf{y}^2 = \begin{pmatrix} -1 \\ -1 \\ 0 \end{pmatrix}, \mathbf{y}^3 = \begin{pmatrix} 0 \\ -2 \\ 0 \end{pmatrix}.$$

Let $Z = \{\mathbf{z}^1, \mathbf{z}^2, \mathbf{z}^3, \mathbf{z}^4\}$, where

$$\mathbf{z}^1 = \begin{pmatrix} 0 \\ 0 \\ -2 \end{pmatrix}, \mathbf{z}^2 = \begin{pmatrix} 0 \\ -2 \\ -4 \end{pmatrix}, \text{ and } \mathbf{z}^3 = \begin{pmatrix} 0 \\ -3 \\ -2 \end{pmatrix}.$$

Then $W \boxtimes (\mathfrak{M}_Z \boxtimes \mathbf{x}^\gamma) = \mathbf{y}^\gamma$ for $\gamma = 1, 2, 3$, where

$$W = W_{ZY} = \begin{pmatrix} -1 & 0 & 2 \\ -1 & 1 & 0 \\ 0 & 0 & 2 \end{pmatrix} \quad \text{and} \quad \mathfrak{M}_Z = \begin{pmatrix} 0 & 3 & 4 \\ 0 & 0 & 2 \\ -2 & 1 & 0 \end{pmatrix}.$$

Obviously, $Z \leq X$ with $Z \neq X$. Thus, according to Definition 5.8, Z is a kernel for (X, Y).

There is an obvious close connection between kernels and minimal representations. If Z is a kernel for X in the sense of Definition 5.3, then $Z \leq X$, $\mathfrak{M}_Z \boxed{\wedge} \mathbf{x}^\gamma = \mathbf{z}^\gamma$, and $W_{ZX} \boxed{\vee} \mathbf{z}^\gamma = \mathbf{x}^\gamma$. Thus, kernels for X satisfy condition 3 of minimal representations. We now consider the converse, namely for what pattern sets do there exist minimal representations that may also serve as kernels. In the remainder of this section we assume that pattern vectors are non-negative! Specifically, a pattern vector $\mathbf{x} \in \mathbb{R}^n$ is said to be *non-negative* if and only if $x_i \geq 0 \; \forall i \in \{1, \ldots, n\}$. If the elements of a set $X \subset \mathbb{R}^n$ are all non-negative pattern vectors, then we say that X is *non-negative*.

By definition, a minimal representations Z of a set X of pattern vectors has the property that $Z \leq X$. Thus any element $\mathbf{z}^\gamma \in Z$ has the property that $\mathbf{z}^\gamma \leq \mathbf{x}^\gamma \in X$. Consequently every element $\mathbf{z}^\gamma$ of Z can be viewed as a noisy eroded version $\tilde{\mathbf{x}}^\gamma$ of $\mathbf{x}^\gamma \in X$. Setting $\tilde{\mathbf{x}}^\gamma = \mathbf{z}^\gamma$ results in $\tilde{x}_j^\gamma \leq x_j^\gamma \; \forall j = 1, \ldots, n$. Hence the condition $\tilde{x}_j^\gamma \leq x_j^\gamma \vee \bigwedge_{i=1}^m [\bigvee_{\xi \neq \lambda}(x_i^\lambda -_i^\xi + x_j^\xi)]$ of Theorem 5.5.(*1*) is automatically satisfied for eroded patterns. This proves the following corollary of Theorem 5.5:

Corollary 5.13 *Suppose that $\tilde{\mathbf{x}}^\gamma$ denotes an eroded version of $\mathbf{x}^\gamma$. The equation $\mathfrak{W}_X \boxed{\vee} \tilde{\mathbf{x}}^\gamma = \mathbf{x}^\gamma$ holds if and only if for each row index $i \in \{1, \ldots, n\}$ there exists a column index $j_i \in \{1, \ldots, n\}$ such that*

$$\tilde{x}_{j_i}^\gamma = x_{j_i}^\gamma \vee \left(\bigvee_{\xi \neq \gamma} \left[x_i^\gamma - x_i^\xi + x_{j_i}^\xi \right] \right). \tag{5.43}$$

Similarly, if $\tilde{\mathbf{x}}^\gamma$ denotes a dilated version of $\mathbf{x}^\gamma$, then the equation $\mathfrak{M}_X \boxed{\wedge} \tilde{\mathbf{x}}^\gamma = \mathbf{x}^\gamma$ holds if and only if for each row index $i \in \{1, \ldots, n\}$ there exists a column index $j_i \in \{1, \ldots, n\}$ such that

$$\tilde{x}_{j_i}^\gamma = x_{j_i}^\gamma \wedge \left(\bigwedge_{\xi \neq \gamma} \left[x_i^\gamma - x_i^\xi + x_{j_i}^\xi \right] \right). \tag{5.44}$$

This corollary is a key component in the proof of the next theorem.

Theorem 5.14 *If X is non-negative and morphologically strongly independent, then there exists a set of patterns $Z \leq X$ with the property that for $\gamma = 1, \ldots, k$*

1. $\mathbf{z}^\gamma \wedge \mathbf{z}^\xi = \mathbf{0}\ \forall \xi \neq \gamma$,

2. $\mathbf{z}^\gamma$ *contains at most one non-zero entry, and*

3. $\mathfrak{W}_X \boxtimes \mathbf{z}^\gamma = \mathbf{x}^\gamma$.

Proof. Since X is strongly independent, by Lemma 5.10 we have that for each $\gamma = 1,\ldots,k$ there exists an index $j_\gamma \in \{1,\ldots,n\}$ such that $x^\xi_{j_\gamma} < x^\gamma_{j_\gamma}\ \forall \xi \neq \gamma$. Define the set Z by defining for each γ the pattern vector $\mathbf{z}^\gamma$ by

$$z^\gamma_i = \begin{cases} x^\gamma_{j_\gamma} & \text{if } i = j_\gamma \\ 0 & \text{if } i \neq j_\gamma \end{cases} \tag{5.45}$$

for $i = 1,\ldots,n$. By construction, each $\mathbf{z}^\gamma$ contains exactly one non-zero entry. Obviously, since $z^\gamma_{j_\gamma} > z^\xi_{j_\gamma}$ and $z^\gamma_{j_\xi} < z^\xi_{j_\xi}\ \forall \xi \neq \gamma$, we must have $j_\gamma \neq j_\xi\ \forall \xi \neq \gamma$ and, hence, $z^\xi_{j_\gamma} = 0\ \forall \xi \neq \gamma$. On the other hand, $z^\gamma_i = 0$ whenever $i \neq j_\gamma$ and $z^\xi_i = 0$ whenever $i \neq j_\xi$. Therefore, $\mathbf{z}^\gamma \wedge \mathbf{z}^\xi = \mathbf{0}\ \forall \xi \neq \gamma$. It remains to be shown that condition *3* of the conclusion holds. By construction, $\mathbf{z}^\gamma \leq \mathbf{x}^\gamma$ for $\gamma = 1,\ldots,k$. We shall now show that given $\gamma \in \{1,\ldots,k\}$, there exists a $j \in \{1,\ldots,n\}$ such that

$$z^\gamma_j = x^\gamma_j \vee \left(\bigvee_{\xi \neq \gamma} \left[x^\gamma_i - x^\xi_i + x^\xi_j \right] \right)\ \forall i = 1,\ldots,n \tag{5.46}$$

Equation 5.46 is verified as follows, let $j = j_\gamma$, then

$$\begin{aligned} x^\gamma_j \vee \left(\bigvee_{\xi \neq \gamma} \left[x^\gamma_i - x^\xi_i + x^\xi_j \right] \right) &= \bigvee_{\xi=1}^{k} \left[x^\gamma_i - x^\xi_i + x^\xi_j \right] \\ &= x^\gamma_i + \bigvee_{\xi=1}^{k} (x^\xi_j - x^\xi_i) = x^\gamma_i - \bigwedge_{\xi=1}^{k} (x^\xi_i - x^\xi_j) \\ &= x^\gamma_i - (x^\gamma_i - x^\gamma_j) = x^\gamma_j = z^\gamma_j. \end{aligned}$$

Therefore, according to Corollary 5.13, condition *3* holds. □

The next corollary is an easy consequence of this theorem.

Corollary 5.15 *If X and Z are as in Theorem 5.14, then Z is a minimal representation of X.*

Proof. Conditions 1 and 2 of the definition of minimal representation are a consequence of Theorem 5.14. Thus, all we have left to show is that $W_{ZX} \boxtimes \mathbf{z}^\gamma = \mathbf{x}^\gamma$ for $\gamma = 1,\ldots,k$. According to Theorem 5.14, $\mathfrak{W}_X$ is a perfect recall memory for the pair (Z,X). Hence, by Theorem 5.2, $\mathfrak{W}_X \leq W_{ZX}$ and $W_{ZX} \boxtimes \mathbf{z}^\gamma = \mathbf{x}^\gamma$ for $\gamma = 1,\ldots,k$. □

Suppose X and Z are as in Theorem 5.14 and $\mathbf{u}^\gamma = \mathfrak{M}_Z \boxtimes \mathbf{x}^\gamma$. Then for each $i = 1,\ldots,n$ we have that

$$u_i^\gamma = (\mathfrak{M}_Z \boxtimes \mathbf{x}^\gamma)_i = \bigwedge_{j=1}^{n} (m_{ij} + x_j^\gamma) \leq m_{ii} + x_i^\gamma = x_i^\gamma \tag{5.47}$$

since $m_{ii} = 0$. Hence $\mathbf{u}^\gamma \leq \mathbf{x}^\gamma$ for each $\gamma = 1,\ldots,k$. Since $\mathbf{z}^\gamma \leq \mathbf{x}^\gamma$, it now follows that

$$\mathbf{z}^\gamma = \mathfrak{M}_Z \boxtimes \mathbf{z}^\gamma \leq \mathfrak{M}_Z \boxtimes \mathbf{x}^\gamma \leq \mathbf{x}^\gamma. \tag{5.48}$$

In view of Theorem 5.14 and equation 5.48 we have

$$\mathbf{x}^\gamma = \mathfrak{W}_X \boxtimes \mathbf{z}^\gamma = \mathfrak{W}_X \boxtimes (\mathfrak{M}_Z \boxtimes \mathbf{z}^\gamma) \leq \mathfrak{W}_X \boxtimes (\mathfrak{M}_Z \boxtimes \mathbf{x}^\gamma) \leq \mathfrak{W}_X \boxtimes \mathbf{x}^\gamma = \mathbf{x}^\gamma.$$

Therefore,

$$\mathfrak{W}_X \boxtimes (\mathfrak{M}_Z \boxtimes \mathbf{x}^\gamma) = \mathbf{x}^\gamma \ \forall \gamma = 1,\ldots,k. \tag{5.49}$$

Equation 5.49 in conjunction with the memory W defined by $W = \mathfrak{W}_X$ satisfy the definition of a kernel for X (Definition 5.8). This verifies the following corollary:

Corollary 5.16 *If X and Z are as in Theorem 5.14, then Z is a kernel for X.*

According to Corollary 5.15, a minimal representation is also a kernel. Hence, for a set of patterns X to be reducible to a kernel, it is sufficient that X is strongly independent. Furthermore, if X is strongly independent, then in order to obtain a kernel one simply selects a minimal representation Z of X using the method given in the proof of Theorem 5.14.

Given a minimal representation Z which is also a kernel for X and a noisy version $\tilde{\mathbf{x}}^\gamma$ of the pattern $\mathbf{x}^\gamma$ having the property that $\mathbf{z}^\gamma \leq \tilde{\mathbf{x}}^\gamma$ and $\mathfrak{M}_Z \boxtimes \tilde{\mathbf{x}}^\gamma \leq \mathbf{x}^\gamma$, then it must follow that

$$\mathfrak{W}_X \boxtimes (\mathfrak{M}_Z \boxtimes \tilde{\mathbf{x}}^\gamma) = \mathbf{x}^\gamma. \tag{5.50}$$

Although the seven exemplar patterns displayed in the top row of Figure 5.10 are morphologically independent, they are not morphologically strongly

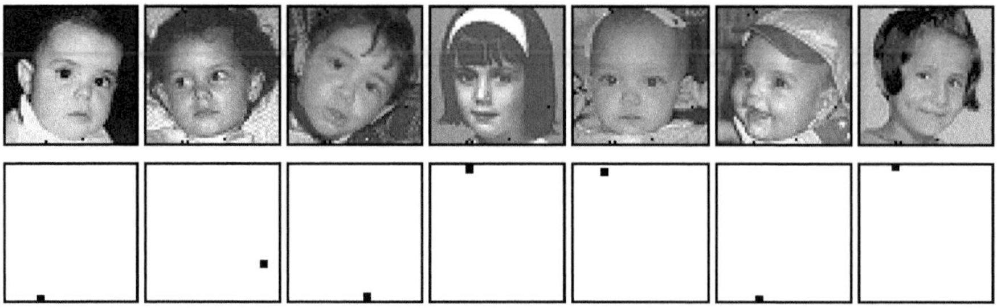

Figure 5.12 Top row: The morphologically strongly independent exemplar patterns. Bottom row: The corresponding non-zero entries of the minimal representation kernel patterns Z [223].

TABLE 5.2 Row index j, pixel position (r,c), and value (underscored) x_j^γ of each pattern γ used to build matrix Z [223].

γ	j	(r,c)	1	2	3	4	5	6	7
1	2463	(50,13)	$\underline{255}$	0	0	0	0	0	0
2	1845	(37,45)	0	$\underline{255}$	0	0	0	0	0
3	2430	(49,30)	0	0	$\underline{255}$	0	0	0	0
4	65	(2,15)	0	0	0	$\underline{255}$	0	0	0
5	112	(3,12)	0	0	0	0	$\underline{255}$	0	0
6	2466	(50,16)	0	0	0	0	0	$\underline{255}$	0
7	14	(1,14)	0	0	0	0	0	0	$\underline{255}$

independent. In order to obtain a morphologically strongly independent set of vectors, the seven exemplar patterns were slightly modified using equation 5.36; the modified exemplar patterns shown in the top row of Figure 5.12 are morphologically strongly independent. An actual algorithm for inducing strong independence on any finite set $X \subset \mathbb{R}^n$ and extracting a set of kernel vectors from the modified set X is given in Section 5.2.6.

Table 5.2 gives the list of the corresponding indexes where the pixel value for each pattern γ in row $j = j_\gamma$ was taken to be the maximum (white) and for $\xi \neq \gamma$ the minimum (black) was assigned. The matrix Z was defined according to equation 5.45. According to Theorem 5.14 and Corollaries 5.15 and 5.16, Z is a minimal representation as well as a kernel respectively; Figure 5.12 shows the morphologically strongly independent set of patterns and the associated minimal representation given by $(\mathbf{z}^1, \ldots \mathbf{z}^7)$.

Randomly corrupting the patterns shown in Figure 5.12 with 30% of noise with an intensity level of 128, and using the minimal representation Z as the kernel set, one obtained the perfect recall $\mathfrak{W}_X \boxed{\vee} (\mathfrak{M}_Z \boxed{\wedge} \tilde{\mathbf{x}}^\gamma) = \mathbf{x}^\gamma$ for

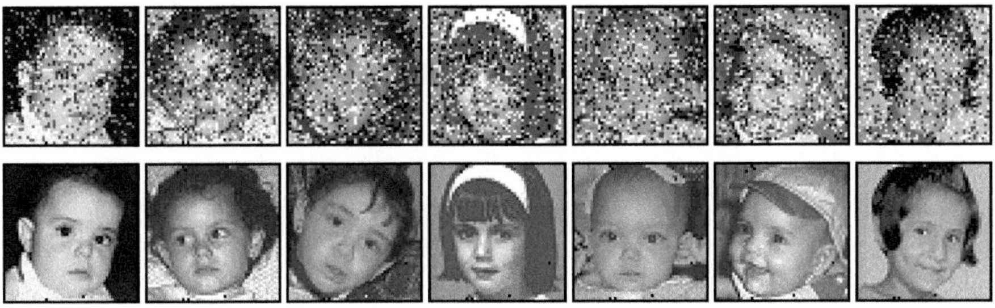

Figure 5.13 Top row: The corrupted (random noise) input patterns. Bottom row: The perfect recall using the kernel Z of Fig. 5.12 for the memory scheme $\rightarrow \mathfrak{M}_Z \rightarrow \mathfrak{W}_X \rightarrow$. [223]

$\gamma = 1, \ldots, 7$ shown in Figure 5.13. Here the memory $\rightarrow \mathfrak{M}_Z \rightarrow W_{ZX} \rightarrow$ works just as well.

As observed earlier, for a set of patterns X to be reducible to a kernel, it is sufficient that X is strongly independent. Strong independence, however, is not a necessary condition. In Example 5.11 it was noted that the set X defined in Example 5.10 is morphologically independent but not strongly independent. By letting Z be as in Example 5.10 one obtains

$$\mathfrak{W}_X \boxed{\vee} (\mathfrak{M}_Z \boxed{\wedge} \mathbf{x}^\gamma) = \mathbf{x}^\gamma \text{ for } \gamma = 1, 2, 3.$$

Hence Z is a kernel for X. Furthermore, the memory W_{ZX}, defined by

$$W_{ZX} = \begin{pmatrix} 2 & -3 & -6 & -2 \\ 2 & 0 & -5 & -4 \\ 1 & -4 & 0 & -1 \\ 4 & -1 & -4 & 0 \end{pmatrix},$$

has the property that $W_{ZX} \boxed{\vee}\, \mathbf{z}^\gamma = \mathbf{x}^\gamma$ for $\gamma = 1, 2, 3$. Thus Z is also a minimal representation for X. It follows that strong independence is not a necessary condition for the existence of kernels or minimal representations.

The question of necessary and sufficient conditions for the existence of kernels remains open. The condition that $\mathfrak{W}_X \boxed{\vee}\, \mathbf{z}^\gamma = \mathbf{x}^\gamma$ is crucial in our proof of the kernel scheme

$$input \rightarrow \mathfrak{M}_Z \rightarrow \mathfrak{W}_X \rightarrow output. \tag{5.51}$$

In order to prove condition (*3*) of Theorem 5.14, we had to use the fact that X is strongly independent. Thus far researchers have been unable to weaken the hypothesis of strong independence.

Even though the set Z in Example 5.10 is a kernel as well as a minimal representation of X, Z does not satisfy property (3) of the conclusion of Theorem 5.14 since

$$\mathfrak{W}_X \boxtimes \mathbf{z}^1 = \begin{pmatrix} 0 & -3 & -6 & -2 \\ -2 & 0 & -5 & -4 \\ -1 & -4 & 0 & -3 \\ 2 & -1 & -4 & 0 \end{pmatrix} \boxtimes \begin{pmatrix} 0 \\ 0 \\ 0 \\ 6 \end{pmatrix} = \begin{pmatrix} 4 \\ 2 \\ 3 \\ 6 \end{pmatrix} \neq \begin{pmatrix} 4 \\ 2 \\ 5 \\ 6 \end{pmatrix} = \mathbf{x}^1.$$

Consequently property (3.) of Theorem 5.14, although sufficient, is not necessary for obtaining a kernel that is also a minimal representation. There is a good reason why minimal representations are the preferred kernels for the recovery of patterns from noisy inputs. Recall that equation 5.50 will be satisfied whenever

$$z_i^\gamma \leq \tilde{x}_i^\gamma \text{ and } (\mathfrak{M}_Z \boxtimes \tilde{\mathbf{x}}^\gamma)_i \leq x_i^\gamma \tag{5.52}$$

$\forall i = 1,\ldots,n$. Now, if for some i the i,jth entry of $\mathfrak{M}_Z$ is zero for every $j = 1,\ldots,n$, then equation 5.52 is satisfied for this particular index i as long as

$$\bigwedge_{j=1}^{n} \tilde{x}_j^\gamma \leq x_i^\gamma. \tag{5.53}$$

This claim follows from the fact that since $m_{ij} = \bigvee_{\xi=1}^{k}(z_i^\xi - z_j^\xi) = 0$ for every $j = 1,\ldots,n$ and Z is a minimal representation, we must have that $z_i^\xi = 0$ for every $\xi = 1,\ldots,k$. Hence, $z_i^\gamma = 0 \leq \tilde{x}_i^\gamma$ and

$$(\mathfrak{M}_Z \boxtimes \tilde{\mathbf{x}}^\gamma)_i = \bigwedge_{j=1}^{n}(m_{ij} + \tilde{x}_j^\gamma) = \bigwedge_{j=1}^{n} \tilde{x}_j^\gamma. \tag{5.54}$$

Due to the lower bound given by equation 5.53, the components of $\mathbf{x}^\gamma$ can be arbitrarily corrupted and still satisfy equation 5.52 for the given index i as long as there exists at least one index $j \in \{1,\ldots,n\}$ such that $\tilde{x}_j^\gamma \leq x_i^\gamma$.

In many cases, equation 5.52 is automatically satisfied for a large number of indices i whenever Z is a minimal representation. These cases occur when $k \ll n$ as, for example, in the case of the seven image patterns shown in the top rows of Figures 5.10 or 5.12, where $k = 7 \ll 2500 = n$. If n is large and $k \ll n$, then $n-k$, which is the cardinality of the set $I = \{i : z_i^\xi = 0 \; \forall \xi = 1,\ldots,k\}$, is also large. Since $m_{ij} = \bigvee_{\xi=1}^{k}(z_i^\xi - z_j^\xi)$ and $z_j^\xi > 0$ for at most one ξ, we have that $\mathfrak{m}_{ij} = 0$ for every $j = 1,\ldots,n$ whenever $i \in I$. This means that $\mathfrak{M}_Z$ contains $n-k$ rows having only zero entries. Hence equation 5.47 is satisfied for at least $n-k$ indices i.

Although equation 5.47 is guaranteed to be satisfied for at least $n-k$ indices i, the likelihood that it is satisfied for the remaining k indices is also very high. Since Z is a minimal representation, the inequality $z_i^\gamma = 0 \leq \tilde{x}_i^\gamma$ is guaranteed for all i except one. The only time the inequality may not hold is in the event that for one single index j_γ, $\tilde{x}_{j_\gamma}^\gamma < x_{j_\gamma}^\gamma = z_{j_\gamma}^\gamma$. The probability of this event occurring becomes small as n increases. Also, since $\mathfrak{M}_Z$ acts as an erosive memory in that $\mathfrak{M}_Z \boxtimes \tilde{\mathbf{x}}^\gamma \leq \mathbf{x}^\gamma$, the expectation that $(\mathfrak{M}_Z \boxtimes \tilde{\mathbf{x}}^\gamma)_i \leq x_i^\gamma$ is dramatically enhanced for large n.

The problem of kernels for pattern pairs (X,Y) where $X \neq Y$ follows from the results established in this section. The next theorem is an easy consequence of Theorem 5.14 and its corollaries.

Theorem 5.17 *If X and Z are as in Theorem 5.14 and W_{XY} is a perfect associative recall memory, then Z is a kernel for (X,Y).*

In order to verify this theorem, simply let $W = W_{XY} \boxdot \mathfrak{W}_X$. Then, for all $\gamma = 1,\ldots,k$

$$\begin{aligned} W \boxdot (\mathfrak{M}_Z \boxtimes \mathbf{x}^\gamma) &= (W_{XY} \boxdot \mathfrak{W}_X) \boxdot (\mathfrak{M}_Z \boxtimes \mathbf{x}^\gamma) \\ &= W_{XY} \boxdot [\mathfrak{W}_X \boxdot (\mathfrak{M}_Z \boxtimes \mathbf{x}^\gamma)] \\ &= W_{XY} \boxdot \mathbf{x}^\gamma = \mathbf{y}^\gamma. \end{aligned} \tag{5.55}$$

The sequence of this associative feed-forward network is given by

$$\mathbf{x}^\gamma \to \mathfrak{M}_Z \to \mathfrak{W}_X \to W_{XY} \to \mathbf{y}^\gamma \text{ or simply by } \mathbf{x}^\gamma \to \mathfrak{M}_Z \to W \to \mathbf{y}^\gamma, \tag{5.56}$$

where $W = W_{XY} \boxdot \mathfrak{W}_X$.

The two major drawbacks of the kernel method are the requirements that $\mathbf{z}^\gamma \leq \tilde{\mathbf{x}}^\gamma$ and that the vector components are non-negative. In the event where $z_{j_\gamma}^\gamma \neq 0$ and $\tilde{x}_{j_\gamma}^\gamma < z_{j_\gamma}^\gamma$, the corrupted pattern will not be recognized as the original pattern. Since all other entries of $\mathbf{z}^\gamma$ are zero, the probability of this event occurring becomes small with increasing dimension n and $k \ll n$. Although Theorem 5.14, in conjunction with Corollary 5.16, provides sufficient conditions for the existence of kernels, the establishment of necessary and sufficient conditions remains an open problem. Several researchers used other approaches to get around this problem. Notable among these are the extensions of the min and max products to fuzzy min and max products as proposed by P. Sussner, et al. [273, 274, 276]. These approaches have proven useful in several applications. Nevertheless, they raise new problems and in various cases do not return the exemplar patterns when confronted with noisy inputs.

Exercises 5.2.4

1. Prove Sussner's theorem (5.9).

2. Use graph paper to create six, 16×16, black-and-white images, using symbols of your choosing. Each square of a 16×16 picture represents a pixel whose value must be either 1 or 0. For each $i = 1, 2, \ldots, 16$, denote the ith row of a picture $\mathbf{p}$ by $\mathbf{p}(i, j)$, where $j = 1, \ldots, 16$, and $\mathbf{p}(1, 1)$ corresponds to the *first* pixel on the *left* side of the *top* row. Denote the images by $\mathbf{p}^1, \ldots, \mathbf{p}^6$ and convert each image $\mathbf{p}^\xi$ into a vector $\mathbf{x}^\xi \in \mathbb{R}^{256}$ using the standard row-scan method (equation 5.35). Find a set Z of kernels for $X = \{\mathbf{x}^1, \ldots, \mathbf{x}^6\}$ in order to construct the matrices $\mathfrak{M}_Z$ and W_{ZX}. Test the performance of $\mathbf{x} \to \mathfrak{M}_Z \to W_{ZX} \to output$ for $\mathbf{x} \in X$.

3. Let $input \to \mathfrak{M}_Z \to W_{ZX} \to output$ represent the memory developed in Exercise 2, and let $\tilde{\mathbf{x}}$ denote a noisy version of $\mathbf{x} \in X$. Test the performance of the memory when the inputs $\tilde{\mathbf{x}}$ are distorted by erosive noise, dilative noise, and by various levels of Boolean mixed noise. Do this testing for different percentages of noise levels.

4. (For readers familiar with image processing) Repeat Exercises 2 and 3 using six 50×50 (or larger) digital grayscale images.

5.2.5 Bidirectional Associative Memories

Unless otherwise stated, in the remainder of this chapter we assume that all pattern vectors under consideration have non-negative components. Let $(X, Y) = \{(\mathbf{x}^\xi, \mathbf{y}^\xi) : \xi = 1, 2, \ldots, k\} \subset \mathbb{R}^n \times \mathbb{R}^m$ be a set of associative pattern vectors. It follows from equations 5.1 and 5.2 that the computation performed by an associative memory for (X, Y) based on these equation can be expressed as

$$f(M \times \mathbf{x}^\xi) = \mathbf{y}_1^\xi, \tag{5.57}$$

where M is a matrix, f denotes the next state activation function, and $\mathbf{y}_1^\xi$ denotes the result of the computation, which may or may not be equal to the exemplar pattern $\mathbf{y}^\xi$. In such cases it may be desirable to synchronously feedback the output $\mathbf{y}_1^\xi$ to another associative memory in order to improve recall accuracy. In classical bidirectional associative memory (BAM) theory as developed by Kosko and later generalized by Grossberg [106, 156, 157], the simplest feedback scheme is to pass $\mathbf{y}_1^\xi$ backwards through M', where M' denotes the transpose of M. If the resulting pattern $\mathbf{x}_1^\xi = f(M' \times \mathbf{y}_1^\xi)$ is fed back through M, a new pattern $\mathbf{y}_2^\xi$ results, which can be fed back through M' to produce $\mathbf{x}_2^\xi$, and so on. This back-and-forth flow will quickly resonate to a fixed data pair $(\mathbf{x}_F^\xi, \mathbf{y}_F^\xi)$ [156]. However, classical BAMs, being generalizations of the Hopfield net, have limitations that are similar to those of the Hopfield net. The number of associations that can be programmed into the memory and effectively recalled is very limited. The storage capacity for reliable recall has to be significantly less than $\min(m,n)$ [156, 157, 220]. Also, BAMs have the tendency to converge to the wrong association pair if components of the two association pair have too many features in common. Thus, it is often likely that at convergence $(\mathbf{x}_F^\xi, \mathbf{y}_F^\xi) \neq (\mathbf{x}^\xi, \mathbf{y}^\xi)$.

Since these early formulations of bidirectional associative memories, numerous advances and improvements in encoding and recall capabilities for traditional BAMs have occurred [50, 51, 200, 252, 258, 305, 306, 308, 309, 323, 331]. Despite the advances made, BAM models based on the classical Kosko paragon have difficulty when it comes to non-linear separable patterns and, in particular, patterns with large overlap. As some later examples in this section will demonstrate, in many cases lattice-based BAMs (LBAMs) allow for heavy overlap of features. Additionally, the notion of kernel patterns discussed in the preceding section can often provide for large increases in storage capacity and robust recall capability.

As in the case of associative memories, in their appearance lattice-based BAMs—also called LBAMs—are surprisingly similar to classical BAMs. The properties and behavior of LBAMs differ drastically from those of the classical BAMs. In the feedback scheme LBAMs employ the dual memory M_{XY}^* of M_{XY}, where M_{XY} denotes the memory defined in equation 5.16. Thus, the i,jth entry of M_{XY}^* is given by

$$m_{ij}^* = -m_{ji} = -\bigvee_{\xi=1}^{k} (y_j^\xi - x_i^\xi) = \bigwedge_{\xi=1}^{k} (x_i^\xi - y_j^\xi), \tag{5.58}$$

where $i = 1, \ldots, n$ and $j = 1, \ldots, m$. Here equation 5.58 follows directly from Theorem 5.2 by simply interchanging $\mathbf{x}^\xi$ with $\mathbf{y}^\xi$ and vice versa. Obviously, applying the same interchange scheme on Theorem 5.3 and Corollary 5.4 results in the following corollary:

Corollary 5.18

1. *$M^*_{XY} \boxed{\vee} Y = X$ if and only if for each row index $i \in \mathbb{N}_n$ and each $\gamma \in \mathbb{N}_k$, there exists a column index $j \in \mathbb{N}_m$—depending on i and γ—such that*

$$m^*_{ij} = x^\gamma_i - y^\gamma_j.$$

2. *$M^*_{XY} \boxed{\vee} Y = X$ if and only if for each row index $i \in \mathbb{N}_n$ and each $\gamma \in \mathbb{N}_k$, there exists a column index $j \in \mathbb{N}_m$—depending on i and γ—such that*

$$y^\gamma_j - x^\gamma_i = \bigvee_{\xi=1}^{k} (y^\xi_j - x^\xi_i).$$

The corollary provides two sufficient conditions for perfect recall of the memory M^*_{XY}. Thus, if either one of these two sufficiencies is satisfied the sufficiency condition of Corollary 5.4(2) also holds, then the memory M_{XY} recalls each pattern $\mathbf{y}^\xi \in Y$ when presented with pattern $\mathbf{x}^\xi \in X$, while the dual memory M^*_{XY} recalls the pattern $\mathbf{x}^\xi$ when presented with the pattern $\mathbf{y}^\xi$. More specifically, when these memories are applied in the sequences

$$\mathbf{x}^\xi \to M_{XY} \to \mathbf{y}^\xi \to M^*_{XY} \to \mathbf{x}^\xi \text{ and } \mathbf{y}^\xi \to M^*_{XY} \to \mathbf{x}^\xi \to M_{XY} \to \mathbf{y}^\xi$$

results in the respective perfect recall auto-associative memories for the pattern sets X and Y as long as the above mentioned sufficiency conditions are satisfied. These observations can be summarized as follows:

Theorem 5.19 *If the following two conditions are satisfied:*

1. for each $i \in \mathbb{N}_m$ and for each $\gamma \in \mathbb{N}_k$ there exists an index $j_1 \in \mathbb{N}_n$—depending on i and γ—such that

$$x^\gamma_{j_1} - y^\gamma_i = \bigwedge_{\xi=1}^{k} (x^\xi_{j_1} - y^\xi_i),$$

and

2. for each $j \in \mathbb{N}_n$ and for each $\gamma \in \mathbb{N}_k$ there exists an index $i_1 \in \mathbb{N}_m$—depending on j and γ—such that

$$y^\gamma_{i_1} - x^\gamma_j = \bigvee_{\xi=1}^{k} (y^\xi_{i_1} - x^\xi_j),$$

then

$$M^*_{XY} \boxed{\vee} (M_{XY} \boxed{\wedge} \mathbf{x}^\xi) = \mathbf{x}^\xi \;\; \forall \xi \in \mathbb{N}_k \;\; \textit{and} \;\; M_{XY} \boxed{\wedge} (M^*_{XY} \boxed{\vee} \mathbf{y}^\xi) = \mathbf{y}^\xi \;\; \forall \xi \in \mathbb{N}_k.$$

Proof. If condition (*1*) is satisfied, then it follows from Corollary 5.4(2) that $M_{XY} \boxtimes \mathbf{x}^\xi = \mathbf{y}^\xi \ \forall \xi \in \mathbb{N}_k$. Similarly, using Corollary 5.18(2), then condition (2) of Theorem 5.19 implies that $M^*_{XY} \boxtimes \mathbf{y}^\xi = \mathbf{x}^\xi \ \forall \xi \in \mathbb{N}_k$. Therefore, $M^*_{XY} \boxtimes (M_{XY} \boxtimes \mathbf{x}^\xi) = M^*_{XY} \boxtimes \mathbf{y}^\xi = \mathbf{x}^\xi$ and $M_{XY} \boxtimes (M^*_{XY} \boxtimes \mathbf{y}^\xi) = M_{XY} \boxtimes \mathbf{x}^\xi = \mathbf{y}^\xi \ \forall \xi \in \mathbb{N}_k$. □

For an associative pair $(\mathbf{x}^\xi, \mathbf{y}^\xi)$, the goal of the bidirectional memory under consideration is to recall the pattern $\mathbf{y}^\xi$ if $\mathbf{x}^\xi$ is fed through the memory M_{XY} and to recall $\mathbf{x}^\xi$ if $\mathbf{y}^\xi$ is fed back through the memory M^*_{XY}. Schematically, we have

$$\left\{ \begin{array}{c} \mathbf{x}^\xi \rightarrow M_{XY} \rightarrow \mathbf{y}^\xi \\ \mathbf{x}^\xi \leftarrow M^*_{XY} \leftarrow \mathbf{y}^\xi \end{array} \right\} \quad \text{or simply} \quad \left\{ \begin{array}{c} \cdots \rightarrow M \rightarrow \cdots \\ \cdots \leftarrow M^* \leftarrow \cdots \end{array} \right\}. \tag{5.59}$$

In this scheme perfect recall is always achieved as long as the conditions of Theorem 5.19 are satisfied. Furthermore, this method is a one-step procedure without the need for thresholding.

Example 5.13 Consider the four associations (a, A), (b, B), (d,D), and (x,X) of letters from the alphabet. Figure 5.14 provides a Boolean image representation of the letters with each letter corresponds to an 18×18 image. The top row of the figure represents the lower case letters while the center row consists of the corresponding upper case letters. Using the standard row-scan method (see equation 5.23), each lower case pattern image $\mathbf{p}^\xi$, convert each pattern image into a Boolean vector $\mathbf{x}^\xi = (x_1^\xi, \ldots, x_{324}^\xi)'$ and each corresponding upper case pattern image $\mathbf{P}^\xi$ into a Boolean vector $\mathbf{y}^\xi = (y_1^\xi, \ldots, y_{324}^\xi)'$. The resulting set of pattern pairs $\{(\mathbf{x}^\xi, \mathbf{y}^\xi) : \xi = 1, \ldots, 4\}$ is then used to construct the two hetero-associative memories M_{XY} and M^*_{XY}. In this particular example, the letters a, b, and d share a large number of pattern pixels (i.e., pixels with value $\mathbf{p}(i, j) = 1$), and so do the letters B and D of the associations (b, B) and (d, D). Nonetheless, the LBAM given by $\mathbf{x}^\xi \rightarrow M_{XY} \rightarrow \mathbf{y}^\xi$ and $\mathbf{x}^\xi \leftarrow M^*_{XY} \leftarrow \mathbf{y}^\xi$ provides for the perfect bidirectional recall shown in Figure 5.14.

In contrast, perfect recall can not be achieved when using the classical semi-linear bidirectional associative memory [219].

In light of this example one may erroneously infer that LBAMs are superior to classical BAMs in recall capability. However, it is important to remember that the conditions (*1*) and (*2*) of Theorem 5.19 are easily violated, in which case perfect recall may not be possible.

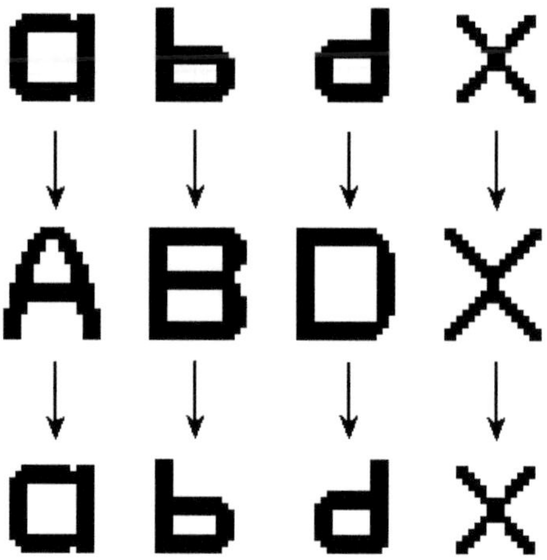

Figure 5.14 The perfect one-step convergence of the LBAM $\{\overset{\cdots\rightarrow M\rightarrow\cdots}{\cdots\leftarrow M^{*}\leftarrow\cdots}\}$ [219].

Example 5.14 Consider the following pairs of associative pattern vectors:

$$\mathbf{x}^1 = \begin{pmatrix} 1 \\ 1 \\ 1 \\ 1 \end{pmatrix}, \mathbf{y}^1 = \begin{pmatrix} 1 \\ -1 \\ 1 \\ -1 \end{pmatrix} \text{ and } \mathbf{x}^2 = \begin{pmatrix} 1 \\ 1 \\ -1 \\ -1 \end{pmatrix}, \mathbf{y}^2 = \begin{pmatrix} 1 \\ -1 \\ -1 \\ 1 \end{pmatrix}.$$

$$\text{Then } M_{XY} = \begin{pmatrix} 0 & 0 & 2 & 2 \\ -2 & -2 & 0 & 0 \\ 0 & 0 & 0 & 0 \\ 0 & 0 & 2 & 2 \end{pmatrix} \text{ and } M^*_{XY} = \begin{pmatrix} 0 & 2 & 0 & 0 \\ 0 & 2 & 0 & 0 \\ -2 & 0 & 0 & -2 \\ -2 & 0 & 0 & -2 \end{pmatrix}.$$

Simple computation shows that $M^*_{XY} \boxed{\vee} \mathbf{y}^\xi = \mathbf{x}^\xi$ for $\xi = 1, 2$ and $M_{XY} \boxed{\wedge} \mathbf{x}^2 = \mathbf{y}^2$, but $M_{XY} \boxed{\wedge} \mathbf{x}^1 = (1, -1, 1, 1)' \neq \mathbf{y}^1$. Starting with $\gamma = 1$ and checking the row indices shows that for $i = 4$ there does not exist an index j such that $x_j^1 - y_4^1 = \bigwedge_{\xi=1}^{2}(x_j^\xi - y_4^\xi) = -2$ since $x_j^1 - y_4^1 = x_j^1 + 1 = 2$ for $j = 1, \ldots, 4$. Thus, condition (*1*) of Theorem 5.19 is not satisfied. However, since these pattern vectors are orthogonal, the associate pairs can be encoded and perfectly recalled using the traditional BAM.

We now turn our attention to noisy patterns. A method for constructing LBAMs that are fairly robust when faced with noisy inputs can be derived from the kernel method discussed in the previous section. Before constructing these LBAMS, we note that since the dual memory $M^*_{XY} = W_{YX}$, the replacements of M_{XY} by M^*_{XY}, $\mathbf{x}^\gamma$ by $\mathbf{y}^\gamma$, and $\tilde{\mathbf{x}}^\gamma$ by $\tilde{\mathbf{y}}^\gamma$ in Theorem 5.5(2) results in the following corollary:

Corollary 5.20 *For $\gamma = 1, \ldots, k$, let $\tilde{\mathbf{y}}^\gamma$ denote a distorted version of the pattern $\mathbf{y}^\gamma$. Then*

$M^*_{XY} \boxed{\vee} \tilde{\mathbf{y}}^\gamma = \mathbf{x}^\gamma \Leftrightarrow \forall\, j \in \mathbb{N}_n$ *and* $i \in \mathbb{N}_m$ *the following two conditions are satisfied:*

$$\tilde{y}^\gamma_j \le y^\gamma_j \vee \bigwedge_{j=1}^{n} [\bigvee_{\xi \neq \lambda} (x^\lambda_i - x^\xi_i + y^\xi_j)] \text{ and } \exists\, j_i \in \mathbb{N}_n \ni \tilde{y}^\gamma_{j_i} = y^\gamma_{j_i} \vee [\bigvee_{\xi \neq \lambda} (x^\lambda_i - x^\xi_i + y^\xi_{j_i})].$$

It follows that if M^*_{XY} is a perfect recall memory, then M^*_{XY} is very robust in the presence of erosive noisy patterns $\tilde{\mathbf{y}}^\xi$. Since both memories M^*_{XY} and M_{XY} are perfect recall memories with respect to their respective pattern inputs from the association patterns defined in Example 5.13, it follows from Theorem 5.5 that the memories $M^*_{XY} = W_{YX}$ and $W^*_{YX} = M_{XY}$ are very robust in the presence of erosive and dilative noisy input patterns, respectively. For example, corrupting the input patterns $\mathbf{x}^\xi$ with 20% randomly generated dilative noise resulted in perfect bidirectional recall as shown in Figure 5.15. Similarly, corrupting the input patterns $\mathbf{y}^\xi$ with 20% randomly generated erosive noise resulted in the perfect bidirectional recall shown in Figure 5.16. However, no sequence of these memories is capable of effectively dealing with input patterns corrupted by random noise. Nonetheless, the particular properties of these memories provide the underlying foundation for the construction of LBAMs that have larger storage capacity and are highly robust in the presence of random noise. Before considering these memories, it will be useful to take another look at the storage and recall limitations of the memories discussed thus far.

Although matrix-based LAMs can often successfully store and retrieve patterns with large feature overlap, the theorems presented in this chapter show that there are certain limitations on perfect recall for perfect input in case the memories are hetero-associative. For instance, increasing the number of patterns given in Example 5.13 by adding the pattern pairs (e,E) and (r,R) to the memories M and M^* results in *almost* perfect performance as illustrated in Figure 5.17. However, the memory M_{XY} derived from the six association fails to satisfy the required condition of Corollary 5.4(2). In particular, for the association $(\mathbf{x}^4, \mathbf{y}^4)$ corresponding to the pair (d,D), we obtain $\mathbf{y}^4 < M_{XY} \boxed{\wedge} \mathbf{x}^4$.

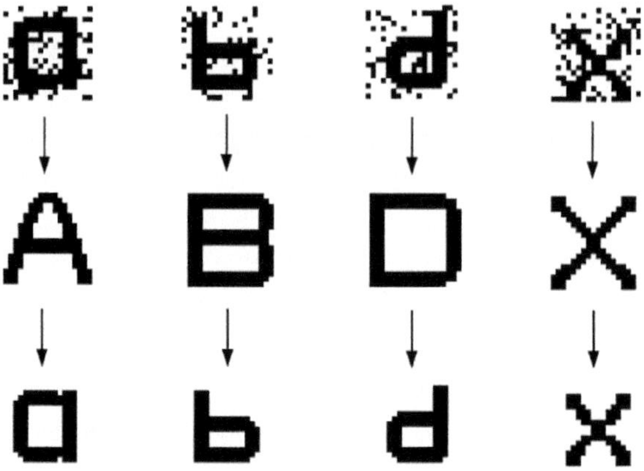

Figure 5.15 One-step convergence of the LBAM $\{\genfrac{}{}{0pt}{}{\cdots\to M\to\cdots}{\cdots\leftarrow M^*\leftarrow\cdots}\}$ using input patterns $\mathbf{x}^\xi$ corrupted by dilative noise [219].

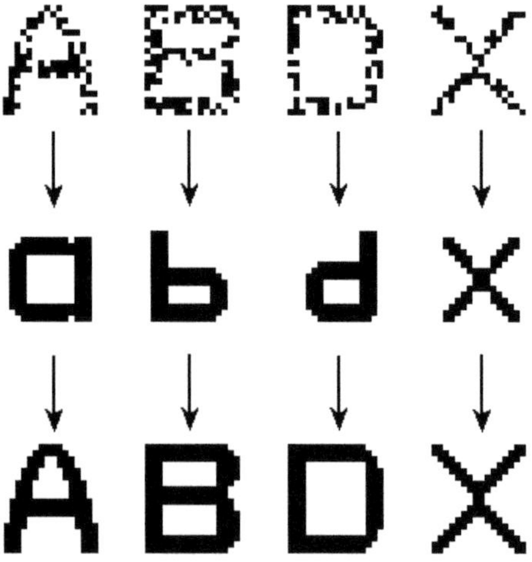

Figure 5.16 One-step convergence of the LBAM $\{\genfrac{}{}{0pt}{}{\cdots\to M^*\to\cdots}{\cdots\leftarrow M\leftarrow\cdots}\}$ using input patterns $\mathbf{y}^\xi$ corrupted by erosive noise [219].

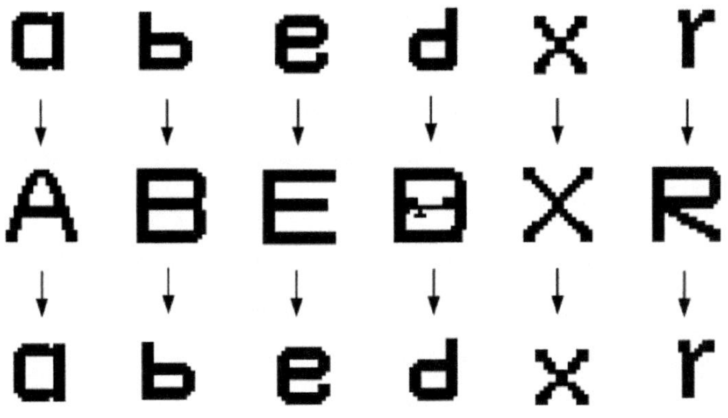

Figure 5.17 Performance of the LBAM $\{\substack{\cdots\rightarrow M\rightarrow\cdots \\ \cdots\leftarrow M^*\leftarrow\cdots}\}$ for the six pattern associations [219].

It follows from Theorem 5.1 that the following inequalities are always satisfied for $\xi = 1, \ldots, k$:

$$W_{XY}\boxed{\vee}\mathbf{x}^\xi \leq \mathbf{y}^\xi \leq M_{XY}\boxed{\wedge}\mathbf{x}^\xi \quad \text{and} \quad M^*_{XY}\boxed{\vee}\mathbf{y}^\xi \leq \mathbf{x}^\xi \leq W^*_{XY}\boxed{\wedge}\mathbf{y}^\xi. \tag{5.60}$$

However, thus far we only used the memories M and M^* to construct the bidirectional memories $\mathbf{x}^\xi \rightarrow M_{XY} \rightarrow \mathbf{y}^\xi \rightarrow M^*_{XY} \rightarrow \mathbf{x}^\xi$ and $\mathbf{y}^\xi \rightarrow M^*_{XY} \rightarrow \mathbf{x}^\xi \rightarrow M_{XY} \rightarrow \mathbf{y}^\xi$. Therefore it may seem reasonable to ask if some other combination that involves the memories W_{XY} and W^*_{XY} may yield a more perfect bidirectional memory. However, since the theorems of this chapter impose the same restrictions on W and W^* the answer to this question will also be negative. For example, adding the additional association (e,E) and (r,R) to the memories W_{XY} and W^*_{XY}, one obtains again near-perfect performance. In this case, perfect recall fails for the associations $(\mathbf{y}^3, \mathbf{x}^3)$ and $(\mathbf{y}^6, \mathbf{x}^6)$ corresponding to the pattern pairs (E,e) and (R,r), respectively. Here we have $\mathbf{x}^3 < W^*_{XY}\boxed{\wedge}\mathbf{y}^3$ and $\mathbf{x}^6 < W^*_{XY}\boxed{\wedge}\mathbf{y}^6$, as illustrated in Figure 5.18

Considering the performance of the two distinct LBAMs illustrated in Figures 5.17 and 5.18, it becomes obvious that a the LBAM defined by $\left\{\begin{array}{l} \mathbf{x}^\xi \rightarrow W_{XY} \rightarrow \mathbf{y}^\xi \\ \mathbf{x}^\xi \leftarrow M^*_{XY} \leftarrow \mathbf{y}^\xi \end{array}\right\}$ exhibits perfect bidirectional recall for perfect inputs for the particular set of six associations. However, the LBAM is practically useless if inputs contain random noise. This weakness can be disposed of by properly choosing a kernel Z for the association (X, Y) and replacing $\{\rightarrow W_{XY} \rightarrow\}$ by $\{\rightarrow \mathfrak{M}_Z \rightarrow W_{ZY} \rightarrow\}$. Since $W_{YX} = M^*_{XY}$ is vulnerable to dilative changes, a properly chosen kernel V for the bi-association is also necessary in order to replace $\{\leftarrow W_{YX} \leftarrow\}$ by $\{\leftarrow W_{VX} \leftarrow \mathfrak{M}_V \leftarrow\}$. The new

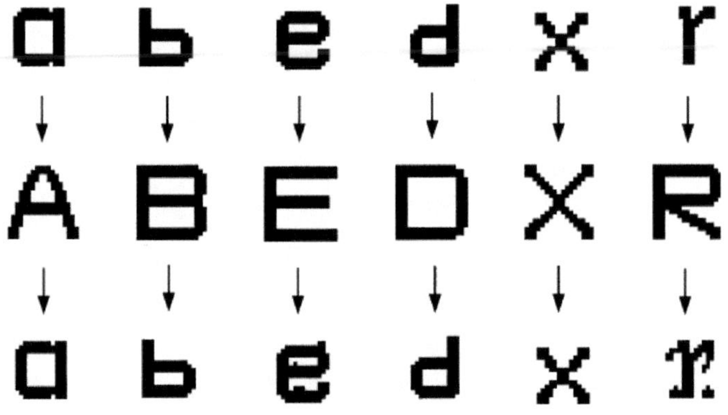

Figure 5.18 Performance of the LBAM $\{\substack{\cdots\to W\to\cdots \\ \cdots\leftarrow W^*\leftarrow\cdots}\}$ for the six pattern associations [219].

LBAM

$$\left\{ \begin{array}{c} \cdots \to \mathfrak{M}_Z \to W_{ZY} \to \cdots \\ \cdots \leftarrow W_{VX} \leftarrow \mathfrak{M}_V \leftarrow \cdots \end{array} \right\} \tag{5.61}$$

will be robust in the presence of random noise in either input X or Y.

Theorem 5.14 and its corollaries (5.15 and 5.16) are usually used as a guide for selecting kernel vectors. For Boolean patterns, the selection of kernels is generally easier because of Theorem 5.9. Additionally, each kernel pattern $\mathbf{z}^\xi$ or $\mathbf{v}^\xi$ contains only one non-zero entry and since $\mathfrak{M}_Z$ as well as $\mathfrak{W}_V$ are perfect recall memories for any finite number of respective kernel vectors, the storage capacity increases dramatically. For example, suppose the preceding example is being increased from six to seven bidirectional associations by adding the pattern pair (i,I) to the memories M_{XY} and M^*_{XY}. Then the output of the LBAM $\{\substack{\cdots\to M\to\cdots \\ \cdots\leftarrow M^*\leftarrow\cdots}\}$ is as shown in Figure 5.19. The performance illustrated in Figure 5.19 shows increased deterioration when compared to the performance of the LBAM $\{\substack{\cdots\to M\to\cdots \\ \cdots\leftarrow M^*\leftarrow\cdots}\}$ shown in Figure 5.17. Even though the output patterns $M_{XY} \boxed{\wedge} \mathbf{x}^\xi$ closely resemble the desired outputs $\mathbf{y}^\xi$, perfect performance was achieved only for the association (x,X). Somewhat surprising is the fact that the feedback through M^*_{XY} provided perfect recall of the pattern $\mathbf{x}^\xi$ even though several of the corresponding patterns $\mathbf{y}^\xi$ were corrupted by *dilative* crosstalk noise.

The kernel vectors $\mathbf{z}^\xi$ were obtained from the 18×18 images shown in the top row of Figure 5.20. Specifically, a black pixel was selected from each image that was unique to that image and did not exist as a black pixel in any of the other images. Converting these kernel images using the row scanning method results in the kernel vectors $\mathbf{z}^\xi$. The kernels are then used in

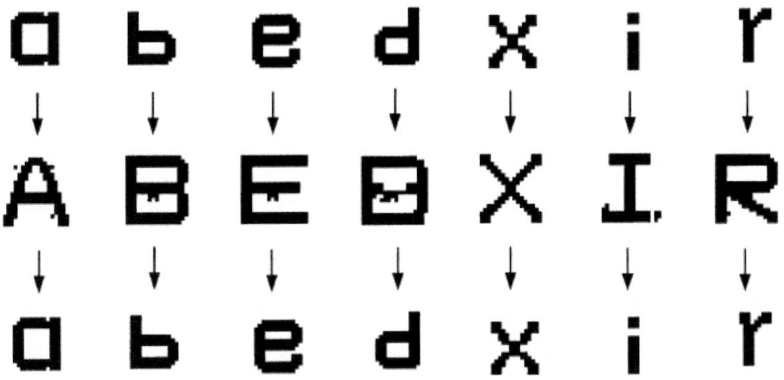

Figure 5.19 Performance of the LBAM $\{\substack{\cdots\rightarrow M\rightarrow\cdots \\ \cdots\leftarrow M^{*}\leftarrow\cdots}\}$ for the six pattern associations [219].

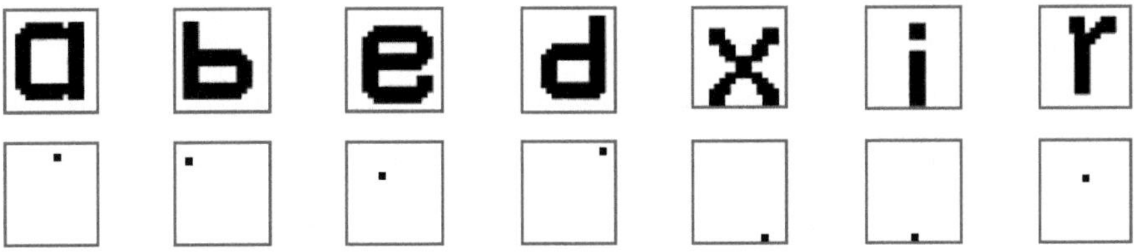

Figure 5.20 Top row: The seven pattern images associated with the pattern vectors $\mathbf{x}^{\xi}$, $\xi = 1,\ldots,7$. Bottom row: The corresponding kernel images associated with the kernel vectors $\mathbf{z}^{\xi}$ [219].

the construction of the memories $\mathfrak{M}_Z$ and W_{ZY}. Using the modified LBAM $\{\substack{\cdots\rightarrow \mathfrak{M}_Z\rightarrow W_{ZY}\rightarrow\cdots \\ \cdots\leftarrow W_{YX}\leftarrow\cdots}\}$ for the seven pattern associations and perfect inputs, perfect bidirectional association was achieved as illustrated in Figure 5.21. This supports the claim of storage increase.

As illustrated in Figure 5.22, corrupting the input patterns $\mathbf{x}^{\xi}$ $\forall \xi \in \mathbb{N}_7$, using 20% random noise did not interfere with the ability of perfect recall. Of course, if the corrupted input patterns violates the inequality of equation 5.52, then perfect recall may not be possible. However, since the single non-zero value is extremely sparse in a large (18^2) kernel vector, chances that this non-zero value being turned off are greatly reduced.

Thus far we have only considered LBAMs for Boolean patterns. Boolean patterns have the advantage that they often satisfy the hypothesis of Theorems 5.9 and 5.14 or can be easily manipulated by changing a few bits in order to satisfy these hypotheses. Additionally, they provide easy methods for selecting kernels which aid in avoiding the strict conditions set by Theo-

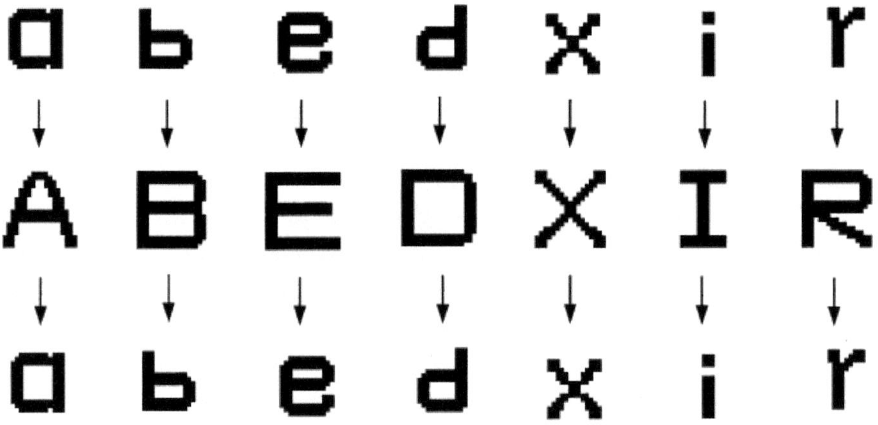

Figure 5.21 Performance of the modified LBAM $\{{}^{\cdots\to\mathfrak{M}_Z\to W_{ZY}\to\cdots}_{\cdots\leftarrow W_{YX}\leftarrow\cdots}\}$ for the six pattern associations [219].

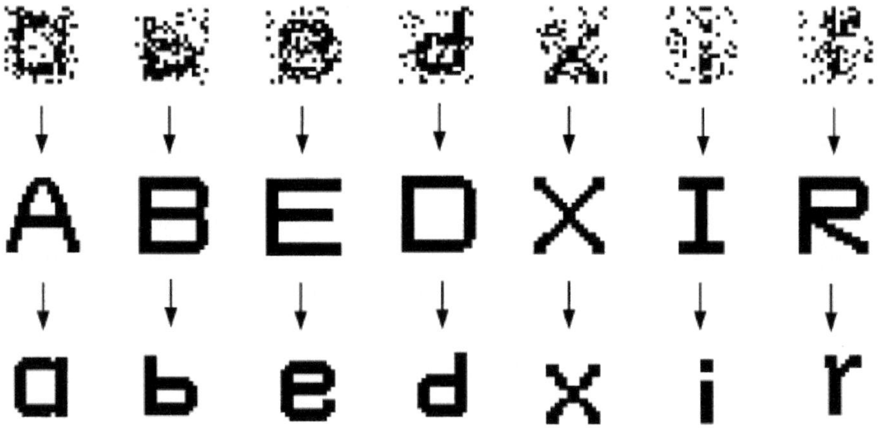

Figure 5.22 Performance of the modified LBAM $\{{}^{\cdots\to\mathfrak{M}_Z\to W_{ZY}\to\cdots}_{\cdots\leftarrow W_{YX}\leftarrow\cdots}\}$ for the six pattern associations [219].

rem 5.5. In contrast, pattern associations for vector patterns with components from $\mathbb{Z}$ or $\mathbb{R}$ seldom satisfy the conditions of Theorem 5.3, let alone Theorem 5.5 for association pairs (X,Y) with $X \neq Y$. Thus, failure of perfect recall is almost guaranteed when using either one of the two LBAMS

$$\left\{\begin{array}{c} \cdots \to M \to \cdots \\ \cdots \leftarrow M^* \leftarrow \cdots \end{array}\right\} \quad \text{or} \quad \left\{\begin{array}{c} \cdots \to W \to \cdots \\ \cdots \leftarrow W^* \leftarrow \cdots \end{array}\right\}. \tag{5.62}$$

However, even in the case that for each pattern $\mathbf{x}^\gamma$ several row indices do exist for which the two hypotheses of Corollary 5.4 are not satisfied, the memories W_{XY} and M_{XY} can still provide a storing mechanism with *almost perfect* recall in terms of a suitable distance measure between the *original* pattern $\mathbf{y}_o^\xi$ and the *recalled* pattern $\mathbf{y}_r^\xi$. One such measure uses the normalized mean square error (NMSE), denoted by $\varepsilon(\mathbf{y}_o^\xi, \mathbf{y}_r^\xi)$, to quantify the difference between $\mathbf{y}_o^\xi$ and $\mathbf{y}_r^\xi$, when recalling stored patterns by means of a specific hetero-associative memory scheme. The following example involving *non-Boolean* patterns of high-dimensionality illustrates this approach.

Consider the five pattern image associations $(\mathbf{p}^1,\mathbf{q}^1),\ldots,(\mathbf{p}^5,\mathbf{q}^5)$ shown in Figure 5.23. Each individual pattern $\mathbf{p}^\xi$ or $\mathbf{q}^\xi$ is a 50×50 pixels 256-grayscale image. For uncorrupted input, *almost* perfect recall is obtained if we use either of the memory schemes given in equation 5.62. Using the standard row-scan method, each pattern image $\mathbf{p}^\xi$ is converted into a pattern vector $\mathbf{x}^\xi = (x_1^\xi,\ldots,x_{2500}^\xi) \in \mathbb{R}^{2500} \in X$. Using the same method generates the pattern vectors $\mathbf{y}^\xi \in Y$ from the pattern images $\mathbf{q}^\xi$ of Y. Figure 5.24 shows the results when applying the canonical memories W_{XY} and M_{XY}. A visual inspection does not reveal immediately the hidden differences that cause the recall to be non-perfect since $\varepsilon(\mathbf{y}_o^\xi, \mathbf{y}_r^\xi) \approx 10^{-4}$ for $\xi = 1,\ldots,5$. Although, for a given arbitrary set (X,Y) of pattern associations, the memories W_{XY} (or W_{YX}) and M_{XY} (or M_{YX}) are not necessarily perfect recall memories, they still can be applied successfully to deal with noisy inputs.

In order to illuminate the problems encountered when trying to obtain perfect recall in the presence of noisy inputs, consider the following corollary of Theorem 5.14:

Corollary 5.21 *If X and Z are as in Theorem 5.14 and W_{XY} is a perfect associative recall memory, then Z is a kernel for (X,Y) with $W = W_{XY} \boxtimes \mathfrak{W}_X$.*

The corollary is a direct consequence of Theorem 5.14 and the fact that equation 5.42 of Definition 5.8 is satisfied for $W = W_{XY} \boxtimes \mathfrak{W}_X$. It follows that a recall mechanism for LBAMs is given by the following feed-forward network

$$\mathbf{x}^\xi \to \mathfrak{M}_Z \to W \to \mathbf{y}^\xi \to \mathfrak{M}_V \to W' \to \mathbf{x}^\xi, \tag{5.63}$$

Figure 5.23 The association (X, Y) that was used in constructing the memories W_{XY} and M_{XY} (of size 2500×2500). First row: patterns of X; second row: patterns of Y [284]

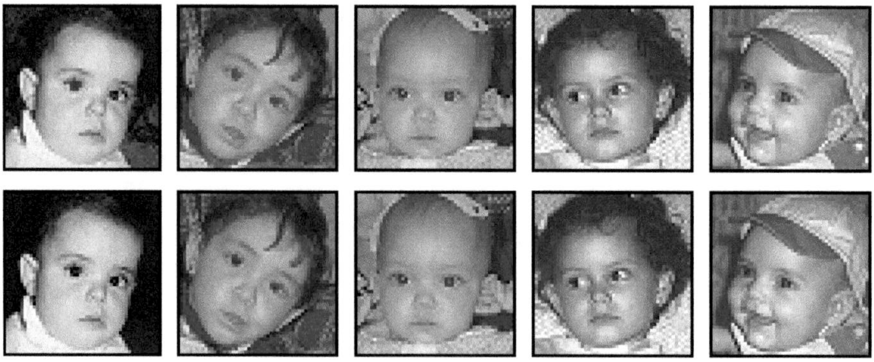

Figure 5.24 The first row displays the associated Y patterns as recalled by the memory W_{XY}; the second row displays the associated Y patterns as recalled by the memory M_{XY} [284]

where $W' = W_{YX} \boxtimes \mathfrak{W}_Y$, Z is a kernel for (X,Y), and V is a kernel for (Y,X). The assumption that W_{XY} and W_{YX} are perfect recall memories—in the proposed memory scheme defined by equation 5.63—is of crucial importance for enhancing the recall capability in the presence of noisy inputs. Given a pair of minimal representations Z and V, which are also the respective kernels for (X,Y) and (Y,X), and a noisy version $(\tilde{\mathbf{x}}^\gamma, \tilde{\mathbf{y}}^\gamma)$ of the pattern association $(\mathbf{x}^\gamma, \mathbf{y}^\gamma)$ having the property that $(\mathbf{z}^\gamma, \mathbf{v}^\gamma) \leq (\tilde{\mathbf{x}}^\gamma, \tilde{\mathbf{y}}^\gamma)$ and $(\mathfrak{M}_Z \boxtimes \tilde{\mathbf{x}}^\gamma, \mathfrak{M}_V \boxtimes \tilde{\mathbf{y}}^\gamma) \leq (\mathbf{x}^\gamma, \mathbf{y}^\gamma)$, then it must follow that

$$\mathfrak{W}_X \boxtimes (\mathfrak{M}_Z \boxtimes \tilde{\mathbf{x}}^\gamma) = \mathbf{x}^\gamma \text{ and } \mathfrak{W}_Y \boxtimes (\mathfrak{M}_V \boxtimes \tilde{\mathbf{y}}^\gamma) = \mathbf{y}^\gamma. \tag{5.64}$$

The requirements that $(\mathbf{z}^\gamma, \mathbf{v}^\gamma) \leq (\tilde{\mathbf{x}}^\gamma, \tilde{\mathbf{y}}^\gamma)$ and $(\mathfrak{M}_Z \boxtimes \tilde{\mathbf{x}}^\gamma, \mathfrak{M}_V \boxtimes \tilde{\mathbf{y}}^\gamma) \leq (\mathbf{x}^\gamma, \mathbf{y}^\gamma)$ are crucial for the recall capability of the memory schemes presented in equation 5.64. Perfect recall of $\mathbf{x}^\gamma$ from input $\tilde{\mathbf{x}}^\gamma$ will fail in the event that for the one single index $i = j_\gamma$, $z_i^\gamma > x_i^\gamma$, or for at most k indexes $i \in \{1, \ldots, n\}$, $(\mathfrak{M}_Z \boxtimes \tilde{\mathbf{x}}^\gamma)_i > x_i^\gamma$. However, even though the performance of the proposed feed-forward LBAM network when presented with noisy inputs can not be assured in a completely deterministic way, experiments have shown that for any pair $(X,Y) \subset \mathbb{R}^n \times \mathbb{R}^m$ of k associated patterns, the expectation of recall capability is greatly enhanced if $\min(n,m) \gg 0$ and $k \ll \min(n,m)$. These experiments used a slight modification of the LBAM scheme based on equation 5.63 which is defined in the next section.

Exercises 5.2.5

1. Prove Corollary 5.18.

2. Use graph paper to create six, 16×16, black-and-white images that are different from the images in Exercise **5.2.4**(2). Using the standard row-scan method, convert the pictures into six 256-dimensional pattern vectors, and let $Y = \{\mathbf{y}^1, \ldots, \mathbf{y}^6\}$ denote the set of these vectors. Construct the bidirectional memories $\mathbf{x}^\xi \to M_{XY} \to \mathbf{y}^\xi$ and $\mathbf{x}^\xi \leftarrow M_{XY}^* \leftarrow \mathbf{y}^\xi$, where X denotes the set of Exercise **5.2.4**(2). Test the performance of the bidirectional associative memory.

3. Test the performance of the bidirectional memory of Exercise 2 in the presence of erosive and dilative noise. Use several different percentages of randomly generated erosive and dilative noise.

4. Let X and Y be as in Exercise 2. Construct kernels Z for (X,Y) and V for (Y,X). Test the performance of the bidirectional memory $\mathbf{x}^\xi \to \mathfrak{M}_Z \to W_{XY} \boxtimes \mathfrak{W}_X \to \mathbf{y}^\xi$ and $\mathbf{y}^\xi \to \mathfrak{M}_V \to W_{YX} \boxtimes \mathfrak{W}_Y \to \mathbf{x}^\xi$ in the

presence of random Boolean noise. Use several different percentages of noise levels.

5. (For readers familiar with image processing) Select a set Q of six grayscale images having the same dimension, but of different context, as the ones defined in Exercise **5.2.4**(4.). Let Y denote the set of pattern vectors obtained when transforming the elements of Q into vectors and let X represent the set of vectors obtained in Exercise **5.2.4**(4.). Test the performance of the associative memories M_{XY} and W_{XY}.

5.2.6 Computation of Kernels

From a theoretical point of view, Theorem 5.14 and its corollaries provide the foundation for the kernel method when applied to *perfect inputs*. In addition, the combined memory scheme suggested by equation 5.62 together with the kernel association shown in equation 5.63 provide a useful mechanism for bidirectional pattern recall of *noisy inputs*. On the other hand, it is clear that the condition of morphological strong independence of the sets X and Y will be rarely satisfied in practical situations and seems to be very restrictive in its possible applications. A practical solution to this dilemma is to induce morphological strong independence (MSI) on patterns that are not morphologically strongly independent. The following algorithm provides a method for realizing this solution.

Algorithm 5.1 (LBAM kernels by induced MSI.)

STEP *1. Compute the global maximum U, and the global minimum L of the input set X that has k patterns of dimension n, i.e., $U = \max(X) = \bigvee_{i=1}^{n} \bigvee_{\xi=1}^{k} x_i^{\xi}$ and $L = \min(X) = \bigwedge_{i=1}^{n} \bigwedge_{\xi=1}^{k} x_i^{\xi}$.*

STEP 2. *Let $I = \{1,\ldots,n\}$. For $\xi = 1,\ldots,k$, compute an index $i_\xi \in I$ where the first available maximum value occurs, i.e.,*

$$x_{i_\xi}^{\xi} = \bigvee_{i\in I} x_i^{\xi}, \tag{5.65}$$

let $I = I \setminus \{i_\xi\}$, $\xi = \xi + 1$, and recompute equation 5.65 for the new pattern $\mathbf{x}^{\xi}$. This assures that $\forall \gamma \neq \xi$ and $i_\gamma \neq i_\xi$.

STEP *3. Change the original pattern set X by replacing all positions i_ξ for*

$\xi = 1,\ldots,k$ with the U and L values determined in Step 1 as follows:

$$x^{\xi}_{i_\gamma} = \begin{cases} U \text{ if } \gamma = \xi, \\ L \text{ if } \gamma \neq \xi. \end{cases} \tag{5.66}$$

It turns out, that the modified pattern set, denoted by $\hat{X}$, is a morphologically strongly independent set.

STEP 4. *Apply to $\hat{X}$ the kernel method and the morphological memory scheme as described in Section 4. The kernel Z of $\hat{X}$ is readily obtained from Step 3, by defining for $i = 1,\ldots,n$ and $\xi = 1,\ldots,k$, $z^{\xi}_{i} = U$ if $i = i_\xi$ otherwise set it to* 0.

STEP 5. *Repeat* STEPS *1–4 for set Y to find the kernel V of $\hat{Y}$. In this final step, a two-way kernel (Z,V) has been determined for (X,Y).*

The following theorem guarantees that the sets $\hat{X}$ and $\hat{Y}$ obtained from the algorithm are morphologically strong independent.

Theorem 5.22 *The sets $\hat{X}$ and $\hat{Y}$ generated by algorithm 5.1 are morphologically strong independent.*

Proof. According to equation 5.66, setting $j_\gamma = i_\gamma$ means that for every $\gamma \in \{1,\ldots,k\}$ there exists an index $j_\gamma \in \{1,\ldots,n\}$ such that $\hat{x}^{\gamma}_{j_\gamma} > \hat{x}^{\xi}_{j_\gamma}$, for all $\xi \neq \gamma$. Therefore, for every $\gamma \in \{1,\ldots,k\}$, $\hat{\mathbf{x}}^{\gamma} \not\leq \hat{\mathbf{x}}^{\xi}$ for all $\xi \neq \gamma$.

We now show that the set $\hat{X}$ is max dominant. Again by the construction of $\hat{X}$ as defined by equation 5.66, for each $\gamma \in \{1,\ldots,k\}$ there exists an index $j_\gamma \in \{1,\ldots,n\}$ such that $U - L = \hat{x}^{\gamma}_{j_\gamma} - \hat{x}^{\xi}_{j_\gamma} \geq \hat{x}^{\gamma}_{i} - \hat{x}^{\xi}_{i}$ $\forall i \in \mathbb{N}_n$. Equivalently, $\hat{x}^{\gamma}_{j_\gamma} - \hat{x}^{\gamma}_{i} \geq \hat{x}^{\xi}_{j_\gamma} - \hat{x}^{\xi}_{i}$ $\forall i \in \mathbb{N}_n$ and $\xi = 1,\ldots,k$. Thus, for each $\gamma \in \{1,\ldots,k\}$ there exists an index $j_\gamma \in \{1,\ldots,n\}$ such that $\hat{x}^{\gamma}_{j_\gamma} - \hat{x}^{\gamma}_{i} = \bigvee_{\xi=1}^{k}(\hat{x}^{\xi}_{j_\gamma} - \hat{x}^{\xi}_{i})$ $\forall i \in \mathbb{N}_n$. This proves that both conditions of Definition 5.7 are satisfied.

The method for proving MSI for $\hat{Y}$ is identical. □

Essentially, the kernel computation suggested by Algorithm 5.1 introduces an alternative to the LBAM scheme based on equation 5.63. The alternative LBAM consists of six distinct memories and is defined by the following diagram:

$$\begin{array}{ll} \mathbf{x}^{\xi} & \rightarrow \hat{\mathbf{x}}^{\xi} \rightarrow \mathfrak{M}_Z \rightarrow \mathfrak{W}_{\hat{X}} \rightarrow W_{\hat{X}Y} \rightarrow \quad \mathbf{y}^{\xi} \\ \mathbf{x}^{\xi} & \leftarrow W_{\hat{Y}X} \leftarrow \mathfrak{W}_{\hat{Y}} \leftarrow \mathfrak{M}_V \leftarrow \hat{\mathbf{y}}^{\xi} \leftarrow \quad \mathbf{y}^{\xi}, \end{array} \tag{5.67}$$

where the recollection mechanism is based on the modified pattern sets $\hat{X}, \hat{Y}$ rather than the original X, Y sets. Observe that induced MSI introduces a negligible amount of deterministic "artificial noise" to the original patterns which does not affect the MBAM performance if $\min(n,m) \gg 0$ and $k \ll \min(n,m)$. For instance, if $X \subset \mathbb{R}^n$ and $k \ll n$, then an element $\mathbf{x}^\xi \in X$ will be modified in k entries in order to guarantee that the conditions for MSI are satisfied. The quality of the content of the derived pattern vector $\hat{\mathbf{x}}^\xi$ can be quantified in a simple manner using the following ratio r_f, defined as the amount of *faithful* vector content,

$$r_f = 1 - \frac{k^2}{n}. \tag{5.68}$$

Example 5.15 Consider the set $X \subset \mathbb{R}^{2500}$ derived from the five images in the top row of Figure 5.23. Then each element of the modified set $\hat{X}$ has 99% of faithful content. The use of this measure implies that the *loss* of image content in each pattern exemplar is given by $r_l = k^2/n = 0.01$. Clearly, $r_f + r_l = 1$, and if the kernel computation described in Algorithm 5.1 is used for pattern recall, then the upper bound to the number of exemplar patterns that can be stored for a fixed amount of allowable loss of pattern content r_l is $k = \sqrt{n \cdot r_l}$. If X is a set of patterns derived from a set of square images of size $s \times s$, then Figure 5.25 depicts the function, $k(s) = s(\sqrt{r_l})$, between the size s of a square image and the number of modified exemplar patterns k, that can be stored for three specific values of the ratio r_l. The functional relation is linear, however a base-2 logarithmic scale has been used to adjust the image size to powers of 2. For a typical square image of size 256×256 pixels ($s = 256$) with a 20% loss of content, the LAM scheme used in Algorithm 5.1 can store up to 114 image patterns.

Recollection of stored images from degraded inputs by means of the LAM scheme of equation 5.67 is subject to the same remarks made after equation 5.64. For instance, suppose that the input pattern $\tilde{\mathbf{x}}^\gamma$, is corrupted with random noise such that the i_γ entry corresponding to the kernel vector $\mathbf{z}^\gamma$ changes its value from $\tilde{x}^\gamma_{i_\gamma} = U$ to $\alpha < U$ with probability p. Clearly, the probability that it does not change value is $1 - p$. Therefore, the experiment of corrupting the modified exemplar pattern $\bar{\mathbf{x}}^\gamma$ a finite number of times and registering exactly how many times the value of $\bar{x}^\gamma_{i_\gamma}$ is changed or not is equivalent to a sequence of Bernoulli trials. Let η be the number of times that the exemplar modified pattern $\bar{\mathbf{x}}^\gamma$ is corrupted with random noise. Then, the probability that exactly μ times, $\bar{x}^\gamma_{i_\gamma}$ changes value to any $\alpha < U$, follows the

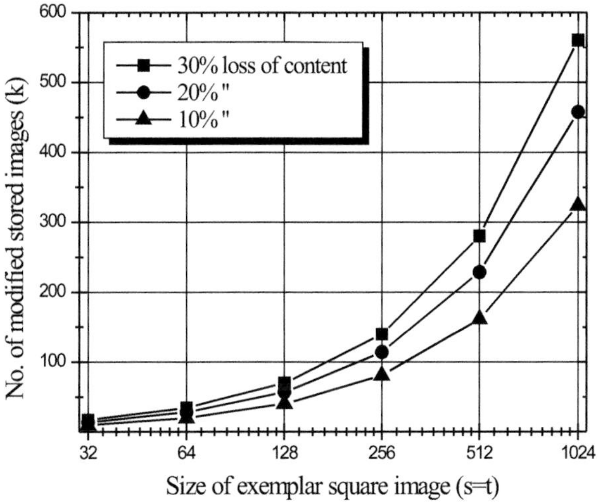

Figure 5.25 Storage capacity of a LAM scheme based on kernel computation for several image loss of content values [283].

binomial distribution

$$\binom{\eta}{\mu} p^{\mu}(1-p)^{\eta-\mu}\,;\, 0 \leq \mu \leq \eta. \tag{5.69}$$

For fixed values of η and p, the expression in equation 5.69 is a function of the number of successes μ; the maximum probability value of exactly μ successes in equation 5.69 occurs for $\mu_{\max} = [(\eta+1)p]$ where $[x]$ denotes the greatest integer function $[x] = inf\{z \in \mathbb{Z} : x \leq z\}$.

Figure 5.26 illustrates the binomial curves $f_p(\mu)$, for the case in which a given input pattern is corrupted 200 times, respectively, with a noise level equal to $10\%, 20\%, 30\%$, and 40%. The most likely number of successes given by $\mu_{\max}$, such as 60 for $p = 0.3$ in Figure 5.26, puts a theoretical limit to the performance of the kernel based LAM scheme used for image recollection. The odds of keeping the integrity of a non-zero kernel entry in a run of η trials are $\eta - \mu_{\max}$ to $\mu_{\max}$. In example 5.15, a run of 200 grayscale images corrupted with 20% random uniform noise, the odds are 4:1 of preserving the kernel association (see the second curve of Figure (5.26).

Exercises 5.2.6

1. Let X and Y denote sets of pattern vectors defined in Exercise **5.2.5**(5). Use Algorithm 5.1 in order to obtain the kernels Z for X and V for Y and construct the alternative LBAM defined in equation 5.67.

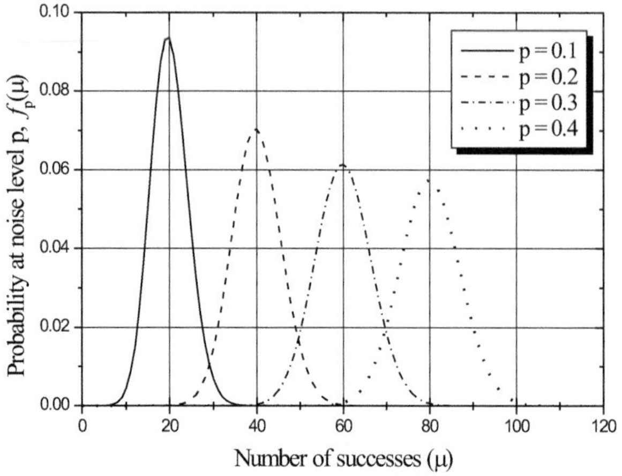

Figure 5.26 Expected probabilistic behavior in a run of 200 trials for an input image $\tilde{\mathbf{x}}^\gamma$ corrupted with noise level p (only 120 trials are shown since $f_p(\mu) \approx 0$ for $\mu > 120$) [283].

2. Test the performance of the LBAM constructed in Example 1 if the input images are corrupted by 5%, 10%, 15%, and 20% of random noise.

5.2.7 Addendum

Although the simple matrix-based approach has several weaknesses—ranging from the fixed point problem to the possible destruction of kernels by noise—it has proven to be a useful tool in several different applications [98, 99, 100, 103, 104, 105, 295, 296, 297]. More importantly, some researchers have provided new and more robust methods for improving the matrix-based approach [114, 115, 273, 274, 275, 276]. An interesting variant, called the *alpha-beta associative memory*, was proposed by Yañez-Márquez [324, 325]. However it too has its inherited weaknesses similar to those of the morphological memories.

Our motive for introducing the matrix-based LAMs is on account that they deal with a given compact data set $X \subset \mathbb{R}^n$ and are directly related to I_X, I_X^*, $\mathfrak{W}_X$, $\mathfrak{M}_X$, and $S(X)$. However, these relations tell us little about the shape of X itself. One important concept for approximating X is its convex hull, a subject discussed in the next chapter, and it is there where the relations discussed in this chapter play an important role.

CHAPTER 6

Extreme Points of Data Sets

VARIOUS concepts of convex analysis are being employed in such diverse fields as pattern recognition, optimization theory, image analysis, computational geometry, and general data analysis. The subsequent applications of the lattice theory presented thus far will require some basic knowledge of convex set theory. The first part of this chapter recollects the necessary relevant concepts from convex set theory.

6.1 RELEVANT CONCEPTS OF CONVEX SET THEORY

Data sets as commonly used in computer science are finite subsets of Euclidean spaces and, generally, not convex sets. One strategy is to consider the smallest convex set containing the data and extrapolating unknown values from the data.

6.1.1 Convex Hulls and Extremal Points

Recall that a set $X \subset \mathbb{R}^n$ is convex if $\mathbf{x}, \mathbf{y} \in X \Rightarrow \langle \mathbf{x}, \mathbf{y} \rangle \subset X$ (subsection 4.4.1). Using induction, it is easy to verify that if X is convex and $\{\mathbf{x}^1, \ldots, \mathbf{x}^k\} \subset X$, then

$$\mathbf{x} \in \mathbb{R}^n \text{ with } \mathbf{x} = \sum_{i=1}^{k} \lambda_i \mathbf{x}^i,\ \sum_{i=1}^{k} \lambda_i = 1, \text{ and } \lambda_i \geq 0\ \forall\, i \in \mathbb{N}_k \Rightarrow \mathbf{x} \in X. \quad (6.1)$$

Definition 6.1 Suppose $X \subset \mathbb{R}^n$ and $\mathfrak{K}(X) = \{C \subset \mathbb{R}^n : C \text{ is convex and } X \subset C\}$.

DOI: 10.1201/9781003154242-6

The *convex hull* of X, denoted by $\mathfrak{H}(X)$, is defined as

$$\mathfrak{H}(X) = \bigcap_{C \in \mathfrak{K}(X)} C.$$

According to this definition, $\mathfrak{H}(X)$ is the smallest convex set possible that contains X.

Since in data analysis one generally deals with finite sets, let us consider the finite set $X = \{\mathbf{x}^1, \ldots, \mathbf{x}^k\} \subset \mathbb{R}^n$ and the set $\mathfrak{X} = \{\mathbf{x} \in \mathbb{R}^n : \mathbf{x} = \sum_{i=1}^k \lambda_i \mathbf{x}^i, \sum_{i=1}^k \lambda_i = 1, \text{ and } \lambda_i \geq 0 \; \forall i \in \mathbb{N}_k\}$. If $\mathbf{x}, \mathbf{y} \in \mathfrak{X}$, then $\mathbf{x} = \sum_{i=1}^k \lambda_i \mathbf{x}^i$ and $\mathbf{y} = \sum_{i=1}^k \gamma_i \mathbf{x}^i$, where $\sum_{i=1}^k \lambda_i = 1 = \sum_{i=1}^k \gamma_i$, with $\lambda_i \geq 0$ and $\gamma_i \geq 0 \; \forall i \in \mathbb{N}_k$. Hence, for any given number $\lambda \in [0, 1]$,

$$\lambda \mathbf{x} + (1-\lambda)\mathbf{y} = \lambda \sum_{i=1}^k \lambda_i \mathbf{x}^i + (1-\lambda) \sum_{i=1}^k \gamma_i \mathbf{x}^i = \sum_{i=1}^k [\lambda \lambda_i + (1-\lambda)\gamma_i] \mathbf{x}^i.$$

Defining $\alpha_i = \lambda \lambda_i + (1-\lambda)\gamma_i$, we obtain $\lambda \mathbf{x} + (1-\lambda)\mathbf{y} = \sum_{i=1}^k \alpha_i \mathbf{x}^i$, where $\alpha_i \geq 0 \; \forall i \in \mathbb{N}_k$ and $\sum_{i=1}^k \alpha_i = \sum_{i=1}^k [\lambda \lambda_i + (1-\lambda)\gamma_i] = \lambda \sum_{i=1}^k \lambda_i + (1-\lambda) \sum_{i=1}^k \gamma_i = \lambda + (1-\lambda) = 1$. Thus the point $\mathbf{z} = \lambda \mathbf{x} + (1-\lambda)\mathbf{y} \in \mathfrak{X}$ and since $\lambda \in [0, 1]$ was arbitrary, $\langle \mathbf{x}, \mathbf{y} \rangle \subset \mathfrak{X}$. This shows that $\mathfrak{X}$ is convex and, obviously, $X \subset \mathfrak{X}$. Thus, according to Definition 6.1, $\mathfrak{H}(X) \subset \mathfrak{X}$. Next we show that $\mathfrak{X} \subset \mathfrak{H}(X)$, which proves that $\mathfrak{H}(X) = \mathfrak{X}$.

Let $\mathbf{x} \in \mathfrak{X}$, then $\mathbf{x}$ satisfies the hypothesis of equation 6.2. Also, since $\{\mathbf{x}^1, \ldots, \mathbf{x}^k\} \subset \mathfrak{H}(X)$, it follows from equation 6.2 that $\mathbf{x} \in \mathfrak{H}(X)$ and, hence, $\mathfrak{H}(X) \subset \mathfrak{X}$. Therefore

$$\mathfrak{H}(\{\mathbf{x}^1, \ldots, \mathbf{x}^k\}) = \{\mathbf{x} \in \mathbb{R}^n : \mathbf{x} = \sum_{i=1}^k \lambda_i \mathbf{x}^i, \sum_{i=1}^k \lambda_i = 1, \text{ and } \lambda_i \geq 0 \; \forall i \in \mathbb{N}_k\} \quad (6.2)$$

The convex hull of a finite set is also known as a *finitely generated* convex hull. In order to simplify notation, we shall use the conventional notation of setting $\mathfrak{H}(\mathbf{x}^1, \ldots, \mathbf{x}^k) = \mathfrak{H}(\{\mathbf{x}^1, \ldots, \mathbf{x}^k\})$. Finitely generated convex hulls are polytopes. The intuitive notions of corner or vertex points associated with polytopes can be rigorously defined in terms of extremal points.

Definition 6.2 A point $\mathbf{z}$ in a convex set $C \subset \mathbb{R}^n$ is called an *extreme* point if and only if $\mathbf{z} \in \langle \mathbf{x}, \mathbf{y} \rangle \subset C \Rightarrow \mathbf{z} \in \{\mathbf{x}, \mathbf{y}\}$. The set of extremal points of C will be denoted by *ext*(C).

We shall use the words *vertices*, *corner points*, and *extremal points* interchangeably. Thus, an n-dimensional interval $[\mathbf{v},\mathbf{u}] \subset \mathbb{R}^n$ has 2^n vertices or corner points since $|ext([\mathbf{v},\mathbf{u}])| = 2^n$. Also, for a finite set $X = \{\mathbf{x}^1,\ldots,\mathbf{x}^k\}$ it is quite possible that there exists a strict subset Y of X such that $\mathfrak{H}(Y) = \mathfrak{H}(X)$. For instance, if $X = \{\mathbf{x}^1,\ldots,\mathbf{x}^5\} \subset \mathbb{R}^2$, where

$$\mathbf{x}^1 = \begin{pmatrix}0\\0\end{pmatrix},\ \mathbf{x}^2 = \begin{pmatrix}\frac{1}{2}\\\frac{1}{2}\end{pmatrix},\ \mathbf{x}^3 = \begin{pmatrix}1\\0\end{pmatrix},\ \mathbf{x}^4 = \begin{pmatrix}0\\1\end{pmatrix},\ \mathbf{x}^5 = \begin{pmatrix}\frac{1}{2}\\\frac{1}{4}\end{pmatrix},\quad \text{and}\quad Y = \{\mathbf{x}^1,\mathbf{x}^3,\mathbf{x}^4\},$$

then $\mathfrak{H}(Y) = \mathfrak{H}(X)$. Clearly, $\mathfrak{H}(X)$ is a triangle with $ext(\mathfrak{H}(X)) = Y$, $\mathbf{x}^2$ is a boundary point of the triangle while $\mathbf{x}^5$ is an interior point of $\mathfrak{H}(Y)$, and the extreme points—which in this case are the vertices of a triangle—are also boundary points. Consequently, every point $\mathbf{x} \in \mathfrak{H}(X)$ if and only if $\mathbf{x} \in \mathfrak{H}(Y)$. More generally, we have that

$$X = \{\mathbf{x}^1,\ldots,\mathbf{x}^k\} \subset \mathbb{R}^n \Rightarrow \mathfrak{H}(X) = \mathfrak{H}(ext(\mathfrak{H}(X))). \tag{6.3}$$

We do not prove the implication in equation 6.3 as this fact is well-known (see for example [81, 292]). A consequence of equation 6.3 is that if $X = \{\mathbf{x}^1,\ldots,\mathbf{x}^{k+1}\}$ is affine independent, then $\mathfrak{H}(\mathbf{x}^1,\ldots,\mathbf{x}^{k+1}) = \langle\mathbf{x}^1,\mathbf{x}^2,\ldots,\mathbf{x}^{k+1}\rangle$ and $ext(\mathfrak{H}(X)) = X$. In most applications of convex analysis, however, the case $ext(\mathfrak{H}(X)) \neq X$ rules and the determination of extremal points of the polytope $\mathfrak{H}(X)$ is generally very laborious. A theorem due to C. Carathéodory shows that $\mathfrak{H}(X)$ can be triangulated, which means that $\mathfrak{H}(X)$ can be expressed as the union of simplexes whose interiors are mutually disjoint and whose vertices are points of X [40]. More precisely, we have

Theorem 6.1 *(Carathéodory) Suppose* $X = \{\mathbf{x}^1,\ldots,\mathbf{x}^k\} \subset \mathbb{R}^n$. *If* $\mathbf{x} \in \mathfrak{H}(X)$, *then there exist indices* $\{i_1,i_2,\ldots,i_m\} \subset \{1,2,\ldots,k\}$ *with* $m \leq n+1$ *such that* $\mathbf{x} \in \langle\mathbf{x}^{i_1},\mathbf{x}^{i_2},\ldots,\mathbf{x}^{i_m}\rangle$.

Proof. Suppose $\mathbf{x} \in \mathfrak{H}(X)$. Then according to equation 6.2

$$\mathbf{x} = \sum_{i=1}^{k} \lambda_i \mathbf{x}^i,\ \sum_{i=1}^{k} \lambda_i = 1,\ \text{with } \lambda_i \geq 0. \tag{6.4}$$

Now if $k > n+1$, then the vectors $\mathbf{x}^2 - \mathbf{x}^1,\ldots,\mathbf{x}^k - \mathbf{x}^1$ are linearly dependent. But this means that there exist scalars $a_2,\ldots,a_k$ not all being zero such that $\sum_{i=2}^k a_i(\mathbf{x}^i - \mathbf{x}^1) = \mathbf{0}$. Setting $a_1 = -\sum_{i=2}^k a_i$, one obtains $\sum_{i=1}^k a_i = 0$ and $\sum_{i=1}^k a_i\mathbf{x}^i = \mathbf{0}$. Note also that not all of the scalars a_i's can be non-positive.

Next, let $\alpha = \inf\{\frac{\lambda_j}{a_j} : a_j > 0\}$. Then

$$\mathbf{x} = \sum_{i=1}^{k} \lambda_i \mathbf{x}^i - \alpha \sum_{i=1}^{k} a_i \mathbf{x}^i = \sum_{i=1}^{k} (\lambda_i - \alpha a_i)\mathbf{x}^i, \ \sum_{i=1}^{k} (\lambda_i - \alpha a_i) = 1 \text{ with } \lambda_i - \alpha a_i \geq 0 \tag{6.5}$$

and $\lambda_i - \alpha a_i = 0$ for at least one $i \in \mathbb{N}_k$. By setting $\beta_i = \lambda_i - \alpha a_i$ we see that equation 6.5 has the same formulation as equation 6.4 for expressing the point $\mathbf{x}$ with the exception that one of the coefficients $\beta_j = 0$. Thus, we can rewrite equation 6.4 using only $k-1$ terms and creating a set $X_1 = X \setminus \{\mathbf{x}^j\} \subset X$ satisfying equation 6.4. If X_1 is not affine independent, the above argument is repeated until for some positive integer $m = k - \ell \leq n+1$, the set X_ℓ is affine independent. □

The proof of Carathéodory's theorem demonstrates the interplay between the notions of affine and linear independence. This interplay becomes an important tool for analyzing and proving properties of the convex hull of a data set. The next lemma corroborates the aforementioned interplay.

Lemma 6.2 *Let $X = \{\mathbf{x}^1, \ldots, \mathbf{x}^k\} \subset \mathbb{R}^n$ and $Y = \{\mathbf{y}^1, \ldots, \mathbf{y}^k\} \subset \mathbb{R}^{n+1}$, where $\mathbf{y}^j = \binom{\mathbf{x}^j}{1} = (x_1^j, \ldots, x_n^j, 1)'$ $\forall j \in \mathbb{N}_k$. The following conditions are equivalent:*

1. *X is affinely independent.*
2. *Y is linearly independent.*
3. *If $\sum_{i=1}^{k} \lambda_i \mathbf{x}^i = \mathbf{0}$ and $\sum_{i=1}^{k} \lambda_i = 0$, then $\lambda_i = 0 \ \forall i \in \mathbb{N}_k$.*

Proof. We show that $1 \Rightarrow 2$. Suppose that $\sum_{i=1}^{k} \lambda_i \mathbf{y}^i = \mathbf{0}$. We need to show that $\lambda_i = 0 \ \forall i \in \mathbb{N}_k$. By our supposition,

$$\sum_{i=1}^{k} \lambda_i \mathbf{y}^i = \sum_{i=1}^{k} \begin{pmatrix} \lambda_i x_1^i \\ \vdots \\ \lambda_i x_n^i \\ \lambda_i \end{pmatrix} = \begin{pmatrix} 0 \\ \vdots \\ 0 \\ 0 \end{pmatrix} \Rightarrow \sum_{i=1}^{k} \lambda_i x_j^i = 0 \text{ for } j = 1, \ldots, n \text{ and } \sum_{i=1}^{k} \lambda_i = 0.$$

It follows that $\sum_{i=1}^{k} \lambda_i \mathbf{x}^i = \mathbf{0}$ so that

$$\mathbf{0} = \sum_{i=1}^{k} \lambda_i \mathbf{x}^i - \sum_{i=1}^{k} \lambda_i \mathbf{x}^1 = \sum_{i=2}^{k} \lambda_i (\mathbf{x}^i - \mathbf{x}^1).$$

Since X is affine independent, $\lambda_2 = \lambda_3 = \ldots = \lambda_k = 0$, and since $\sum_{i=1}^{k} \lambda_i = 0$, we also have that $\lambda_1 = 0$. The proofs of $2 \Rightarrow 3$ and $3 \Rightarrow 1$ are just as easy and left as exercises. □

Exercises 6.1.1

1. Verify equation 6.1.

2. Complete the proof of Lemma 6.2 by verifying $2 \Rightarrow 3$ and $3 \Rightarrow 1$.

6.1.2 Lattice Polytopes

A *lattice polytope* is simply a polytope that is also a complete lattice. For instance, the closed interval $[\mathbf{a},\mathbf{b}] \subset \mathbb{R}^n$ with $\mathbf{a} < \mathbf{b}$ is an n-dimensional lattice polytope. The main emphasis of this chapter is on the smallest lattice polytope containing a given data set X.

Suppose $X \subset \mathbb{R}^n$ is compact, then according to Theorem 2.34 X is bounded. Thus, there exists an interval $[\mathbf{a}_1,\mathbf{b}_1]$ such that $X \subset [\mathbf{a}_1,\mathbf{b}_1]$. Now let $\mathbf{v} = \sup\{\mathbf{a} \in \mathbb{R}^n : \mathbf{a}_1 \leq \mathbf{a}$ and $X \subset [\mathbf{a},\mathbf{b}_1]\}$. Similarly, let $\mathbf{u} = \inf\{\mathbf{b} \in \mathbb{R}^n : \mathbf{b} \leq \mathbf{b}_1$ and $X \subset [\mathbf{a}_1,\mathbf{b}]\}$. Then $[\mathbf{v},\mathbf{u}]$ is the smallest interval with the property $X \subset [\mathbf{v},\mathbf{u}]$. As an aside, the interval $[\mathbf{v},\mathbf{u}]$ in higher-dimensional spaces is also referred to as a *hyperbox*. In case X is finite, say $X = \{\mathbf{x}^1,\ldots,\mathbf{x}^k\}$, the points $\mathbf{v}$ and $\mathbf{u}$ can be computed directly by using the formulae

$$\mathbf{u} = \bigvee_{j=1}^{k} \mathbf{x}^j \quad \text{and} \quad \mathbf{v} = \bigwedge_{j=1}^{k} \mathbf{x}^j. \tag{6.6}$$

Since $X \subset [\mathbf{v},\mathbf{u}]$ and $X \subset S(X)$, where $S(X)$ denotes the ℓ-span of X, the polytope $\mathfrak{P}(X)$, defined by

$$\mathfrak{P}(X) = [\mathbf{v},\mathbf{u}] \cap S(X), \tag{6.7}$$

has the property that $X \subset \mathfrak{H}(X) \subset \mathfrak{P}(X)$. Furthermore, since both $[\mathbf{v},\mathbf{u}]$ and $S(X)$ are convex, $\mathfrak{P}(X)$ is also convex. Moreover, it is easy to show that $\mathfrak{P}(X)$ is indeed a polytope as well as the smallest complete lattice with universal bounds $\mathbf{u}$ and $\mathbf{v}$ containing X. For the remainder of this subsection we establish a series of theorems and equations that are relevant in the analysis of the lattice polytope $\mathfrak{P}(X)$ and its extreme points.

As can be ascertained from Chapter 7, the importance of the extreme points of $\mathfrak{P}(X)$ lies in their applications to real-world problems. Since the data used in that chapter consists of non-negative real-valued vectors, our current focus will be on finite subsets of $\mathbb{R}^n$ with the property that if $X \subset \mathbb{R}^n$ and $x \in X$, then $x_i \geq 0$. Under these conditions, the sets $\mathfrak{W}$ and $\mathfrak{M}$ are the collection of vectors obtained from X as defined by equation 4.46. We will also reserve the bold letters $\mathbf{u}$ and $\mathbf{v}$ for the maximum and minimum vectors associated with $X = \{\mathbf{x}^1, \ldots, \mathbf{x}^k\}$ as defined by equation 6.9. The j-th coordinates of $\mathbf{u}$ and $\mathbf{v}$ will be denoted by u_j and v_j, respectively.

Recall that the vectors in $\mathfrak{W}$ and $\mathfrak{M}$ correspond to the respective column vectors of the identity matrices $I_X = \mathfrak{W}_X$ and $I_X^* = \mathfrak{M}_X$. With the collections $\mathfrak{W}$ and $\mathfrak{M}$ we associate two respective sets of vectors W and M defined by

$$W = \{\mathbf{w}^j : \mathbf{w}^j = u_j + \mathfrak{w}^j, j \in \mathbb{N}_n\} \text{ and } M = \{\mathbf{m}^j : \mathbf{m}^j = v_j + \mathfrak{m}^j, j \in \mathbb{N}_n\}. \quad (6.8)$$

It is possible that $\mathbf{w}^j = \mathbf{w}^\ell$ or $\mathbf{m}^j = \mathbf{m}^\ell$ with $j \neq \ell$, hence the number of elements of W and M may be less than the number of columns of $\mathbf{W}$ or $\mathbf{M}$. We shall reserve the bold letters $\mathbf{W}$ and $\mathbf{M}$ to denote the matrices whose columns correspond to the vectors of the collections W and M, respectively. More precisely, $\mathbf{W} = (w_{ij})_{n\times n}$ and $\mathbf{M} = (m_{ij})_{n\times n}$, where $w_{ij} = \mathbf{w}_i^j$ and $m_{ij} = \mathbf{m}_i^j$. Equivalently, $\mathbf{W} = (\mathbf{w}^1, \ldots, \mathbf{w}^n)$ and $\mathbf{M} = (\mathbf{m}^1, \ldots, \mathbf{m}^n)$. For $i \in \{1, \ldots, n\}$, the respective ith row vectors of the matrices $\mathbf{W}$ and $\mathbf{M}$ will be denoted by $\mathbf{w}_i$ and $\mathbf{m}_i$.

By definition (see equation 6.11), the min and max vectors $\mathbf{u}$ and $\mathbf{v}$ are vital in defining the collections W and M and their associated matrices $\mathbf{W}$ and $\mathbf{M}$. The following two equalities are a direct consequence of the strong bonds between W and $\mathbf{u}$, and M and $\mathbf{v}$:

Theorem 6.3

$$\mathbf{u} = \bigvee_{j=1}^{n} \mathbf{w}^j \quad \textit{and} \quad \mathbf{v} = \bigwedge_{j=1}^{n} \mathbf{m}^j.$$

Proof. We only prove that $\mathbf{u} = \bigvee_{j=1}^n \mathbf{w}^j$ as the proof for $\mathbf{v} = \bigwedge_{j=1}^n \mathbf{m}^j$ is analogous. Arbitrarily choose $i \in \{1, \ldots, n\}$. Then $u_i = u_i + \mathfrak{w}_i^i \leq \bigvee_{j=1}^n (u_j + \mathfrak{w}_i^j) = \bigvee_{j=1}^n \mathbf{w}_i^j$. Thus,

$$u_i \leq \bigvee_{j=1}^{n} \mathbf{w}_i^j. \quad (6.9)$$

We also have

$$\begin{aligned}
\bigvee_{j=1}^{n} \mathbf{w}_i^j &= \bigvee_{j=1}^{n} (u_j + \mathfrak{w}_i^j) = \bigvee_{j=1}^{n} [(\bigvee_{\xi=1}^{k} x_j^\xi) + \mathfrak{w}_i^j] = \bigvee_{j=1}^{n} [\bigvee_{\xi=1}^{k} (x_j^\xi + \mathfrak{w}_i^j)] \\
&= \bigvee_{j=1}^{n} \{\bigvee_{\xi=1}^{k} [x_j^\xi + \bigwedge_{\gamma=1}^{k} (x_i^\gamma - x_j^\gamma)]\} \leq \bigvee_{j=1}^{n} \{\bigvee_{\xi=1}^{k} [x_j^\xi + (x_i^\xi - x_j^\xi)]\} = \bigvee_{j=1}^{n} (\bigvee_{\xi=1}^{k} x_i^\xi) = u_i
\end{aligned}$$

and, therefore, $\bigvee_{j=1}^{n} \mathbf{w}_i^j \leq u_i$. In view of equation 6.9 this proves that $u_i = \bigvee_{j=1}^{n} \mathbf{w}_i^j$. Since i was arbitrary, this equality holds for $i = 1,\ldots,n$. Therefore, $\mathbf{u} = \bigvee_{j=1}^{n} \mathbf{w}^j$. □

Corollary 6.4 $L(\mathbf{w}^\ell) = L(\mathbf{w}^j) \Leftrightarrow \mathbf{w}^\ell = \mathbf{w}^j$, *and* $L(\mathbf{m}^\ell) = L(\mathbf{m}^j) \Leftrightarrow \mathbf{m}^\ell = \mathbf{m}^j$.

Proof. To avoid trivialities, assume that $\ell \neq j$, $\mathbf{w}^\ell \neq \mathbf{w}^j$ and $\mathbf{w}^\ell \in L(\mathbf{w}^j)$. Then by definition of $L(\mathbf{x})$ in equation 4.51, $\mathbf{w}^\ell = a + \mathbf{w}^j$ with $a \neq 0$ by our assumption. Thus, either $a > 0$ or $a < 0$. Assuming $a > 0$, then according to Theorem 6.3, $w_{j\ell} \leq \bigvee_{h=1}^{n} w_{jh} = u_j < a + u_j = a + w_{jj} = w_{j\ell}$, which contradicts Theorem 6.3. If $a < 0$, then setting $b = -a > 0$ one obtains the contradiction $w_{\ell j} \leq \bigvee_{h=1}^{n} w_{\ell h} = u_\ell = u_\ell < b + u_\ell = -a + w_{\ell\ell} = w_{\ell j}$. Therefore $a = 0$. Proving the validity of $L(\mathbf{m}^\ell) = L(\mathbf{m}^j) \Leftrightarrow \mathbf{m}^\ell = \mathbf{m}^j$ is analogous. □

According to the definition of W and M the vectors $\mathbf{w}^j$ and $\mathbf{m}^j$ are just translates of the respective basis vectors $\mathfrak{w}^j$ and $\mathfrak{m}^j$ in the direction $\mathbf{e}$ so that $L(\mathbf{w}^j) = L(\mathfrak{w}^j)$ and $L(\mathbf{m}^j) = L(\mathfrak{m}^j)$. This relationship between $\mathbf{w}^j$ and $\mathfrak{w}^j$, or $\mathbf{m}^j$ and $\mathfrak{m}^j$, is a valuable tool in the proofs of several theorems associated with the polytope $\mathfrak{P}(X)$. A *constant real-valued row vector* is a vector of form $\mathbf{c} = (c_1,\ldots,c_n)$, where $c_1 = \cdots = c_n$ with $c_1 \in \mathbb{R}$.

Theorem 6.5 *Suppose* $\mathbf{W} = (w_{ij})_{n\times n}$, $\mathbf{M} = (m_{ij})_{n\times n}$, *and* $j,\ell \in \{1,\ldots,n\}$, *then*

$$\mathbf{w}^\ell = \mathbf{w}^j \Leftrightarrow \mathbf{m}_j - \mathbf{m}_\ell = \mathbf{c}^1 \quad \textit{and} \quad \mathbf{m}^\ell = \mathbf{m}^j \Leftrightarrow \mathbf{w}_j - \mathbf{w}_\ell = \mathbf{c}^2,$$

where $\mathbf{c}_i^1 = u_j - u_\ell$ *and* $\mathbf{c}_i^2 = v_j - v_\ell$ *for* $i = 1,\ldots,n$.

Proof. According to the definition of the elements of W and M given in equation 6.8), we have

$$\begin{aligned}
\mathbf{w}^\ell = \mathbf{w}^j \quad &\Leftrightarrow \quad w_{i\ell} = w_{ij}\, \forall\, i \in \mathbb{N}_n \Leftrightarrow u_\ell + \mathfrak{w}_{i\ell} = u_j + \mathfrak{w}_{ij}\, \forall\, i \in \mathbb{N}_n \\
&\Leftrightarrow \quad \mathfrak{w}_{i\ell} - \mathfrak{w}_{ij} = u_j - u_\ell\ \forall\, i \in \mathbb{N}_n \Leftrightarrow -\mathfrak{m}_{\ell i} + \mathfrak{m}_{ji} = u_j - u_\ell\ \forall\, i \in \mathbb{N}_n \\
&\Leftrightarrow \quad v_i + -\mathfrak{m}_{\ell i} + \mathfrak{m}_{ji} - v_i = u_j - u_\ell\ \forall\, i \in \mathbb{N}_n \\
&\Leftrightarrow \quad (v_i + \mathfrak{m}_{ji}) - (v_i + \mathfrak{m}_{\ell i}) = u_j - u_\ell\ \forall\, i \in \mathbb{N}_n \\
&\Leftrightarrow \quad m_{ji} - m_{\ell i} = u_j - u_\ell\, \forall\, i \in \mathbb{N}_n \Leftrightarrow \mathbf{m}_j - \mathbf{m}_\ell = \mathbf{c}^1. \ \square
\end{aligned}$$

The proof of the second assertion is much the same and left as an exercise.

Example 6.1 Let $X = \{\mathbf{x}^1, \mathbf{x}^2, \mathbf{x}^3\} \subset \mathbb{R}^4$, where $\mathbf{x}^1 = (4,8,10,5)'$, $\mathbf{x}^2 =$

$(7,11,13,8)'$, and $\mathbf{x}^3 = (7,4,8,8)'$. Then, $\mathbf{v} = (4,4,8,5)'$ and $\mathbf{u} = (7,11,13,8)'$. Computing $\mathfrak{W}_X$ and setting $\mathfrak{M}_X = \mathfrak{W}^*_X$, one obtains

$$\mathfrak{W}_X = \begin{pmatrix} 0 & -4 & -6 & -1 \\ -3 & 0 & -4 & -4 \\ 1 & 2 & 0 & 0 \\ 1 & -3 & -5 & 0 \end{pmatrix}, \mathbf{W} = \begin{pmatrix} 7 & 7 & 7 & 7 \\ 4 & 11 & 9 & 4 \\ 8 & 13 & 13 & 8 \\ 8 & 8 & 8 & 8 \end{pmatrix}, \mathbf{M} = \begin{pmatrix} 4 & 7 & 7 & 4 \\ 8 & 4 & 6 & 8 \\ 10 & 8 & 8 & 10 \\ 5 & 8 & 8 & 5 \end{pmatrix}.$$

Consequently $\mathbf{w}^1 = \mathbf{w}^4$ and $\mathbf{m}_4 - \mathbf{m}_1 = \mathbf{c}^1$, where $c_i^1 = u_4 - u_1 = 1$ for $i = 1,\ldots,4$. In this simplistic example we also have $\mathbf{m}^4 = \mathbf{m}^1$ and $\mathbf{w}_1 - \mathbf{w}_4 = \mathbf{c}^2$, where $c_i^2 = v_1 - v_4 = -1$ for $i = 1,\ldots,4$.

Theorem 6.5 plays a role in data reduction, which is often necessary in such areas as signal and image processing. Considering the two vectors $\mathbf{w}^1$ and $\mathbf{w}^4$ in the above example, they are just duplicates of the same information. Thus it makes sense to discard one of them. Also, if the rows of the matrix W or M represent different frequency bands and two rows differ only by a constant, then the two rows have the same profile. In such cases one of the rows can be eliminated. Specifically, eliminating $\mathbf{w}^4$ and removing $\mathbf{w}_1$ in the above example results in a 3×3 matrix that contains the same basic information as the original 4×4 matrix.

Since $[\mathbf{v},\mathbf{u}] = \{\mathbf{x} \in \mathbb{R}^n : v_i \leq x_i \leq u_i \text{ for } i = 1,\ldots,n\}$, it follows that

$$\begin{aligned} \partial[\mathbf{v},\mathbf{u}] &= \bigcup_{i=1}^{n}\{\mathbf{x} \in [\mathbf{v},\mathbf{u}] : x_i = v_i\} \cup \bigcup_{i=1}^{n}\{\mathbf{x} \in [\mathbf{v},\mathbf{u}] : x_i = u_i\} \qquad (6.10) \\ &= \bigcap_{i=1}^{n}[\overline{E^-_{\mathbf{w}^i}(\mathbf{e}^i)} \cap \overline{E^+_{\mathbf{m}^i}(\mathbf{e}^i)}] \cap (\bigcup_{i=1}^{n}[E_{\mathbf{w}^i}(\mathbf{e}^i) \cup E_{\mathbf{m}^i}(\mathbf{e}^i)]) \end{aligned}$$

and

$$W \cup M \subset \partial[\mathbf{v},\mathbf{u}] \cap \bigcup_{i=1}^{n}[L(\mathbf{w}^i) \cup L(\mathbf{m}^i)]. \qquad (6.11)$$

This last equation emphasizes the fact that the set $W \cup M$ is generally only a subset of the intersection of $\bigcup_{i=1}^{n}[L(\mathbf{w}^i) \cup L(\mathbf{m}^i)]$ with the boundary of the interval $[\mathbf{v},\mathbf{u}]$. On the other end of the spectrum it may happen that $[\mathbf{v},\mathbf{u}] \subset \mathbb{R}^n$, but $dim[\mathbf{v},\mathbf{u}] < n$. For instance, if $u_i = v_i$ for some $i \in \mathbb{N}_n$, then $[\mathbf{v},\mathbf{u}] \subset E_{\mathbf{x}}(\mathbf{e}^i)$ $\forall \mathbf{x} \in [\mathbf{v},\mathbf{u}]$, and the dimension of $[\mathbf{v},\mathbf{u}]$ is reduced accordingly. The notion of boundary—i.e., $\partial[\mathbf{v},\mathbf{u}]$—is still well-defined in the subspace topology for $[\mathbf{v},\mathbf{u}]$. In conjunction with these observations, we have the following results:

Theorem 6.6 *For* $i \in \mathbb{N}_n$*,* $E_{\mathbf{w}^i}(\mathbf{e}^i) \cap E_{\mathbf{m}^i}(\mathbf{e}^i) = \emptyset \Leftrightarrow v_i < u_i$*.*

.

Theorem 6.7 *For* $i, j \in \mathbb{N}_n$ *with* $i \neq j$*,* $\mathbf{m}^j \in E_{\mathbf{w}^i}(\mathbf{e}^i) \Leftrightarrow m_i^j = u_i$ *and* $w_j^i = v_j$*.*

Proof. We only prove Theorem 6.7 as the proof of Theorem 6.6 is trivial. Since $E_{\mathbf{w}^i}(\mathbf{e}^i) = \{\mathbf{x} \in \mathbb{R}^n : x_i = u_i\}$, and using the definition of W and M given in terms of equation 6.8, we have

$$\mathbf{m}^j \in E_{\mathbf{w}^i}(\mathbf{e}^i) \Leftrightarrow m_i^j = u_i \Leftrightarrow v_j + \mathfrak{m}_i^j = u_i \Leftrightarrow v_j = u_i - \mathfrak{m}_i^j = u_i + \mathfrak{w}_j^i = w_j^i. \quad \square$$

This also establishes the relationship

$$\mathbf{m}^j \in E_{\mathbf{w}^i}(\mathbf{e}^i) \Leftrightarrow \langle \mathbf{m}^j, \mathbf{w}^i \rangle \subset E_{\mathbf{m}^j}(\mathbf{e}^j) \cap E_{\mathbf{w}^i}(\mathbf{e}^i), \tag{6.12}$$

which implies that if $\mathbf{m}^j \in E_{\mathbf{w}^i}(\mathbf{e}^i)$, then $\langle \mathbf{m}^j, \mathbf{w}^i \rangle$ is an edge in $\mathfrak{P}(X)$ with $\mathbf{w}^i, \mathbf{m}^j \in ext(\mathfrak{P}(X))$ (see Figures 6.2 and 6.3).

The next result shares some similarities with Theorem 6.7.

Theorem 6.8 *Suppose* $i, j \in \mathbb{N}_n$ *with* $i < j$*. Then*

1. $E_{\mathbf{w}^i}(\mathbf{d}_{ij}) = E_{\mathbf{m}^j}(\mathbf{d}_{ij})$ *and* $E_{\mathbf{m}^i}(\mathbf{d}_{ij}) = E_{\mathbf{w}^j}(\mathbf{d}_{ij})$*.*
2. $E_{\mathbf{w}^j}(\mathbf{d}_{ij}) = E_{\mathbf{w}^i}(\mathbf{d}_{ij}) \Leftrightarrow w_i^j = p + m_i^j$*, where* $p = u_j - v_j$*.*

.

Proof. Part *1* is again a direct consequence of equation 6.8 and left as an exercise.

According to part *1* $E_{\mathbf{w}^i}(\mathbf{d}_{ij}) = E_{\mathbf{m}^j}(\mathbf{d}_{ij})$. Therefore

$$E_{\mathbf{w}^j}(\mathbf{d}_{ij}) = E_{\mathbf{w}^i}(\mathbf{d}_{ij}) \Leftrightarrow w_i^j - u_j = u_i - w_j^i = m_i^j - v_j \Leftrightarrow w_i^j = (u_j - v_j) + m_i^j. \quad \square$$

A consequence of Theorems 6.7 and 6.8 is that if $u_j = v_j$ and $w_i^j = m_i^j$, then $E_{\mathbf{w}^j}(\mathbf{e}^j) = E_{\mathbf{m}^j}(\mathbf{e}^j)$ and $E_{\mathbf{w}^j}(\mathbf{d}_{ij}) = E_{\mathbf{w}^i}(\mathbf{d}_{ij})$.

The interval $[\mathbf{v}, \mathbf{u}]$ can also be viewed as the intersection of the half-spaces

$$[\mathbf{v}, \mathbf{u}] = \{\mathbf{x} \in \mathbb{R}^n : \mathbf{v} \leq \mathbf{x} \leq \mathbf{u}\} = \bigcap_{i=1}^{n} (\overline{E_{\mathbf{v}}^{+}}(\mathbf{e}^i) \cap \overline{E_{\mathbf{u}}^{-}}(\mathbf{e}^i)). \tag{6.13}$$

Consequently,

$$E_{\mathbf{v}}(\mathbf{e}^i) = E_{\mathbf{m}^i}(\mathbf{e}^i) \text{ and } E_{\mathbf{u}}(\mathbf{e}^i) = E_{\mathbf{w}^i}(\mathbf{e}^i) \text{ with } E_{\mathbf{m}^i}(\mathbf{e}^i) \parallel E_{\mathbf{w}^i}(\mathbf{e}^i). \tag{6.14}$$

.

Exercises 6.1.2

1. Prove the assertion $\mathbf{v} = \bigwedge_{j=1}^{n} \mathbf{m}^j$ of Theorem 6.3.

2. Complete the proof of Corollary 6.4 by verifying the claim that $L(\mathbf{m}^\ell) = L(\mathbf{m}^j) \Leftrightarrow \mathbf{m}^\ell = \mathbf{m}^j$.

3. Verify the claim of Theorem 6.5 that $\mathbf{m}^\ell = \mathbf{m}^j \Leftrightarrow \mathbf{w}_j - \mathbf{w}_\ell = \mathbf{c}^2$.

6.2 AFFINE SUBSETS OF $EXT(\mathfrak{P}(X))$

The main goal of this section is the extraction of affine subsets of $ext(\mathfrak{P}(X))$. In order to accomplish this goal, it is imperative to establish methods for effectively identifying the elements of $ext(\mathfrak{P}(X))$. These methods are based on the structure of $\mathfrak{P}(X)$ and various concepts from the geometry of n-dimensional spaces.

6.2.1 Simplexes and Affine Subspaces of $\mathbb{R}^n$

Suppose that $\sigma^m \subset \mathbb{R}^n$ is an m-simplex with vertices $\mathbf{v}^0, \mathbf{v}^1, \ldots, \mathbf{v}^m$. Then it follows from equation 6.3 that $\mathfrak{H}(\{\mathbf{v}^0, \mathbf{v}^1, \ldots, \mathbf{v}^m\}) = \sigma^m$. If $\{\mathbf{v}^{i_0}, \mathbf{v}^{i_1}, \ldots, \mathbf{v}^{i_k}\} \subset \{\mathbf{v}^0, \mathbf{v}^1, \ldots, \mathbf{v}^m\}$, then the simplex $\sigma^k = \langle \mathbf{v}^{i_0}, \mathbf{v}^{i_1}, \ldots, \mathbf{v}^{i_k} \rangle$ is called a *k-dimensional face* of σ^m. According to this definition, σ^m is an m-dimensional face of σ^m. Since the vertices of a simplex are affine independent, any subset is of $ext(\sigma^m)$ is affine independent. In particular, if $m = n$, and $\sigma^n = \langle \mathbf{v}^0, \mathbf{v}^1, \ldots, \mathbf{v}^n \rangle$, then the set of vectors $\{\mathbf{v}^1 - \mathbf{v}^0, \ldots, \mathbf{v}^n - \mathbf{v}^0\}$ is linearly independent and, hence, a basis for $\mathbb{R}^n$. Consequently, any vector of form $(\mathbf{x} - \mathbf{v}^0) \in \mathbb{R}^n$ can be written as $\mathbf{x} - \mathbf{v}^0 = \sum_{i=1}^{n} r_i(\mathbf{v}^i - \mathbf{v}^0)$ with $r_i \in \mathbb{R}$. Since this equation is equivalent to $\mathbf{x} = (1 - \sum_{i=1}^{n} r_i)\mathbf{v}^0 + \sum_{i=1}^{n} r_i(\mathbf{v}^i - \mathbf{v}^0)$, $\mathbf{x}$ can be uniquely expressed as $\mathbf{x} = \sum_{i=0}^{n} \lambda_i \mathbf{v}^i$, where $\lambda_0 = (1 - \sum_{i=1}^{n} r_i)$ and $\lambda_i = r_i$ for $1 \leq i \leq n$, with $\sum_{i=0}^{n} \lambda_i = 1$. Therefore

$$\{\mathbf{x} \in \mathbb{R}^n : \mathbf{x} = \sum_{i=0}^{n} \lambda_i \mathbf{v}^i \text{ and } \sum_{i=0}^{n} \lambda_i = 1\} = \mathbb{R}^n. \tag{6.15}$$

If $m < n$, then the set $\{\mathbf{x} \in \mathbb{R}^n : \mathbf{x} = \sum_{i=0}^{m} \lambda_i \mathbf{v}^i \text{ and } \sum_{i=0}^{m} \lambda_i = 1\} \subset \mathbb{R}^n$ is an m-dimensional affine subspace of $\mathbb{R}^n$ denoted by L^m. Various authors use the term *linear subspace* or *flat* instead of affine subspace. In this treatise we may also use the notation $L^m(\sigma^m)$ if it useful to clarify the specific vertices $\langle \mathbf{v}^0, \mathbf{v}^1, \ldots, \mathbf{v}^m \rangle = \sigma^m$ that generated the affine space L^m. For a given $\sigma^n = \langle \mathbf{v}^0, \mathbf{v}^1, \ldots, \mathbf{v}^n \rangle \subset \mathbb{R}^n$, every vertex $\mathbf{v}^i$ is a 0-dimensional face of σ^n and every line segment $\langle \mathbf{v}^i, \mathbf{v}^j \rangle$ is a 1-dimensional face of σ^n. Furthermore, $L^0(\sigma^0) = \{\mathbf{x} \in \mathbb{R}^n : \mathbf{x} = \lambda_0 \mathbf{v}^i \text{ and } \lambda_0 = 1\} = \{\mathbf{v}^i\}$ while $L^1(\sigma^1) = \{\mathbf{x} \in \mathbb{R}^n : \mathbf{x} =$

$\lambda_1\mathbf{v}^i + \lambda_2\mathbf{v}^j$ and $\lambda_1 + \lambda_2 = 1\} = L(\mathbf{v}^1, \mathbf{v}^2)$, where $L(\mathbf{v}^1, \mathbf{v}^2)$ is the line defined in equation 4.42. Similarly, if σ^2 denotes the 2-dimensional face defined by the triangle $\langle \mathbf{v}^i, \mathbf{v}^j, \mathbf{v}^k \rangle$, then $L^2(\sigma^2)$ represents the plane defined by $L^2(\sigma^2) = \{\mathbf{x} \in \mathbb{R}^n : \mathbf{x} = \lambda_0\mathbf{v}^i + \lambda_1\mathbf{v}^j + \lambda_2\mathbf{v}^k$ and $\lambda_0 + \lambda_1 + \lambda_2 = 1\}$. Continuing in this fashion until one obtains the $(n-1)$-dimensional face $\sigma^{n-1} = \langle \mathbf{v}^{i_0}, \mathbf{v}^{i_1}, \dots, \mathbf{v}^{i_{n-1}} \rangle$, where $\{\mathbf{v}^{i_0}, \mathbf{v}^{i_1}, \dots, \mathbf{v}^{i_{n-1}}\} \subset \{\mathbf{v}^0, \mathbf{v}^1, \dots, \mathbf{v}^n\}$ and $L^{n-1}(\sigma^{n-1})$ correspond to the hyperplane $E = L^{n-1} = \{\mathbf{x} \in \mathbb{R}^n : \mathbf{x} = \sum_{j=0}^{n-1} \lambda_j \mathbf{v}^{i_j}$ and $\sum_{j=0}^{n-1} \lambda_j = 1\} \subset \mathbb{R}^n$. The number of $(n-1)$-dimensional faces of σ^n is n which also coincides with number of vertices, of σ^{n-1}. In general, the number of m-dimensional faces of an n-simplex is equal to the binomial coefficient $\binom{n+1}{m+1}$.

In a similar fashion one can define faces for polytopes. In the case of the polytopic beam $S(X) \subset \mathbb{R}^n$, a line $L(\mathbf{w}^i)$ is a 1-dimensional face. If $L(\mathbf{w}^i) \cap L(\mathbf{m}^j) = \varnothing$, and P denotes the plane defined by two distinct points $\mathbf{x}, \mathbf{y} \in L(\mathbf{w}^i)$ and the point $\mathbf{m}^j \notin L(\mathbf{w}^i)$, then $P \cap S(X)$ will be a 2-dimensional face and so on, until one reaches the possible case where $E_{\mathbf{w}^i}(\mathbf{d}_{ij}) \cap E_{\mathbf{w}^k}(\mathbf{d}_{k\ell}) = L^{n-2}$, which in turn implies that the lattice space $F_{\mathbf{w}^i}(\mathbf{d}_{ij}) = E_{\mathbf{w}^i}(\mathbf{d}_{ij}) \cap S(X)$ is an $(n-1)$-dimensional face of $S(X)$. Note that polytope the $S(X)$ has no 0-dimensional faces.

In contrast to the polytope $S(X)$, the polytope $\mathfrak{P}(X)$ will always have 0-dimensional faces. This is due to the fact that the geometry of $\mathfrak{P}(X)$ is determined by $ext(\mathfrak{P}(X))$ since $\mathfrak{H}(ext(\mathfrak{P}(X))) = \mathfrak{P}(X)$. It follows from Definition 6.2 and the definition of $\mathfrak{P}(X)$ that the elements of the set $W \cup M \cup \{\mathbf{v}, \mathbf{u}\}$ are vertices of the polytope $\mathfrak{P}(X)$. More explicitly, since $L(\mathbf{w}^\ell) = L(\mathfrak{w}^\ell)$ and $L(\mathbf{m}^\ell) = L(\mathfrak{m}^\ell)$, $L(\mathbf{w}^\ell)$ and $L(\mathbf{m}^\ell)$ are edges of $S(X)$. Furthermore, equations 4.62 and 4.63 imply that

$$L(\mathbf{w}^\ell) = \begin{cases} [\bigcap_{j=2}^{n} E_{\mathbf{w}^1}(\mathbf{d}_{1j})] & \text{if } \ell = 1 \\ [(\bigcap_{i=1}^{\ell-1} E_{\mathbf{w}^\ell}(\mathbf{d}_{i\ell})) \cap (\bigcap_{j=\ell+1}^{n} E_{\mathbf{w}^\ell}(\mathbf{d}_{\ell j}))] & \text{if } 1 < \ell < n \\ [\bigcap_{i=1}^{n-1} E_{\mathbf{w}^n}(\mathbf{d}_{in})] & \text{if } \ell = n \end{cases} \tag{6.16}$$

and

$$L(\mathbf{m}^\ell) = \begin{cases} [\bigcap_{j=2}^{n} E_{\mathbf{m}^1}(\mathbf{d}_{1j})] & \text{if } \ell = 1 \\ [(\bigcap_{i=1}^{\ell-1} E_{\mathbf{m}^\ell}(\mathbf{d}_{i\ell})) \cap (\bigcap_{j=\ell+1}^{n} E_{\mathbf{m}^\ell}(\mathbf{d}_{\ell j}))] & \text{if } 1 < \ell < n, \\ [\bigcap_{i=1}^{n-1} E_{\mathbf{m}^n}(\mathbf{d}_{in})] & \text{if } \ell = n \end{cases} \tag{6.17}$$

Also, the edge $L(\mathbf{w}^\ell)$ of $S(X)$ is being cut by $E_u(\mathbf{e}^\ell)$ at the edge point $\mathbf{w}^\ell$ of $S(X)$ since $\mathbf{w}^\ell_\ell = u_\ell$, $\{\mathbf{w}^\ell\} = L(\mathbf{w}^\ell) \cap E_{\mathbf{u}}(\mathbf{e}^\ell)$. It is now easy to verify that $\mathbf{w}^\ell$ is a vertex point of $\mathfrak{P}(X)$ (see Exercise **6.2.1**(4.)). A similar argument shows that $\{\mathbf{m}^\ell\} = L(\mathbf{m}^\ell) \cap E_{\mathbf{v}}(\mathbf{e}^\ell)$. This testifies that $\mathbf{w}^\ell$ and $\mathbf{m}^\ell$ are vertex points. Finally, since $\mathbf{u}$ and $\mathbf{v}$ are two of the 2^n vertices—also called 0-dimensional faces—of the hyperbox $[\mathbf{v}, \mathbf{u}]$ and since $\mathbf{v}, \mathbf{u} \in S(X)$, they are vertices of $\mathfrak{P}(X)$.

Exercises 6.2.1

1. Determine the number of 1-dimensional faces of an n-dimensional interval $[\mathbf{v}, \mathbf{u}] \subset \mathbb{R}^n$.
2. For $1 < k \leq n - 1$, determine the number of k-dimensional faces of the interval in Exercise 1.
3. Determine the number vertices if $[\mathbf{v}, \mathbf{u}] \subset \mathbb{R}^n$ is a k-dimensional interval.
4. Prove that $\mathbf{w}^\ell$ and $\mathbf{m}^\ell$ are vertex points of $\mathfrak{P}(X)$.

6.2.2 Analysis of $ext(\mathfrak{P}(X)) \subset \mathbb{R}^n$

It follows from the discussion in Section 6.2.1 that if $V = W \cup M \cup \{\mathbf{v}, \mathbf{u}\}$, then $V \subset ext(\mathfrak{P}(X))$ and $\mathfrak{H}(V) \subset \mathfrak{P}(X)$. The method for constructing the set V is simple, fast, and consists of the following steps:

Algorithm 6.1 (Constructing the elements of V)

STEP *1. Given* $X = \{\mathbf{x}^1, \ldots, \mathbf{x}^k\}$, *compute the vectors* $\mathbf{v} = \bigwedge_{j=1}^k \mathbf{x}^j$ *and* $\mathbf{u} = \bigvee_{j=1}^k \mathbf{x}^j$.
STEP *2. Compute the identity matrix* $\mathfrak{W}_X$ *by setting* $\mathfrak{w}_{ij} = \bigwedge_{\xi=1}^k (x_i^\xi - x_j^\xi)$ *and set* $\mathfrak{M}_X = \mathfrak{W}_X^*$.
STEP *3. For* $j = 1$ *to* n, *use equation 6.8 to construct the vectors* $\mathbf{w}^j$ *and* $\mathbf{m}^j$.
STEP *4. Remove any duplicates from the collection* $\{\mathbf{w}^1, \ldots, \mathbf{w}^n, \mathbf{m}^1, \ldots, \mathbf{m}^n, \mathbf{v}, \mathbf{u}\}$ *and let* V *denote the set of points thus obtained.*

Observe that the algorithm computes the set $V \subset ext(\mathfrak{P}(X)) \subset \mathbb{R}^n$ for any positive integer $n \geq 2$.

6.2.2.1 *The case* $n = 2$

If $n = 2$ and $X \subset \mathbb{R}^2$ is compact, then $S(X)$ is either a line with direction $\mathbf{e}$ or a region bounded by two support lines of X with each having direction $\mathbf{e}$. Furthermore $V = ext(\mathfrak{P}(X))$.

Example 6.2

Let $X = \{\mathbf{x}^1, \dots, \mathbf{x}^{12}\} \subset \mathbb{R}^2$, where

$$\mathbf{x}^1 = \begin{pmatrix}2.5\\3.5\end{pmatrix}, \mathbf{x}^2 = \begin{pmatrix}2\\2\end{pmatrix}, \mathbf{x}^3 = \begin{pmatrix}2.5\\1\end{pmatrix}, \mathbf{x}^4 = \begin{pmatrix}4\\2\end{pmatrix}, \mathbf{x}^5 = \begin{pmatrix}5\\4\end{pmatrix}, \mathbf{x}^6 = \begin{pmatrix}4.5\\5\end{pmatrix}, \mathbf{x}^7 = \begin{pmatrix}4\\3\end{pmatrix},$$
$$\mathbf{x}^8 = \begin{pmatrix}4.5\\3.5\end{pmatrix}, \mathbf{x}^9 = \begin{pmatrix}3.5\\2\end{pmatrix}, \mathbf{x}^{10} = \begin{pmatrix}3\\3\end{pmatrix}, \mathbf{x}^{11} = \begin{pmatrix}4\\3.5\end{pmatrix}, \text{ and } \mathbf{x}^{12} = \begin{pmatrix}2.5\\1.5\end{pmatrix}.$$

Step 1. of algorithm 6.1 yields $\mathbf{v} = \bigwedge_{j=1}^{12} \mathbf{x}^j = \binom{2}{1}$ and $\mathbf{u} = \bigvee_{j=1}^{12} \mathbf{x}^j = \binom{5}{5}$ and, hence, the interval $[\mathbf{v}, \mathbf{u}]$. Applying Step 2. results in the identity matrices

$$\mathfrak{W}_X = \begin{pmatrix} 0 & -1 \\ -2 & 0 \end{pmatrix} \text{ and } \mathfrak{M}_X = \begin{pmatrix} 0 & 2 \\ 1 & 0 \end{pmatrix}$$

for the lattice $S(X)$. The first two steps of the algorithm provide the input parameters for Step 3, which constructs the vectors $\mathbf{w}^1 = \binom{5}{3}$, $\mathbf{w}^2 = \binom{4}{5}$, $\mathbf{m}^1 = \binom{2}{3}$, and $\mathbf{m}^2 = \binom{3}{1}$. Figure 6.1 illustrates the subset relationship $X \subset \mathfrak{H}(X) \subset \mathfrak{P}(X)$ as well as the relationships of the elements of $ext(\mathfrak{P}(X))$ and the elements of the standard dual basis of $S(X)$.

It is apparent that if $X \subset \mathbb{R}^2$ is as in Example 6.2, then $V = ext(\mathfrak{P}(X))$ so that $\mathfrak{H}(V) = \mathfrak{P}(X)$. These two equations are easily verifiable for any compact set $X \subset \mathbb{R}^2$. The dimension of $\mathfrak{P}(X)$, however, can be 0-dimensional, 1-dimensional, or 2-dimensional. The first case occurs when X is trivial, namely $|X| = 1$. Figure 6.2 displays various possible examples of $\mathfrak{P}(X)$ for $|X| > 1$. As can be ascertained from the figure, if $dim(\mathfrak{P}(X)) = 1$, then $\mathfrak{P}(X)$ is either a horizontal line segment, a vertical line segment, or a line segment with slope equal to 1. In the case $\mathfrak{P}(X)$ is 2-dimensional, the shape of $\mathfrak{P}(X)$ is triangular, quadrilateral, pentagonal, or hexagonal, as illustrated in Figure 6.2. The basic shapes of $\mathfrak{P}(X) \subset \mathbb{R}^2$ shown in the figure demonstrate that any collection of $k \leq 3$ extreme points of $\mathfrak{P}(X)$ is affine independent and that $V = ext(\mathfrak{P}(X))$ for any compact subset X of $\mathbb{R}^2$.

Before examining the shape of $\mathfrak{P}(X)$ for finite or compact subsets of $\mathbb{R}^n$ with $n > 2$, we note that if $\ell \in \mathbb{N}_0$, then there exist compact sets $Z \subset \mathbb{R}^n$ such that $dim(\mathfrak{P}(Z)) = \ell$. These sets can be constructed inductively from lower dimensional examples as follows. Given $\mathfrak{P}(X) \subset \mathbb{R}^n$, let $Y = \{\mathbf{y} \in \mathbb{R}^{n+1} : \mathbf{y} = a + (\mathbf{x}, 0)', \mathbf{x} \in \mathfrak{P}(X), a \in \mathbb{R}\}$, where $(\mathbf{x}, 0)' = (x_1, \dots, x_n, 0)'$. Note that if $dim(\mathfrak{P}(X)) = m \leq n$, then $dim(Y) = m + 1$. Thus, if Z is any compact subset of Y and $dim(\mathfrak{H}(Z)) = m + 1$, then $dim(\mathfrak{P}(Z)) = m + 1$. For instance, for any set $\mathfrak{P}(X)$ in the first column in Figure 6.2, $dim(Y) = 2$ while for any set $\mathfrak{P}(X)$ in columns 2 and 3, $dim(Y) = 3$. If $Z \subset Y$ is chosen so that $dim(\mathfrak{H}(Z)) = k \leq m + 1$, then of course $dim(\mathfrak{P}(Z)) \leq m + 1$.

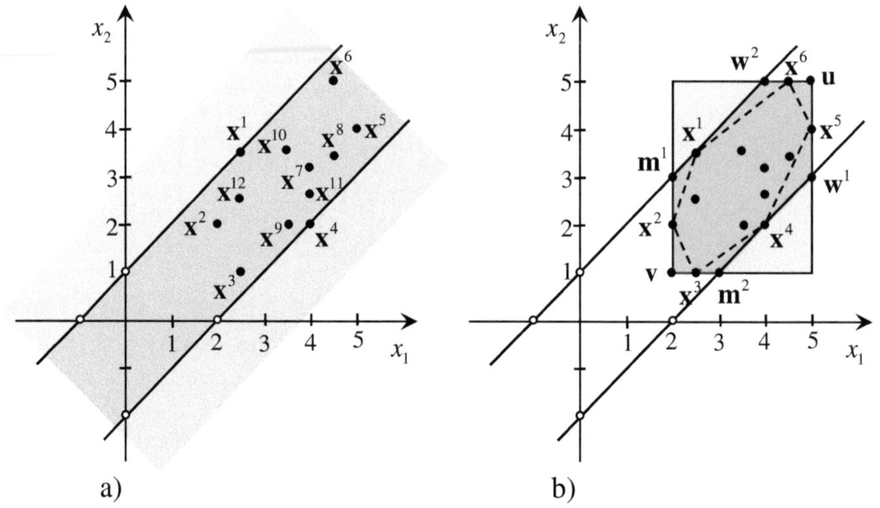

Figure 6.1 (a) shows the data set X of Example 6.2, the associated half-spaces $\overline{E^-_{\mathfrak{w}^1}}(\mathbf{d}_{12})$, $\overline{E^+_{\mathfrak{w}^2}}(\mathbf{d}_{12})$, and the dark shaded set $S(X) = \overline{E^-_{\mathfrak{w}^1}}(\mathbf{d}_{12}) \cap \overline{E^+_{\mathfrak{w}^2}}(\mathbf{d}_{12})$. Here the base points $\mathfrak{w}^1 = (0,-2)'$, $\mathfrak{m}^2 = (-1,0)'$, etc., derived from the matrices $\mathfrak{W}_X$ and $\mathfrak{M}_X$ are depicted by the open circles on the x_1 and x_2 axes. The dark shaded region in illustration (b) represents the ℓ-polytope $\mathfrak{P}(X)$ and the extremal points of $V = W \cup M \cup \{\mathbf{v},\mathbf{u}\}$. The dotted lines indicate the boundary of $\mathfrak{H}(X)$ [229].

Example 6.3 Suppose $X = \{\mathbf{x} \in \mathbb{R}^2 : \mathbf{x} = \lambda\mathbf{x}^1 + (1-\lambda)\mathbf{x}^2\}$, where $\mathbf{x}^1 = \binom{4}{4}$ and $\mathbf{x}^2 = \binom{4}{2}$. Then $W = \{\mathbf{w}^1, \mathbf{w}^2\} = M$, where $\mathbf{w}^1 = \mathbf{x}^2 = \mathbf{m}^2 = \mathbf{v}$ and $\mathbf{w}^2 = \mathbf{x}^1 = \mathbf{m}^1 = \mathbf{u}$. It follows that $\mathfrak{P}(X) = X$ with $dim(\mathfrak{P}(X)) = 1$ and $Y = \{\mathbf{y} \in \mathbb{R}^3 : \mathbf{y} = (a+\mathbf{x}, a)', \mathbf{x} \in X, a \in \mathbb{R}\}$ with $dim(Y) = 2$. Letting $Z = \{\mathbf{z}^1, \mathbf{z}^2, \mathbf{z}^3, \mathbf{z}^4\}$, where

$$\mathbf{z}^1 = \begin{pmatrix} 10 \\ 10 \\ 6 \end{pmatrix}, \mathbf{z}^2 = \begin{pmatrix} 10 \\ 8 \\ 6 \end{pmatrix}, \mathbf{z}^3 = \begin{pmatrix} 4 \\ 4 \\ 0 \end{pmatrix}, \text{ and } \mathbf{z}^4 = \begin{pmatrix} 4 \\ 2 \\ 0 \end{pmatrix},$$

then $\mathfrak{P}(Z)$ is a rectangle (see Figure 4.7) whose corners are the elements of Z. Thus $dim(\mathfrak{P}(Z)) = 2$. Had we chosen $Z = \{\mathbf{z}^1, \mathbf{z}^2\}$, where $\mathbf{z}^1$ and $\mathbf{z}^2$ are as above, then $dim(\mathfrak{P}(Z)) = 1$.

Since $\mathfrak{P}(X) \subset [\mathbf{v},\mathbf{u}]$, $dim(\mathfrak{P}(X)) \leq dim[\mathbf{v},\mathbf{u}]$. Consequently, if $A = \{u_i : u_i = v_i\}$ and $0 \leq |A| = m \leq n$, then $dim(\mathfrak{P}(X)) \leq dim[\mathbf{v},\mathbf{u}] = n-m$. Thus, if $m = n$, then $dim(\mathfrak{P}(X)) = 0$ and $\mathbf{u} = \mathbf{v}$, which happens only when $|X| = 1$. For instance, in Example 6.3 since $\mathbf{u} = \mathbf{z}^1$ and $\mathbf{v} = \mathbf{z}^4$, $dim[\mathbf{v},\mathbf{u}] = 3$ while $dim(\mathfrak{P}(Z)) = 2$, which also shows that strict inequality $dim(\mathfrak{P}(X)) < dim[\mathbf{v},\mathbf{u}]$ is possible.

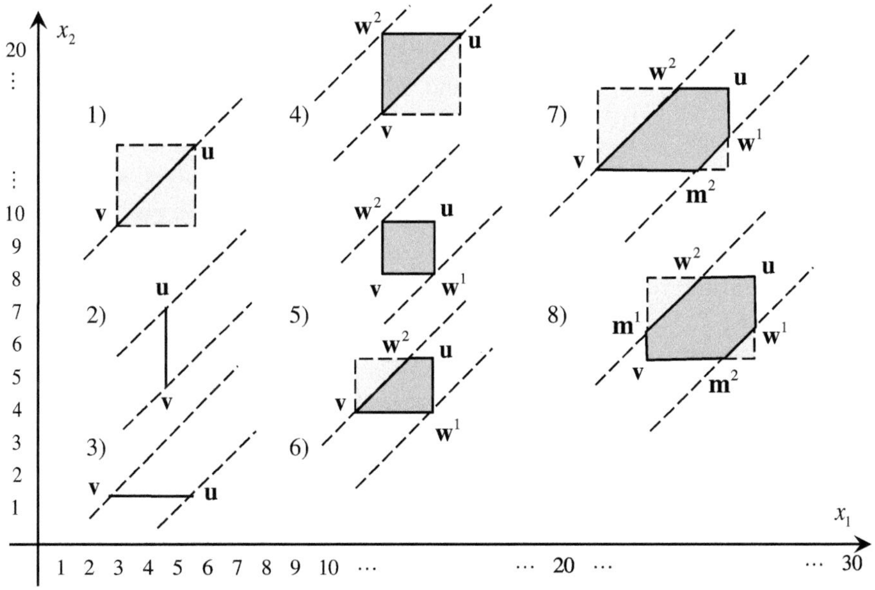

Figure 6.2 Examples of various configurations of the polyhedron $\mathfrak{P}(X)$ in the (x_1, x_2) plane. In the first column $dim(\mathfrak{P}(X)) = 1$, while in the remaining examples $dim(\mathfrak{P}(X)) = 2$. For the graphic representations (2), (3), and (5), $[\mathbf{v}, \mathbf{u}] = \mathfrak{P}(X)$, while for (1), (2), and (3) we also have $\langle \mathbf{v}, \mathbf{u} \rangle = \mathfrak{P}(X)$.

6.2.2.2 The case $n = 3$

If $X \subset \mathbb{R}^3$ is compact but the set $A = \{u_i : u_i = v_i\} \neq \varnothing$, then $0 \leq dim(\mathfrak{P}(X)) \leq dim[\mathbf{v}, \mathbf{u}] < 3$ and we are back to a k-dimensional problem with $0 \leq k \leq 2$. For this reason we shall assume that $dim(\mathfrak{P}(X)) = 3$.

We begin our discussion by constructing the 3-dimensional analogue of the example illustrated in Figure 6.2(5).

Example 6.4 Let $X = \{\mathbf{x}^1, \mathbf{x}^2, \mathbf{x}^3, \mathbf{x}^4, \mathbf{x}^5, \mathbf{x}^6\} \subset \mathbb{R}^3$, where

$$\mathbf{x}^1 = \begin{pmatrix} 14 \\ 10 \\ 12 \end{pmatrix}, \mathbf{x}^2 = \begin{pmatrix} 10 \\ 14 \\ 12 \end{pmatrix}, \mathbf{x}^3 = \begin{pmatrix} 12 \\ 14 \\ 10 \end{pmatrix}, \mathbf{x}^4 = \begin{pmatrix} 14 \\ 12 \\ 10 \end{pmatrix}, \mathbf{x}^5 = \begin{pmatrix} 12 \\ 10 \\ 14 \end{pmatrix}, \mathbf{x}^6 = \begin{pmatrix} 10 \\ 12 \\ 14 \end{pmatrix}.$$

Then

$$\mathfrak{W}_X = \begin{pmatrix} 0 & -4 & -4 \\ -4 & 0 & -4 \\ -4 & -4 & 0 \end{pmatrix}, \mathfrak{M}_X = \begin{pmatrix} 0 & 4 & 4 \\ 4 & 0 & 4 \\ 4 & 4 & 0 \end{pmatrix}, \mathbf{u} = \begin{pmatrix} 14 \\ 14 \\ 14 \end{pmatrix} \text{ and } \mathbf{v} = \begin{pmatrix} 10 \\ 10 \\ 10 \end{pmatrix}.$$

while

$$\mathbf{W} = \begin{pmatrix} 14 & 10 & 10 \\ 10 & 14 & 10 \\ 10 & 10 & 14 \end{pmatrix} \text{ and } \mathbf{M} = \begin{pmatrix} 10 & 14 & 14 \\ 14 & 10 & 14 \\ 14 & 14 & 10 \end{pmatrix},$$

In this particular example $[\mathbf{v},\mathbf{u}] \cap F_{\mathbf{w}^1}(\mathbf{d}_{12}) = ([\mathbf{v},\mathbf{u}] \cap F_{\mathbf{w}^1}(\mathbf{e}^1)) \cap F_{\mathbf{w}^1}(\mathbf{d}_{12}) = \langle \mathbf{w}^1, \mathbf{m}^2 \rangle$, $[\mathbf{v},\mathbf{u}] \cap F_{\mathbf{w}^2}(\mathbf{d}_{12}) = ([\mathbf{v},\mathbf{u}] \cap F_{\mathbf{w}^2}(\mathbf{e}^2)) \cap F_{\mathbf{w}^2}(\mathbf{d}_{12}) = \langle \mathbf{w}^2, \mathbf{m}^1 \rangle$, and $[\mathbf{v},\mathbf{u}] \cap F_{\mathbf{w}^3}(\mathbf{d}_{13}) = ([\mathbf{v},\mathbf{u}] \cap F_{\mathbf{w}^3}(\mathbf{e}^3)) \cap F_{\mathbf{w}^3}(\mathbf{d}_{13}) = \langle \mathbf{w}^3, \mathbf{m}^1 \rangle$. We encourage the reader to find the remaining edges generated by the intersections $F_{\mathbf{w}^1}(\mathbf{d}_{13})$, $F_{\mathbf{w}^2}(\mathbf{d}_{23})$, and $F_{\mathbf{w}^3}(\mathbf{d}_{23})$ with $[\mathbf{v},\mathbf{u}]$ and prove that $[\mathbf{v},\mathbf{u}] \subset S(X)$. Since $[\mathbf{v},\mathbf{u}] \subset S(X)$, we have $[\mathbf{v},\mathbf{u}] = [\mathbf{v},\mathbf{u}] \cap S(X) = \mathfrak{P}(X)$.

The location of the extreme points of $\mathfrak{P}(X)$ are shown in Figure 6.3. If $\mathbf{x} \in \{\mathbf{v},\mathbf{u}\}$, then each of the sets $\{\mathbf{m}^1,\mathbf{m}^2,\mathbf{m}^3,\mathbf{x}\}$, $\{\mathbf{w}^1,\mathbf{w}^2,\mathbf{w}^3,\mathbf{x}\}$, $\{\mathbf{m}^1,\mathbf{m}^2,\mathbf{w}^1,\mathbf{x}\}$, $\{\mathbf{m}^1,\mathbf{m}^2,\mathbf{w}^3,\mathbf{v}\}$, and $\{\mathbf{w}^1,\mathbf{m}^1,\mathbf{m}^2,\mathbf{m}^3\}$ is affine independent. On the other hand, the elements of each of the following sets $\{\mathbf{m}^1,\mathbf{m}^2,\mathbf{w}^3,\mathbf{u}\}$, $\{\mathbf{m}^1,\mathbf{w}^2,\mathbf{w}^3,\mathbf{v}\}$, or $\{\mathbf{w}^1,\mathbf{m}^1,\mathbf{m}^2,\mathbf{w}^2\}$ are coplanar. Thus each one of these sets is affine dependent. To recapitulate, we can take any three extreme points in one of the planar faces $E_{\mathbf{w}^i}(\mathbf{e}^i) \cap [\mathbf{v},\mathbf{u}]$ of $[\mathbf{v},\mathbf{u}]$ and any point in the opposite parallel planar face $E_{\mathbf{m}^i}(\mathbf{e}^i) \cap [\mathbf{v},\mathbf{u}]$ in order to obtain four affine independent points of $\mathfrak{P}(X)$.

As noted earlier, if $n = 2$, then any three points of $ext(\mathfrak{P}(X))$ are affine independent. Since $\mathfrak{P}(X)$ is convex and no three extreme points can be elements of a line segment $\langle \mathbf{x},\mathbf{y} \rangle \subset \mathfrak{P}(X)$, any three points of $ext(\mathfrak{P}(X)) \subset \mathbb{R}^n$ are affine independent for any integer $n \geq 2$. Additionally, we observed that $V = ext(\mathfrak{P}(X))$. The next example answers the question as to whether or not this equality holds in general.

Example 6.5 Let $X = \{\mathbf{x}^1,\mathbf{x}^2,\mathbf{x}^3\} \subset \mathbb{R}^3$, with

$$\mathbf{x}^1 = \begin{pmatrix} 6 \\ 1 \\ 14 \end{pmatrix}, \mathbf{x}^2 = \begin{pmatrix} 1 \\ 6 \\ 14 \end{pmatrix}, \mathbf{x}^3 = \begin{pmatrix} 5 \\ 5 \\ 20 \end{pmatrix}, \text{ thus } \mathbf{v} = \begin{pmatrix} 1 \\ 1 \\ 14 \end{pmatrix} \text{ and } \mathbf{u} = \begin{pmatrix} 6 \\ 6 \\ 20 \end{pmatrix}.$$

Here the corresponding matrices $\mathfrak{W}_X$ and $\mathfrak{M}_X$ are given by

$$\mathfrak{W}_X = \begin{pmatrix} 0 & -5 & -15 \\ -5 & 0 & -15 \\ 8 & 8 & 0 \end{pmatrix} \text{ and } \mathfrak{M}_X = \begin{pmatrix} 0 & 5 & -8 \\ 5 & 0 & -8 \\ 15 & 15 & 0 \end{pmatrix}.$$

Employing equation 6.8, one obtains the sets W and M or, equivalently, the

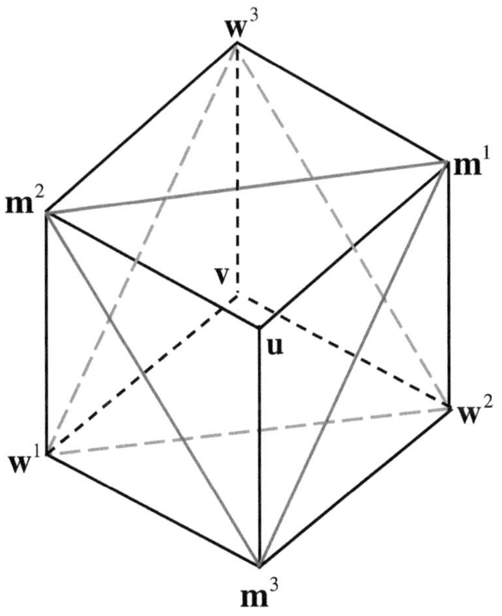

Figure 6.3 The ℓ-polytope $\mathfrak{P}(X)$ with the properties $\mathfrak{P}(X) = [\mathbf{v}, \mathbf{u}] \subset S(X)$ and $V = ext[\mathbf{v}, \mathbf{u}] = ext(\mathfrak{P}(X))$.

matrices

$$\mathbf{W} = \begin{pmatrix} 6 & 1 & 5 \\ 1 & 6 & 5 \\ 14 & 14 & 20 \end{pmatrix} \text{ and } \mathbf{M} = \begin{pmatrix} 1 & 6 & 6 \\ 6 & 1 & 6 \\ 16 & 16 & 14 \end{pmatrix}.$$

It follows that $W = X$ and the set $V = W \cup M \cup \{\mathbf{u}, \mathbf{v}\}$. However, as shown in Figure 6.4, $V \neq ext(\mathfrak{P}(X))$.

The polyhedron $\mathfrak{P}(X)$ shown in Figure 6.4 illustrates the fact that in dimensions $n > 2$ the equation $V = ext(\mathfrak{P}(X))$ is generally not achievable. For the set X of Example 6.5, $V \neq ext(\mathfrak{P}(X))$ since the extreme points $\mathbf{s}^1$, $\mathbf{s}^2$, and $\mathbf{p}^3$ of $\mathfrak{P}(X)$ are not elements of V. Even if the points $\mathbf{s}^1$ and $\mathbf{s}^2$ are included as elements of the set X, the sets W and M in the example remain the same and the additional elements $\mathbf{s}^1$ and $\mathbf{s}^2$ of X will not be elements of $\mathfrak{H}(V)$.

A consequence of equation 6.3 is that $\mathfrak{H}(ext(\mathfrak{P}(X))) = \mathfrak{P}(X)$. Thus, for a given data set $X \subset \mathbb{R}^n$, the geometry of $\mathfrak{P}(X)$ is completely dependent on the set $ext(\mathfrak{P}(X))$. It is therefore imperative to determine procedures for obtaining all extremal points of $\mathfrak{P}(X)$. To establish this goal we begin by examining Figure 6.4(b) as it provides a simple example of a 3-dimensional ℓ-polytope. The first step in constructing the polytope $\mathfrak{P}(X)$, where X is de-

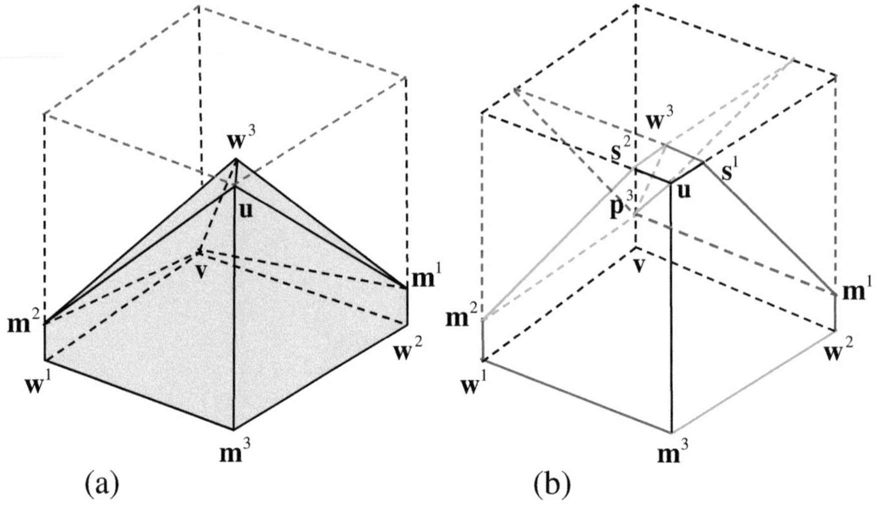

Figure 6.4 The shaded region in figure (a) indicates the polyhedron $\mathfrak{H}(V)$. Figure (b) shows the intersection of $[\mathbf{v},\mathbf{u}]$ with $S(X)$; here the red lines indicate the intersections of $E_{\mathbf{w}^1}(\mathbf{d}_{13})$ and $E_{\mathbf{w}^3}(\mathbf{d}_{13})$ with the cube $[\mathbf{v},\mathbf{u}]$, the blue lines correspond to the intersections of $E_{\mathbf{w}^1}(\mathbf{d}_{12})$ and $E_{\mathbf{w}^2}(\mathbf{d}_{12})$ with the cube, and the green lines mark the intersections of $E_{\mathbf{w}^2}(\mathbf{d}_{23})$ and $E_{\mathbf{w}^3}(\mathbf{d}_{23})$ with $[\mathbf{v},\mathbf{u}]$. Since $ext(\mathfrak{P}(X)) = V \cup \{\mathbf{s}^1,\mathbf{s}^2,\mathbf{p}^3\}, \mathfrak{P}(X) \neq \mathfrak{H}(V)$ [229].

fined in Example 6.5, is to compute the sets W, M, and $\{\mathbf{v},\mathbf{u}\}$. These three sets make up the *basic* extreme points that are used in deriving the additional extreme points $\mathbf{s}^1$, $\mathbf{s}^2$, and $\mathbf{p}^3$. The basic extremal points are the points used in establishing the boundaries of the sets $S(X)$ and $[\mathbf{v},\mathbf{u}]$, which in turn define the boundary pieces of $S(X) \cap [\mathbf{v},\mathbf{u}]$. As illustrated in the figure, the corner point $\mathbf{s}^1$ of $\mathfrak{P}(X)$ is due to the intersection of $F_{\mathbf{w}^3}(\mathbf{d}_{13})$ with the line $E_{\mathbf{w}^2}(\mathbf{e}^2) \cap E_{\mathbf{w}^3}(\mathbf{e}^3)$. Similarly, $\{\mathbf{s}^2\} = F_{\mathbf{w}^3}(\mathbf{d}_{23}) \cap (E_{\mathbf{w}^1}(\mathbf{e}^1) \cap E_{\mathbf{w}^3}(\mathbf{e}^2))$, and $\{\mathbf{p}^3\} = F_{\mathbf{m}^2}(\mathbf{d}_{23}) \cap (E_{\mathbf{m}^1}(\mathbf{e}^1) \cap E_{\mathbf{m}^2}(\mathbf{e}^2))$. What can also be deduced from the drawing—and easily calculated—is that

$$\mathbf{w}^2 \vee \mathbf{w}^3 = \begin{pmatrix} 5 \\ 6 \\ 20 \end{pmatrix} = \mathbf{s}^1,\ \mathbf{w}^1 \vee \mathbf{w}^3 = \begin{pmatrix} 6 \\ 3 \\ 20 \end{pmatrix} = \mathbf{s}^2 \text{ and } \mathbf{m}^1 \wedge \mathbf{m}^2 = \begin{pmatrix} 1 \\ 1 \\ 18 \end{pmatrix} = \mathbf{p}^3. \tag{6.18}$$

Another pertinent observation is that $\mathbf{p}^3 \in F_{\mathbf{m}^1}(\mathbf{d}_{13}) \cap F_{\mathbf{m}^2}(\mathbf{d}_{23}) = L(\mathbf{w}^3)$ and $\mathbf{p}^3 = a + \mathbf{w}^3$, where $a = -4$ so that $\mathbf{p}^3 < \mathbf{w}^3$. It follows that $\langle \mathbf{p}^3, \mathbf{w}^3 \rangle \subset L(\mathbf{w}^3)$ is an edge of $\mathfrak{P}(X)$ with endpoints $\{\mathbf{p}^3, \mathbf{w}^3\} = L(\mathbf{w}^3) \cap \partial[\mathbf{v},\mathbf{u}]$.

More generally, suppose $X \subset \mathbb{R}^n$ is a finite set, $\mathbf{v} = \bigwedge_{\mathbf{x}\in X} \mathbf{x}$, and $\mathbf{u} = \bigvee_{\mathbf{x}\in X} \mathbf{x}$.

If $\mathbf{w} \in \mathbb{R}^n$ and $L(\mathbf{w}) \cap int[\mathbf{v},\mathbf{u}] \neq \varnothing$, then $|L(\mathbf{w}) \cap \partial[\mathbf{v},\mathbf{u}]| = 2$. Thus, if $\{\mathbf{p},\mathbf{q}\} = L(\mathbf{w}) \cap \partial[\mathbf{v},\mathbf{u}]$, then since $\mathbf{p}, \mathbf{q} \in L(\mathbf{w})$, either $\mathbf{p} < \mathbf{q}$ or $\mathbf{q} < \mathbf{p}$. The lesser point is called the *entry point* of $L(\mathbf{w})$ and the larger point is called the *exit point* of the line $L(\mathbf{w})$ with respect to the hyperbox $[\mathbf{v},\mathbf{u}]$. Assuming that $L(\mathbf{w}^i) \cap int[\mathbf{v},\mathbf{u}] \neq \varnothing$, then since $\{\mathbf{w}^i\} = E_{\mathbf{u}}(\mathbf{e}^i) \cap L(\mathbf{w}^i)$ and $\forall\ \mathbf{x} \in [\mathbf{v},\mathbf{u}] \cap L(\mathbf{w}^i)$ with $\mathbf{x} \neq \mathbf{w}^i$, $\mathbf{x} < \mathbf{w}^i$, it follows that $\mathbf{w}^i$ is the exit point of $L(\mathbf{w}^i)$. Henceforth we will denote the corresponding entry point of $L(\mathbf{w}^i)$ by $\mathbf{p}^i$.

An analogous argument shows that if $L(\mathbf{m}^j) \cap int[\mathbf{v},\mathbf{u}] \neq \varnothing$, then $\mathbf{m}^j$ is an entry point of $L(\mathbf{m}^j)$ into $[\mathbf{v},\mathbf{u}]$. The corresponding exit point of $L(\mathbf{m}^j)$ will be denoted by $\mathbf{q}^j$. Since $L(\mathbf{w}^i) \cap int[\mathbf{v},\mathbf{u}] = \varnothing$ for $i = 1, 2$ in Example 6.5, neither $\mathbf{w}^1$ or $\mathbf{w}^2$ is an exit or entry point. The same is true for $\mathbf{m}^i$ with $i \in \{1,2,3\}$.

Although $|ext(\mathfrak{P}(X))| = 11 > 8 = |W \cup M \cup \{\mathbf{v},\mathbf{u}\}|$ in Example 6.5, the number of extreme points can be much higher. This is due to the observation that for large randomly generated data sets, the sets M and W are generally well-separated. Here the notion of separation does not refer to topological separation or disjointness, but is based on disjointness in terms of lattice structures.

Definition 6.3 Suppose $(L, \vee, \wedge)$ is a lattice and $X, Y \subset L$. The sets X and Y are said to be *lattice disjoint*, or simply *ℓ-disjoint* if and only if $X \cap Y = \varnothing$ and there does not exist a subset $A \subset X$ or a subset $B \subset Y$, or both, such that

$$y = \bigvee_{x \in A} x \text{ or } y = \bigwedge_{x \in A} x \text{ for some } y \in Y, \text{ or } x = \bigvee_{y \in B} y \text{ or } x = \bigwedge_{y \in B} y \text{ for some } x \in X.$$

Sets that are ℓ-disjoint are also called *lattice-separated* or simply *ℓ-separated* while sets that are not ℓ-separated are said to be *lattice-connected* or *ℓ-connected*.

If L is a lattice vector space and X and Y are ℓ-separated, then X and Y are *strong lattice-separated* if and only if $L(x) \neq L(y)\ \forall\ x \in X$ and $y \in Y$.

For instance, the sets W and M in Example 6.2 are lattice disjoint. In contrast, the sets W and M in Example 6.5 are ℓ-connected since $\mathbf{m}^3 \wedge \mathbf{m}^2 = \mathbf{w}^1$ and $\mathbf{w}^1 \vee \mathbf{w}^2 = \mathbf{m}^3$ even though $W \cap M = \varnothing$. The following is an example of a set $X \subset \mathbb{R}^3$ for which the sets W and M are lattice-separated.

Example 6.6 Suppose $X = \{\mathbf{x}^1,\mathbf{x}^2,\mathbf{x}^3,\mathbf{x}^4,\mathbf{x}^5,\mathbf{x}^6\} \subset \mathbb{R}^3$, where

$$\mathbf{x}^1 = \begin{pmatrix} 10 \\ 7 \\ 10 \end{pmatrix}, \mathbf{x}^2 = \begin{pmatrix} 8 \\ 12 \\ 9 \end{pmatrix}, \mathbf{x}^3 = \begin{pmatrix} 9 \\ 8 \\ 10 \end{pmatrix}, \mathbf{x}^4 = \begin{pmatrix} 1 \\ 3 \\ 1 \end{pmatrix}, \mathbf{x}^5 = \begin{pmatrix} 2 \\ 1 \\ 1 \end{pmatrix}, \mathbf{x}^6 = \begin{pmatrix} 1 \\ 2 \\ 2 \end{pmatrix}.$$

The corresponding matrices $\mathfrak{W}_X$, $\mathfrak{M}_X$, and the vectors $\mathbf{u}$ and $\mathbf{v}$ are given by

$$\mathfrak{W}_X = \begin{pmatrix} 0 & -4 & -1 \\ -3 & 0 & -3 \\ -1 & -3 & 0 \end{pmatrix}, \ \mathfrak{M}_X = \begin{pmatrix} 0 & 3 & 1 \\ 4 & 0 & 3 \\ 1 & 3 & 0 \end{pmatrix}, \ \mathbf{u} = \begin{pmatrix} 10 \\ 12 \\ 10 \end{pmatrix}, \ \mathbf{v} = \begin{pmatrix} 1 \\ 1 \\ 1 \end{pmatrix}.$$

Using Algorithm 6.1 results in

$$\mathbf{w}^1 = \begin{pmatrix} 10 \\ 7 \\ 9 \end{pmatrix}, \mathbf{w}^2 = \begin{pmatrix} 8 \\ 12 \\ 9 \end{pmatrix}, \mathbf{w}^3 = \begin{pmatrix} 9 \\ 7 \\ 10 \end{pmatrix}, \mathbf{m}^1 = \begin{pmatrix} 1 \\ 5 \\ 2 \end{pmatrix}, \mathbf{m}^2 = \begin{pmatrix} 4 \\ 1 \\ 4 \end{pmatrix}, \mathbf{m}^3 = \begin{pmatrix} 2 \\ 4 \\ 1 \end{pmatrix}.$$

It is now an easy exercise to check that the sets $W = \{\mathbf{w}^1, \mathbf{w}^2, \mathbf{w}^3\}$ and $M = \{\mathbf{m}^1, \mathbf{m}^2, \mathbf{m}^3\}$ are not ℓ-connected.

Following the rational that resulted in equation 6.10, we note that since $\mathbf{m}^i \in E_{\mathbf{m}^i}(\mathbf{e}^i) = E_{\mathbf{v}}(\mathbf{e}^i)$ and $\mathbf{m}^j \in E_{\mathbf{m}^j}(\mathbf{e}^j) = E_{\mathbf{v}}(\mathbf{e}^j)$, it follows that $\mathbf{m}^i \wedge \mathbf{m}^j \in E_{\mathbf{v}}(\mathbf{e}^i) \cap E_{\mathbf{v}}(\mathbf{e}^j)$. Thus, $\mathbf{m}^i \wedge \mathbf{m}^j$ is always an edge point of the hyperbox $[\mathbf{v}, \mathbf{u}]$. The notation $\mathbf{r}^\ell = \mathbf{m}^i \wedge \mathbf{m}^j$ is a reminder that $i \neq \ell \neq j$ and important in the generalization of this process. Similarly, $\mathbf{s}^\ell = \mathbf{w}^i \vee \mathbf{w}^j \in E_{\mathbf{u}}(\mathbf{e}^i) \cap E_{\mathbf{u}}(\mathbf{e}^j)$. However, an edge point of $[\mathbf{v}, \mathbf{u}]$ is not necessarily an extremal point of $\mathfrak{P}(X)$. That the points $\mathbf{w}^i \vee \mathbf{w}^j$ and $\mathbf{m}^i \wedge \mathbf{m}^j$ are indeed extremal points of $\mathfrak{P}(X)$ is a consequence of the following equations:

$$\mathbf{w}^1 \vee \mathbf{w}^2 = (E_{\mathbf{u}}(\mathbf{e}^1) \cap E_{\mathbf{u}}(\mathbf{e}^2)) \cap (F_{\mathbf{w}^1}(\mathbf{d}_{13}) \cap F_{\mathbf{w}^2}(\mathbf{d}_{23})) \tag{6.19}$$
$$\mathbf{w}^1 \vee \mathbf{w}^3 = (E_{\mathbf{u}}(\mathbf{e}^1) \cap E_{\mathbf{u}}(\mathbf{e}^3)) \cap (F_{\mathbf{w}^1}(\mathbf{d}_{12}) \cap F_{\mathbf{w}^3}(\mathbf{d}_{23})) \tag{6.20}$$
$$\mathbf{w}^2 \vee \mathbf{w}^3 = (E_{\mathbf{u}}(\mathbf{e}^2) \cap E_{\mathbf{u}}(\mathbf{e}^3)) \cap F_{\mathbf{w}^3}(\mathbf{d}_{13}) \tag{6.21}$$
$$\mathbf{m}^1 \wedge \mathbf{m}^2 = (E_{\mathbf{v}}(\mathbf{e}^1) \cap E_{\mathbf{v}}(\mathbf{e}^2)) \cap F_{\mathbf{m}^1}(\mathbf{d}_{13}) \tag{6.22}$$
$$\mathbf{m}^1 \wedge \mathbf{m}^3 = (E_{\mathbf{v}}(\mathbf{e}^1) \cap E_{\mathbf{v}}(\mathbf{e}^3)) \cap F_{\mathbf{m}^3}(\mathbf{d}_{23}) \tag{6.23}$$
$$\mathbf{m}^2 \wedge \mathbf{m}^3 = (E_{\mathbf{v}}(\mathbf{e}^2) \cap E_{\mathbf{v}}(\mathbf{e}^3)) \cap F_{\mathbf{m}^3}(\mathbf{d}_{13}) \tag{6.24}$$

where

$$\mathbf{w}^1 \vee \mathbf{w}^2 = \begin{pmatrix} 10 \\ 12 \\ 9 \end{pmatrix} = \mathbf{s}^3, \ \mathbf{w}^1 \vee \mathbf{w}^3 = \begin{pmatrix} 10 \\ 7 \\ 10 \end{pmatrix} = \mathbf{s}^2, \ \mathbf{w}^2 \vee \mathbf{w}^3 = \begin{pmatrix} 9 \\ 12 \\ 10 \end{pmatrix} = \mathbf{s}^1 \text{ and}$$

$$\mathbf{m}^1 \wedge \mathbf{m}^2 = \begin{pmatrix} 1 \\ 1 \\ 2 \end{pmatrix} = \mathbf{r}^3, \ \mathbf{m}^1 \wedge \mathbf{m}^3 = \begin{pmatrix} 1 \\ 4 \\ 1 \end{pmatrix} = \mathbf{r}^2, \ \mathbf{m}^2 \wedge \mathbf{m}^3 = \begin{pmatrix} 2 \\ 1 \\ 1 \end{pmatrix} = \mathbf{r}^1.$$

Accordingly, we now have a set $V \cup \{\mathbf{s}^1, \mathbf{s}^2, \mathbf{s}^3, \mathbf{r}^1, \mathbf{r}^2, \mathbf{r}^3\}$ of 14 easily computable extremal points of $\mathfrak{P}(X)$. However these are not all of the extremal points of $\mathfrak{P}(X)$. Additional extremal points that need to be considered are the entry and exit points generated by the lines $L(\mathbf{w}^i)$ and $L(\mathbf{m}^j)$.

Although the coordinate values of the points in set $\{\mathbf{s}^1,\mathbf{s}^2,\mathbf{s}^3,\mathbf{r}^1,\mathbf{r}^2,\mathbf{r}^3\}$ can be quickly computed in terms of the respective max and min operation $\mathbf{w}^i \vee \mathbf{w}^j$ and $\mathbf{m}^i \wedge \mathbf{m}^j$, their more complex formulations expressed in equations 6.19 through 6.24 provide additional geometric information. Note that equations 6.19 and 6.20 represents the intersection of two lines, while equations 6.21 through 6.24 represents the intersection of a line with a plane. Also, since $E_{\mathbf{w}^1}(\mathbf{d}_{13}) \cap E_{\mathbf{w}^2}(\mathbf{d}_{23}) = E_{\mathbf{m}^3}(\mathbf{d}_{13}) \cap E_{\mathbf{m}^3}(\mathbf{d}_{23}) = L(\mathbf{m}^3)$ and $E_{\mathbf{w}^1}(\mathbf{d}_{12}) \cap E_{\mathbf{w}^3}(\mathbf{d}_{23}) = E_{\mathbf{m}^2}(\mathbf{d}_{12}) \cap E_{\mathbf{m}^2}(\mathbf{d}_{23}) = L(\mathbf{m}^2)$, it follows that $\mathbf{s}^3 = a+\mathbf{m}^3$ and $\mathbf{s}^2 = b+\mathbf{m}^2$ with $a = 8$ and $b = 6$. Since the two points $\mathbf{s}^3$ and $\mathbf{s}^2$ are just translates of $\mathbf{m}^3$ and $\mathbf{m}^2$, they correspond to the exit points $\mathbf{q}^3 = 8+\mathbf{m}^3$ and $\mathbf{q}^2 = 6+\mathbf{m}^2$. We also have $L(\mathbf{m}^1) \cap \partial[\mathbf{v},\mathbf{u}] = \{\mathbf{m}^1,\mathbf{w}^2\}$ since $7+\mathbf{m}^1 = \mathbf{w}^2$ so that $\mathbf{q}^1 = \mathbf{w}^2$ and $\mathbf{p}^2 = \mathbf{m}^1$. This holds whenever $L(\mathbf{m}^i) = L(\mathbf{w}^j)$ or, equivalently, whenever W and M are not strong ℓ-separated.

We now know the extreme points generated by $L(\mathbf{m}^i)$ for $i = 1,2,3$, as well as those associated with $L(\mathbf{w}^2)$, but we have not yet explored possible extreme points generated by $L(\mathbf{w}^1)$ and $L(\mathbf{w}^3)$. Considering the line $L(\mathbf{w}^1)$, the first step is to find the shortest distance from $\mathbf{w}^1$ along the line $L(\mathbf{w}^1)$ to the set of planes $E_{\mathbf{v}}(\mathbf{e}^i)$ for $i = 1,2$, and 3. This can be achieved by setting $a = \bigvee_{i=1}^{3}(v_i - w_i^1)$ and computing $a+\mathbf{w}^1$. In this case one obtains $a = -6$ and $-6+\mathbf{w}^1 = (4,1,3)' = \mathbf{p}^1$. In a similar fashion one obtains $-6+\mathbf{w}^3 = (3,1,4)' = \mathbf{p}^3$. This shows that $ext(\mathfrak{P}(X)) = W \cup M \cup \{\mathbf{p}^1,\mathbf{p}^3,\mathbf{q}^2,\mathbf{q}^3,\mathbf{r}^1,\mathbf{r}^2,\mathbf{r}^3,\mathbf{s}^1,\mathbf{v},\mathbf{u}\}$ so that $|ext(\mathfrak{P}(X))| = 16$. The shaded region in Figure 6.4 represents the polytope $\mathfrak{P}(X)$ and shows the locations of the various extremal points. It is also constructive to note that while $|W \cup M| = 6$, the polytope $S(X)$ has only five one-dimensional faces and five two-dimensional faces.

Example 6.6 demonstrates that lattice separation forces an increase in distance between the sets W and M, culminating in the inequality $\mathbf{m}^j \leq \mathbf{w}^i$. The downside is that a large increase in the number of extremal points can be expected whenever W and M are lattice disjoint. More important is the method employed in Example 6.6 for determining any remaining extreme points in $ext(\mathfrak{P}(X)) \setminus (W \cup M \cup \{\mathbf{v},\mathbf{u}\})$ in case $n = 3$. This method reduces to three simple steps and can also be employed to compute large subsets of $ext(\mathfrak{P}(X))$ in case $n > 3$.

Algorithm 6.2 (Computing the Elements of $ext(\mathfrak{P}(X))$)

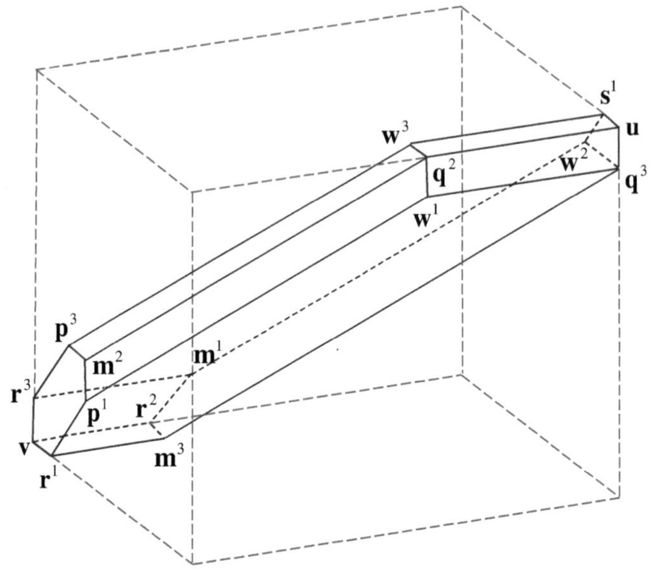

Figure 6.5 The set $\mathfrak{P}(X) \subset [\mathbf{v}, \mathbf{u}]$ and its extreme points [229].

STEP *1. Given* $X \subset \mathbb{R}^n$, *apply algorithm 6.1 in order to obtain the set* V *from the collection* $\{\mathbf{w}^1, \ldots, \mathbf{w}^n, \mathbf{m}^1, \ldots, \mathbf{m}^n, \mathbf{v}, \mathbf{u}\}$.
STEP *2. For* $\mathbf{w}^j$ *and* $\mathbf{m}^j \in V$, *define*

$$a_j = \bigvee_{i=1}^{n} (v_i - \mathbf{w}_i^j),\ b_j = \bigwedge_{i=1}^{n} (u_i - \mathbf{m}_i^j),\ \textit{and set}\ \mathbf{p}^j = a_j + \mathbf{w}^j\ \textit{and}\ \mathbf{q}^j = b_j + \mathbf{m}^j.$$

Let $\Psi = \{\mathbf{p}^j : \mathbf{w}^j \in V\}$, $Q = \{\mathbf{q}^j : \mathbf{m}^j \in V\}$, *and set* $U = (\Psi \cup Q) \setminus V$.
STEP *3. For* $i, j \in \mathbb{N}_n$ *set*

$$\mathbf{r}^j = \bigwedge_{i \neq j} \mathbf{m}^i\ \textit{and}\ \mathbf{s}^j = \bigvee_{i \neq j} \mathbf{w}^i,$$

Let $R_1 = \{\mathbf{r}^j : j \in \mathbb{N}_n\}$, $S_1 = \{\mathbf{s}^j : j \in \mathbb{N}_n\}$, *and set* $T_1 = (R_1 \cup S_1) \setminus (V \cup U)$.

The computation of the points $\mathbf{r}^\ell$ and $\mathbf{s}^\ell$ in $\mathbb{R}^3$ can also be accomplished by defining $\mathbf{r}(i,j) = \mathbf{m}^i \wedge \mathbf{m}^j$ and $\mathbf{s}(i,j) = \mathbf{w}^i \vee \mathbf{w}^j$. Then $\mathbf{r}^\ell = \mathbf{r}(i,j)$ and $\mathbf{s}^\ell = \mathbf{s}(i,j)$, where $\ell \notin \{i,j\}$. The notation $\mathbf{r}(i,j)$ and $\mathbf{s}(i,j)$ is convenient when dealing with extreme points generated by lower-dimensional faces of $S(X)$ in dimension $n \geq 4$.

The points $\mathbf{p}^j$ and $\mathbf{q}^j$ generated in STEP 2 are the entry points and exit points of the lines $L(\mathbf{w}^j)$ and $L(\mathbf{m}^j)$, respectively. The exit point of the line $L(\mathbf{w}^j)$ is $\mathbf{w}^j$, while the entry point of $L(\mathbf{m}^j)$ is $\mathbf{m}^j$. In view of Example 6.6, the points $\mathbf{s}^\ell$ and $\mathbf{r}^\ell$ obtained from STEP *3* are the intersection of the respective lines $E_{\mathbf{u}}(\mathbf{e}^i) \cap E_{\mathbf{u}}(\mathbf{e}^j)$ with 2-dimensional faces of $S(X) \subset \mathbb{R}^3$, i.e., satisfying equations 6.21 through 6.24 but not 6.19 and 6.20. Any point $\mathbf{s}^\ell$ or $\mathbf{r}^\ell$ that is in the intersection involving more than one 2-dimensional face of $S(X)$ is superfluous as it is already an element of $U \cup V$.

Clearly, if $n = 2$, then only STEP *1* is necessary and $V = ext(\mathfrak{P}(X))$. Furthermore, if $n \geq 2$, then STEP *1* of the algorithm can result in a maximum of $2n+2$ vectors. If $n \geq 3$, STEP *2* as well as STEP *3* may each produce an additional $2n$ distinct extremal points. Thus, the maximal number of extreme points possible using this algorithm is $3(2n)+2 = 2(3n+1)$.

As can be deduced from equations 6.19 through 6.24, the elements of T_1 represent the points where the faces $F_{\mathbf{w}^i}(\mathbf{d}_{ij})$ or $F_{\mathbf{m}^i}(\mathbf{d}_{ij})$ cut an edge of $[\mathbf{v},\mathbf{u}]$. These points could also be exit or entry points of a line $L(\mathbf{w}^i)$ or $L(\mathbf{m}^i)$, while the elements of the set U obtained from Step *2* are all entry and/or exit points. Note that if for some j, $a_j = 0$ (or $b_j = 0$), then $\mathbf{p}^j = \mathbf{w}^j$ (or $\mathbf{q}^j = \mathbf{m}^j$). However, these points will not be members of U or T_1 since by definition of U and T_1 we have that $V \cap U = T_1 \cap V = T_1 \cap U = T_1 \cap (V \cup U) = \varnothing$. Consequently, if $n = 3$, then

$$V \cup U \cup T_1 = ext(\mathfrak{P}(X)) \text{ and } |V|+|U|+|T_1| = |ext(\mathfrak{P}(X))|. \qquad (6.25)$$

The case $n = 4$ verifies that for $n > 3$ it is very likely that $V \cup U \cup T_1 \neq ext(\mathfrak{P}(X)$.

Definition 6.4 If $n \geq 3$ and $\mathfrak{P}(X) \subset \mathbb{R}^n$, then $\mathfrak{P}(X)$ is said to be *maximal* if and only if $|W \cup M| = |U| = 2n$.

It follows that if $\mathfrak{P}(X)$ is maximal, then for $i \in \mathbb{N}_n$, $L(\mathbf{w}^i) \neq L(\mathbf{m}^j)\ \forall\, j = 1, 2, \ldots, n$. However, as Example 6.4 shows, the converse is not true since $|U| = 0$. Nevertheless, $\mathfrak{P}(X)$ is maximal if and only if $|W \cup M| = 2n$ and W and M are strong ℓ-separated.

Example 6.7 Suppose $X = \{\mathbf{x}^1, \ldots, \mathbf{x}^6\} \subset \mathbb{R}^3$, where

$$\mathbf{x}^1 = \begin{pmatrix} 10 \\ 7 \\ 5 \end{pmatrix}, \mathbf{x}^2 = \begin{pmatrix} 9 \\ 10 \\ 4 \end{pmatrix}, \mathbf{x}^3 = \begin{pmatrix} 5 \\ 4 \\ 10 \end{pmatrix}, \mathbf{x}^4 = \begin{pmatrix} 3 \\ 5 \\ 7 \end{pmatrix}, \mathbf{x}^5 = \begin{pmatrix} 3 \\ 1 \\ 3 \end{pmatrix}, \mathbf{x}^6 = \begin{pmatrix} 1 \\ 3 \\ 2 \end{pmatrix},$$

$$\mathbf{x}^7 = \begin{pmatrix} 6 \\ 2 \\ 1 \end{pmatrix}, \text{ and } \mathbf{x}^8 = \begin{pmatrix} 2 \\ 9 \\ 4 \end{pmatrix}. \text{ Thus } \mathbf{v} = \begin{pmatrix} 1 \\ 1 \\ 1 \end{pmatrix}, \text{ and } \mathbf{u} = \begin{pmatrix} 10 \\ 10 \\ 10 \end{pmatrix}.$$

STEP *1* of Algorithm 6.2 results in the set $W \cup M$ consisting of the vectors

$$\mathbf{w}^1 = \begin{pmatrix} 10 \\ 6 \\ 5 \end{pmatrix}, \mathbf{w}^2 = \begin{pmatrix} 3 \\ 10 \\ 4 \end{pmatrix}, \mathbf{w}^3 = \begin{pmatrix} 5 \\ 4 \\ 10 \end{pmatrix}, \mathbf{m}^1 = \begin{pmatrix} 1 \\ 8 \\ 6 \end{pmatrix}, \mathbf{m}^2 = \begin{pmatrix} 5 \\ 1 \\ 7 \end{pmatrix}, \mathbf{m}^3 = \begin{pmatrix} 6 \\ 7 \\ 1 \end{pmatrix}.$$

Thus, $|V| = 8$.

The first computation of STEP *2* yields the following additive coefficients:

$$a_1 = -4,\ a_2 = -2,\ a_3 = -3, \text{ and } b_1 = 2,\ b_2 = 3,\ b_3 = 3.$$

Using these values in order to compute the points $\mathbf{p}^j = a_j + \mathbf{w}^j$ and $\mathbf{q}^j = b_j + \mathbf{m}^j$, results in

$$\mathbf{p}^1 = -4 + \mathbf{w}^1 = \begin{pmatrix} 6 \\ 2 \\ 1 \end{pmatrix}, \mathbf{p}^2 = -2 + \mathbf{w}^2 = \begin{pmatrix} 1 \\ 8 \\ 2 \end{pmatrix}, \mathbf{p}^3 = -3 + \mathbf{w}^3 = \begin{pmatrix} 2 \\ 1 \\ 7 \end{pmatrix} \text{ and}$$

$$\mathbf{q}^1 = 2 + \mathbf{m}^1 = \begin{pmatrix} 3 \\ 10 \\ 8 \end{pmatrix}, \mathbf{q}^2 = 3 + \mathbf{m}^2 = \begin{pmatrix} 8 \\ 4 \\ 10 \end{pmatrix}, \mathbf{q}^3 = 3 + \mathbf{m}^3 = \begin{pmatrix} 9 \\ 10 \\ 4 \end{pmatrix}.$$

In this case there are no duplicates, which means that $U = \Psi \cup Q$ and $|U| = 6$. Also, since $|W \cup M| = |U| = 6$, $\mathfrak{P}(X)$ is maximal.

The final STEP *3* produces

$$\mathbf{r}^1 = \mathbf{m}^2 \wedge \mathbf{m}^3 = \begin{pmatrix} 5 \\ 1 \\ 1 \end{pmatrix}, \mathbf{r}^2 = \mathbf{m}^1 \wedge \mathbf{m}^3 = \begin{pmatrix} 1 \\ 7 \\ 1 \end{pmatrix}, \mathbf{r}^3 = \mathbf{m}^1 \wedge \mathbf{m}^2 = \begin{pmatrix} 1 \\ 1 \\ 6 \end{pmatrix} \text{ and}$$

$$\mathbf{s}^1 = \mathbf{w}^2 \vee \mathbf{w}^3 = \begin{pmatrix} 5 \\ 10 \\ 10 \end{pmatrix}, \mathbf{s}^2 = \mathbf{w}^1 \vee \mathbf{w}^3 = \begin{pmatrix} 10 \\ 6 \\ 10 \end{pmatrix}, \mathbf{s}^3 = \mathbf{w}^1 \vee \mathbf{w}^2 = \begin{pmatrix} 10 \\ 10 \\ 5 \end{pmatrix}.$$

Again, there are no duplicates so that $|T_1| = 6$. Therefore $|ext(\mathfrak{P}(X))| = |V| + |U| + |T_1| = 20$, which is the maximal number of extreme points possible for an ℓ-polytope $\mathfrak{P}(X) \subset \mathbb{R}^3$.

Although algorithm 6.2 computes all the extremal points of a given ℓ-polytope $\mathfrak{P}(X) \subset \mathbb{R}^3$, the algorithm does not directly provide a method for selecting affine independent subsets of $ext(\mathfrak{P}(X))$. For $n = 2$, this task is trivial, while for $n = 3$ and $dim(\mathfrak{P}(X)) = 3$, the task is still easy. Here we use the simple fact that any three elements of $ext(\mathfrak{P}(X))$ are affine independent and therefore define a plane. Thus, once three extreme points $\mathbf{x}$, $\mathbf{y}$, and $\mathbf{z}$ have been selected, the remaining objective is the identification of a fourth extremal point

$\mathbf{w}$ that is not an element of the plane generated by $\{\mathbf{x}, \mathbf{y}, \mathbf{z}\}$. Of course, this objective depends on the selection of the initial three points. An easy method is to simply choose a plane $E_{\mathbf{w}^i}(\mathbf{d}_{ij})$ such that $L(\mathbf{w}^i) \neq L(\mathbf{m}^j)$ and select the two extreme points $\mathbf{w}^i$ and $\mathbf{p}^i$ from $\{\mathbf{w}^i, \mathbf{p}^i\} = L(\mathbf{w}^i) \cap ext(\mathfrak{P}(X))$ and one extreme point $\mathbf{z} \in \{\mathbf{m}^j, \mathbf{q}^j\} = L(\mathbf{m}^j) \cap ext(\mathfrak{P}(X))$, or vice versa. Suppose P corresponds to the plane generated by $\{\mathbf{w}^i, \mathbf{p}^i, \mathbf{z}\}$. Since $E_{\mathbf{w}^i}(\mathbf{d}_{ij}) \cap E_{\mathbf{w}^j}(\mathbf{d}_{ij}) = \varnothing$, it follows that for any point $\mathbf{w} \in E_{\mathbf{w}^j}(\mathbf{d}_{ij}) \cap ext(\mathfrak{P}(X))$, $\mathbf{w} \notin P$.

This simple method of extracting four affine independent points can be generalized by selecting any three points from $F_{\mathbf{w}^i}(\mathbf{d}_{ij}) \cap ext(\mathfrak{P}(X))$ and one point from $ext(\mathfrak{P}(X)) \setminus [F_{\mathbf{w}^i}(\mathbf{d}_{ij}) \cap ext(\mathfrak{P}(X))]$. For instance, given the set $ext(\mathfrak{P}(X))$ of Example 6.6 and choosing any three elements from $\{\mathbf{r}^1, \mathbf{m}^3, \mathbf{q}^3, \mathbf{w}^1, \mathbf{p}^1\} = F_{\mathbf{w}^1}(\mathbf{d}_{13}) \cap ext(\mathfrak{P}(X))$ and one from $ext(\mathfrak{P}(X)) \setminus \{\mathbf{r}^1, \mathbf{m}^3, \mathbf{q}^3, \mathbf{w}^1, \mathbf{p}^1\}$, results in a set of four affine independent points.

As demonstrated in the next section, the method just described can be easily generalized for extracting affine independent points from $ext(\mathfrak{P}(X)) \subset \mathbb{R}^n$ for $n > 3$. Nonetheless, the method does not cover all possible combinations of finding four extreme points that are affine independent. For example, the set $\{\mathbf{w}^1, \mathbf{w}^2, \mathbf{w}^3, \mathbf{v}\}$ shown in Figure 6.5 is affine independent but would not be extracted from $ext(\mathfrak{P}(X))$ when using this method alone.

6.2.2.3 The case $n \geq 4$

As the preceding section corroborates, for $2 \leq n \leq 3$ it is relatively easy to visualize the various faces of $\mathfrak{P}(X)$ in terms of the intersections of the faces of $[\mathbf{v}, \mathbf{u}]$ with those of $S(X)$. However when $n \geq 4$ we leave the plane and solid geometry of our early education. Although analogies with lower dimensional spaces can be useful, they can also lead us astray. For instance, any two distinct non-parallel planes in $\mathbb{R}^3$ intersect in a line. Now consider the two non-parallel planes $P_1, P_2 \subset \mathbb{R}^4$, defined by $P_1 = \{\mathbf{x} \in \mathbb{R}^4 : \mathbf{x} = (x_1, x_2, 0, 0)'$ and $x_1, x_2 \in \mathbb{R}\}$ and $P_2 = \{\mathbf{x} \in \mathbb{R}^4 : \mathbf{x} = (0, 0, x_3, x_4)'$ and $x_3, x_4 \in \mathbb{R}\}$. Then $\mathbf{0} \in P_1 \cap P_2$ with $\{\mathbf{0}\} = P_1 \cap P_2$. Thus $P_1 \cap P_2$ is a point and not a line. However, any two distinct planes that are contained in a hyperplane $E \subset \mathbb{R}^4$ will intersect in a line.

Similarly we have the fact that a line can intersect a hyperplane in a single point. For instance, if $\mathbf{a} = (-1, 0, 0, 0)'$, $\mathbf{b} = (1, 0, 0, 0)'$, $L = L(\mathbf{a}, \mathbf{b})$, and $E = \{\mathbf{x} \in \mathbb{R}^4 : \mathbf{x} = (0, x_2, x_3, x_4)'$ and $x_i \in \mathbb{R}$ for $i = 2, 3, 4\}$, then $L \cap E = \{\mathbf{0}\}$.

In the remainder of this chapter we reserve the letters L, P, L^m and E for denoting a line, a plane, an m-dimensional affine space and a hyperplane, respectively. Care must be taken when considering intersections of two linear spaces in high dimensions. While properties of linear subspaces of $\mathbb{R}^n$

for $n = 2$ and $n = 3$ rest on Hilbert's axiomatic foundation of solid geometry [116], this section relies on the extended axioms of higher-dimensional geometry (e.g., [243, 298, 299, 319, 181]). Thus, for instance, if $n = 3$, then the intersection of two distinct planes is either empty or a line while in $\mathbb{R}^4$ the intersection can be either empty, a point, or a line. Nonetheless, if E_1 and E_2 are two distinct hyperplanes in $\mathbb{R}^n$ with $n > 3$ and $E_1 \cap E_2 \neq \varnothing$, then $E_1 \cap E_2 = L^{n-2}$.

Algorithm 6.1 computes the elements of the set $V = W \cup M \cup \{\mathbf{v}, \mathbf{u}\}$ for any dimension $n \geq 2$, and the results serve as input to Algorithm 6.2. STEP *2* of Algorithm 6.2 computes the entry and exit points (if they exist) of the respective lines $L(\mathbf{w}^i)$ and $L(\mathbf{m}^j)$. The final STEP *3* checks for possible intersection points of the $(n-1)$-dimensional faces $F_{\mathbf{w}^i}(\mathbf{d}_{ij})$ of $S(X)$ with the one-dimensional faces of $[\mathbf{v}, \mathbf{u}]$. Thus, if $n = 3$, then Algorithm 6.2 accounts for all extreme points of $\mathfrak{P}(X)$. However, for $n > 3$ the algorithm ignores intersections of k-dimensional faces of $S(X)$ with $[\mathbf{v}, \mathbf{u}]$ when $1 < k < n-1$. Hence, if $n = 4$, then the extreme points that are due to the intersections of two-dimensional faces of $S(X)$ with $\partial[\mathbf{v}, \mathbf{u}]$ will be disregarded by Algorithm 6.2. A straightforward method for identifying additional extreme points in $\mathbb{R}^4$ that are due to the intersections of two-dimensional faces of $S(X)$ with $\partial[\mathbf{v}, \mathbf{u}]$, is simply to set $\mathbf{s}(i, j) = \mathbf{w}^i \vee \mathbf{w}^j$ and $\mathbf{r}(i, j) = \mathbf{m}^i \wedge \mathbf{m}^j$, where $i, j \in \mathbb{N}_4$ and $i < j$. Then eliminate all points $\mathbf{s}(i, j)$ and $\mathbf{r}(i, j)$ that are superfluous, i.e., already computed by Algorithm 6.2.

Example 6.8 Suppose $X = \{\mathbf{x}^1, \ldots, \mathbf{x}^{10}\} \subset \mathbb{R}^4$, where

$$\mathbf{x}^1 = \begin{pmatrix} 22 \\ 18 \\ 24 \\ 26 \end{pmatrix}, \mathbf{x}^2 = \begin{pmatrix} 1 \\ 5 \\ 13 \\ 11 \end{pmatrix}, \mathbf{x}^3 = \begin{pmatrix} 20 \\ 18 \\ 6 \\ 8 \end{pmatrix}, \mathbf{x}^4 = \begin{pmatrix} 31 \\ 22 \\ 9 \\ 12 \end{pmatrix}, \mathbf{x}^5 = \begin{pmatrix} 14 \\ 20 \\ 26 \\ 3 \end{pmatrix},$$

$$\mathbf{x}^6 = \begin{pmatrix} 5 \\ 4 \\ 8 \\ 12 \end{pmatrix}, \mathbf{x}^7 = \begin{pmatrix} 8 \\ 17 \\ 28 \\ 9 \end{pmatrix}, \mathbf{x}^8 = \begin{pmatrix} 15 \\ 25 \\ 20 \\ 18 \end{pmatrix}, \mathbf{x}^9 = \begin{pmatrix} 10 \\ 2 \\ 5 \\ 13 \end{pmatrix}, \mathbf{x}^{10} = \begin{pmatrix} 3 \\ 5 \\ 4 \\ 6 \end{pmatrix}.$$

Applying STEP *1* of Algorithm 6.2 one obtains the following matrices $\mathfrak{W}_X$,

$\mathfrak{M}_X$, and the vectors $\mathbf{u}$ and $\mathbf{v}$:

$$\mathfrak{W}_X = \begin{pmatrix} 0 & -10 & -20 & -10 \\ -9 & 0 & -11 & -11 \\ -22 & -13 & 0 & -8 \\ -19 & -17 & -23 & 0 \end{pmatrix}, \mathfrak{M}_X = \begin{pmatrix} 0 & 9 & 22 & 19 \\ 10 & 0 & 13 & 17 \\ 20 & 11 & 0 & 23 \\ 10 & 11 & 8 & 0 \end{pmatrix}, \mathbf{u} = \begin{pmatrix} 31 \\ 25 \\ 28 \\ 26 \end{pmatrix}, \mathbf{v} = \begin{pmatrix} 1 \\ 2 \\ 4 \\ 3 \end{pmatrix}.$$

Computing the respective sets $W = \{\mathbf{w}^1, \ldots, \mathbf{w}^4\}$ and $M = \{\mathbf{m}^1, \ldots, \mathbf{m}^4\}$ results in

$$\mathbf{w}^1 = \begin{pmatrix} 31 \\ 22 \\ 9 \\ 12 \end{pmatrix}, \mathbf{w}^2 = \begin{pmatrix} 15 \\ 25 \\ 12 \\ 8 \end{pmatrix}, \mathbf{w}^3 = \begin{pmatrix} 8 \\ 17 \\ 28 \\ 5 \end{pmatrix}, \mathbf{w}^4 = \begin{pmatrix} 16 \\ 15 \\ 18 \\ 26 \end{pmatrix}, \text{ and}$$

$$\mathbf{m}^1 = \begin{pmatrix} 1 \\ 11 \\ 21 \\ 11 \end{pmatrix}, \mathbf{m}^2 = \begin{pmatrix} 11 \\ 2 \\ 13 \\ 13 \end{pmatrix}, \mathbf{m}^3 = \begin{pmatrix} 26 \\ 17 \\ 4 \\ 12 \end{pmatrix}, \mathbf{m}^4 = \begin{pmatrix} 22 \\ 20 \\ 26 \\ 3 \end{pmatrix}.$$

In this case we have that $V = W \cup M \cup \{\mathbf{v}, \mathbf{u}\}$, resulting in $|V| = 10$.
Step *2* computes the additive coefficients

$$a_1 = -5,\ a_2 = -5,\ a_3 = -2,\ a_4 = -13 \text{ and } b_1 = 7,\ b_2 = 13,\ b_3 = 5,\ b_4 = 2.$$

Using these values in order to compute the points $\mathbf{p}^j = a_j + \mathbf{w}^j$ and $\mathbf{q}^j = b_j + \mathbf{m}^j$, results in

$$\mathbf{p}^1 = \begin{pmatrix} 26 \\ 17 \\ 4 \\ 7 \end{pmatrix}, \mathbf{p}^2 = \begin{pmatrix} 10 \\ 20 \\ 7 \\ 3 \end{pmatrix}, \mathbf{p}^3 = \begin{pmatrix} 6 \\ 15 \\ 26 \\ 3 \end{pmatrix}, \mathbf{p}^4 = \begin{pmatrix} 3 \\ 2 \\ 5 \\ 13 \end{pmatrix}, \text{ and}$$

$$\mathbf{q}^1 = \begin{pmatrix} 8 \\ 18 \\ 28 \\ 18 \end{pmatrix}, \mathbf{q}^2 = \begin{pmatrix} 24 \\ 15 \\ 26 \\ 26 \end{pmatrix}, \mathbf{q}^3 = \begin{pmatrix} 31 \\ 22 \\ 9 \\ 17 \end{pmatrix}, \mathbf{q}^4 = \begin{pmatrix} 24 \\ 22 \\ 28 \\ 5 \end{pmatrix}.$$

It follows that $U = \Psi \cup Q$ with $(\Psi \cup Q) \cap V = \varnothing$ so that $|U| = 8$.

Step *3* computes the points $\mathbf{r}^j = \bigwedge_{i \neq j} \mathbf{m}^i$ and $\mathbf{s}^j = \bigvee_{i \neq j} \mathbf{w}^i$, which in this case are

$$\mathbf{r}^1 = \begin{pmatrix} 11 \\ 2 \\ 4 \\ 3 \end{pmatrix}, \mathbf{r}^2 = \begin{pmatrix} 1 \\ 11 \\ 4 \\ 3 \end{pmatrix}, \mathbf{r}^3 = \begin{pmatrix} 1 \\ 2 \\ 13 \\ 3 \end{pmatrix}, \mathbf{r}^4 = \begin{pmatrix} 1 \\ 2 \\ 4 \\ 11 \end{pmatrix}, \text{ and}$$

$$\mathbf{s}^1 = \begin{pmatrix} 16 \\ 25 \\ 28 \\ 26 \end{pmatrix}, \mathbf{s}^2 = \begin{pmatrix} 31 \\ 22 \\ 28 \\ 26 \end{pmatrix}, \mathbf{s}^3 = \begin{pmatrix} 31 \\ 25 \\ 18 \\ 26 \end{pmatrix}, \mathbf{s}^4 = \begin{pmatrix} 31 \\ 25 \\ 28 \\ 12 \end{pmatrix}.$$

In this case we have $|T_1| = 8$ and $|V| + |U| + |T_1| = 26$ and $V \cup U \cup T_1 \subset ext(\mathfrak{P}(X))$.

Next compute $\mathbf{r}(i, j) = \mathbf{m}^i \wedge \mathbf{m}^j$ for $i = 1, \ldots, 3$ and $j = i+1, \ldots, 4$, and set $R_2 = \{\mathbf{r}(1,2), \ldots, \mathbf{r}(3,4)\}$. In a like manner let $S_2 = \{\mathbf{s}(1,2), \ldots, \mathbf{s}(3,4)\}$. For this example we obtain

$$\mathbf{r}(1,2) = \begin{pmatrix} 1 \\ 2 \\ 13 \\ 11 \end{pmatrix}, \mathbf{r}(1,3) = \begin{pmatrix} 1 \\ 11 \\ 4 \\ 11 \end{pmatrix}, \mathbf{r}(1,4) = \begin{pmatrix} 1 \\ 11 \\ 21 \\ 3 \end{pmatrix}, \mathbf{r}(2,3) = \begin{pmatrix} 11 \\ 2 \\ 4 \\ 12 \end{pmatrix},$$

$$\mathbf{r}(2,4) = \begin{pmatrix} 11 \\ 2 \\ 13 \\ 4 \end{pmatrix}, \mathbf{r}(3,4) = \begin{pmatrix} 22 \\ 17 \\ 4 \\ 3 \end{pmatrix}, \mathbf{s}(1,2) = \begin{pmatrix} 31 \\ 25 \\ 12 \\ 12 \end{pmatrix}, \mathbf{s}(1,3) = \begin{pmatrix} 31 \\ 22 \\ 28 \\ 12 \end{pmatrix},$$

$$\mathbf{s}(1,4) = \begin{pmatrix} 31 \\ 22 \\ 18 \\ 26 \end{pmatrix}, \mathbf{s}(2,3) = \begin{pmatrix} 15 \\ 25 \\ 28 \\ 8 \end{pmatrix}, \mathbf{s}(2,4) = \begin{pmatrix} 16 \\ 25 \\ 18 \\ 26 \end{pmatrix}, \mathbf{s}(3,4) = \begin{pmatrix} 16 \\ 17 \\ 28 \\ 26 \end{pmatrix}.$$

Since none of the points $\mathbf{r}(i, j)$ and $\mathbf{s}(i, j)$ are elements of $V \cup U \cup T_1$, the set $R_2 \cup S_2$ provides for 12 additional extreme points.

To show that the elements of $R_2 \cup S_2$ in Example 6.8 are indeed extreme points, note that $\mathbf{s}(i, j)$ is an element of the plane $P_{\mathbf{u}}(i, j) = \{\mathbf{x} \in \mathbb{R}^4 : x_i = u_i \text{ and } x_j = u_j\}$ and $\mathbf{r}(i, j)$ is an element of the plane $P_{\mathbf{v}}(i, j) = \{\mathbf{x} \in \mathbb{R}^4 : x_i = v_i \text{ and } x_j = v_j\}$, i.e., $\mathbf{s}(i, j) \in E_{\mathbf{u}}(\mathbf{e}^i) \cap E_{\mathbf{u}}(\mathbf{e}^j)$ and $\mathbf{r}(i, j) \in E_{\mathbf{v}}(\mathbf{e}^i) \cap E_{\mathbf{v}}(\mathbf{e}^j)$. In fact, these points are points of the two-dimensional faces of $[\mathbf{v}, \mathbf{u}]$, e.g. $\mathbf{s}(i, j) \in F_u(i, j) = P_{\mathbf{u}}(i, j) \cap [\mathbf{v}, \mathbf{u}]$. Given $\mathbf{s}(i, j)$ and $P_{\mathbf{u}}(i, j)$ we can determine the two-dimensional face of $S(X)$ that contains the point $\mathbf{s}(i, j)$. Specifically, the face will be given by $F_{\mathbf{s}}(i, j) = P_{\mathbf{s}}(i, j) \cap S(X)$, where $P_{\mathbf{s}}(i, j)$ denotes the plane to be determined. For a point $\mathbf{x} \in \mathbb{R}^4$ to be a point of $P_{\mathbf{s}}(i, j)$, it must first have the property that $x_i, x_j \in \mathbb{R}$ are the two variables of $\mathbf{x}$. The two remaining coordinates x_k and x_ℓ of $\mathbf{x}$ are determined as follows: If the kth coordinate of $\mathbf{s}(i, j)$ is w^i_k, set $x_k = x_i - (u_i - w^i_k)$, and if the kth coordinate of $\mathbf{s}(i, j)$ is w^j_k, set $x_k = x_j - (u_j - w^j_k)$. Similarly, we construct the plane $P_{\mathbf{r}}(i, j)$ by using

$\mathbf{r}(i, j)$ as the starting point. In this case, set the kth coordinate of $\mathbf{x} \in P_{\mathbf{r}}(i, j)$ to $x_k = x_i - (v_i - m_k^i)$ if the kth coordinate of $\mathbf{r}(i, j)$ is m_k^i and if the kth coordinate of $\mathbf{r}(i, j)$ is m_k^j, set $x_k = x_j - (v_j - m_k^j)$.

Example 6.9 Let $\mathbf{s}(i, j)$ and $\mathbf{r}(i, j)$ be the points computed in Example 6.8 and consider $\mathbf{s}(1,2) = (31,25,12,12)' = (u_1,u_2,w_3^2,w_4^1)' \in P_{\mathbf{u}}(1,2) = \{\mathbf{x} \in \mathbb{R}^4 : x_1 = u_1, x_2 = u_2 \text{ and } x_3, x_4 \in \mathbb{R}\}$. Following the above mentioned rules in order to describe the elements $\mathbf{x} \in \mathbb{R}^4$ that define $P_{\mathbf{s}}(1,2)$, we need to have $x_1, x_2 \in \mathbb{R}$ to be the variables. Since the third component of $\mathbf{s}(1,2)$ is $x_3 = w_3^2$, the third component of $\mathbf{x} \in P_{\mathbf{s}}(1,2)$ must be $x_3 = x_2 - (u_2 - w_3^2) = x_2 - 13$. Similarly, $x_4 = x_1 - (u_1 - w_4^1) = x_1 - 19$. This shows that

$$\mathbf{x} \in P_{\mathbf{s}}(1,2) \Leftrightarrow \mathbf{x} = \begin{pmatrix} x_1 \\ x_2 \\ x_2 - 13 \\ x_1 - 19 \end{pmatrix}, \text{ where } x_1, x_2 \in \mathbb{R}.$$

Thus, for $x_1 = u_1$ and $x_2 = u_2$ results in $s(1,2) \in P_{\mathbf{s}}(1,2) = E_{\mathbf{w}^1}(\mathbf{d}_{14}) \cap E_{\mathbf{w}^2}(\mathbf{d}_{23})$ and $\{\mathbf{s}(1,2)\} = P_{\mathbf{s}}(1,2) \cap P_{\mathbf{u}}(1,2)$ since no other point is common to both planes. Note also that $E_{\mathbf{w}^1}(\mathbf{d}_{14}) \perp E_{\mathbf{w}^2}(\mathbf{d}_{23})$.

Using the same approach to find the plane $P_{\mathbf{r}}(1,2)$ given $s(1,2) = (1,2,13,11)' = (v_1, v_2, m_3^2, m_4^1)'$, one obtains

$$\mathbf{x} \in P_{\mathbf{r}}(1,2) \Leftrightarrow \mathbf{x} = \begin{pmatrix} x_1 \\ x_2 \\ x_2 + 11 \\ x_1 + 10 \end{pmatrix},$$

where $x_1, x_2 \in \mathbb{R}$ and $P_{\mathbf{r}}(1,2) = E_{\mathbf{m}^1}(\mathbf{d}_{14}) \cap E_{\mathbf{m}^2}(\mathbf{d}_{23})$

Again, $\{\mathbf{r}(1,2)\} = P_{\mathbf{r}}(1,2) \cap P_{\mathbf{v}}(1,2)$ and $E_{\mathbf{m}^1}(\mathbf{d}_{14}) \perp E_{\mathbf{m}^2}(\mathbf{d}_{23})$.

Continuing this approach results in

$$\mathbf{x} \in P_{\mathbf{s}}(1,3) \Leftrightarrow \mathbf{x} = \begin{pmatrix} x_1 \\ x_1 - 9 \\ x_3 \\ x_1 - 19 \end{pmatrix} \qquad \mathbf{x} \in P_{\mathbf{s}}(1,4) \Leftrightarrow \mathbf{x} = \begin{pmatrix} x_1 \\ x_1 - 9 \\ x_4 - 8 \\ x_4 \end{pmatrix}$$

$$\mathbf{x} \in P_{\mathbf{s}}(2,3) \Leftrightarrow \mathbf{x} = \begin{pmatrix} x_2 - 10 \\ x_2 \\ x_3 \\ x_2 - 17 \end{pmatrix} \qquad \mathbf{x} \in P_{\mathbf{s}}(2,4) \Leftrightarrow \mathbf{x} = \begin{pmatrix} x_4 - 10 \\ x_2 \\ x_4 - 8 \\ x_4 \end{pmatrix}$$

$$\mathbf{x} \in P_{\mathbf{s}}(3,4) \Leftrightarrow \mathbf{x} = \begin{pmatrix} x_4 - 10 \\ x_3 - 11 \\ x_3 \\ x_4 \end{pmatrix} \qquad \mathbf{x} \in P_{\mathbf{r}}(1,3) \Leftrightarrow \mathbf{x} = \begin{pmatrix} x_1 \\ x_1 + 10 \\ x_3 \\ x_1 + 10 \end{pmatrix}, \text{ etc.}$$

Each plane $P_{\mathbf{s}}(i,j)$ and $P_{\mathbf{r}}(i,j)$ represents the intersection of two hyperplanes. These hyperplanes can be readily determined from the respective points $\mathbf{s}(i,j)$ and $\mathbf{r}(i,j)$. For instance, since $\mathbf{s}(3,4) = (w_1^4, w_2^3, u_3, u_4)'$ it follows that $(w_1^4, x_2, x_3, u_4)' \in E_{\mathbf{w}^4}(\mathbf{d}_{14})$ and $(x_1, w_2^3, u_3, x_4) \in E_{\mathbf{w}^3}(\mathbf{d}_{23})$. Hence $(w_1^4, w_2^3, u_3, u_4)' \in E_{\mathbf{w}^4}(\mathbf{d}_{14}) \cap E_{\mathbf{w}^3}(\mathbf{d}_{23})$ and, more to the point, $\mathbf{x} \in P_{\mathbf{s}}(3,4) \Leftrightarrow \mathbf{x} \in E_{\mathbf{w}^4}(\mathbf{d}_{14}) \cap E_{\mathbf{w}^3}(\mathbf{d}_{23}) \Leftrightarrow P_{\mathbf{s}}(3,4) = E_{\mathbf{w}^4}(\mathbf{d}_{14}) \cap E_{\mathbf{w}^3}(\mathbf{d}_{23})$. Also, since $\mathbf{x} \in P_{\mathbf{s}}(i,j) \Rightarrow x_i, x_j \in \mathbb{R}$ and $\mathbf{x} \in P_{\mathbf{u}}(i,j)$ whenever $x_i = u_i$ and $x_j = u_j$, we have that $\mathbf{x} \in P_{\mathbf{s}}(i,j) \cap P_{\mathbf{u}}(i,j) \Leftrightarrow \mathbf{x} = \mathbf{s}(i,j)$. An analogous argument proves that $\mathbf{x} \in P_{\mathbf{r}}(i,j) \cap P_{\mathbf{v}}(i,j) \Leftrightarrow \mathbf{x} = \mathbf{r}(i,j)$. Therefore $\mathbf{s}(i,j)$ is the *only* point in the intersection $P_{\mathbf{r}}(i,j) \cap P_{\mathbf{v}}(i,j)$. An analogous argument shows that $\mathbf{x} \in P_{\mathbf{r}}(i,j) \cap P_{\mathbf{v}}(i,j) \Leftrightarrow \mathbf{x} = \mathbf{r}(i,j)$.

The intersections of the two hyperplanes that correspond to the planes $P_{\mathbf{s}}(i,j)$ and $P_{\mathbf{r}}(i,j)$ are as follows:

$$\begin{aligned}
P_{\mathbf{s}}(1,2) &= E_{\mathbf{w}^1}(\mathbf{d}_{14}) \cap E_{\mathbf{w}^1}(\mathbf{d}_{23}) & P_{\mathbf{s}}(1,3) &= E_{\mathbf{w}^1}(\mathbf{d}_{12}) \cap E_{\mathbf{w}^1}(\mathbf{d}_{14}) \\
P_{\mathbf{s}}(1,4) &= E_{\mathbf{w}^1}(\mathbf{d}_{12}) \cap E_{\mathbf{w}^4}(\mathbf{d}_{34}) & P_{\mathbf{s}}(2,3) &= E_{\mathbf{w}^2}(\mathbf{d}_{12}) \cap E_{\mathbf{w}^2}(\mathbf{d}_{24}) \\
P_{\mathbf{s}}(2,4) &= E_{\mathbf{w}^4}(\mathbf{d}_{14}) \cap E_{\mathbf{w}^4}(\mathbf{d}_{34}) & P_{\mathbf{s}}(3,4) &= E_{\mathbf{w}^3}(\mathbf{d}_{23}) \cap E_{\mathbf{w}^4}(\mathbf{d}_{14}) \\
P_{\mathbf{r}}(1,2) &= E_{\mathbf{m}^1}(\mathbf{d}_{14}) \cap E_{\mathbf{m}^1}(\mathbf{d}_{23}) & P_{\mathbf{r}}(1,3) &= E_{\mathbf{m}^1}(\mathbf{d}_{12}) \cap E_{\mathbf{m}^1}(\mathbf{d}_{14}) \\
P_{\mathbf{r}}(1,4) &= E_{\mathbf{m}^1}(\mathbf{d}_{12}) \cap E_{\mathbf{m}^1}(\mathbf{d}_{13}) & P_{\mathbf{r}}(2,3) &= E_{\mathbf{m}^2}(\mathbf{d}_{12}) \cap E_{\mathbf{m}^3}(\mathbf{d}_{34}), \text{ etc.}
\end{aligned}$$

We leave the construction of the remaining planes $P_{\mathbf{r}}(i,j)$ as an exercise for the reader.

Example 6.8 introduced two new sets R_2 and S_2 consisting of the points $\mathbf{r}(i,j)$ and $\mathbf{s}(i,j)$, respectively. Example 6.9 demonstrates that these points are indeed extreme points. It also follows from Example 6.8 that these points are not elements of $V \cup U \cup T_1$. This means that for $n \geq 3$ Algorithm 6.2 is incapable of discovering all extreme points of $\mathfrak{P}(X)$ for various compact sets $X \subset \mathbb{R}^n$. Nevertheless, Examples 6.8 and 6.9 provide for a simple expansion of Algorithm 6.2 for finding all extreme points of $\mathfrak{P}(X) \subset \mathbb{R}^4$.

Algorithm 6.3 (Computing the Elements of $ext(\mathfrak{P}(X))$**)**

Step *1. Given $X \subset \mathbb{R}^n$ apply Algorithm 6.2.*
Step 2. *For each sequence $i_1, i_2, \ldots, i_{n-2} \in \mathbb{N}_n$ with $1 \leq i_1 < i_2 < \cdots < i_{n-2} \leq n$, set*

$$\mathbf{r}(i_1, i_2, \ldots, i_{n-2}) = \bigwedge_{j=1}^{n-2} \mathbf{m}^{i_j} \text{ and } \mathbf{s}(i_1, i_2, \ldots, i_{n-2}) = \bigvee_{j=1}^{n-2} \mathbf{w}^{i_j}.$$

Let $R_2 = \{\mathbf{r}(i_1, i_2, \ldots, i_{n-2}) : 1 \leq i_1 < i_2 < \cdots < i_{n-2} \leq n\}$, $S_2 = \{\mathbf{s}(i_1, i_2, \ldots, i_{n-2}) : 1 \leq i_1 < i_2 < \cdots < i_{n-2} \leq n\}$, and set $T_2 = (R_2 \cup S_2) \setminus T_1$.

In this case we have that for $n = 4$, $V \cup U \cap T_1 \cup T_2 \subset ext(\mathfrak{P}(X))$.

We may now straightforwardly extend the three algorithms to the n-dimensional case.

Algorithm 6.4 (Computing the Elements of $ext(\mathfrak{P}(X))$**)**
Step *1. Given $X \subset \mathbb{R}^n$ with $n > 4$, apply Algorithm 6.3.*
Step 2. *For $\ell = 3, 4, \ldots, n-2$ and every sequence $\{i_1, i_2, \ldots, i_{n-\ell}\} \subset \mathbb{N}_n$ with $1 \leq i_1 < i_2 < \cdots < i_{n-\ell} \leq n$, set*

$$\mathbf{r}(i_1, i_2, \ldots, i_{n-\ell}) = \bigwedge_{j=1}^{n-\ell} \mathbf{m}^{i_j} \text{ and } \mathbf{s}(i_1, i_2, \ldots, i_{n-\ell}) = \bigvee_{j=1}^{n-\ell} \mathbf{w}^{i_j}.$$

Let $R_\ell = \{\mathbf{r}(i_1, i_2, \ldots, i_{n-\ell}) : 1 \leq i_1 < i_2 < \cdots < i_{n-\ell} \leq n\}$, $S_\ell = \{\mathbf{s}(i_1, i_2, \ldots, i_{n-\ell}) : 1 \leq i_1 < i_2 < \cdots < i_{n-\ell} \leq n\}$, and set $T_\ell = (R_\ell \cup S_\ell) \setminus T_{\ell-1}$.

In view of Algorithms 6.2 through 6.4, it becomes obvious that the number of extreme points of type **r** and **s** increases dramatically as the dimension n increases. For instance, for $n = 3$ the total number of points of type $\mathbf{r}(i_1, i_2)$ and $\mathbf{s}(i_1, i_2)$ is 6 while for dimension $n = 6$ the maximal number is 30. More generally, the maximal number of points of type $\mathbf{r}(i_1, i_2)$ and $\mathbf{s}(i_1, i_2)$ for any dimension $n \geq 3$ is $\frac{n(n-1)}{2} + \frac{n(n-1)}{2} = n(n-1)$. Similar observations can be made for the number of elements of R_ℓ and S_ℓ for any $\ell \in \{3, 4, \ldots, n-2\}$. It is also noteworthy that for $n \geq 4$, perpendicular hyperplanes entered the computations.

The method for selecting $n+1$ affine independent points from $ext(\mathfrak{P}(X)) \subset \mathbb{R}^n$, where $n = 3$ and $n < |U| \leq 2n$, easily extends to $n = 4$. Here one chooses any face $F_{\mathbf{w}^i}(\mathbf{d}_{ij}) = E_{\mathbf{w}^i}(\mathbf{d}_{ij}) \cap S(X)$ satisfying the property that $L(\mathbf{w}^i) \neq L(\mathbf{m}^j)$. The next step is to consider the plane

$$P(i,j) = \{\mathbf{x} \in \mathbb{R}^4 : \mathbf{x} = \lambda_1 \mathbf{w}^i + \lambda_2 \mathbf{p}^i + \lambda_3 \mathbf{m}^j \text{ and } \sum_{\ell=1}^{3} \lambda_\ell = 1\} \subset E_{\mathbf{w}^i}(\mathbf{d}_{ij}),$$

generated by the three points $\mathbf{w}^i, \mathbf{p}^i \in L(\mathbf{w}^i)$ and $\mathbf{m}^j \in L(\mathbf{m}^j)$. In order to obtain another extremal point of $F_{\mathbf{w}^i}(\mathbf{d}_{ij})$ that is not an element of $P(i,j)$, simply choose a plane $P_{\mathbf{s}}(i,\ell) \subset E_{\mathbf{w}^i}(\mathbf{d}_{ij})$. Then $P_{\mathbf{s}}(i,\ell) \cap P(i,j) = L(\mathbf{w}^i)$ and $\mathbf{w}^i \vee \mathbf{w}^\ell = \mathbf{s}(i,\ell) \notin P(i,j)$. Consequently, the set of points $\{\mathbf{w}^i, \mathbf{p}^i, \mathbf{m}^j, \mathbf{s}(i,\ell)\}$ is affine independent.

To illustrate this process, suppose that X is as in Example 6.8 and consider the plane $P(1,2) \subset E_{\mathbf{w}^1}(\mathbf{d}_{12})$ generated by the three points $\mathbf{w}^1, \mathbf{p}^1 \in L(\mathbf{w}^1)$ and $\mathbf{m}^2 \in L(\mathbf{m}^2)$. In this case $P_{\mathbf{s}}(1,3) \subset E_{\mathbf{w}^1}(\mathbf{d}_{12})$ and $P_{\mathbf{s}}(1,4) \subset E_{\mathbf{w}^1}(\mathbf{d}_{12})$. Choosing $P_{\mathbf{s}}(1,3)$ we have $P(1,2) \cap P_{\mathbf{s}}(1,3) = L(\mathbf{w}^1)$ and since $\mathbf{s}(1,3) = (31,22,28,12)' \notin L(\mathbf{w}^1)$, $\mathbf{s}(1,3) \notin P(1,2)$ Therefore the set of points $A = \{\mathbf{w}^1, \mathbf{p}^1, \mathbf{m}^2, \mathbf{s}(1,3)\}$ is affine independent. An analogous argument shows that the set of points $B = \{\mathbf{w}^1, \mathbf{p}^1, \mathbf{m}^2, \mathbf{s}(1,4)\}$ is affine independent had we chosen the plane $P_{\mathbf{s}}(1,4)$ instead. Note also that the 3-dimensional affine space generated by the points of either A or B is $E_{\mathbf{w}^1}(\mathbf{d}_{12})$. Furthermore, since $E_{\mathbf{w}^2}(\mathbf{d}_{12}) \cap E_{\mathbf{w}^1}(\mathbf{d}_{12}) = \varnothing$, we can choose either $\mathbf{w}^2$ or $\mathbf{m}^1$ and add it to the set A (or B) in order to have five affine independent extreme points.

The above example reveals the utility of the elements of R_ℓ and S_ℓ in quickly finding sets of $n+1$ affine independent extreme points. Another revelation is the fact that the intersection $E_{\mathbf{w}^1}(\mathbf{d}_{12}) \cap E_{\mathbf{w}^1}(\mathbf{d}_{13})$ does not appear in the calculation of the various extreme points of Examples 6.8 and 6.9. Since $Ł(\mathbf{w}^1) \subset E_{\mathbf{w}^1}(\mathbf{d}_{12}) \cap E_{\mathbf{w}^1}(\mathbf{d}_{13})$, the intersection is not empty and, therefore, must be a plane. Letting $P_{\mathbf{s}}$ denote this plane, then $P_{\mathbf{s}} = E_{\mathbf{w}^1}(\mathbf{d}_{12}) \cap E_{\mathbf{w}^1}(\mathbf{d}_{13}) = \{\mathbf{x} \in \mathbb{R}^4 : x_1 - x_2 = u_1 - w_2^1 \text{ and } x_1 - x_3 = u_1 - w_3^1\}$. Using the data from Example 6.8, we obtain

$$\mathbf{x} \in P_{\mathbf{s}} \Leftrightarrow \mathbf{x} = \begin{pmatrix} x_1 \\ x_1 - 9 \\ x_1 - 22 \\ x_4 \end{pmatrix}, \text{ where } x_1, x_4 \in \mathbb{R}.$$

The extreme points $\mathbf{w}^1$, $\mathbf{p}^1$, $\mathbf{m}^3$, and $\mathbf{q}^3$ all satisfy the equation for elements of $P_{\mathbf{s}}$. By definition of $P(1,2)$, the planar strip $F^2(1,2) = P(1,2) \cap S(X)$ is a

2-dimensional face of $S(X)$ with $\partial F^2(1,2) = L(\mathbf{w}^1) \cup L(\mathbf{m}^2)$. Similarly, letting $F^2_{\mathbf{s}} = P_{\mathbf{s}} \cap S(X)$, then $\partial F^2_{\mathbf{s}} = L(\mathbf{w}^1) \cup L(\mathbf{m}^3)$ and $L(\mathbf{w}^1) = P(1,2) \cap P_{\mathbf{s}} = F^2(1,2) \cap F^2_{\mathbf{s}}$. Obviously, $\mathbf{m}^2 \notin P_{\mathbf{s}}$. Therefore $\mathbf{m}^2$ (or $\mathbf{q}^2$) and any three points from the set $\{\mathbf{w}^1, \mathbf{p}^1, \mathbf{m}^3, \mathbf{q}^3\}$ yields a set of four affine independent points. For the same reason the point $\mathbf{m}^3$ (or $\mathbf{q}^3$) and any three points from the set $\{\mathbf{w}^1, \mathbf{p}^1, \mathbf{m}^2, \mathbf{q}^2\}$ yields a set of four affine independent points. Finally, since $E_{\mathbf{w}^2}(\mathbf{d}_{12}) \cap E_{\mathbf{w}^1}(\mathbf{d}_{12}) = \emptyset$, adding the point $\mathbf{w}^2$ (or $\mathbf{m}^1$) to the set of the four selected affine points results in five affine independent points.

Setting $x_1 = u_1 = 31$ one obtains $(31,22,9,x_4)' \in P_{\mathbf{s}}\ \forall x_4 \in \mathbb{R}$ and, consequently, $(31,22,9,26)' \in P_{\mathbf{s}}$. Since $(31,22,9,26)' \in E_{\mathbf{u}}(\mathbf{e}^1) \cap E_{\mathbf{u}}(\mathbf{e}^4) = P_{\mathbf{u}}(1,4)$, it follows that $\{(31,22,9,26)'\} = P_{\mathbf{s}} \cap P_{\mathbf{u}}(1,4)$, which verifies that $(31,22,9,26)'$ is an extreme point. This extreme point is not accounted for when using Algorithms 6.3 or 6.4. Since $\partial F^2_{\mathbf{s}} = L(\mathbf{w}^1) \cup L(\mathbf{m}^3)$ and $(31,22,9,26)' \notin L(\mathbf{w}^1) \cup L(\mathbf{m}^3)$, $(31,22,9,26)'$ is an interior point of the 2-dimensional face $F^2_{\mathbf{s}}$ of $S(X)$. This example shows that Algorithms 6.3 and 6.4 do not extract all extreme points from $ext(\mathfrak{P}(X))$. In order to extract all extreme points, all intersections of the different k-dimensional faces that were not in play (i.e., have not been used) when employing the algorithms need to be examined. However, care must be taken when determining lower-dimensional faces of $S(X)$ derived from the intersections of higher-dimensional affine spaces. As illustrated in Figure 6.6, $E_{\mathbf{w}^i}(\mathbf{d}_{ij}) \cap E_{\mathbf{w}^\ell}(\mathbf{d}_{\ell k}) \neq \emptyset$ whenever $\{i,j\} \neq \{\ell,k\}$ does not imply that $F^{n-1}_{\mathbf{w}^i}(\mathbf{d}_{ij}) \cap F^{n-1}_{\mathbf{w}^\ell}(\mathbf{d}_{\ell k}) \neq \emptyset$.

The points $\mathbf{w}^i$ and $\mathbf{m}^i$ are translates of foundational elements $\mathfrak{w}^i$ and $\mathfrak{m}^i$ that determine the structure of $S(X)$ and $[\mathbf{v},\mathbf{u}]$ and, hence, of $\mathfrak{P}(X)$. For this reason, $\mathbf{w}^i$, $\mathbf{m}^i$ and their respective entry and exit points are considered the main information carriers among the extreme points of $\mathfrak{P}(X)$. For this reason it is often preferable to establish sets of affine independent points consisting of these extreme points.

We conclude this section with three theorems and a couple of examples that are instructive when assembling affine independent sets of extreme points.

Theorem 6.9 *Suppose that* $X \subset \mathbb{R}^n$ *is compact and the set of points* $\{\mathbf{v}^0, \mathbf{v}^1, \ldots, \mathbf{v}^m\} \subset E_{\mathbf{w}^i}(\mathbf{e}^i) \cap [\mathbf{v},\mathbf{u}]$ *is affine independent. If* $v_i < u_i$, *then the set of points* $\{\mathbf{v}^0, \mathbf{v}^1, \ldots, \mathbf{v}^m, \mathbf{w}\}$ *is affine independent for any point* $\mathbf{w} \in E_{\mathbf{m}^i}(\mathbf{e}^i) \cap [\mathbf{v},\mathbf{u}]$.

Proof. To prove the theorem we must show that $\mathbf{w}$ is not a point of the affine subspace generated by $\{\mathbf{v}^0, \mathbf{v}^1, \ldots, \mathbf{v}^m\}$. Since $E_{\mathbf{w}^i}(\mathbf{e}^i) \parallel E_{\mathbf{m}^i}(\mathbf{e}^i)$ and $v_i < u_i$, $E_{\mathbf{w}^i}(\mathbf{e}^i) \cap E_{\mathbf{m}^i}(\mathbf{e}^i) = \emptyset$. Thus, $\mathbf{w} \notin E_{\mathbf{w}^i}(\mathbf{e}^i)$. Also, since $\{\mathbf{v}^0, \mathbf{v}^1, \ldots, \mathbf{v}^m\} \subset E_{\mathbf{w}^i}(\mathbf{e}^i)$

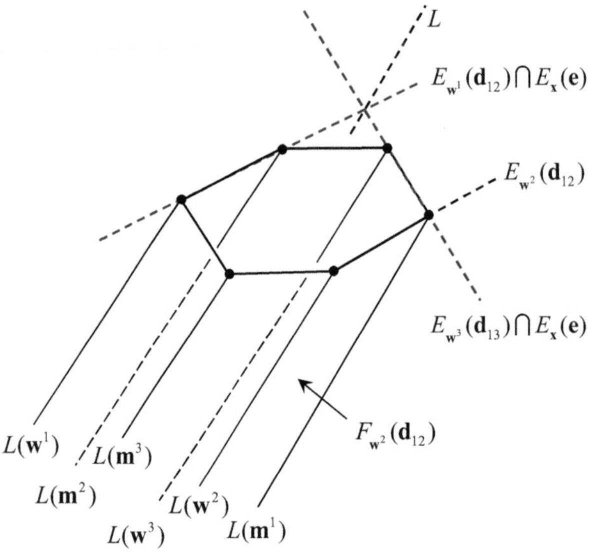

Figure 6.6 The cut of the ℓ-span $S(X)$ by the plane $E_{\mathbf{x}}(\mathbf{e})$, where $\mathbf{x} \in \mathbb{R}^3$. The blue line denotes the intersection of $E_{\mathbf{w}^3}(\mathbf{d}_{13})$ with $E_{\mathbf{x}}(\mathbf{e})$ and the red line denotes the intersection $E_{\mathbf{w}^1}(\mathbf{d}_{12}) \cap E_{\mathbf{x}}(\mathbf{e})$. The line L represents the intersection $E_{\mathbf{w}^3}(\mathbf{d}_{13}) \cap E_{\mathbf{w}^1}(\mathbf{d}_{12})$. Since $L \cap S(X) = \emptyset$, L is not a one-dimensional face of $S(X)$.

is affine independent, the $(m-1)$-dimensional affine space $A = \{\mathbf{x} \in \mathbb{R}^n : \mathbf{x} = \sum_{i=0}^{m} \lambda_i \mathbf{v}^i \text{ and } \sum_{i=0}^{m} \lambda_i = 1\}$, generated by $\{\mathbf{v}^0, \mathbf{v}^1, \ldots, \mathbf{v}^m\}$, must also be a subset of $E_{\mathbf{w}^i}(\mathbf{e}^i)$. Therefore, $\mathbf{w} \notin A$. □

Note that the assumption $\{\mathbf{v}^0, \mathbf{v}^1, \ldots, \mathbf{v}^m\} \subset E_{\mathbf{w}^i}(\mathbf{e}^i) \cap [\mathbf{v}, \mathbf{u}]$ is affine independent implies that $m < n$, and if $m = n-1$, then $A = E_{\mathbf{w}^i}(\mathbf{e}^i)$. It may also be obvious that an analogous theorem exists for the planes of type $E_{\mathbf{w}^i}(\mathbf{d}_{ij})$.

Theorem 6.10 *Suppose that $X \subset \mathbb{R}^n$ is compact and the set of points $\{\mathbf{v}^0, \mathbf{v}^1, \ldots, \mathbf{v}^m\} \subset E_{\mathbf{w}^i}(\mathbf{d}_{ij}) \cap [\mathbf{v}, \mathbf{u}]$ is affine independent. If $E_{\mathbf{w}^i}(\mathbf{d}_{ij}) \cap E_{\mathbf{w}^j}(\mathbf{d}_{ij}) = \emptyset$, then the set of points $\{\mathbf{v}^0, \mathbf{v}^1, \ldots, \mathbf{v}^m, \mathbf{w}\}$ is affine independent for any point $\mathbf{w} \in E_{\mathbf{m}^i}(\mathbf{d}_{ij}) \cap [\mathbf{v}, \mathbf{u}]$.*

We dispense with the proof of this theorem as it is analogous to the proof of Theorem 6.9.

Since any three points of $ext(\mathfrak{P}(X))$ are affine independent, we also have the following theorem that does not rely on the parallel planes approach.

Theorem 6.11 *Suppose that X is a finite subset of $\mathbb{R}^n$, $\mathbf{u} = \bigvee_{\mathbf{x}\in X}\mathbf{x}$, and $\{\mathbf{w}^r, \mathbf{w}^s, \mathbf{w}^t\} \subset W$. If for some pair of distinct indices $i, j \in \{r,s,t\}$, $\mathbf{w}_i^j \neq u_i$, then the set $\{\mathbf{u}, \mathbf{w}^r, \mathbf{w}^s, \mathbf{w}^t\}$ is affine independent.*

Similarly, suppose that $\mathbf{v} = \bigwedge_{\mathbf{x}\in X}\mathbf{x}$, and $\{\mathbf{m}^r, \mathbf{m}^s, \mathbf{m}^t\} \subset M$. If for each pair of distinct indices $i, j \in \{r,s,t\}$, $\mathbf{m}_i^j \neq v_i$, then the set $\{\mathbf{v}, \mathbf{m}^r, \mathbf{m}^s, \mathbf{m}^t\}$ is affine independent.

Proof. To prove affine independence, we will show that the set $\{\mathbf{w}^r - \mathbf{u}, \mathbf{w}^s - \mathbf{u}, \mathbf{w}^t - \mathbf{u}\}$ is linearly independent. Suppose $a(\mathbf{w}^r - \mathbf{u}) + b(\mathbf{w}^s - \mathbf{u}) + c(\mathbf{w}^t - \mathbf{u}) = \mathbf{0}$. Then

$$\begin{array}{lcl} a(\mathbf{w}_r^r - u_r) + b(\mathbf{w}_r^s - u_r) + c(\mathbf{w}_r^t - u_r) = 0 & & b(\mathbf{w}_r^s - u_r) + c(\mathbf{w}_r^t - u_r) = 0 \\ a(\mathbf{w}_s^r - u_s) + b(\mathbf{w}_s^s - u_s) + c(\mathbf{w}_s^t - u_s) = 0 & \text{or,} & a(\mathbf{w}_s^r - u_s) + c(\mathbf{w}_s^t - u_s) = 0 \\ a(\mathbf{w}_t^r - u_t) + b(\mathbf{w}_t^s - u_t) + c(\mathbf{w}_t^t - u_t) = 0. & & a(\mathbf{w}_t^r - u_t) + b(\mathbf{w}_t^s - u_t) = 0. \end{array}$$

It follows from the theorem's hypothesis that for $i, j \in \{r,s,t\}$ with $i \neq j$, $(\mathbf{w}_i^j - u_i) < 0$. Obviously, if any one of the coefficients a, b, or c is equal to zero, then the remaining two coefficients must also be equal to zero. Also, if $a > 0$, then $a(\mathbf{w}_s^r - u_s) < 0$ and $a(\mathbf{w}_t^r - u_t) < 0$, which implies that $b < 0$ and $c < 0$. But then $b(\mathbf{w}_r^s - u_r) + c(\mathbf{w}_r^t - u_r) > 0$, contradicting the assumption that $a(\mathbf{w}_r^r - u_r) + b(\mathbf{w}_r^s - u_r) + c(\mathbf{w}_r^t - u_r) = 0$. The case for $a < 0$ is analogous and also leads to a contradiction. Thus $a = b = c = 0$.

The proof of the second part of the theorem is almost the same and left as an exercise. □

The next example shows that the restriction $i, j \in \{r,s,t\}$ with $\mathbf{w}_i^j \neq u_i$ is necessary.

Example 6.10 Suppose $X = \{\mathbf{x}^1, \ldots, \mathbf{x}^6\} \subset \mathbb{R}^3$, where

$$\mathbf{x}^1 = \begin{pmatrix} 10 \\ 7 \\ 10 \end{pmatrix}, \mathbf{x}^2 = \begin{pmatrix} 8 \\ 12 \\ 10 \end{pmatrix}, \mathbf{x}^3 = \begin{pmatrix} 8 \\ 7 \\ 10 \end{pmatrix}, \mathbf{x}^4 = \begin{pmatrix} 1 \\ 3 \\ 1 \end{pmatrix}, \mathbf{x}^5 = \begin{pmatrix} 1 \\ 5 \\ 3 \end{pmatrix}, \mathbf{x}^6 = \begin{pmatrix} 1 \\ 1 \\ 3 \end{pmatrix}.$$

The set $W \cup \{\mathbf{u}\} \subset V$ consists of the vectors

$$\mathbf{w}^1 = \begin{pmatrix} 10 \\ 7 \\ 10 \end{pmatrix}, \mathbf{w}^2 = \begin{pmatrix} 8 \\ 12 \\ 10 \end{pmatrix}, \mathbf{w}^3 = \begin{pmatrix} 8 \\ 7 \\ 10 \end{pmatrix}, \text{ and } \mathbf{u} = \begin{pmatrix} 10 \\ 12 \\ 10 \end{pmatrix}.$$

These four vertices are not affine independent since they are points of the same plane $E_u(\mathbf{e}^3)$. However, $\mathbf{v} \notin E_u(\mathbf{e}^3)$ which proves that $\{\mathbf{w}^1, \mathbf{w}^2, \mathbf{w}^3, \mathbf{v}\}$ is affine independent.

Theorem 6.11 relies on the fact that any three elements of $W \subset ext(\mathfrak{P}(X))$ are affine independent. Ergo, it is reasonable to ask if the theorem can be generalized to more than three elements of W. The next example answers this question in the negative.

Example 6.11 Suppose $X = \{\mathbf{x}^1, \mathbf{x}^2, \mathbf{x}^3, \mathbf{x}^4\} \subset \mathbb{R}^4$, where

$$\mathbf{x}^1 = \begin{pmatrix} 1 \\ \frac{1}{3} \\ \frac{2}{3} \\ 0 \end{pmatrix}, \mathbf{x}^2 = \begin{pmatrix} \frac{1}{3} \\ 1 \\ 0 \\ \frac{2}{3} \end{pmatrix}, \mathbf{x}^3 = \begin{pmatrix} \frac{2}{3} \\ 0 \\ 1 \\ \frac{1}{3} \end{pmatrix}, \mathbf{x}^4 = \begin{pmatrix} 0 \\ \frac{2}{3} \\ \frac{1}{3} \\ 1 \end{pmatrix}.$$

In this case we have $W = X$, $\mathbf{v} = \mathbf{0}$, and $\mathbf{u} = \mathbf{1}$, where $\mathbf{1} = (1,1,1,1)'$. By definition of affine independence (Section 4.4.1), W is affine independent if and only if the set $V = \{\mathbf{w}^2 - \mathbf{w}^1, \mathbf{w}^3 - \mathbf{w}^1, \mathbf{w}^4 - \mathbf{w}^1\}$ is linearly independent. Setting $\mathbf{v}^i = \mathbf{w}^{i+1} - \mathbf{w}^1$ for $i = 1, 2$, and 3, and solving the equation $a\mathbf{v}^1 + b\mathbf{v}^2 + c\mathbf{v}^3 = \mathbf{0}$, one obtains $a = 1 = b$ and $c = -1$. Thus V is linearly dependent since $\mathbf{v}^1 + \mathbf{v}^2 = \mathbf{v}^3$.

Exercises 6.2.2

1. Construct the remaining hyperplanes $P_r(i, j)$ of Example 6.9.

2. Prove Theorem 6.10.

3. Prove the second part of Theorem 6.11.

4. Suppose $[\mathbf{a}, \mathbf{b}] \subset \mathbb{R}^3$, the interval defined by $\mathbf{a} = (10, 4, 8)'$ and $\mathbf{b} = (20, 15, 16)'$. Randomly choose twelve points from $[\mathbf{a}, \mathbf{b}]$ and let $X = \{\mathbf{x}^1, \ldots, \mathbf{x}^{12}\}$ denote the set of these randomly chosen elements. Use Algorithms **6.1** and **6.2** in order to obtain the set $V \cup U \cup T_1$ of extreme points of $\mathfrak{P}(X)$.

5. Suppose $[\mathbf{a}, \mathbf{b}] \subset \mathbb{R}^4$ is an interval defined by $\mathbf{a} = (10, 4, 8, 5)'$ and $\mathbf{b} = (20, 15, 16, 25)'$. Randomly choose twelve points from $[\mathbf{a}, \mathbf{b}]$ and let $X = \{\mathbf{x}^1, \ldots, \mathbf{x}^{12}\}$ denote the set of these randomly chosen elements. Use Algorithms 6.3 and 6.4 in order to obtain the set $V \cup U \cup T_2$ of extreme points of $\mathfrak{P}(X)$.

CHAPTER 7

Image Unmixing and Segmentation

THIS chapter presents a lattice algebra approach to hyperspectral image unmixing and to color image segmentation. The first three subsections are devoted to hyperspectral image unmixing and the last section covers color image segmentation.

7.1 SPECTRAL ENDMEMBERS AND LINEAR UNMIXING

Due to their optimal absolute storage capacity and one step convergence, applications of matrix-based LAMs commenced rapidly after their initial specification in 1998 [218]. These applications varied widely and ranged from the reconstruction of grayscale images, to databases, and hyperspectral image segmentation [5, 84, 98, 99, 100, 213, 277, 288, 291, 295, 296, 307]. Most of the early applications were concerned with the recall of objects associated with exemplar patterns. The focus of this chapter will use the lattice auto-associative memories $\mathfrak{W}_X$ and $\mathfrak{M}_X$ to unmix a hyperspectral image, a process that may be considered a specific approach to the more general problem of image segmentation.

Image segmentation refers to subdividing an image into regions of particular interest. The idea of employing the memories $\mathfrak{W}_X$ and $\mathfrak{M}_X$ in image segmentation originated with M. Grãna et al. [99, 100] in their efforts of hyperspectral image unmixing. Since segmentation is data-dependent, it became clear that a more thorough understanding of the distribution of various data values within the ℓ-span generated by the data was a necessity. The relationships between the data set X, the smallest interval (hyperbox) containing X, and the dual standard bases for the ℓ-span $S(X)$, established in [226, 231],

DOI: 10.1201/9781003154242-7

are essential for this understanding. Shortly thereafter, G. Urcid et al. applied these new insights to hyperspectral image unmixing as well as to color image segmentation [230, 288, 289].

Advances in remote sensing has resulted in imaging devices with ever-growing spectral resolution. A pixel of a multispectral or hyperspectral image can be expressed as a pair $(\mathbf{p}, \mathbf{x}) \in \mathbb{R}^2 \times \mathbb{R}^n$, where $\mathbf{p} \in \mathbb{R}^2$ denotes a position or location on the ground, $\mathbf{x} = (x_1, \ldots, x_n)' \in \mathbb{R}^n$, and x_i denotes the measured reflected radiation, also called the *reflectance*, of the ith wavelength at location $\mathbf{p}$. The vector $\mathbf{x}$ is called the *spectrum* of the pixel. For a multispectral image, the number n of spectral bands is usually small, with $n \leq 20$, while the number of bands in hyperspectral images may consist of several hundred bands. Figure 7.1 illustrates the fact that a hyperspectral image can be viewed as a 3-dimensional rectangular box or as a "cube".

The high spectral resolution produced by current hyperspectral imaging devices facilitates identification of fundamental materials that make up a remotely sensed scene. In other words, hyperspectral resolution can improve the discrimination between various different ground-based materials. However, a typical pixel of a multispectral image generally represents a region on the ground consisting of several square meters. For example, each Landsat Thematic Mapper pixel represents a 30×30 m^2 footprint on the ground. Consequently, a hyperspectral image pixel can contain various different materials or objects and its spectrum is, therefore, usually a mixture of various different reflective objects. This raises the question as to whether or not it is possible to know the percentage of objects that are most represented in the spectrum of a given pixel. One widely used method for estimating these percentages from the pixel's spectrum is known as *linear spectral unmixing* (LSU). Linear spectral unmixing is based on the *constrained linear mixing* (CLM) model that in turn is based on the fact that points on a simplex can be represented as a linear sum of the vertices that determine the simplex [83, 140, 141]. Therefore the LSU model assumes that the spectrum of a pixel $(\mathbf{p}, \mathbf{x})$ is a linear combination of the endmembers present in $\mathbf{x}$. The mathematical equations of the model and its constraints are given by

$$\mathbf{x} = S\,\mathbf{a} + \mathbf{n} = \sum_{j=1}^{m} a_j \mathbf{s}^j + \mathbf{n}, \tag{7.1}$$

$$\sum_{j=1}^{m} a_j = 1 \quad \text{and} \quad a_j \geq 0 \; \forall i, \tag{7.2}$$

where $\mathbf{x} \in \mathbb{R}^n$ is the measured spectrum over n bands of an image pixel,

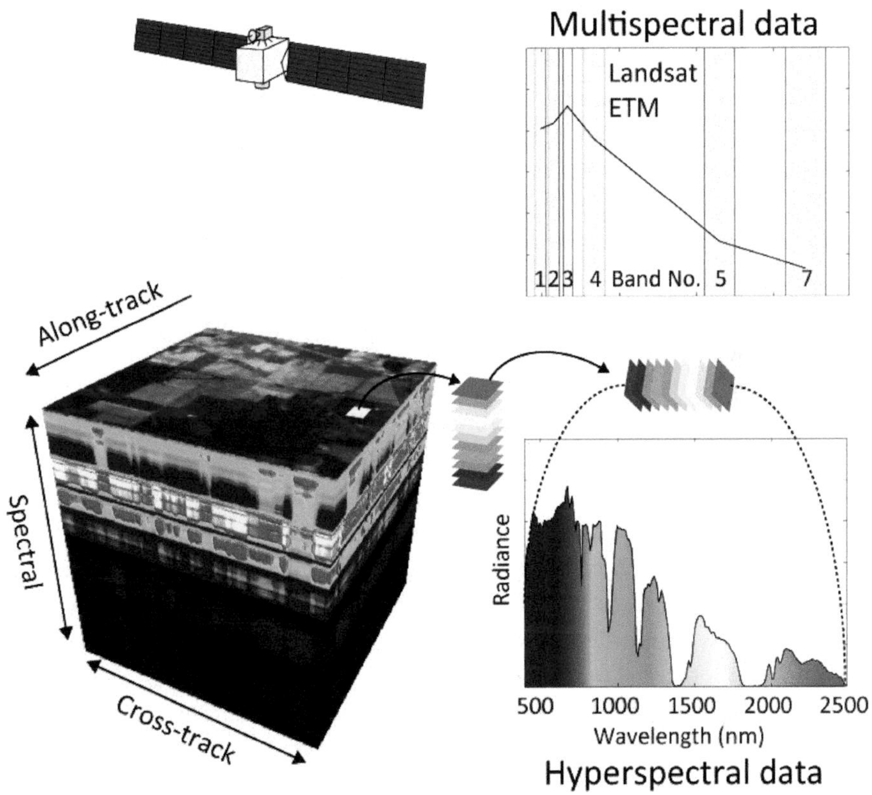

Figure 7.1 Counterclockwise from the top-left image: satellite imaging device for push-broom image generation; the push-broom generated hyperspectral (or multispectral) image cube; the spectrum of a hyperspectral pixel; the corresponding spectral bands of a multispectral image, where the pixel value of the band is derived from the intensity of the radiance detected within the band. The actual radiance at location **p** within the visible spectrum (from near-ultraviolet to near-infrared) would be a continuous curve. The figure between the image cube and the hyperspectral data is the spectrum **x** of the pixel with location indicated by the white square on top of the cube.

$S = (\mathbf{s}^1, \ldots, \mathbf{s}^m)$ is an $n \times m$ matrix whose columns are the m endmember spectra assumed to be affinely independent, the entries of m-dimensional column vector, $\mathbf{a} = (a_1, \ldots, a_m)'$, are the corresponding abundances or fractions of endmember spectra present in $\mathbf{x}$, and $\mathbf{n} \in \mathbb{R}^n$ represents an additive noise vector.

This model has been used by a multitude of researchers ever since Adam et al [2] analyzed an image of Mars using four endmembers. In the cited reference and various other applications, hyperspectral image segmentation and analysis takes the form of a pattern recognition problem, as the segmentation problems reduces to matching the spectra of the hyperspectral image to predetermined spectra stored in a library. In many cases, however, endmembers cannot be determined in advance and must be selected from the image directly by identifying the pixel spectra that are most likely to represent the fundamental materials. This comprises the *autonomous endmember detection* problem. Unfortunately, the spatial resolution of a sensor makes it often unlikely that any pixel is composed of a single endmember. Thus, the determination of endmembers becomes a search for image pixels with the least contamination from other endmembers. These are also referred to as *pure* pixels. The pure pixels exhibit *maximal* or *minimal* reflectance in certain spectral bands and correspond to vertices of a high dimensional simplex that, hopefully, encloses most if not all pixel spectra.

There exist a multitude of different approaches to construct endmembers and endmember abundances (percentages) in mixed pixels [18, 31, 47, 48, 57, 126, 197, 317]. A large number of these approaches are based on the *convex polytope model*. This model is derived from the physical fact that radiance is a non-negative quantity. Therefore, the vectors formed by discrete radiance spectra are linear combinations of non-negative components and as such must lie in a convex region located in the non-negative orthant $\mathbb{R}^n_+ = \{\mathbf{x} \in \mathbb{R}^n : \mathbf{x} \geq \mathbf{0}\}$. The vertices of this convex region, which can be elements of X or vectors that are physically related to the elements of X, have proven to be excellent endmember candidates. The reason for this is based on the observation that endmembers exhibit maximal and minimal reflectances within certain bands and correspond to vertices of a high-dimensional polyhedron. The N-FINDR and MVT (minimal value transform) algorithms are two classic examples of the convex polytope based endmember selection approach [315, 57]. Since these earlier methods, researchers have developed various sophisticated endmember extraction and endmember generation techniques based on the convex polytope assumption [90, 91, 197, 317]. The WM-method described here differs from those described by Graña [101, 102] and Myers [194] as the endmembers we obtain have a physical relationship to the pixels of the hyperspectral image under consideration. The WM-method will always provide the same sets of candidate endmembers based on theoretical facts already exposed in Chapter 6. A brief comparison is made against two approaches based on convex optimization, namely vertex component analysis [197] and the minimal volume enclosing simplex [46].

7.1.1 The Mathematical Basis of the WM-Method

Here, we recall the mathematical concepts and expressions that serve as the foundation to the $\mathfrak{W}_X \mathfrak{M}_X$ method used to determine a set of endmembers in a given hyperspectral image specified by the data set X. We denote by k the number of pixel spectra belonging to X and display its elements as $\{\mathbf{x}^1, \ldots, \mathbf{x}^k\}$ where each $\mathbf{x}^\xi \in \mathbb{R}^n_+$. The reader is invited to review equation 4.46 of Section 4.3.3 and the beginning paragraphs of Section 4.4.1 before equation 4.50. From Section 5.2.2 about lattice auto-associative memories, equation 5.20 is rewritten equivalently as follows,

$$\mathfrak{m}_{ij} = \bigvee_{\mathbf{x}^\xi \in X} (x_i^\xi - x_j^\xi) = \bigvee_{\xi=1}^{k} (x_i^\xi - x_j^\xi) \tag{7.3}$$
$$\mathfrak{w}_{ij} = \bigwedge_{\mathbf{x}^\xi \in X} (x_i^\xi - x_j^\xi) = \bigwedge_{\xi=1}^{k} (x_i^\xi - x_j^\xi).$$

The corner points $\mathbf{v} = \min(X)$ and $\mathbf{u} = \max(X)$ of the n-dimensional hyperbox $[\mathbf{v}, \mathbf{u}]$ enclosing X can be computed directly by using the formulae given in equation 6.6 of Section 6.1.2 and repeated here for convenience in developed form (cf. Step 1 of Algorithm 5.1),

$$\mathbf{v} = \bigwedge_{j=1}^{k} \mathbf{x}^j = \bigwedge_{i=1}^{n} \bigwedge_{\xi=1}^{k} x_i^\xi \quad \text{and} \quad \mathbf{u} = \bigvee_{j=1}^{k} \mathbf{x}^j = \bigvee_{i=1}^{n} \bigvee_{\xi=1}^{k} x_i^\xi. \tag{7.4}$$

Again, from Section 6.1.2 on lattice polytopes, we will use the collections $\mathfrak{W}$ and $\mathfrak{M}$ with corresponding *shifted* or *translated* vector sets, respectively, W and M defined as in equation 6.8,

$$W = \{\mathbf{w}^j : \mathbf{w}^j = u_j + \mathfrak{w}^j, j = 1, \ldots, n\}, \tag{7.5}$$
$$M = \{\mathbf{m}^j : \mathbf{m}^j = v_j + \mathfrak{m}^j, j = 1, \ldots, n\}.$$

From these equations, it is not difficult to see that the main diagonal of

W equals $\mathbf{u}$ and, similarly, the main diagonal of M equals $\mathbf{v}$. In relation to equation 6.7, the reader should have in mind that $X \subset [\mathbf{v},\mathbf{u}]$ and $X \subset S(X)$, where $S(X)$ denotes the ℓ-span of X. Hence, the polytope $\mathfrak{P}(X) = [\mathbf{v},\mathbf{u}] \cap S(X)$ has the property that $X \subset \mathfrak{H}(X) \subset \mathfrak{P}(X)$. In addition, convexity of $[\mathbf{v},\mathbf{u}]$ and $S(X)$ imply the convexity of $\mathfrak{P}(X)$, and $\mathfrak{P}(X)$ is the smallest complete lattice polytope containing X.

In various applications examples, one often obtains an overlap in the elements of W and M in that $\mathbf{w}^\ell = \mathbf{m}^j$ without affecting the affine independence of either set W or M. Furthermore, in order to ascertain that W or M is affinely independent, all one has to do is to check that no two vectors of W or M are identical. Thus, if $\mathbf{w}^j \neq \mathbf{w}^\ell$ for any pair $\{j,\ell\} \subset J$, then any non-empty subset of W can serve as a possible set of endmembers. If set X is such that $\mathbf{w}^j = \mathbf{w}^\ell$ for several pairs $\{j,\ell\}$, then an elimination algorithm of columns and rows from W results in an affinely independent set of vectors as described in [231]. So far, in our tests with hyperspectral image data, the equalities $\mathbf{w}^j = \mathbf{w}^\ell$ or $\mathbf{m}^j = \mathbf{m}^\ell$ for any $j \neq \ell$ *do not* occur and therefore we can select subsets of W or M as candidate sets of endmembers which are affinely independent.

In most endmember detection schemes, endmembers form a linearly independent set of m points in $\mathbb{R}^n_+$ and a *dark point* is chosen to obtain a large m-simplex containing most or all of X. In order for the $m+1$ points to be affinely independent, one must assure that the dark point is not an element of the hyperplane spanned by the $(m-1)$-simplex defined by the m affine independent points. Generally, the subsets $\mathcal{W} = W \cup \mathbf{u}$ and $\mathcal{M} = M \cup \mathbf{v}$ of set $V = W \cup M \cup \{\mathbf{v},\mathbf{u}\}$ are affinely independent. Hence, the choice of $\mathbf{v}$ for the dark point remains a good one, especially in view of the fact that it represents the low values of actual data. Also, note that a vector $\mathbf{w}^j \in W$ has the property that its j-th coordinate corresponds to the *maximum measured reflectance* within the j-th band of the data set X. In this sense, the elements of W can be viewed as excellent representatives of endmembers of the data cube X. However, it is important to remark that for any pair $\{\mathbf{w}^j,\mathbf{m}^i\}$ the inequalities $\mathbf{w}^j \leq \mathbf{m}^i$ or $\mathbf{m}^i \leq \mathbf{w}^j$ are generally false even though $m^i_j \leq w^j_j = u_j = \bigvee_{\xi=1}^k x^\xi_j$ and $\bigwedge_{\xi=1}^k x^\xi_i = v_i = m^i_i \leq w^j_i$. In fact, when using hyperspectral data, one usually obtains $w^j_\ell < m^i_\ell$ for several indices ℓ. For this reason various elements of M may represent important endmembers as demonstrated in the test example described in the next subsection. Geometrically, the elements of V are vertices of the convex polytope $\mathfrak{P}(X)$. Thus, the image data cube X, contained in $\mathfrak{P}(X)$, lends itself to convex hull analysis using the elements of V.

All previous mathematical expressions reviewed here that appear in different theoretical developments treated before in Chapters 4, 5, and 6, are

TABLE 7.1 Representative minerals in the Cuprite site [233].

Description	**Color**	**Area %**
Alunite GDS96 (250C)	Red	4.2
Calcite CO 2004	Pink	22.6
Kaolinite KGa-2 (pxl)	Green	10.8
Montmorillonite SCa-2.b	Seagreen	16.1
Muscovite CU93-1 low-Al	Blue	27.1
Unclassified	Black	19.2

combined in Algorithm 6.1 (Constructing the elements of V) that computes the fundamental set of extreme points of the lattice convex polytope enclosing X as derived from the LAAMs, $\mathfrak{W}_X$ and $\mathfrak{M}_X$. Therefore, computing set V with at most $2(n+1)$ elements gives all potential endmembers and the steps of Algorithm 6.1 are the core of what we call the *WM-method* used to find endmembers in spectral imagery.

7.1.2 A Validation Test of the WM-Method

The aim of the following validation test, based on a hyperspectral image remotely acquired with the *Airborne Visible and Infrared Imaging Spectrometer* (AVIRIS) of NASA's Jet Propulsion Laboratory, is to provide enough detail to demonstrate the effectiveness of using the dual lattice auto-associative memories, $\mathfrak{W}_X$ and $\mathfrak{M}_X$, to determine sets of endmembers from which a subset of *final endmembers* is selected to accomplish hyperspectral image segmentation.

One of the most studied cases in mineralogical and chemical composition has been the mining site of Cuprite, Nevada. The United States Geographical Survey (USGS) Laboratory has produced detailed maps of mineral and chemical compound distribution at the Cuprite site. The left part of Figure 7.2, shows an approximate true color RGB image with wavelengths of 0.67 μm (red), 0.56 μm (green) and 0.48 μm (blue). The right image in the same figure displays a simplified 6-false color map, built with five representative minerals chosen from the mineral groups present in the original USGS image map produced by Clark and Swayze in 1995. Table 7.1 lists specific minerals whose assigned color is a visual aid to distinguish its distribution on site.

The USGS source image map shows only the geographical region of Cuprite where mineral distribution is prominent. It has a width of 534 pixels out of 614 pixels available in a scan line and is formed with the lower

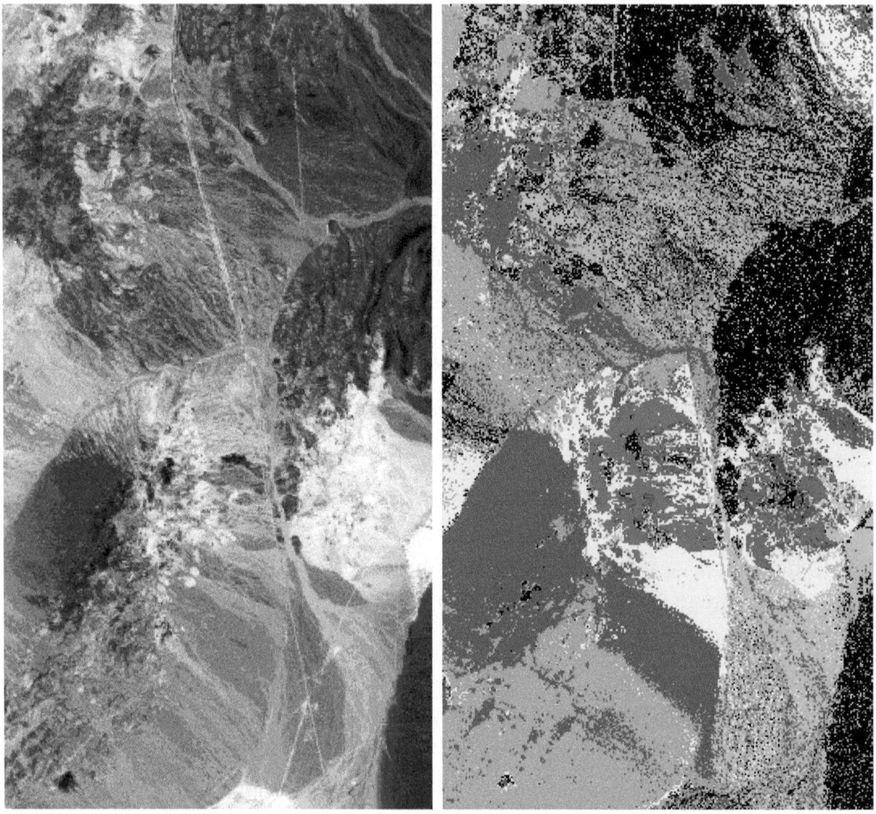

Figure 7.2 Left, RGB composite image of the mining site of Cuprite, Nevada. Right, simplified 6-false color mineral distribution map. True size of each image is 534 × 972 pixels [233].

TABLE 7.2 AVIRIS & USGS spectrometers wavelength scales (microns). AVIRIS channels not used for spectral sample matching are marked with a cross "×" [233].

Channel	AVIRIS	USGS
212	2.388	2.386
213	2.398	2.400
214	2.408	×
215	2.418	2.418
216	2.427	×
217	2.437	2.440
218	2.447	×
219	2.457	×
220	2.467	2.466

358 scan lines of scene no. 3, followed by all 512 scan lines of scene no. 4, and ends with the upper 102 scan lines of scene no. 5 as registered in flight (1997). The first 80 pixels were dropped from each scan line to obtain the region studied by USGS scientists. The AVIRIS device acquires hyperspectral images in 224 channels; however, only 52 noiseless channels that fall within the short wavelength infrared band are considered for mineral detection. Specifically, channels no. 169 (1.95 μm) to 220 (2.47 μm) are used to match remote spectra against subsampled ground or laboratory spectra, since these last ones are obtained with higher spectral resolution [55]. However, only 48 AVIRIS channels were selected for spectral matching due to the existing difference in wavelength scale between ground and remote spectrometers. Table 7.2 gives the correspondence between the wavelength scale used by AVIRIS imaging spectrometer and the reference scale of a USGS laboratory spectrometer, showing explicitly which channels at the end of the scale were not considered. Therefore, a *test data cube* associated with the 6-false color map is composed of "pure" pixel spectra corresponding to the minerals listed in Table 7.1 and illustrated in Figure 7.3 where each pixel spectra is a 48-dimensional vector.

Computation of the lattice auto-associative memories $\mathfrak{W}_X$ and $\mathfrak{M}_X$ is quite simple in this case. From equation 7.3, e.g., for $i, j = 1, \ldots, n$ with $n = 48$,

$$\mathfrak{w}_{ij} = \bigwedge_{\xi=1}^{k} (x_i^\xi - x_j^\xi) = \bigwedge_{\zeta=1}^{m} \bigwedge_{\xi\in\Omega_\zeta} (x_i^\xi - x_j^\xi) = \bigwedge_{\zeta=1}^{m} (x_i^{\xi_\zeta} - x_j^{\xi_\zeta}), \tag{7.6}$$

where $k = \sum_{\zeta=1}^{m} |\Omega_\zeta| = 419,390$ (out of $519,048 = 534 \times 972$ vectors) is the number of *non-zero* vectors in the data cube, $m = 5$ is the number of "pure"

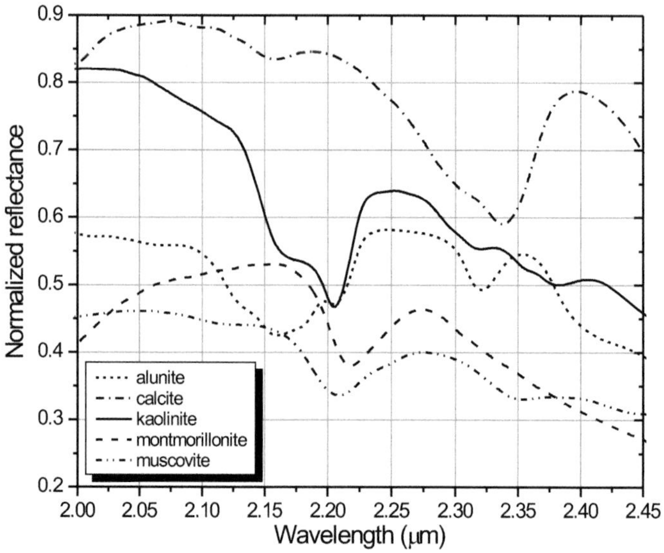

Figure 7.3 Spectral curves of the minerals given in Table 7.1 (true endmembers).

pixels, each Ω_ζ is the family of indices that correspond to pixels describing the same mineral, and $\xi_\zeta \in \{1,\ldots,k\}$ is the first sequential index in the data cube equal to a different spectral vector. The first equality in Eq. (7.6) is a consequence of the associative property of the minimum binary operation, and the second equality follows by *idempotency*. With X reduced to $\{\mathbf{x}^{\xi_1},\ldots,\mathbf{x}^{\xi_5}\}$, the shifted min- and max-memories are calculated in milliseconds, and the 48 columns of each memory matrix give all candidate endmembers. Since this is a test data set for which *a priori knowledge* is available about the hyperspectral image, a simple matching procedure applied to the column vectors $\mathbf{w}^j \in W$ and $\mathbf{m}^j \in M$, together with the $\mathbf{v}$ and $\mathbf{u}$ bounds, immediately yields a subset of final endmembers.

Consider $\mathcal{W} \subset V$ and, for $p = 1,\ldots,5$ and $q = 1,\ldots,49$, let $c_{pq} = \rho(\mathbf{x}^{\xi_p},\mathbf{y}^q)$, where $\mathbf{x}^{\xi_p} \in X$ and $\mathbf{y}^q \in \mathcal{W}$, be the linear correlation coefficients between true and LAAM determined endmembers. In addition, for all p, let $\mu_p = \max_{1\leq q\leq 49}\{c_{pq}\}$ be the *maximum* correlation coefficient of each pure pixel against all potential endmembers in Y, and compute

$$b_{pq} = \mathbf{if}(c_{pq} = \mu_p, \mathbf{if}(c_{pq} \geq \alpha, 1, 0), 0), \tag{7.7}$$

where α is a parameter used to threshold correlation coefficient values and the three argument function, **if**(*condition, true, false*) is the usual **if-then-else**

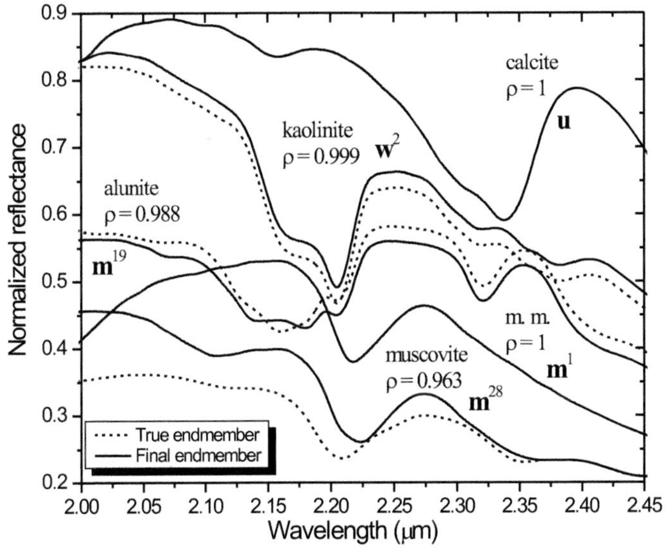

Figure 7.4 The solid lines represent the final endmembers determined from set V and the dashed lines correspond to the given true endmembers; the mineral montmorillonite is abbreviated as m.m. [233]

programming construct. A binary value, $b_{pq} = 1$, gives the row and column indices of a "very good" match between a true endmember (row index) and the potential endmember marked as final (column index). A similar spectral matching procedure is applied to $\mathcal{M} \subset V$.

It turns out that *all five* "pure" pixels used to build the test data cube are correctly determined from both LAAMs as final endmembers, a result that is in agreement with the theoretical background developed earlier. Figure 7.4 displays the set of true and final endmembers, where the curve labels give the specific element in V whose *correlation coefficient*, denoted by ρ, is highest if compared to each true endmember ("pure" spectra). Final and true endmember curves are superimposed if a perfect match exists ($\rho = 1$) and, to avoid signature overlapping, the bottom two curves (muscovite) are displaced -0.1 off their original normalized reflectances. As a final remark, the set $\{\mathbf{v}, \mathbf{m}^1, \mathbf{m}^{19}, \mathbf{m}^{28}, \mathbf{w}^2, \mathbf{u}\}$, where $\mathbf{v}$ is taken as the dark point, is affinely independent and its convex hull forms a 5-simplex that encloses X.

7.1.3 Candidate and Final Endmembers

In order to solve the constrained linear unmixing problem established in Eqs. (7.1) and (7.2) without a heavy computational burden, different techniques have been developed to find a representative set of endmembers, i.e., "the purest pixels", or otherwise reducing the number of potential endmembers to a smaller subset. The benefit of having a small collection of final spectral signatures lies in significant savings of computational effort to determine the fractional coefficients, $a_j \geq 0$ for $j = 1, \ldots, m$, by means of numerical matrix inversion operations or least square estimation methods. In the case of hyperspectral imagery acquired with hundreds of bands sampled from the continuous electromagnetic spectrum, the goal of any endmember detection method is to find a subset $S \subset X$ with m pure pixels such that $m \leq 15$.

In the case of the WM-method, the vector sets W and M derived from $\mathfrak{W}_X$ and $\mathfrak{M}_X$, contain always a fixed but large number of endmembers. As shown before, $|V| = 2(n+1)$; however, several techniques can be applied to V or selected proper subsets S of V to have $m \ll 2(n+1)$ if $n \geq 100$. Prior knowledge of the desired number of endmembers or use of alternative endmember reduction methods can be applied to eliminate unimportant candidate endmembers. Another possibility consists of a preprocessing stage where the hyperspectral data cube dimensionality is diminished by specific conditions associated with the type of resources to be extracted from a hyperspectral image or by the application of known techniques such as *Principal Component Analysis* (PCA), *Minimum Noise Fraction Transform* (MNFT), or *Adjacent Band Removal* (ABR) of highly correlated bands [99, 141, 230]. Such reductions are often necessary to eliminate undesirable effects produced during data acquisition, as well as lowering computational requirements such as storage space, available main memory, and processing hardware speed. For example, the *maximal* storage space required for a single hyperspectral image scene, of size $614 \times 512 \times 224$ (pixels, lines & bands), captured by NASA-JPL's Airborne Visible and Infrared Imaging Spectrometer (AVIRIS) is 134.3125 Mb (megabytes) [294].

Other methods for autonomous endmember determination have been described elsewhere, for example, hierarchical Bayesian models [73], minimum mutual information [101], and pattern elimination based on minimal Chebyshev distance or angle between vector pairs [230]. At the end of the following section, linear unmixing based on the vertex component analysis [197] and the minimum volume enclosing simplex [46] approaches are described briefly, including a comparison of algorithm characteristics between them and the proposed WM-method.

Exercises 7.1.3

1. From the viewpoint of linear algebra, describe the type of system of linear equations in several unknowns that correspond to equation. 7.1, representing the LSU model, for a) $n > m$, b) $n < m$, and c) $n = m$; explain the relationship between n (vector dimensionality) and m (number of endmembers).

2. In relation to equations 7.1 and 7.2 representing the CLM model, let $J = \{1,\ldots,m\}$ be an index set and let $K \subset J$ such that, after solving the given equations subject to the restrictions imposed, it turns out that $a_j = 0$ for all $j \in K$. What is the physical meaning in that case?

3. Equation 7.1 can be simplified by considering that the noise vector $\mathbf{n} = \mathbf{0}$. Otherwise, what physical factors may contribute if $\mathbf{n} \geq \mathbf{0}$? If, $I = \{1,\ldots,n\}$ and $H \subset I$ is such that $n_i > (S\,a)_i$ for all $i \in H$, how can the CLM model be solved to avoid numerical instabilities?

4. Computational complexity of sets W and M in equation 7.5 is linear in n (in a sequential machine). Discuss the *mean* computational complexity for equations 7.3 and 7.4.

5. Search the AVIRIS website for free data hyperspectral "cubes" and download the Cuprite Nevada data set. a) Describe the data structure of the file, and b) propose a low-level programming pseudo-code to visualize, interactively as a 256-grayscale image, the i-th spectral band where $i \in \{1,\ldots,n\}$.

6. With respect to equation 7.6, how much computer memory space occupies the test data cube with size $534 \times 972 \times 48$ (pixels, scan lines, and spectral bands)?

7. For the channels listed in Table 7.2, a) compute the wavelength differences in microns between consecutive channels for both AVIRIS and USGS columns. b) Search the USGS website for the USGS spectral library, download the spectral data of some mineral which has both remotely and laboratory (or field) sensed signatures, and compare them graphically. Include in your comparison the number of wavelength samples and the wavelength step size(s) between samples in each scale (remote vs. laboratory).

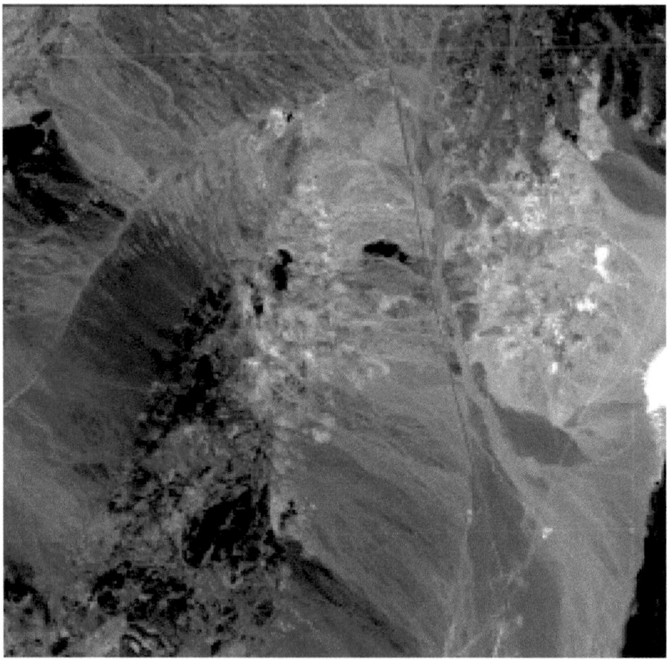

Figure 7.5 Scene no. 4 of the Cuprite Nevada site. Enhanced grayscale image obtained from an RGB pseudocolor composition of bands 207 (red), 123 (green), and 36 (blue) [230].

7.2 AVIRIS HYPERSPECTRAL IMAGE EXAMPLES

For our first application example, we consider a subcube corresponding to scene no. 4 of the 1997 AVIRIS flight over the Cuprite mining site in Nevada, displayed in Figure 7.2. With respect to the image format used by AVIRIS, pixels 81 to 614 and bands 169 to 220, covering the wavelength range from 1.95 to 2.47 microns, where selected to compare our results with the USGS (United States Geographical Survey) Cuprite 1995 mineral distribution made by Clark and Swayze. Thus, the hyperspectral image subcube is of size $534 \times 512 \times 52$ and the data set X has 273,408 vectors of dimension 52.

The first step is to form the memories $\mathfrak{W}_X$ and $\mathfrak{M}_X = \mathfrak{W}_X^* = -\mathfrak{W}_X'$ based on equation 7.3. Using the vectors $\mathbf{v}$ and $\mathbf{u}$ computed with equation 7.4, the elements of $\mathfrak{W}_X$ and $\mathfrak{M}_X$ are then shifted, respectively, by $\mathbf{u}$ and $\mathbf{v}$ to obtain the column vector sets W and M. Since W and M are square matrices of size 52×52, each provides 52 "candidate" endmembers. Since contiguous columns are highly correlated, most of these potential endmembers can be discarded

TABLE 7.3 Potential endmember groups and representatives (Rep). Final endmembers are marked with an underline [230].

W columns	Rep. $\mathbf{w}^j$	M columns	Rep . $\mathbf{m}^j$
$[1,\ldots,15]$	$\{\underline{2},5\}$	$[1,\ldots,10]$	$\{1\}$
$[16,\ldots,27]$	$\{\underline{25}\}$	$[11,\ldots,20]$	$\{\underline{16}\}$
$[28,\ldots,35]$	$\{\underline{33}\}$	$[21,\ldots,32]$	$\{26,\underline{27}\}$
$[36,\ldots,41]$	$\{39\}$	$[34,\ldots,41]$	$\{38\}$
$[42,\ldots,48]$	$\{45\}$	$[42,\ldots,48]$	$\{47\}$
$[49,\ldots,52]$	$\{51\}$	$[49,\ldots,52]$	$\{51\}$

using appropriate techniques. Here, column vector elimination is realized by computing the minimal Chebyshev distances or the angles between pairs of vectors as explained next.

Given any two vectors $\mathbf{x}^\xi, \mathbf{x}^\gamma \in \mathbb{R}^n$, the expressions given by

$$\tau(\mathbf{x}^\xi, \mathbf{x}^\gamma) = \bigvee_{i=1}^{n} |x_i^\xi - x_i^\gamma| \quad \text{and} \quad \theta(\mathbf{x}^\xi, \mathbf{x}^\gamma) = \cos^{-1} \frac{\mathbf{x}^\xi \cdot \mathbf{x}^\gamma}{\| \mathbf{x}^\xi \| \| \mathbf{x}^\gamma \|}, \tag{7.8}$$

where $\| \cdot \|$ denotes Euclidean norm, define, respectively the *Chebyshev distance* (left expression in equation 7.8), and the *angle* (in radians) between the two vectors (right expression). Each formula is symmetric in its arguments for $\gamma \neq \xi$ and, if $\gamma = \xi$, then $\tau = 0$ and $\theta = 0$. Two vectors are said to be "similar" if τ or θ are less than a given tolerance small value. In our application example, to discard correlated potential endmembers, equation 7.8 was computed between every pair of column vectors of W as well as between columns of M, for $\xi = 1, \ldots, 51$ and $\gamma = \xi + 1, \ldots, 52$.

Vector grouping based on distance values and calculated angles is shown in Table 7.3, where each set of indices indicates strong correlation between corresponding columns. In a second step, representative endmembers were selected from each group based on additional criteria such as, overall graphical shape, number of local minima and maxima, and amplitude range. The third and last step involves a matching process between available library spectra and representative endmembers from which the set of final endmembers is obtained. As an example, Figure 7.6 displays the Chebyshev distance curves of each final endmember selected from W. Notice that $\tau(\mathbf{w}^2, \mathbf{w}^j)$ for $j \in [1, 15]$, $\tau(\mathbf{w}^{25}, \mathbf{w}^j)$ for $j \in [16, 27]$, and $\tau(\mathbf{w}^{33}, \mathbf{w}^j)$ for $j \in [28, 52]$, are all less than 0.4. Similarly, $\tau(\mathbf{m}^{16}, \mathbf{m}^j)$ for $j \in [2, 20]$ and $\tau(\mathbf{m}^{27}, \mathbf{m}^j)$ for $j \in [21, 46]$, are also less than 0.4.

Another aspect to note is that the scaling based on equation 7.5 generates

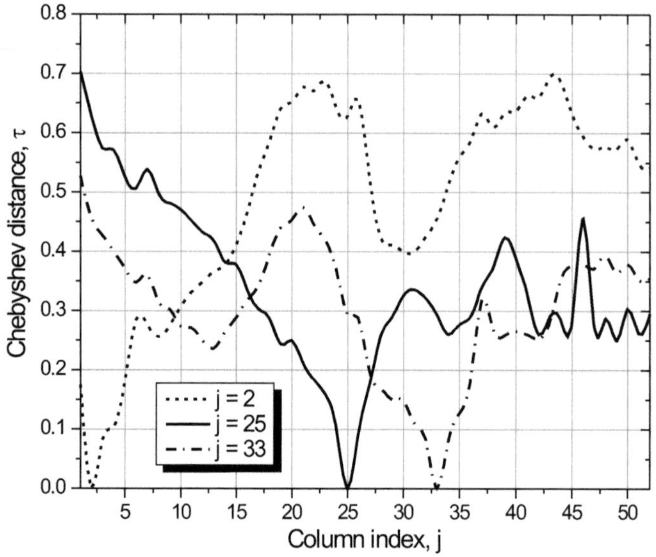

Figure 7.6 Chebyshev distance curves for $\mathbf{w}^2$, $\mathbf{w}^{25}$, and $\mathbf{w}^{33}$ [230].

an "upward spike" in endmembers selected from W since $w_{ii} = u_i$, or a "downward spike" if endmembers come from M because $m_{ii} = v_i$. The anomalous spikes can be smoothed so that, the global shape of each final endmember agrees with available library reference spectra. A simple *smoothing* procedure [194] considers the nearest one or two spectral samples next to w_{ii} or m_{ii}, i.e., for any $i \in \{1,\ldots,n\}$, we have

$$s_{ii} = \begin{cases} s_{21} & \Leftrightarrow i = 1, \\ \frac{1}{2}\left[s_{(i-1)i} + s_{(i+1)i}\right] & \Leftrightarrow 1 < i < n, \\ s_{(n-1)n} & \Leftrightarrow i = n, \end{cases} \tag{7.9}$$

where the endmember vector $\mathbf{s}$ can be a selected column vector $\mathbf{w}$ or $\mathbf{m}$.

Figure 7.7 displays five final endmembers extracted from scene no. 4 of Cuprite. Note that a correspondence between endmembers and laboratory spectra cannot be exact, since some wavelength values over the entire spectral range are different between airborne imaging spectrometers and ground instrumentation. Thus, small "displacements" of the absorption bands characteristic of each mineral may be noticed in endmember curves. The approximate mineral tags have the following endmember column correspondence: $\mathbf{w}^{33}$ is alunite, $\mathbf{m}^{16}$ is buddingtonite, $\mathbf{w}^{25}$ is calcite, $\mathbf{w}^2$ is kaolinite, and $\mathbf{m}^{27}$ is muscovite. In comparison, Figure 7.8 shows the subsampled spectra in the range 2.0 to 2.5 microns of similar minerals obtained from the USGS spectral library (splib06a, 2007) corresponding to: Alunite-al706, Buddingtonite-gds85, Calcite-hs483b, Kaolinite-kga2, and Muscovite-gds108. We remark that spectral characteristics are common to a family of minerals that are variants or mixtures, and hence, are not restricted to a single material. Therefore, an abundance map calculated from a set of final endmembers usually represents a class of minerals with analogous spectral behavior.

For the unmixing stage, recall that equation 7.1 is an *overdetermined* system of linear equations since $n > m$, subject to the restrictions of full additivity and non-negativity of abundance coefficients as stated in equation 7.2. In the present case, both matrices W and M have full rank, thus their columns are linearly independent vectors. Also, it happens that, the set of final endmembers, $S = \{\mathbf{w}^{33}, \mathbf{m}^{16}, \mathbf{w}^{25}, \mathbf{w}^{2}, \mathbf{m}^{27}\} \subset V$, is a linearly independent set whose pseudo-inverse matrix is unique. Although the *unconstrained* solution corresponding to equation 7.1, where $n = 52 > 5 = m$, has a single solution, some coefficients turn out to be negative for many pixel spectra and do not sum up to unity. If full additivity is enforced, again negative coefficients appear. Therefore, the best approach consists of imposing non-negativity for the abundance proportions and simultaneously relaxing full additivity by considering the inequality $\sum_{j=1}^{m} a_j < 1$. Specifically, we use the *Non Negative Least Squares* (NNLS) algorithm that solves the problem of minimizing $\|S\mathbf{a} - \mathbf{x}\|_2$ (Euclidean norm) subject to the condition $\mathbf{a} \geq \mathbf{0}$. The details related to the NNLS algorithm can be found in [125, 160]. Figure 7.9 illustrates the abundance maps corresponding to three of the five approximate endmembers. The maps shown were obtained with the NNLS method as implemented in Matlab. They have been enhanced for visual clarity by incrementing their brightness and contrast in 15%. The maps correlate well with the standard USGS reference map (after color thresholding) and with the results presented in [316].

Our second example, Moffett Field, is a remote sensing test site on the bay of San Francisco, a few kilometers north of the city of Mountain View, California. The site includes the Naval Air Station Moffet Field, agriculture fields, water ponds, salt banks, and man-made constructions including several airplane hangars which today are museums. The hyperspectral data collected by AVIRIS is ideal for water variability, vegetation and urban studies [8, 158, 215]. Figure 7.10, shows an RGB color image of scene no. 3 with wavelengths of 0.693 μm (red, channel 36), 0.557 μm (green, channel 20) and 0.458 μm (blue, channel 10). Several artificial and natural resources can be extracted from the scene using only 103 noiseless channels that fall within the visible and first half of the short wavelength infrared bands. Specifically,

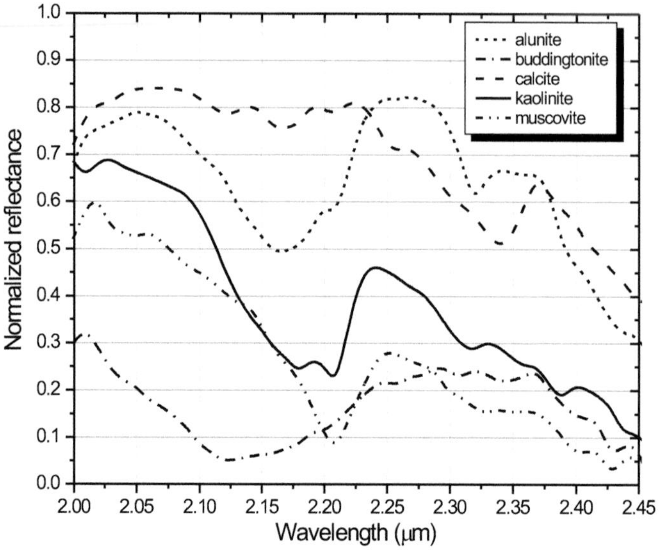

Figure 7.7 Final endmembers determined from sets W and M, computed with data of the Cuprite site, scene 4 [230].

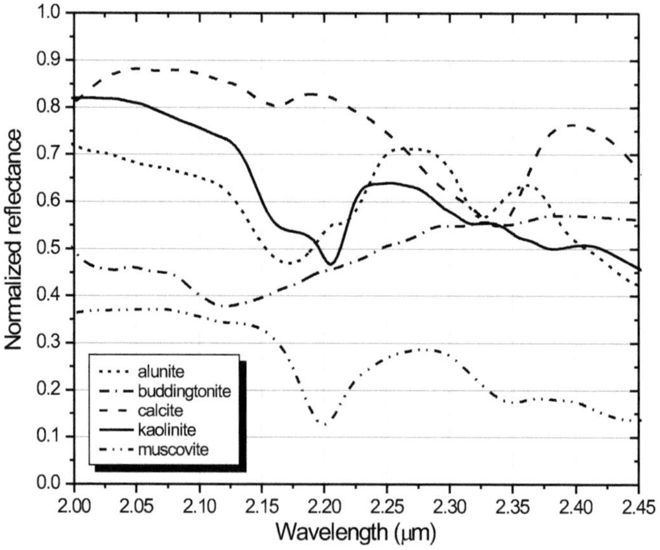

Figure 7.8 Mineral reference spectra similar to the final endmembers shown in Figure 7.7 [230].

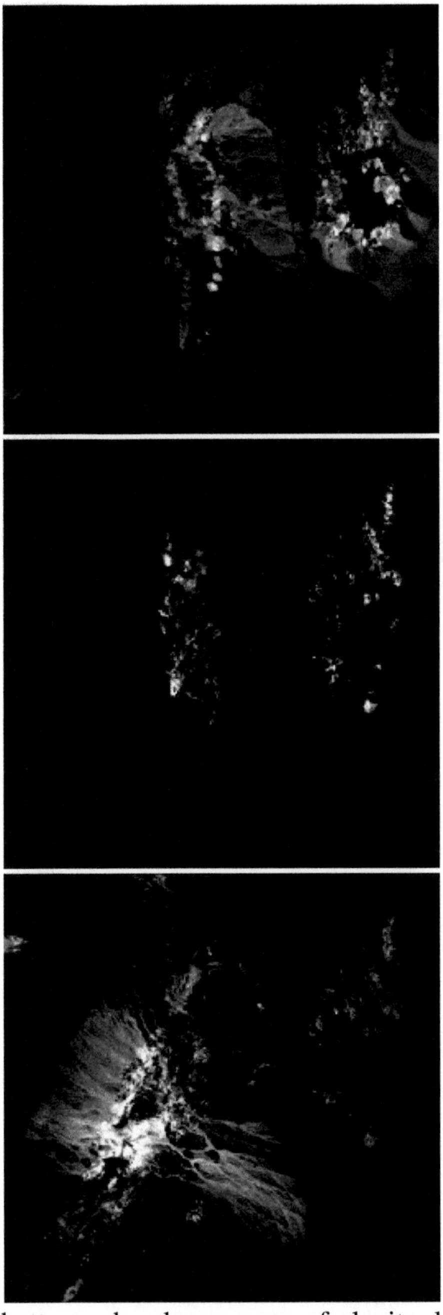

Figure 7.9 Top to bottom: abundance maps of alunite, kaolinite, and muscovite. Grayscale values indicate relative abundance proportion; brighter zones signal high mineral content, darker zones signal low content or mineral mixtures [230].

Figure 7.10 RGB composite "true" color image of scene no. 3 of the Moffett Field site in San Francisco, California. Real size of image is 614×512 pixels [233].

channels no. 4 (0.40 μm) to 106 (1.34 μm) can be used for endmember determination. Thus, the hyperspectral cube of reflectance data is formed by 314,368 pixel spectra (614×512) and each pixel is a 103-dimensional vector.

In this case, the specific format of the AVIRIS data file requires as a first step the extraction of all pixel spectra line by line to form set $X = \{\mathbf{x}^1, \ldots, \mathbf{x}^k\} \subset \mathbb{R}^n$, where $k = 314,368$ and $n = 103$. The second step performs the computation of the vector bounds $\mathbf{v}, \mathbf{u}$ of X as well as the entries of the shifted matrices W and M for $i, j = 1, \ldots, n$ in developed form (cf. equation 7.5),

$$w_{ij} = \bigvee_{\xi=1}^{k} x_j^\xi + \bigwedge_{\xi=1}^{k} (x_i^\xi - x_j^\xi), \tag{7.10}$$

$$m_{ij} = \bigwedge_{\xi=1}^{k} x_j^\xi + \bigvee_{\xi=1}^{k} (x_i^\xi - x_j^\xi). \tag{7.11}$$

As mentioned in the Cuprite example, the shifting operation that generates an "upward spike" in endmembers selected from W since $w_{ii} = u_i$, or a

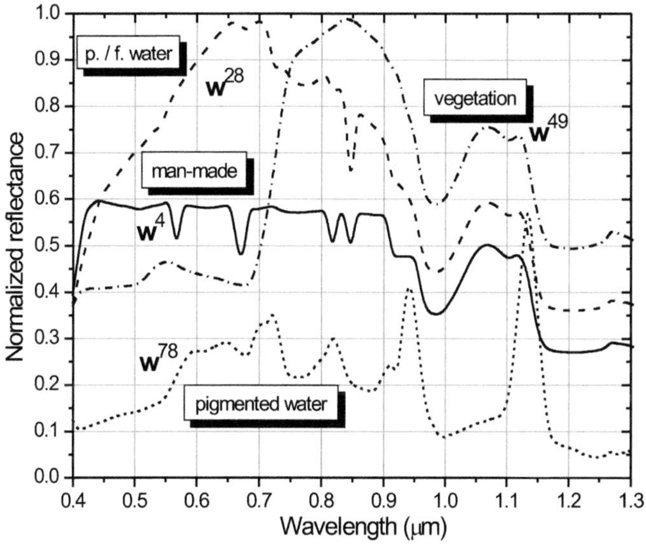

Figure 7.11 Spectral curves of final endmembers determined from $\mathcal{W}$, with corresponding resource classification; p./f. means pigmented or fresh [233].

"downward spike" if endmembers come from M because $m_{ii} = v_i$, needs to be smoothed to suppress the anomalous spikes using equation 7.9. Since $\mathcal{W}$ or $\mathcal{M}$ result in matrices of size 103×104, each one gives *all* 104 column vectors as possible endmembers. The LAAMs method always gives a number of "candidate" endmembers which is either equal or slightly less than the spectral dimensionality. However, several contiguous columns are highly correlated and most of these potential endmembers can be discarded using appropriate metric techniques.

As an alternative way to automate endmember screening, we calculate a matrix whose entries are given by the linear correlation coefficients, $c_{pq} = \rho(\mathbf{s}^p, \mathbf{s}^q)$ for $p, q = 1, \ldots, n$, where $\mathbf{s}^p, \mathbf{s}^q \in \mathcal{W}$ (resp. in $\mathcal{M}$). A threshold τ value is then applied on c_{pq} to get a subset of selected endmember pairs with *low* correlation coefficients. For scene no. 3 of the Moffett Field data, $\tau = 0.005$ was applied to $\mathcal{W}$ (resp., $\tau = 0.0005$ for $\mathcal{M}$). The resulting set is first sorted by ascending column index and after elimination of repeated or contiguous indices, the number of potential endmembers is decreased from 104 to 10. On physical ground, it should be clear that a low correlation value between potential endmembers does not necessarily guarantee a clear cut criteria to obtain "good" endmembers useful to match high-resolution laboratory spectra. Nevertheless, the reduction in the number of LAAM column vectors

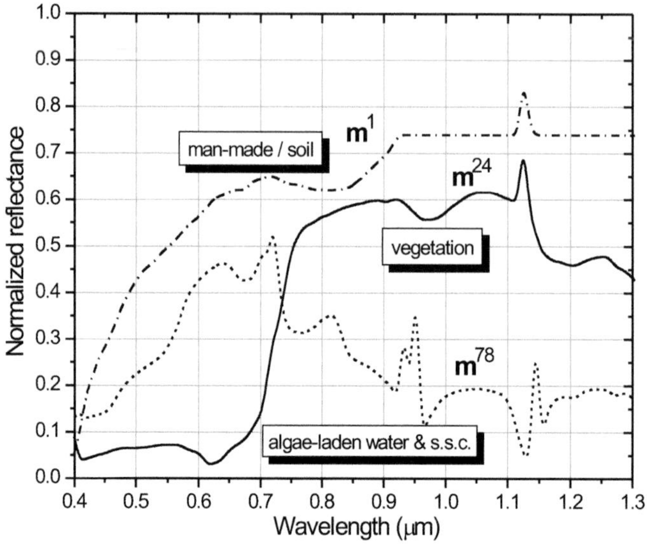

Figure 7.12 Spectral curves of final endmembers determined from $\mathcal{M}$, with corresponding resource classification; s.s.c. means suspended sediment concentration [233].

is quite useful to find a smaller subset of *final* endmembers. Another simpler but supervised technique for the lattice-based approach forms $\lfloor \sqrt{n+1} \rfloor$ subsets, each with $\lfloor \sqrt{n+1} \rfloor$ column vectors taken from $\mathcal{W}$ (resp. $\mathcal{M}$), and then, a representative from each group is selected at random as a final endmember. Thus, in this scheme, the number of possible endmembers is always diminished one order of magnitude. Both techniques provide a reasonable number of approximate true endmembers *without* sacrificing spectral resolution, which, from a physical point of view, is an advantage if compared to data dimensionality reduction techniques such as principal component analysis [76, 154] or alternative neural network techniques [196].

Figures 7.11 and 7.12 display the final endmember curves for scene no. 3 of the Moffett Field site determined, respectively, from the translated versions of $\mathfrak{W}_X$ and $\mathfrak{M}_X$. Normalization of reflectance data values in spectral distributions are linearly scaled from the range $[-50, 12000]$ to the unit interval $[0, 1]$. For the problem at hand, the vector bounds $\mathbf{v}$ and $\mathbf{u}$ as well as many other potential endmembers were rejected by the correlation coefficient technique described earlier.

The unmixing procedure represented by equatiion 7.1 corresponds again to an *overdetermined* system of linear equations with $n > m$, subject to the

Figure 7.13 Combined abundances obtained from $\mathbf{w}^4$, $\mathbf{w}^{28}$, and $\mathbf{w}^{49}$. Cyan colors indicate artificial or man-made resources, e.g., urban settlements. Blue colors show pigmented and fresh water distribution. The upper left bright blue regions correspond to evaporation ponds pigmented by red brine shrimp. Yellow colors indicate vegetation distribution, e. g., shore-line and in-land vegetation [233].

Figure 7.14 Combined abundances obtained from $\mathbf{m}^1$, $\mathbf{m}^{24}$, and $\mathbf{m}^{78}$. Cyan and sea-green colors reveal both man-made resources and different kinds of soil. Large irregular bright green regions with grooves correspond to golf courses. Red colors show possibly algae-laden water with some concentration of suspended sediment [233].

restrictions of full additivity and non-negativity of abundance coefficients. Here, in matrix W (resp. M), any two column vectors are distinct and therefore its columns form a linearly independent set of vectors. It turns out that the set of final endmembers, $S = \{\mathbf{w}^4, \mathbf{w}^{28}, \mathbf{w}^{49}, \mathbf{w}^{78}, \mathbf{m}^1, \mathbf{m}^{24}, \mathbf{m}^{78}\} \subset W \cup M$, is a linearly independent set whose pseudo-inverse matrix is unique. In similar fashion to the Cuprite example, the *unconstrained* solution corresponding to equation 7.1, where $n = 103 > m = |S| = 7$, has a single solution but some coefficients are negative for many pixel spectra and their sum is not equal to one. If full additivity is enforced, again negative coefficients appear. Therefore, we chose to impose non-negativity for the abundance proportions and simultaneously relaxing the full additivity constraint. Again, the NNLS algorithm solves the problem of minimizing the Euclidean norm expressed by $||S\mathbf{a} - \mathbf{x}||_2$, subject to the condition $\mathbf{a} \geq \mathbf{0}$. Thus, abundance maps were obtained using the NNLS numerical method as implemented in Matlab and, for visual clarity, each image map has been contrast enhanced by the application of a non-linear increasing function. Additionally, abundance fractions below 0.2625 were set to zero. Figures 7.13 and 7.14 display as false RGB color images the combined abundance distribution associated to the final endmembers shown in Figures 7.11 and 7.12. The abundance map obtained from $\mathbf{w}^{78}$ was not incorporated in Figure 7.13 due to its similarity to the abundance map generated from endmember $\mathbf{w}^{28}$. In this case, both spectral distributions provide information about the presence of pigmented water in wet soil or wet vegetation areas.

In this example, there is *no a priori knowledge* about the hyperspectral image content, and therefore, approximate resource classification is possible after the abundance maps are computed. However, a minimal base knowledge of hyperspectral image analysis as well as a fundamental background in spectral analysis is required for proper recognition of spectral signature characteristics in single, multiple or mixed resources. For example, in the abundance distribution shown as cyan colors in Figure 7.14 (endmember $\mathbf{m}^1$), the presence of man-made or soil is clearly differentiated. Thus, the identification labels that appear in Figures 7.11 and 7.12 are the result of an overall spectral scrutiny, aided by a visual match between an RGB "true" color reference image and each final endmember corresponding abundance map obtained by constrained linear unmixing. Precise resource identification would require an exhaustive matching procedure between each final endmember that belongs to S against high-resolution signatures available in professional spectral libraries [55, 120]. Table 7.4 shows the time spent by a standard personal computer (2.5 GHz quad-processor, 4 GB RAM, 8 GB virtual memory space, and 500 GB hard disk) for each computational task performed in the analysis of

TABLE 7.4 Processing times for Moffett Field hyperspectral image.

Task	Minutes
1 Pixel spectra extraction	1.522
2 Vector bounds & LAMs computation	0.717
3 LAAMs linear independence test	0.005
4 Correlated endmembers elimination	0.358
5 Final endmember selection	–
6 NNLS abundance map generation	3.864
7 General resource classification	–

scene no. 3 of the Moffet Field hyperspectral image. Note that tasks 5 and 7 in the analysis process are of an interactive nature but rely, respectively, on tasks 4 and 6 which are completely automatic. Specifically, tasks 2, 3, and 4 are the fundamental steps of the lattice auto-associative memories based technique whose computational effort is minimum in comparison to the processing times needed to perform the other tasks.

For a hyperspectral image of size $k = p \times q$ pixels acquired over n spectral bands, the computational effort required for pixel spectra extraction (task 1) is linear in k since $n \ll k$. The overall computational complexity of the WM based technique (tasks 2,3,4) is $n^2(k+3)$, which for values of n of a few hundreds is in the order of few minutes (see Table 7.4). Endmember determination (task 5) relies on a subset of $2\lfloor \sqrt{n+1} \rfloor$ candidate extremal vectors generated after task 4 is completed and, although interactive in nature, $m \propto \lfloor \sqrt{n} \rfloor$ "final" endmembers can be selected in a lapse of minutes. On the other hand, the NNLS method needs about nm^3 arithmetical operations to find a unique set of abundance coefficients for each pixel spectra. Since m represents the number of final endmembers and $m \ll k$, it turns out that for hyperspectral image segmentation the computational complexity of the non-negative least square method (task 6) is nm^3 per pixel. Finally, general resource classification (task 7) may also be accomplished within minutes whenever a working knowledge in spectral identification or prior experience in hyperspectral image analysis is available for the problem at hand [129].

Two recent approaches to linear spectral unmixing, comparable in performance to N-FINDR and based on the geometry of convex sets and convex optimization algorithms are *Vertex Component Analysis* (VCA) [197] and the *Minimum-Volume Enclosing Simplex* (MVES) [46]. Vertex component analysis is an unsupervised technique that relies on singular value decomposition (SVD) and principal component analysis (PCA) as subprocedures, and assumes the existence of pure pixels. Specifically, VCA exploits the fact that

TABLE 7.5 Five characteristics of three autonomous linear unmixing techniques: 1) input data, 2) endmember search numerical methods, 3) endmember search computational complexity, 4)abundance map generation algorithm, and 5) system type [233].

	VCA	MVES	WM
1)	X, m	X, m	X
2)	SVD & PCA	APS & LPs	LAAMs & CMs
3)	$2m^2k$	$\alpha m^2 k^{1.5}$	$n^2(k+3)$
4)	M-P inverse	M-P inverse	LNNS
5)	unconstrained	unconstrained	constrained

endmembers are vertices of a simplex and that the affine transformation of a simplex is again a simplex. This algorithm iteratively projects data onto a direction orthogonal to the subspace generated by the endmembers already determined. The new endmember spectrum is the extreme of the projection and the main loop continues until all given endmembers are exhausted. Similarly, the minimum-volume enclosing simplex is an autonomous technique supported on a linear programming (LP) solver but does not require the existence of pure pixels in the hyperspectral image. However, when pure pixels exist, the MVES technique leads to unique identification of endmembers. In particular, dimension reduction is accomplished by affine set fitting, and Craig's unmixing criterion [57] is applied to formulate hyperspectral unmixing as an MVES optimization problem. Table 7.5 gives the main characteristics of the VCA and MVES convex geometry based algorithms as well as the lattice algebra approach based on the W & M vector sets derived respectively from the $\mathfrak{W}_X$ and $\mathfrak{M}_X$ LAAMs.

Note that the proposed lattice algebra based method *does not* require to know in advance the number of endmembers as specified in row 1 of Table 7.5. In row 2 of the same table, the numerical procedures used by VCA are SVD for projections onto a subspace of dimension m and PCA for projections onto a subspace of dimension $m-1$. Algorithm MVES first determines the affine parameters set (APS), solves by LP an initial feasibility problem with linear convex constraints, and iteratively optimizes two LP problems with non-convex objective functions. The WM algorithm computes first the scaled min- and max LAMs and their corresponding correlation coefficient matrices (CMs) for further endmember discrimination. Abundance coefficients are determined in the VCA and MVES algorithms using the Moore-Penrose (M-P) pseudoinverse and, as explained earlier, the WM algorithm makes use of the non-negative least squares method (LNNS). Computational complexity given in row 3 is expressed as a function of the number of endmembers m, the number of pixels or observed spectra k, the number of iterations α, and the number of spectral bands n. These expressions assume that $k \gg n \approx m^2$ and consider only the computations required by the numerical endmember search methods. Although the M-P inverse operation is faster than the LNNS numerical method for the calculation of the abundance coefficients for all pixels in the input image X, we favored the LNNS algorithm since it enforces fraction positivity and consequently allows for a better rendering of the corresponding abundance maps.

We remark that since VCA and MVES require to know in advance the number m of endmembers to be found, their application to a real hyperspectral image must probe all values of m in a specified interval, e. g., from 1 to $m_{\max}$. Hence, if *no a priori* information is known about the number of pure pixels existing in a hyperspectral image, the computational performance for finding endmembers and determining their abundance fractions increases respectively, to $2k\sum_{m=1}^{m_{\max}} m^2$ and $\alpha k^{1.5}\sum_{m=1}^{m_{\max}} m^2$, which are proportional to $m_{\max}^3 k$ and $\alpha m_{\max}^3 k^{1.5}$. Therefore, in practical situations, overall complexity of our proposed WM-method is lower by one order of magnitude than VCA, though better in performance to MVES. Furthermore, the VCA and MVES Matlab codes provided by their respective authors were applied to specific subimages of the given examples producing similar results to those obtained by the WM technique.

Exercises 7.2

1. In reference to Figure 7.5, showing the 4th scene of the mining site in Cuprite Nevada, a) reproduce the results listed in Table 7.3, b) reproduce the distance curve graphs displayed in Figure 7.6, and c) build a figure that displays the selected final endmembers from M.

2. Following the explanation after equation 7.9 of the unmixing stage based on the NNLS algorithm, complete Figure 7.9 with the abundance maps corresponding to minerals buddingtonite and calcite.

3. Equations (7.10) and (7.11) show directly the dependence of w_{ij} and m_{ij} in terms of data entries x_j^ξ and x_i^ξ, for all $\xi = 1,\ldots,k$. Explain why both expressions, as written, are *not* computationally efficient and give an already stated equation that is a better way to calculate the W and M matrices.

4. Based on the Cuprite Nevada and the Moffett Field site examples, make

a similar hyperspectral image analysis based on the W-M method followed by the application of the NNLS algorithm for the linear unmixing stage using the data of Indian Pines. Specifically, a) Find, download, and prepare the numerical data extracted from the image "cube", b) compute W and M and determine an adequate set of final endmembers, and c) produce an image panel displaying the corresponding resource abundance maps in grayscale.

7.3 ENDMEMBERS AND CLUSTERING VALIDATION INDEXES

The lattice algebra approach to hyperspectral imagery described in the previous two sections, mainly the WM-method [102, 230, 231, 232], always gives a fixed set of potential endmembers. Specifically, for an n-band hyperspectral image X, m is equal to $2(n+1)$, which in most cases is redundant since column vectors of W and M are highly correlated. Measures between n-dimensional vectors such as the spectral angle, the Chebyshev distance, or the correlation coefficient matrix, can be considered *supervised* or *semisupervised* techniques since these numerical criteria were oriented towards spectral identification of selected endmembers against known spectral libraries. These tools have seen useful in the examples provided to obtain a smaller set of representative endmembers without sacrificing spectral resolution.

Another possibility for reducing the cardinality of V is the application of well-established *unsupervised clustering* techniques such as the crisp and fuzzy c-means algorithms. It is generally recognized that c-means clustering is fundamental to pattern recognition where, as a rule of thumb, it has been accepted that an upper bound to the number of clusters specified by c is given by $\lfloor \sqrt{n} \rfloor$, where n denotes feature space dimensionality. In fact, supervised extraction of final endmembers obtained with the WM-method suggested the same criterion as an alternative way to reduce the cardinality of the vector sets W and M equal to $n+1$ to $\lfloor \sqrt{n+1} \rfloor$. Therefore, given a hyperspectral image data cube X, computation of the augmented matrices $\mathcal{W}$ and $\mathcal{M}$ lowers data size from kn to $2(n+1)n$, (where $k \gg n$), and application of c-means clustering to V reduces data size from $2(n+1)n$ to $2cn$, where $c \in [2, c_{\max}]$ and $c_{\max} = \lfloor \sqrt{n+1} \rfloor$.

Without any a priori knowledge concerning possible data structure present in pixel spectra, such as those drawn from particular joint probability distributions or specific n-dimensional cluster shapes, crisp and fuzzy partitions of $\mathcal{W}$ and $\mathcal{M}$ were randomly seeded. All c values in the interval $c \in [2, c_{\max}]$ were used to generate partitions to which selected validation indexes were computed to draw quantitative information about the intrin-

TABLE 7.6 Crisp and fuzzy validation indexes; $\|\cdot\|$ denotes Euclidean norm [290].

Name	Definition	Related quantities
Dunn (crisp)	$\mathcal{V}_{\text{D}}(c) = (1/\bigvee_{k=1}^{c} \Delta_k)\times \bigwedge_{i=1}^{c} \bigwedge_{j\neq i} \delta_{ij}$	$\delta_{ij} = \bigwedge_{\mathbf{a}\in X_i, \mathbf{b}\in X_j} \delta(\mathbf{a},\mathbf{b})$ $\Delta_k = \bigvee_{\mathbf{a},\mathbf{b}\in X_k} \delta(\mathbf{a},\mathbf{b})$ $\delta(\mathbf{a},\mathbf{b}) = \|\mathbf{a}-\mathbf{b}\|$
Davies-Bouldin (crisp)	$\mathcal{V}_{\text{DB}}(c) = (1/c)\times \sum_{i=1}^{c} \bigvee_{j\neq i}(\alpha_i+\alpha_j)/\|\mathbf{v}_i-\mathbf{v}_j\|$	$\alpha_i = \sqrt{\sum_{\mathbf{x}\in X_i} \|\mathbf{x}-\mathbf{v}_i\|^2/\|X_i\|}$ $\mathbf{v}_i = \sum_{\mathbf{x}\in X_i} \mathbf{x}/\|X_i\|$
Araki et al. (fuzzy)	$\mathcal{V}_{\text{ANW}}(c) = (1/c)\times \sum_{i=1}^{c} \bigvee_{j\neq i}(\alpha_i+\alpha_j)/\|\mathbf{v}_i-\mathbf{v}_j\|$	$\beta_i = \sum_{k=1}^{n} u_{ik}^2$ $\alpha_i = \left(\sum_{k=1}^{n} u_{ik}^2\|\mathbf{x}_k-\mathbf{v}_i\|^2\right)/\beta_i$ $\mathbf{v}_i = (1/\beta_i)\times\sum_{k=1}^{n} u_{ik}^2\mathbf{x}_k$
Xie-Beni (fuzzy)	$\mathcal{V}_{\text{XB}}(c) = (1/n)\times \sum_{i=1}^{c} \alpha_i / \bigwedge_{i\neq j}\|\mathbf{v}_i-\mathbf{v}_j\|^2$	$\alpha_i = \sum_{k=1}^{n} u_{ik}^2\|\mathbf{x}_k-\mathbf{v}_i\|^2$

sic agglomeration structure for potential endmembers determined from the scaled lattice memories W and M. The *clustering validation indexes* serve the purpose of finding an "optimal" value of c and hence suggest the best possible choice for correctly grouping the data patterns under analysis. The validation indexes used in the present study and defined in Table 7.6 include Dunn's ($\mathcal{V}_{\text{D}}$) and Davies-Bouldin ($\mathcal{V}_{\text{DB}}$) indexes for crisp partitions [77, 67], and Araki-Nomura-Wakami ($\mathcal{V}_{\text{ANW}}$) and Xie-Beni ($\mathcal{V}_{\text{XB}}$) quantitative criteria for fuzzy partitions [11, 322]. An optimal value of c occurs in a subinterval of $[2, c_{\max}]$ if $\mathcal{V}_{\text{D}}$ is maximum and any of the other three validity indexes is minimum. It is not surprising that different validity indexes can give different values for c and this is enough reason to consider not just a single value of c but possibly a neighborhood of c or even other relative extrema.

In reference to Table 7.6, $\{X_1,\ldots,X_c\}$ denotes the collection of subsets forming a *c-partition* of X, $\{\mathbf{v}_1,\ldots,\mathbf{v}_c\}$ denotes the set of *prototype* or *centroid* vectors (crisp or fuzzy), $u_{ik}\in[0,1]$ denotes the membership value of the k-th sample vector with respect to the i-th cluster, X_i and X_j denote two different clusters, Δ_k stands for the *diameter* of the k-th cluster, and δ_{ij} is the *inter-cluster distance* matrix. Also, memberships appear squared since we used a fuzzy exponent equal to 2 in the fuzzy c-means algorithm. Further details on c-means algorithms and validity measures can be found in [24, 25]. More recent developments on cluster validation and comparison appear in [201, 142, 321].

TABLE 7.7 Validation index values for AVIRIS hyperspectral images [290].

V **subset – Image**	$\mathcal{V}_{\mathrm{D}}$	$\mathcal{V}_{\mathrm{DB}}$	$\mathcal{V}_{\mathrm{ANW}}$	$\mathcal{V}_{\mathrm{XB}}$	c **– values**
$\mathcal{W}$ – Cuprite	0.156	1.334	0.036	0.939	6,6,6,6
$\mathcal{M}$ – Cuprite	0.200	1.578	0.026	0.328	7,6,3,3
$\mathcal{W}$ – Moffett Field	0.141	1.496	0.175	0.319	6,10,4,3
$\mathcal{M}$ – Moffett Field	0.115	1.512	0.163	0.358	7,10,4,3
$\mathcal{W}$ – Jasper Ridge	0.141	1.825	0.037	0.512	4,10,6,4
$\mathcal{M}$ – Jasper Ridge	0.111	1.539	0.046	0.220	3,13,3,3

In our computational experiments, we have used particular scenes (512 scan lines and 614 pixel spectra per line) of several hyperspectral images acquired by the AVIRIS sensor such as the mining site of Cuprite (scene 4), urban settlements in Moffett Field (scene 3), and biological preserve at Jasper Ridge (subimage of scene 3). Although the AVIRIS sensor captures spectral information in 224 bands, analysis of each image requires specific wavelength ranges, e.g., Cuprite needs only 52 bands and Moffett Field as well as Jasper Ridge use only 194 bands. Hence, the empirical interval for c in the case of Cuprite is $[2,7]$ since $c_{\max} = 7 = \lfloor \sqrt{52} \rfloor$, and for the other two hyperspectral images is $[2,13]$ since $c_{\max} = 13 = \lfloor \sqrt{194} \rfloor$. Nevertheless, to exhibit the behavior of the selected validation indexes, we perform clustering of $\mathcal{W}$ and $\mathcal{M}$ for all three images in the common interval $c \in [2,15]$ because $\sqrt{224} \approx 15$. As an example, Figures 7.15 and 7.16 display the validation index graphs for crisp and fuzzy partitions of $\mathcal{W}$ and $\mathcal{M}$ of Cuprite's hyperspectral image. For $\mathcal{W}$ observe that, on $[2,7]$, $\mathcal{V}_{\mathrm{D}}(c)$ is maximum at $c = 6$ and $\mathcal{V}_{\mathrm{DB}}(c)$ is minimum at the same c value. However, for $\mathcal{M}$, $\mathcal{V}_{\mathrm{D}}(c)$ is maximum at $c = 7$ whereas $\mathcal{V}_{\mathrm{DB}}(c)$ is minimum at $c = 6$. On the other hand, for $\mathcal{W}$, the minimum values of the fuzzy indexes $\mathcal{V}_{\mathrm{ANW}}(c)$ and $\mathcal{V}_{\mathrm{XB}}(c)$ occur at $c = 2$ and the next relative minimum of both indexes occur at $c = 6$ matching what the crisp indexes give. In the case of $\mathcal{M}$, both fuzzy indexes point to $c = 3$ as the best number of clusters.

It turns out that the prototype vectors, $\mathbf{v}_i$ for $i = 1,\ldots,c$, of the partition suggested by each validation index are approximations to true endmembers as shown in Figure 7.17. From the centroid graphs displayed in Figure 7.17, it can be seen that further simplification is possible if additional knowledge is available regarding the nature of some spectral signatures, as is the case for the dashed curves, which are highly correlated with the corresponding solid line curve of the same color. Thus, the complete set or a proper subset of the $\mathbf{v}_i$ can be used as final endmembers for unmixing a hyperspectral image. We close this section with Table 7.7 that gives the number of clusters c suggested

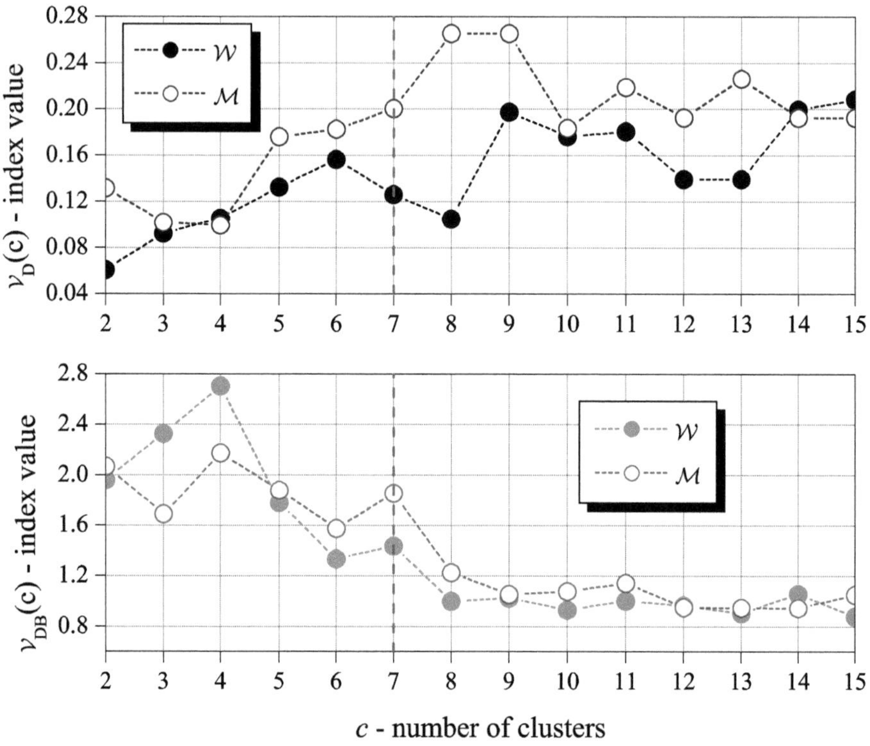

Figure 7.15 Crisp index graphs obtained from Cuprite hyperspectral image data. Top: Dunn's index for LAAMs derived subsets $\mathcal{W}$ and $\mathcal{M}$; bottom: Davies-Bouldin index for the same subsets. Empirical interval for c is $[2,7]$ (vertical dashed line) [290].

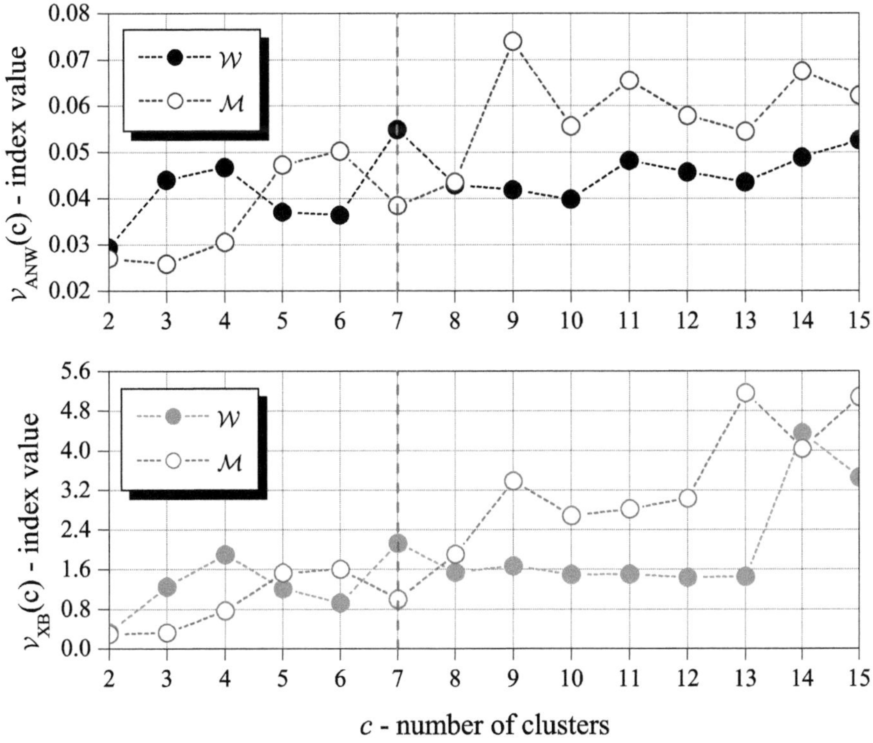

Figure 7.16 Fuzzy index graphs obtained from Cuprite hyperspectral image data. Top: Araki's et al. index for LAAMs derived subsets $\mathcal{W}$ and $\mathcal{M}$; bottom: Xie-Beni index for the same subsets. Empirical interval for c is $[2, 7]$ (vertical dashed line) [290].

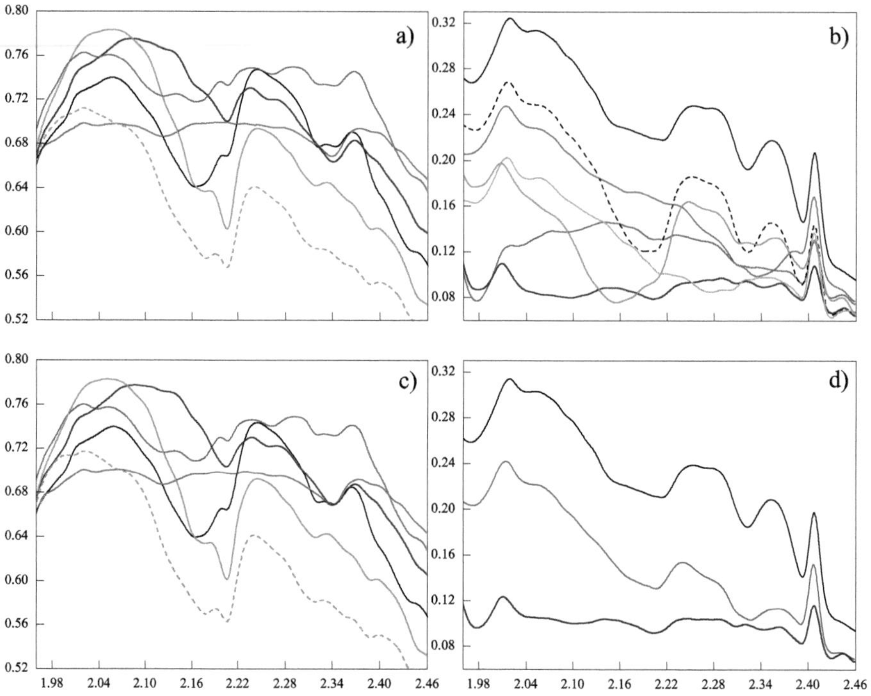

Figure 7.17 Approximate endmembers of the Cuprite hyperspectral image suggested by a) the max and min values of $\mathcal{V}_{\mathrm{D}}$ and $\mathcal{V}_{\mathrm{DB}}$, resp., from $\mathcal{W}$, b) the max and min values of $\mathcal{V}_{\mathrm{D}}$ and $\mathcal{V}_{\mathrm{DB}}$, resp., from $\mathcal{M}$, c) the next lowest rel. min of $\mathcal{V}_{\mathrm{ANW}}$ and $\mathcal{V}_{\mathrm{XB}}$, resp., from W', d) the min values of $\mathcal{V}_{\mathrm{ANW}}$ and $\mathcal{V}_{\mathrm{XB}}$ from M' (axes: wavelength vs. reflectance) [290]

by the crisp and fuzzy validation indexes found by partitioning the $\mathcal{W}$ and $\mathcal{M}$ subsets for three AVIRIS example hypespectral images. We point out that variations of c values for the same image clearly depend on the way validation indexes are defined.

Exercises 7.3

1. Using the mathematical expressions given in Table 7.6 for crisp and fuzzy validation indexes, draw the corresponding graphs related to $\mathcal{W}$ and $\mathcal{M}$, as shown in Figures 7.15 and 7.16, based on the hyperspectral image analysis realized for the Indian Pines site (see last exercise of Section 7.2).

2. Discuss the use of clustering validation indexes for datasets X of low

to medium dimensionality, i.e., where $2 \leq n \leq 25$, which includes color or multispectral images, or in a more general setting for problems in pattern recognition.

7.4 COLOR IMAGE SEGMENTATION

In several image processing and analysis applications, image segmentation is a preliminary step in the description and representation of regions of interest [13, 16, 109, 128]. Segmentation techniques, first developed for grayscale images [45, 202, 255, 330], have been extended, enhanced or changed to deal efficiently with color images coded in different color spaces. Color image segmentation has been approached from several perspectives that currently are categorized as *pixel*, *area*, *edge*, and *physics* based segmentation, for which early compendiums appeared in [211, 261] and state-of-the-art surveys are given in [53, 173]. For example, pixel based segmentation includes histogram techniques and cluster analysis in color spaces. Optimal thresholding [42] and the use of a perceptually uniform color space [247] are examples of histogram-based techniques. Area-based segmentation contemplates region growing as well as split-and-merge techniques, whereas edge-based segmentation embodies local methods and extensions of the morphological watershed transformation. This transformation and the flat-zone approach to color image segmentation were originally developed, respectively, in [189] and [58]. A seminal work employing Markov random fields for splitting and merging color regions was proposed in [172]. Other recent developments contemplate the fusion of various segmentation techniques, such as the application of morphological closing and adaptive dilation to color histogram thresholding [207] or the use of the watershed algorithm for color clustering with Markovian labeling [92]. Physics-based segmentation relies on adequate reflection models of material objects such as inhomogeneous dielectrics, plastics, or metals [111, 144]. Nevertheless, its applicability has been limited to finding changes in materials whose reflection properties are well-studied and modeled properly.

Recently, soft computing techniques [262] or fuzzy principal component analysis coupled with clustering based on recursive one-dimensional histogram analysis [82], suggest alternative ways to segment a color image. In order to quantify the results obtained from different segmentation schemes, the subject of color image segmentation evaluation has been briefly exposed in [205]. Basic treatment of image segmentation performed in both Hue-Saturation-Intensity (HSI) and RGB color spaces is given in [96, 327];

for a more complete and systematic exposition of color image segmentation methods see [206] or [155].

In this chapter, we present a lattice algebra based technique for image segmentation applied to RGB (Red-Green-Blue) color images transformed to other representative systems, such as the HSI (Hue-Saturation-Intensity), the $I_1I_2I_3$ (principal components approximation), and the L*a*b* (Luminance - redness/greenness - yellowness/blueness) color spaces. The proposed method relies on the min $\mathfrak{W}_X$ and max $\mathfrak{M}_X$ lattice auto-associative memories (LAAMs), where X is the set formed by 3D pixel vectors or colors. The scaled column vectors of any memory together with the minimum or maximum bounds of X form the vertices of tetrahedra enclosing subsets of X, and will correspond to the most saturated color pixels in the image. Image partition into regions of similar colors is realized by linearly unmixing pixels belonging to tetrahedra determined by the columns of the scaled lattice auto-associative memories W and M, and then by thresholding and scaling pixel color fractions obtained numerically by applying a least squares method, such as the *linear least squares* (LLS) method also known as *generalized matrix inversion* [108], or the *non-negative least squares* (NNLS) method [160]. In the final step, segmentation results are displayed as grayscale images. The lattice algebra approach to color image segmentation can be categorized as a pixel based unsupervised technique. Preliminary research and computational experiments on the proposed method for segmenting color images were reported in [286, 287].

7.4.1 About Segmentation and Clustering

Although there are several approaches to segment a color image, as briefly described before, a mathematical description of the segmentation process, common to all approaches, can be given using set theory [16, 128, 13, 96]. In this framework, to segment an image is to divide it into a finite set of disjoint regions whose pixels share well-defined attributes. We recall from basic set theory that a partition of a set is a family of pairwise disjoint subsets covering it. Mathematically, we have

Definition 7.1 Let X be a finite set with k elements. A *partition* of X is a family $\mathcal{P} = \{R_i\}$ of subsets R_i of X, each with k_i elements for $i = 1,\ldots,q$, that satisfy the following conditions:

1. $R_i \cap R_j = \varnothing$ for $i \neq j$ (pairwise disjoint subsets) and
2. $\bigcup_{i=1}^{q} R_i = X$ where $\sum_{i=1}^{q} k_i = k$ (whole-set covering).

Note that the only attribute shared between any two elements of X with respect to a given partition $\mathcal{P}$ is their membership to a single subset R_i of X. Unfortunately, the simple attribute of sharing the same membership is not enough to distinguish or separating objects of interest in a given image. Therefore, Definition 7.1 must be enriched by imposing other conditions required for image segmentation. Additional attributes shared between pixels or elements of X can be, for example, spatial contiguity, similar intensity or color, and type of discrete connectedness. All or some of these quantifiable attributes can be gathered into a single *uniformity* criterion specified by a logical predicate. A mathematical statement of our intuitive notion of segmentation follows next.

Definition 7.2 Let X be a finite set with k elements. A *segmentation* of X is a pair $(\{R_i\}, p)$ composed of a family $\{R_i\}$ of subsets of X each with k_i elements for $i = 1, \ldots, q$, and a logical predicate p specifying a uniformity criterion between elements of X satisfying the following conditions:

1. The family $\{R_i\}$ is a partition $\mathcal{P}$ of X.

2. $\forall i$, R_i is a connected subset of X.

3. $\forall i$, $p(R_i)$ is true; i.e., elements in a single subset share the same attributes.

4. For $i \neq j$, $p(R_i \cup R_j)$ is false or in words, elements in a pairwise union of subsets do not share the same attributes.

Since finite sets are topologically totally disconnected, the reader needs to be mindful that with respect to condition 2.) of Definition 7.2, a discrete connected subset R_i is a set where every pair of elements $\{x_s, x_t\} \in R_i$ is connected in the sense that a sequence of elements, denoted by $(x_s, \ldots, x_r, x_{r+1}, \ldots, x_t)$, exists such that $\{x_r, x_{r+1}\}$ belong to the same spatial neighborhood and all points belong to R_i. A weaker but still useful version of Condition 4) in Definition 7.2, requires that R_i and R_j should be neighbors. Loosely speaking, a subset $R_i \subset X$ is commonly referred to as an image *region*. Whether regions can be disconnected (2nd condition of Definition 7.2 is not imposed), multi-connected (with holes), should have smooth boundaries, and so forth, depends on the application's domain, segmentation technique, and goals. Perceptually, the segmentation process must convey the necessary information to visually recognize or identify the prominent features contained in the image such as color hue, brightness or texture. Hence, adequate segmentation is essential for further description and representation of regions of interest suitable for image analysis or image understanding.

Here, we consider only images coded in the *Red-Green-Blue* (RGB) color space, and segmentation of an image in that space is performed in stages including: 1) computation of the scaled lattice auto-associative memories, 2) linear unmixing of color pixels using least square methods, and 3) thresholding color fractions to produce color segmentation maps represented as grayscale images. These stages are explained in detail in the following paragraphs.

Given a color image A consisting of $p \times q$ pixels, we build a set X containing all different colors (3-dimensional vectors) present in A. If $|X| = k$ denotes the number of elements in set X, then $k \leq pq = |A|$, where pq is the maximum number of colors available in A. Then, using the right expressions of equation 7.3, the memory matrices min-$\mathfrak{W}_X$ and max-$\mathfrak{M}_X$ are computed and to make explicit their respective column vectors, we rewrite them, respectively, as $\mathfrak{W} = (\mathfrak{w}^1, \mathfrak{w}^2, \mathfrak{w}^3)$ and $\mathfrak{M} = (\mathfrak{m}^1, \mathfrak{m}^2, \mathfrak{m}^3)$. By definition, the column vectors of $\mathfrak{W}$ may not necessarily belong to the space $[0, 255]^3$ since $\mathfrak{W}$ usually has negative entries. The general transformation given in Eqs. (7.4) and (7.5) will translate the column vectors of $\mathfrak{W}$ within the color cube. After additive scaling, we obtain the matrices $W = (\mathbf{w}^1, \mathbf{w}^2, \mathbf{w}^3)$ and $M = (\mathbf{m}^1, \mathbf{m}^2, \mathbf{m}^3)$. Note that for $j = 1, \ldots, 3$, $\mathbf{w}_{jj} = u_j$ and $\mathbf{m}_{jj} = v_j$. Hence, $\text{diag}(W) = \mathbf{u}$ and $\text{diag}(M) = \mathbf{v}$.

Thus, the first stage of the segmentation process is completed by applying Eqs. (7.4) and (7.5) to X, $\mathfrak{W}_X$, and $\mathfrak{M}_X$. Continuing with the description of the proposed segmentation procedure, use is made of the underlying sets W and M of scaled columns, respectively, $\{\mathbf{w}^1, \mathbf{w}^2, \mathbf{w}^3\}$ and $\{\mathbf{m}^1, \mathbf{m}^2, \mathbf{m}^3\}$, including the extreme vector bounds $\mathbf{v}$ and $\mathbf{u}$. Note that, the vectors belonging to the set $W \cup M \cup \{\mathbf{v}, \mathbf{u}\}$ provide a way to determine several tetrahedra enclosing specific subsets of X such as, e. g., $W \cup \{\mathbf{u}\}$ and $M \cup \{\mathbf{v}\}$.

The second stage in the segmentation process is accomplished using concepts from convex set geometry. These concepts make it possible to mix colors in any color space. Recall that X is said to be a *convex set* if the straight line joining any two points in X lies completely within X; also, an n-dimensional *simplex* is the minimal convex set or convex hull whose $n+1$ vertices (extreme points) are affinely independent vectors in $\mathbb{R}^n$. Since the color cube is a subspace of $\mathbb{R}^3$, a 3-dimensional simplex will correspond to a tetrahedron. Thus, considering pixel vectors in a color image enclosed by some tetrahedron, whose base face is determined by its most saturated colors, an estimation of the fractions in which they appear at any other color pixel can be made. Observe that the model commonly used for the analysis of spectral mixtures in hyperspectral images, the *constrained linear mixing* (CLM) model [141], can readily be adapted to segment noiseless color images by representing each pixel vector as a convex linear combination of the most saturated colors. As a particular case of Eqs. (7.1) and (7.2), its mathematical representation is given by

$$\mathbf{x} = S\,\mathbf{c} = c_1\mathbf{s}^1 + c_2\mathbf{s}^2 + c_3\mathbf{s}^3,\ c_1, c_2, c_3 \geq 0 \quad \text{and} \quad c_1 + c_2 + c_3 = 1. \qquad (7.12)$$

where, $\mathbf{x}$ is a 3×1 pixel vector, $S = (\mathbf{s}^1, \mathbf{s}^2, \mathbf{s}^3)$ is a square matrix of size 3×3 whose columns are the extreme colors, and $\mathbf{c}$ is the 3×1 vector of "saturated color fractions" present in $\mathbf{x}$. Notice that the most saturated colors in a given image may easily be equal to the set of primary colors (red, green, blue) or to the set of complementary colors (cyan, magenta, yellow). Therefore, the *linear unmixing* step consists of solving equation 7.12 to find vector $\mathbf{c}$ given that $S = W$ or $S = M$ for every $\mathbf{x} \in X$. As mentioned earlier, in the case of hyperspectral images, to solve the constrained linear system displayed in equation 7.12, one can employ the LLS or NNLS methods imposing the full additivity or the positivity constraint, respectively.

In the third and last stage of the segmentation process, once equation 7.12 is solved for every color pixel $\mathbf{x}^\xi \in X$, all c^j_ξ fraction values are assembled to form a vector associated with the saturated color $\mathbf{s}^j$, and the final step is carried out by applying a threshold value, in most cases, between 0.3 and 1 to obtain an adequate segmentation depicting the corresponding image partition (see Definition 7.2).

Of the many existing approaches to image segmentation [261, 211, 173, 53], clustering techniques such as c-means and fuzzy c-means can be applied to color images provided the number of clusters is known beforehand. When using any of these techniques a cluster is interpreted as the mean or average color assigned to an iteratively determined subset of color pixels belonging to X. For an explanation of the basic theory and algorithmic variants concerning the c-means clustering technique cf. [178, 76, 80] and similarly, for the fuzzy c-means clustering technique see [23, 169]. In relation to our proposed method based on LAAMs, a comparison with both clustering techniques is immediate since the maximum number of saturated colors determined from $\mathfrak{W}_X$, $\mathfrak{M}_X$, and possibly $\{\mathbf{v}, \mathbf{u}\}$ is always 8, thus the number of clusters is bounded by the interval $[1, 8]$. Furthermore, since any member in the set $W \cup M \cup \{\mathbf{v}, \mathbf{u}\}$ is an extreme point, we are able to select any two disjoint subsets of three column vectors to form a 3×3 system in order to obtain unique solutions to equation 7.12. Therefore, once a pair of triplets is fixed, the number of clusters c can be restricted to the interval $[6, 8]$.

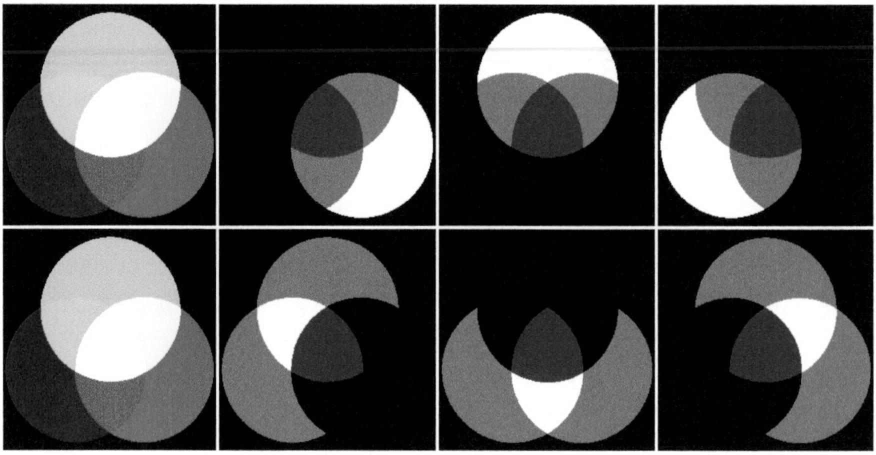

Figure 7.18 1st column: test RGB color image; 1st row, 2nd to 4th cols.: grayscale images depicting segmented regions containing proportions of red ($\mathbf{w}^1$), green ($\mathbf{w}^2$), and blue ($\mathbf{w}^3$) colors; 2nd row, 2nd to 4th cols.: grayscale images with regions composed of cyan ($\mathbf{m}^1$), magenta ($\mathbf{m}^2$), and yellow ($\mathbf{m}^3$) colors. Brighter gray tones correspond to high fractions of saturated colors [289].

7.4.2 Segmentation Results and Comparisons

Example 7.1 (Flat color image) Figure 7.18 shows in the left column a test RGB color image (primary colors additive mixtures) of size 256×256 pixels that has only 8 different colors. Hence, $X = \{\mathbf{x}^1, \ldots, \mathbf{x}^8\}$ out of a total of 65,536 pixel vectors. The minimum and maximum vector bounds are $\mathbf{v} = (0,0,0)'$ and $\mathbf{u} = (255,255,255)'$, respectively, whereas the scaled lattice memory matrices given by

$$W = \begin{pmatrix} 255 & 0 & 0 \\ 0 & 255 & 0 \\ 0 & 0 & 255 \end{pmatrix} \quad \text{and} \quad M = \begin{pmatrix} 0 & 255 & 255 \\ 255 & 0 & 255 \\ 255 & 255 & 0 \end{pmatrix}. \tag{7.13}$$

For this trivial color image, a simple algebraic analysis yields a closed solution for unmixing color pixels obeying equation 7.12. In this case we have

$$W^{-1} = \frac{1}{255}\begin{pmatrix} 1 & 0 & 0 \\ 0 & 1 & 0 \\ 0 & 0 & 1 \end{pmatrix} \quad \text{and} \quad M^{-1} = \frac{1}{510}\begin{pmatrix} -1 & 1 & 1 \\ 1 & -1 & 1 \\ 1 & 1 & -1 \end{pmatrix}. \tag{7.14}$$

From equation 7.14, $W^{-1} = I/255$ where I is the 3×3 identity matrix and

TABLE 7.8 Fraction values for unmixing pixels of the test RGB color image [289].

Saturated Color	**Pixel Values** (x_1, x_2, x_3)	**From** W (c_1, c_2, c_3)	**From** M (c_1, c_2, c_3)
Black	$(0,0,0)$	$(0,0,0)$	$(0,0,0)$
Red	$(255,0,0)$	$(1,0,0)$	$(0,\frac{1}{2},\frac{1}{2})$
Green	$(0,255,0)$	$(0,1,0)$	$(\frac{1}{2},0,\frac{1}{2})$
Blue	$(0,0,255)$	$(0,0,1)$	$(\frac{1}{2},\frac{1}{2},0)$
Cyan	$(0,255,255)$	$(0,\frac{1}{2},\frac{1}{2})$	$(1,0,0)$
Magenta	$(255,0,255)$	$(\frac{1}{2},0,\frac{1}{2})$	$(0,1,0)$
Yellow	$(255,255,0)$	$(\frac{1}{2},\frac{1}{2},0)$	$(0,0,1)$
White	$(255,255,255)$	$(\frac{1}{3},\frac{1}{3},\frac{1}{3})$	$(\frac{1}{3},\frac{1}{3},\frac{1}{3})$

considering that $x_i^\xi \in \{0,255\}$, $c_i = x_i/255$ verifies trivially the inequalities $0 \leq c_i \leq 1$ for all $i = 1,2,3$ and $\xi \in \{1,\ldots,8\}$. Full additivity is satisfied if $\sum_{i=1}^{3} c_i = \sum_{i=1}^{3} x_i/255 = 1$, therefore color pixel values x_1, x_2, and x_3 lie in the plane $x_1 + x_2 + x_3 = 255$ which occurs only at the points $(255,0,0)$, $(0,255,0)$, and $(0,0,255)$. However, letting $s = x_1 + x_2 + x_3$, the color fractions obtained from the scaled min memory W are readily specified by the simple formula

$$c_i = \frac{x_i}{s} = \frac{x_i}{x_1 + x_2 + x_3} \Leftrightarrow s \neq 0, \tag{7.15}$$

otherwise if $s = 0$ let $c_i = 0$. Similarly, from the inverse matrix M^{-1} given in equation 7.14, one finds that $c_i = [(\sum_{j \neq i} x_j) - x_i]/510$ for $i = 1,2,3$. However, since $x_i^\xi \in \{0,255\}$ we have $c_i \in \{-0.5, 0, 0.5, 1\}$; thus, non-negativity is not satisfied for all i. Full additivity is verified if $\sum_{i=1}^{3} c_i = \sum_{i=1}^{3} [(\sum_{j \neq i} x_j) - x_i]/510 = 1$, implying that color pixel values x_1, x_2, and x_3 belong to the plane $x_1 + x_2 + x_3 = 510$, and this can occur only at the points $(0,255,255)$, $(255,0,255)$, and $(255,255,0)$. Therefore, making $s = x_1 + x_2 + x_3$, the color fractions obtained from the scaled max memory M are given by the formula

$$c_i = \frac{(\sum_{j \neq i} x_j) - x_i}{s} = \frac{(\sum_{j \neq i} x_j) - x_i}{x_1 + x_2 + x_3} \Leftrightarrow s \neq 0, \tag{7.16}$$

otherwise if $s = 0$, then $c_i = 0$; also, if $c_i = -1$ (for some i), then set $c_i = 0$ and change c_j to $c_j/2$ for $j \neq i$. Table 7.8 displays the correspondence between pixel color values and color fractions derived from the scaled LAAMs.

Using the mapping established in Table 7.8, the color fraction solution vector '**c**' is quickly determined for each one of the 65,536 pixels forming the image, using for S, first the W matrix that unmixes the primary colors, then the M matrix that unmixes the secondary colors. To the right of the test RGB color image in Figure 7.18, the color fraction maps displayed as grayscale images are associated to the saturated colors derived from the column vectors of the scaled LAAMs, except black, considered the image background, and white that results from additive mixture of the three primary colors. Each color fraction segmented image $\mathbf{s}^j$ is visible after a linear scaling from the interval $[0,1]$ to the grayscale dynamic range $[0,255]$.

Example 7.2 (Gradient color image) In Figure 7.19, the left column shows a synthetic RGB color image composed by a gradient of primary and secondary colors of size 256×256 pixels with 2,400 different colors. Thus, $X = \{\mathbf{x}^1, \ldots, \mathbf{x}^{2400}\}$ (again, from a total of 65,536 color pixels). It turns out that the scaled LAAM matrices and the minimum-, maximum vector bounds are almost the same as those computed in the previous example, equation 7.13, except that the numeric value 255 is replaced by 254. Although the given image is rather simple, an algebraic analysis would be impractical for finding a color fractions formula applicable for unmixing every different color present in the image. However, *fast* pixel linear decomposition can be realized, e.g., by generalized matrix inversion (LLS) enforcing full additivity and adequate thresholding of numerical values.

From equation 7.12 any $c_q = 1 - c_p - c_r$, where $q = 1,2,3$ and $q \neq p < r \neq q$, can be selected to reduce the size of matrix S and vector $\mathbf{c}$. Consequently, computations are simplified by solving for each color pixel the linear system given by $\mathbf{x}_q = S_q \mathbf{c}_q$, where $\mathbf{c}_q = (c_p, c_r)^t$, $S_q = W_q$ or $S_q = M_q$, and

$$S_q = \begin{pmatrix} s_{1p} - s_{1q} & s_{1r} - s_{1q} \\ s_{2p} - s_{2q} & s_{2r} - s_{2q} \\ s_{3p} - s_{3q} & s_{3r} - s_{3q} \end{pmatrix}, \quad \vec{x}_q = \begin{pmatrix} x_1 - s_{1q} \\ x_2 - s_{2q} \\ x_3 - s_{3q} \end{pmatrix}. \tag{7.17}$$

In this example we let $q = 1$ and equation 7.17 is solved only for $S_1 = M_1$. Hence, $c_1 = 1 - c_2 - c_3$ and $\mathbf{c}_1 = (c_2, c_3)^t$. Also, each i-th row of S_1 and entries of the transformed input color vector $\mathbf{x}_1$, for $i = 1,2,3$, are given by $(m_{i2} - m_{i1}, m_{i3} - m_{i1})$ and $x_i - m_{i1}$, respectively. Thresholds applied to fractions values for generating segmented images were computed as

$$u_j = \frac{\tau_j}{256} \bigvee_{\xi=1}^{k} c_\xi^j, \tag{7.18}$$

where $k = 2,400$ and by setting the user-defined grayscale threshold $\tau_j = 85$ for all j. The first row in Figure 7.19 shows the segmentation produced using

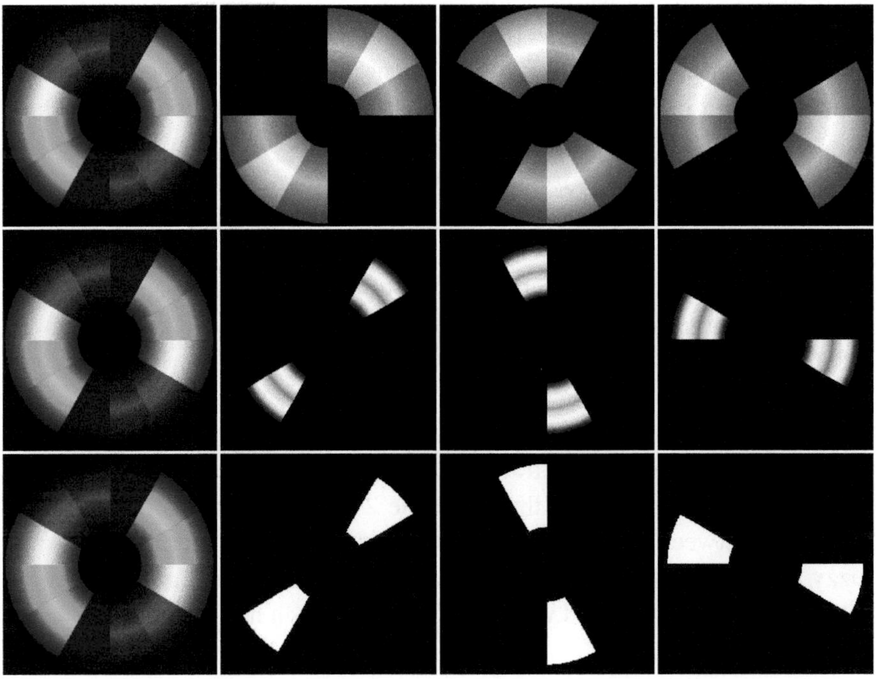

Figure 7.19 1st column: test RGB color image; 1st row, 2nd to 4th cols.: grayscale images of color fractions obtained by linear unmixing showing the segmentation of cyan ($\mathbf{m}^1$), magenta ($\mathbf{m}^2$), and yellow ($\mathbf{m}^3$) (CMY) colors; 2nd row, 2nd to 4th cols.: fuzzy c-means grayscale images depicting membership distribution in regions of CMY color gradients; 3rd row, 2nd to 4th cols.: c-means binary images depicting uniform segmented regions labeled from CMY centroids [289].

M (secondary colors), where the brighter gray tones correspond to high fractions of saturated colors. Hence color gradients are preserved as grayscale gradients. Additionally, original color regions composed of some proportion of the saturated colors $\mathbf{m}^1$, $\mathbf{m}^2$, and $\mathbf{m}^3$ appear as middle or dark gray tones. The second row displays the results obtained by applying the fuzzy c-means technique with $c = 7$ and the thresholds values u_j used to cut fuzzy memberships were calculated with equation 7.18 setting $\tau_j = 64$ for all $j = 1,\ldots,7$. Observe that the brighter gray tones are associated with pixels near to fuzzy centroids (high-membership values) whereas darker gray tones correspond to pixels far from fuzzy centroids (low-membership values); note that original color gradients are not preserved. The third row depicts as black and white binary images the clusters found using the c-means algorithm with $c = 7$ and initial centroids given by the set $W \cup M \cup \{\mathbf{v}\}$. In this last case, thresholds are not needed since the c-means algorithm is a labeling procedure that assigns to all similar colors belonging to a cluster the color value of its centroid. Consequently, a simple labeling procedure is implemented to separate regions of different color.

If W_1 is selected instead of M_1 for the system matrix S_1 in equation 7.17, similar segmentation results are obtained except that, in this case, red, green, and blue regions would be extracted from the corresponding saturated colors $\mathbf{w}^1$, $\mathbf{w}^2$, and $\mathbf{w}^3$. We remark that Example 2 clearly shows the fundamental difference between the three segmentation methods compared: c-means and fuzzy c-means clustering are statistical and iterative in nature whereas the LAAM's approach coupled with the CLM model is a non-iterative geometrical procedure.

Example 7.3 (Real color images) Next, we provide additional segmentation results for three realistic RGB color images of size 256×256 pixels displayed in the first column of Figure 7.20 (see Table 7.9 for image information). For each of these color images, we create a set $X_\ell = \{\mathbf{x}^1,\ldots,\mathbf{x}^{k_\ell}\} \subset [0,255]^3$ where $\ell = \alpha,\beta,\gamma$, and each vector $\mathbf{x}^\xi \in X_\ell$ is distinct from the others, i.e., $\mathbf{x}^\xi \neq \mathbf{x}^\zeta$ whenever $\xi \neq \zeta$. This is achieved by eliminating pixel vectors of the same color (k_ℓ is given in Table 7.9). After application of Eqs. (7.3) (LAAMs) and (7.4) (vector bounds), the scaled matrices W and M are computed. The numerical entries for the scaled LAAM matrices (3rd column, Table 7.9) of the sample images are explicitly given below:

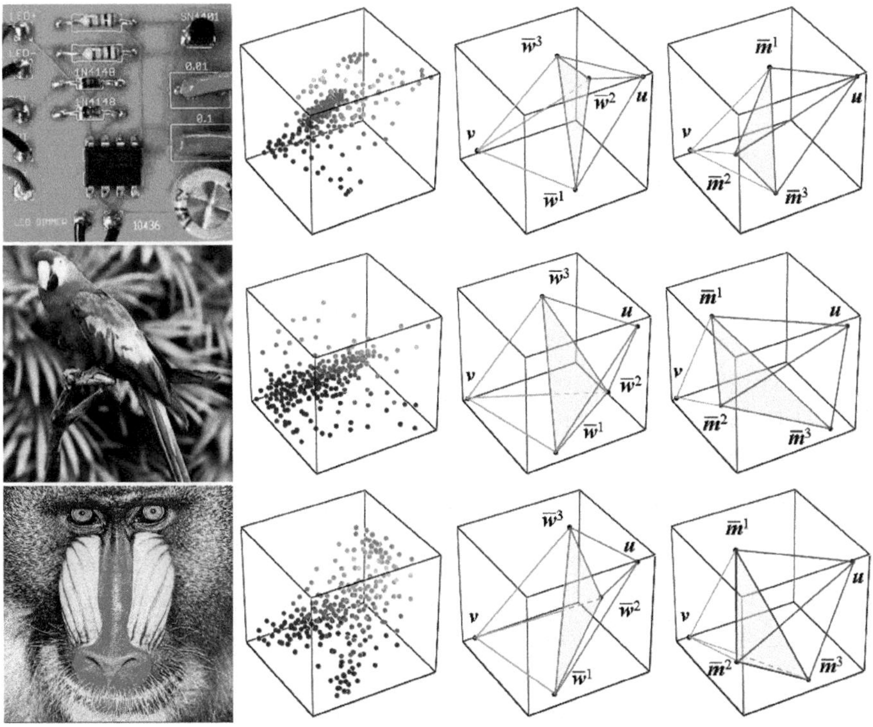

Figure 7.20 1st column: sample RGB color images; 2nd col.: scatter plot of a subset of X showing 256 different colors including the most saturated colors determined from W and M; 3rd and 4th cols.: tetrahedra determined from proper subsets of $W \cup M \cup \{\mathbf{v}, \mathbf{u}\}$ [289].

TABLE 7.9 Information of sample real RGB color images [289].

Image	**Pixels, pq**	**Colors, $\lvert X_\ell \rvert = k_\ell$**	**Scaled LAAMs**
circuit	65536	35932	W_α, M_α
parrot	65536	55347	W_β, M_β
baboon	65536	63106	W_γ, M_γ

$$W_\alpha = \begin{pmatrix} 255 & 80 & 101 \\ 71 & 255 & 135 \\ 46 & 154 & 255 \end{pmatrix}, \quad M_\alpha = \begin{pmatrix} 19 & 203 & 228 \\ 194 & 19 & 120 \\ 173 & 139 & 19 \end{pmatrix},$$

$$W_\beta = \begin{pmatrix} 255 & 121 & 35 \\ 55 & 251 & 128 \\ 1 & 23 & 255 \end{pmatrix}, \quad M_\beta = \begin{pmatrix} 0 & 200 & 254 \\ 130 & 0 & 228 \\ 220 & 127 & 0 \end{pmatrix},$$

$$W_\gamma = \begin{pmatrix} 255 & 129 & 72 \\ 55 & 255 & 156 \\ 0 & 90 & 255 \end{pmatrix}, \quad M_\gamma = \begin{pmatrix} 0 & 200 & 255 \\ 126 & 0 & 165 \\ 183 & 99 & 0 \end{pmatrix}.$$

Notice that the corresponding minimum and maximum vector bounds $\{\mathbf{v}_\ell, \mathbf{u}_\ell\}$ for $\ell = \alpha, \beta, \gamma$ are readily available from the main diagonals of the corresponding LAAM matrices. A 3D scatter plot of each set X showing only 256 different colors, including the extreme points of the set $W \cup M \cup \{\mathbf{v}, \mathbf{u}\}$, is depicted in the second column of Figure 7.20 for each sample image. Two tetrahedra enclosing points of X are illustrated in the third column of the same figure. The vertices of the left tetrahedron belong to the set $W \cup \{\mathbf{v}\}$ and those of the right tetrahedron are in $W \cup \{\mathbf{u}\}$; similarly, in the fourth column of Figure 7.20, the left tetrahedron has its vertices in the set $M \cup \{\mathbf{v}\}$ and the right tetrahedron is formed with the points of $M \cup \{\mathbf{u}\}$.

Again, for each RGB color image in Figure 7.20, equation 7.12 was simplified to equation 7.17 setting $q = 1$ and solving it using LLS for each $\mathbf{x} \in X_\ell$, by taking first W_ℓ and then M_ℓ as the S matrix for $\ell = \alpha, \beta, \gamma$. It turns out that for the sample images selected, the corresponding 3×3 computed scaled LAAMs are non-singular matrices (full rank) and, therefore, the solutions found by the linear unmixing scheme are unique. Since the minimum and maximum bounds $\{\mathbf{v}_\ell, \mathbf{u}_\ell\}$ correspond, respectively, to a "dark" color near black and to a "bright" color near white, it is possible to replace a specific column in W or M with one of these extreme bounds in order to obtain segmentations of dark or bright regions. Thus, final satisfactory segmentation results are produced by an adequate selection of saturated colors $\mathbf{s}^j$ from the set $W \cup M \cup \{\mathbf{v}, \mathbf{u}\}$. Figure 7.21 displays the segmentation produced by applying the clustering techniques of c-means, fuzzy c-means, and our proposed LAAMs plus linear unmixing based technique. Results are shown as quantized grayscale images where specific gray tones are associated with selected colors corresponding to cluster centers or extreme points. Table 7.10 provides the technical information relative to each segmenting algorithm; for example, "runs" is the number of times an algorithm is applied to a given

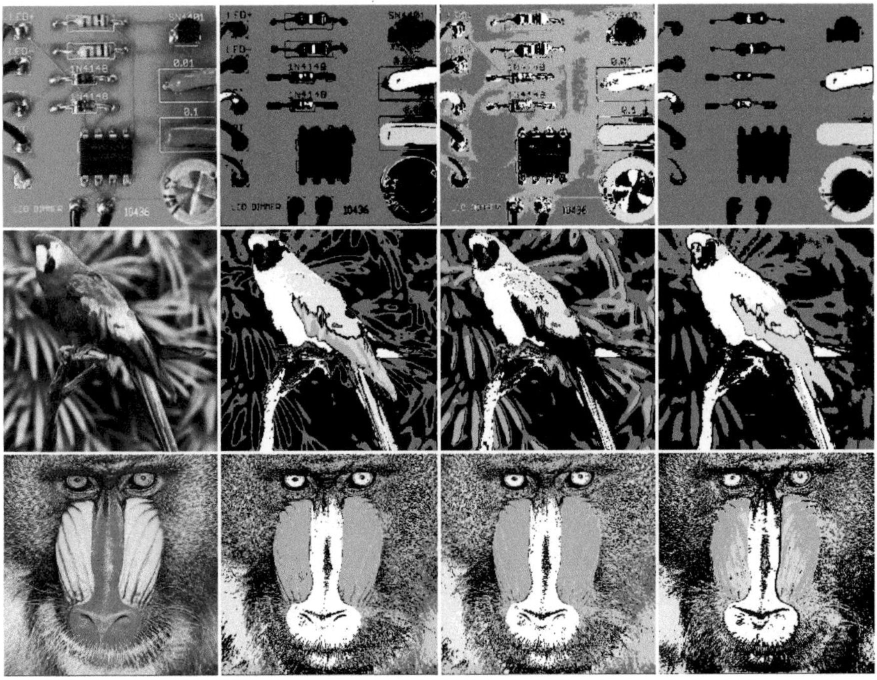

Figure 7.21 1st column: sample RGB color images; 2nd, 3rd, and 4th cols.: quantized grayscale segmented images composed from results obtained, respectively, with c-means clustering, fuzzy c-means clustering, and scaled LAAMs & LLS linear unmixing techniques [289].

TABLE 7.10 Technical data used for RGB color image segmentation [289].

Image	***c*-means**	**fuzzy *c*-means**	**LAAMs & LLS**
	kmeans (Matlab)	**fcm (Matlab)**	**geninv (Mathcad)**
circuit	c=8, runs=5 squared Euclidean 5th run, 58 iterations 57 seconds RGB values: 255,128,192	c=7, runs=3 exp(U)=2, OFC=10^{-5} 3rd run, 108 iterations 720 seconds RG_1G_2B values: 255,128,160,200	c=6, runs=1 $W_\alpha, M_\alpha, q = 1$ non-iterative 30 seconds RGB values: 255,128,192
parrot	c=8, runs=5 city block 5th run, 20 iterations 32 seconds RGBCY values: 255,128,160,192,216	c=6, runs=3 exp(U)=2, OFC=10^{-2} 3rd run, 134 iterations 238 seconds RG_1G_2Y values: 255,128,160,216	c=6, runs=1 $W_\beta, M_\beta, q = 1$ non-iterative 30 seconds RGB values: 255,128,192
baboon	c=7, runs=5 city block 5th run, 39 iterations 38 seconds RGB_1B_2Y values: 255,128,160,176,216	c=6, runs=3 exp(U)=2, OFC=10^{-2} 3rd run, 94 iterations 270 seconds RGB_1B_2Y values: 255,128,160,176,216	c=6, runs=1 $W_\gamma, M_\gamma, q = 1$ non-iterative 30 seconds RGBCY values: 255,128,160,192,216

image. Specifically, in the Matlab environment, "runs" is equivalent to the "replicates" parameter used for c-means clustering; exp(U) and OFC refer to, respectively, the partition matrix exponent and the minimum amount of improvement needed for the objective function to converge in fuzzy c-means clustering. Also, e.g., "RGB values: 255,128,192", gives the gray levels assigned to the red, green, and blue colors or additional colors.

Exercises 7.4.2

1. Given a color image X of size $p \times q$ pixels coded in RGB space, a) propose an algorithm that counts the different colors contained in X, knowing that each color pixel is a three-dimensional vector. Hence, if k denotes the number of different colors, then $k \leq pq$, though for many images, $k < pq$. b) As a way to speed up color counting, define two maps, the first map must assign a 3D vector to a unique positive integer and the second map assigns to a given positive integer, belonging to the range of the first map, a unique 3D vector in RGB space (also belonging to X). Once you find both mappings, use the first one to map X to a list of integers that you can sort with an algorithm such as heapsort or quicksort. Then, eliminate from the ordered list of integers those

that are repeated, and finally remap the reduced list to X' containing all different colors from X. Note that, in general, $X' \subset X$ and that X' can be interpreted as the *color palette* of X (without quantization).

2. Search and select three public domain 24-bit color images of size 256×256 coded in RGB space. Based on the WM-method repeat the steps explained in Example 7.3 and reproduce for the chosen images the results presented in Figure 7.20, Table 7.9 (show the numerical values for the scaled LAAMs), Figure 7.21, and Table 7.10.

CHAPTER 8

Lattice-Based Biomimetic Neural Networks

CURRENT artificial neural network (ANN) models are intimately associated with a particular learning algorithm or learning rule. Thus, we have multi-layer perceptrons (MLPs), kernel function based learning such as radial bases function (RBF) neural nets, support vector machines (SVM), kernel Fisher discriminant (KFD) neural networks, Boltzmann machines, and many others. The latest ANNs in vogue are *deep* neural networks (DNNs), that is networks with more than one hidden layer between the input and output layer. The word "deep" refers to the learning algorithms affecting the hidden layers, with each layer training on a distinct set of features. In addition, learning at a given hidden layer also depends on the output of the preceding layer as is the case for the aforementioned ANNs. As a consequence, the biological approach to ANNs has been largely abandoned for a more practical approach based on well-known mathematical, statistical, and signal processing methods. One basic goal of dendritic computing is the return of ANNs to its roots in neurobiology and neuro-physics.

8.1 BIOMIMETIC ARTIFICIAL NEURAL NETWORKS

The term *biomimetic* refers to man-made systems of processes that imitate nature. Accordingly, biomimetic artificial neurons are man-made models of biological neurons, while biomimetic computational systems deal with information processing in the brain. More specifically, biomimetic computational

DOI: 10.1201/9781003154242-8

systems are concerned with such questions as how do neurons encode, transform and transfer information, and how can this encoding and transfer of information be expressed mathematically.

8.1.1 Biological Neurons and Their Processes

In order to imitate biological neural structures, one has to first understand some very basic concepts concerning the morphology and function of the fundamental component of the structure, namely the neuron. A neuron (or nerve cell) is a cell in the animal kingdom and as such contains numerous components common to all animal cells. These include a cell membrane, a cell nucleus, mitochondria, Golgi apparatus, ribosomes, and so on. Just as there are many different type of cells making up the overall structure of an animal, there are many different types of nerve cells making up the nervous system of an animal. These different types of neurons are classified according to their morphological differences such as their dendritic structures as well as their functionality. Nevertheless, every neuron consists of a cell body, called *soma*, and several processes. These processes are of two kinds and are called, respectively, *dendrites* and *axons*. The dendrites, which are usually multiple, conduct impulses toward the body of the cell; the axon conducts from the cell body. Dendrites typically have many branches that create large and complicated trees. Many (but not all) types of dendrites are studded with large numbers of tiny branches called *spines*. Dendritic spines, when present, are the major *postsynaptic* target for the synaptic input. The soma and the dendrites constitute the input surface of the neuron. When a neuron fires, then all neurons receiving the fired signal are called the *postsynaptic* neurons while the firing neuron is called the *presynaptic* neuron. When the voltage profile of a fired signal is recorded, it usually consists of a sequence of short electrical pulses known as a *spike train*. It has been conjectured that the number of and distances between the spikes in a train represents the encoded information that the neuron is transmitting to the recipient postsynaptic neuron [68]. In humans, and most vertebrates, spike trains initiate in the axonal initial segment (AIS) near the soma. For our purposes a spike train is simply a combination of spikes and silences which can be expressed in binary form, namely a one for a spike and a zero for the absence of a spike. In this interpretation a spike train may look like 11110011111000001.

The axon, which usually arises from the opposite pole of the cell at a point called the *axon hillock*, consists of a long fiber whose branches form the *axonal arborization* or *axonal tree*. For some neurons the axon may have branches at intervals along its length in addition to its terminal arborization.

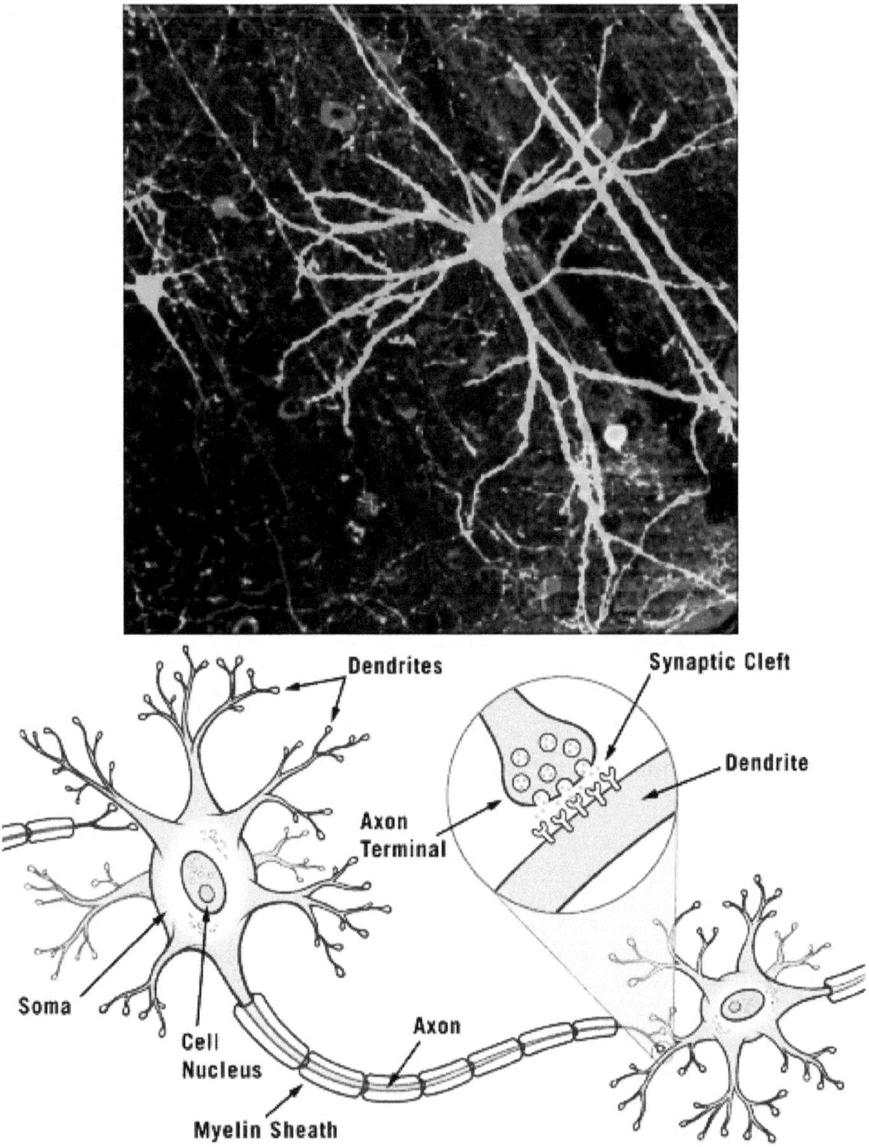

Figure 8.1 Merged color image of a biological neuron cell showing dendrites, dendritic trees, soma, axon, and thin terminal branches [162], and schematic drawing of biological neuron cells (pre- and postsynaptic) showing dendrites, dendritic trees, soma with nucleus, axon, myelin sheath, synaptic cleft, and terminal branches with boutons [164]

The tips of the branches of the axon are called *nerve terminals* or *boutons* or *synaptic knobs*. The axon is the principal fiber branch of the neuron for the transmission of signals to other neurons. Figure 8.1 shows an image and a typical schematic representation of a biological neuron with its branching processes. An impulse traveling along an axon from the axon hillock propagates through the axonal tree all the way to the nerve terminals. The terminals of the branches make contact with the soma and the many dendrites of other neurons. The sites of contact are the *synaptic sites* where the synapses take place. The *synapse* is a specialized structure whereby neurons communicate, but there is no actual structural union of the two neurons at the synaptic site. The synaptic knob is separated from the surface of the dendrite or soma by an extremely narrow space called the *synaptic cleft*. The exact mechanism of synaptic structures is fairly well-understood and there exist two kinds of synapses; *excitatory synapses*, which tend to depolarize the postsynaptic membrane and consequently excite the postsynaptic cell to fire impulses, and *inhibitory synapses* that try to prevent the neuron from firing impulses in response to excitatory synapses. In brief, it is at the synaptic cleft where the presynaptic neuron communicates with the postsynaptic neuron. This communication takes place via neurotransmitters. Neurotransmitters are small molecules that are released by the axon terminal of the presynaptic neuron after an action potential reaches the synapse. There exist various different types of neurotransmitters that have been identified [198, 301]. These transmitter molecules can bind with the dendritic receptors and create an excitatory electrical potential that is then transmitted down the cell membrane or it may block (inhibit) the signal from being carried to the soma.

The common belief that only one type neurotransmitter (i.e. either excitatory or inhibitory, but not both) is released at the various axon terminals of a single presynaptic neuron was challenged in 1976 [34]. It is now widely accepted that cotransmission is an integral feature of neurotransmission. Here cotransmission refers to the release of several different types of neurotransmitters from a single nerve terminal. Recent research has demonstrated that most, if not all, neurons release different neurotransmitters [35, 199, 281]. Another interesting synapse discovered in 1972 is the *autapse*. The autapse is a synapse formed by the axon of a neuron on its own dendrites [21, 300].

8.1.2 Biomimetic Neurons and Dendrites

The number of neurons in the human brain lies somewhere around 85 billion, with each of the neurons having an average of 7,000 synaptic connections to other neurons [74, 78, 117, 302]. The number of synapses on a *single* neuron in the cerebral cortex ranges between 500 and 200,000, while the cerebral cortex of an adult has an estimated range of 100 to 500 trillion (10^{14} to 5×10^{14}) synapses [74, 133, 183, 332]. Dendrites make up the largest component in both surface area and volume of the brain. Part of this is due to the fact that pyramidal cell dendrites span all cortical layers in all regions of the cerebral cortex [79, 148, 244]. Thus, when attempting to model artificial neural networks that bear more than just a passing resemblance to biological brain networks, one cannot ignore dendrites (and their associated spines) which can make up more than 50% of the neuron's membrane. This is especially true in light of the fact that some brain researchers have proposed that dendrites and not the neuron are the elementary computing devices of the brain. Neurons with dendrites can function as many, almost independent, functional subunits with each unit being able to implement a rich repertoire of logical operations [33, 148, 188, 214, 244, 312]. Possible mechanisms for dendritic computation of such logical functions as XOR, AND, and NOT have been proposed by several researchers [12, 119, 148, 150, 180, 187, 214, 244, 251].

In account of the above observations it is apparent that a more realistic biomimetic model of a neuron needs to include both dendrites and an axon with arborization. Also, the operations of AND, OR, NOT, and XOR are operations common to lattice theory and can be achieved in the dendrites starting at the synapses and accumulating in branches of the dendritic tree. They are just as easy to implement on the gate array level and therefore provide for fast computational results. Additionally, for additive lattice groups the operation of multiplication is generally absent and thus yields extremely fast convergence in most lattice-based learning algorithms.

8.2 LATTICE BIOMIMETIC NEURAL NETWORKS

In the dendritic lattice-based model of ANNs [54, 221, 222, 224], a set of presynaptic neurons $N_1,\ldots,N_n$ provides information through its axonal arborization to the dendritic trees of some other set of postsynaptic neurons $M_1,\ldots,M_m$. Figure 8.2 illustrates the neural pathways from the presynaptic neurons to the postsynaptic neuron M_j, whose dendritic tree is assumed to consist of K_j branches $\tau_1^j,\ldots,\tau_{K_j}^j$ that contain the synaptic sites upon which the axonal fibers of the presynaptic neurons terminate. The *location* or *address* of a specific synapse is defined by the quintuple (i,j,k,h,ℓ), where $i\in\{1,\ldots,n\}$, $j\in\{1,\ldots,m\}$, and $k\in\{1,\ldots,K_j\}$ that a terminal axonal branch of N_i has a bouton on the kth dendritic branch τ_k^j of M_j. The index $h\in\{1,\ldots,\rho\}$ denotes the hth synapse of N_i on τ_k^j since there may be more terminal axonal branches of N_i synapsing on τ_k^j. The index $\ell \in \{0,1\}$ classifies the type of the synapse, where $\ell = 0$ indicates that the synapse at (i,j,k,h,ℓ) is inhibitory (i.e., releases inhibitory neurotransmitters) and $\ell = 1$ indicates that the synapse is excitatory (releases excitatory neurotransmitters).

The *strength* of the synapse (i,j,k,h,ℓ) corresponds to a real number, commonly referred to as the *synaptic weight* and customarily denoted by w_{ijkh}^{ℓ}. Thus, if S denotes the set of synapses on the dendritic branches of the set of the postsynaptic neurons $M_1,\ldots,M_m$, then w can be viewed as the function $w : S \to \mathbb{R}$ defined by $w : (i,j,k,h,\ell) \to w_{ijkh}^{\ell} \in \mathbb{R}$ or, equivalently, by $w(i,j,k,h,\ell) = w_{ijkh}^{\ell}$. In order to reduce notational overhead, we simplify the synapse location and type as follows: (1) (j,k,h,ℓ) if $n = 1$ and set $N = N_1$, (2) (i,k,h,ℓ) if $m = 1$, set $M = M_1$ and denote its dendritic branches by $\tau^1, \tau^2, \ldots, \tau^K$ or simply τ if $K = 1$, and (3) (i,j,k,ℓ) if $\rho = 1$.

The axon terminals of different presynaptic biological neurons that have synaptic sites on a single branch of the dendritic tree of a postsynaptic neuron may release dissimilar neurotransmitters which, in turn, affect the receptors of the branch. Since the receptors serve as storage sites of the synaptic strengths, the resulting electrical signal generated by the branch is the result of the combination of the output of all its receptors. As the signal travels toward the cell's body it again combines with signals generated in the other branches of the dendritic tree. In the lattice-based biomimetic model the various biological synaptic processes due to dissimilar neurotransmitters are replaced by different operations of a lattice group. More specifically, if $O = \{\vee, \wedge, +\}$ represents the operations of a lattice group $\mathbb{F}$, then the symbols $\oplus$, $\otimes$, and $\odot$ will mean that $\oplus, \otimes, \odot \in O$ but are not explicitly specified operations. For instance, if $\bigoplus_{i=1}^{n} a_i = a_1 \oplus \cdots \oplus a_n$ and $\oplus = \vee$, then $\bigoplus_{i=1}^{n} a_i = \bigvee_{i=1}^{n} a_i = a_1 \vee \cdots \vee a_n$, and if $\oplus = +$, then $\bigoplus_{i=1}^{n} a_i = \sum_{i=1}^{n} a_i = a_1 + \cdots + a_n$.

The total response (or output) of a branch τ_k^j to the received input at its synaptic sites is given by the general formula

$$\tau_k^j(\mathbf{x}) = p_{jk} \bigoplus_{i \in I(k)} \bigotimes_{h=1}^{r} (-1)^{1-\ell}(x_i \odot w_{ijkh}^{\ell}), \tag{8.1}$$

where $\mathbf{x} = (x_1,\ldots,x_n) \in \mathbb{F}^n$, $x_i \in \mathbb{F}$ denotes the information propagated by N_i via its axon and axonal branches, $I(k) \subseteq \{1,\ldots,n\}$ corresponds to the set of all presynaptic neurons with terminal axonal fibers that synapse on the k-th dendritic branch of M_j, and r denotes the number of synoptic knobs of N_i on the branch d_{jk}. The value $p_{jk} \in \{-1,1\}$ marks the final signal outflow from the k-th branch as inhibitory if $p_{jk} = -1$ and excitatory if $p_{jk} = 1$. The value $\tau_k^j(\mathbf{x})$ is passed to the cell body of M_j and the state of M_j is a function of the

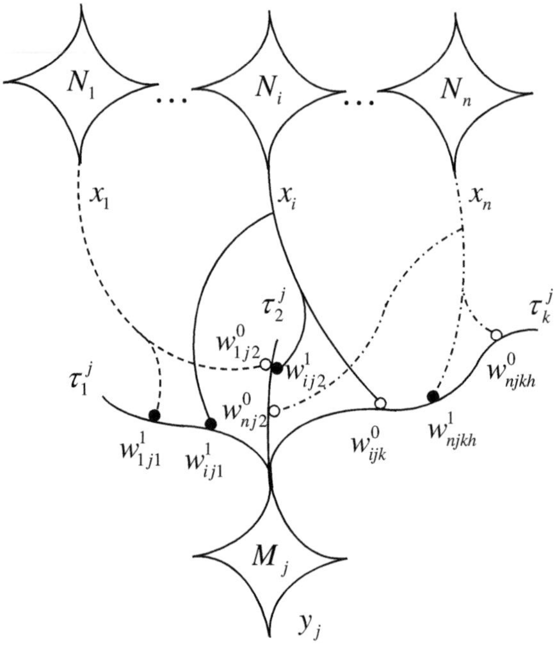

Figure 8.2 Illustration of the neural pathways from the presynaptic neurons N_i to the postsynaptic neuron M_j. An open circle ○ means that the synaptic weight is inhibitory, while a solid circle • indicates an excitatory synapse. The value x_i denotes information transferred from neuron N_i to the synaptic sites of neuron M_j. Terminal branches of axonal fibers originating from the presynaptic neurons make contact with synaptic sites on dendritic branches of M_j.

combined values received from its dendritic structure and is computed as

$$\tau^j(\mathbf{x}) = p_j \bigodot_{k=1}^{K_j} \tau_k^j(\mathbf{x}), \tag{8.2}$$

where $p_j = \pm 1$ denotes the response of the cell to the received input. Here again $p_j = -1$ means rejection (inhibition) and $p_j = 1$ means acceptance (excitation) of the received input. This mimics the *summation* that occurs in the axonal hillock of biological neurons.

In many applications of LNNs, the presynaptic neurons have at most one axonal button synapsing on any given dendritic branch τ_k^j. In these cases equation 8.1 simplifies to

$$\tau_k^j(\mathbf{x}) = p_{jk} \bigoplus_{i \in I(k)} (-1)^{1-\ell}(x_i \odot w_{ijk}^{\ell}). \tag{8.3}$$

As in most ANNs, the next state of M_j is determined by an activation function f_j, which—depending on the problem domain—can be the identity function, a simple hard-limiter, or a more complex function. The *next state* refers to the information being transferred via M_j's axon to the next level neurons or the output if M_j is an output neuron. Any ANN that is based on dendritic computing and employs equations of type 8.1 and 8.2, or 8.3 and 8.2, will be called a *lattice biomimetic neural network* or simply an LBNN. In the technical literature there exist a multitude of different models of lattice-based neural networks, usually abbreviated as LNNs. The matrix-based lattice-associative memories discussed in Chapter 5 and LBNNs are just two examples of LNNs. What sets LBNNs apart from current ANNs are the inclusion of the following processes employed by biological neurons:

1. The use of dendrites and their synapses.
2. A presynaptic neuron N_i can have more than one terminal branch on the dendrites of a postsynaptic neuron M_j.
3. If the axon of a presynaptic neuron N_I has two or more terminal branches that synapse on different dendritic locations of the postsynaptic neuron M_j, then it is possible that some of the synapses are excitatory and others are inhibitory to the same information received from N_i.
4. The basic computations resulting from the information received from the presynaptic neurons takes place in the dendritic tree of M_j.
5. As in standard ANNs, the number of input and output neurons are problem dependent. However in contrast to standard ANNs where the number of neurons in a hidden layer as well as the number of hidden layers are pre-set by the user or an optimization process, hidden layer neurons, dendrites, synaptic sites and weights, and axonal structures are grown during the learning process.

8.2.1 Simple Examples of Lattice Biomimetic Neural Networks

Substituting specific lattice operations in the general equations 8.1 and 8.2 results in a specific model of the computations performed by the postsynaptic neuron M_j. For instance,

$$\tau_k^j(\mathbf{x}) = p_{jk} \sum_{i\in I(k)} \left(\bigwedge_{h=1}^{\rho} (-1)^{1-\ell}(x_i + w_{ijkh}^{\ell})\right) \quad ; \quad \tau^j(\mathbf{x}) = p_j \bigvee_{k=1}^{K_j} \tau_k^j(\mathbf{x}), \tag{8.4}$$

or

$$\tau_k^j(\mathbf{x}) = p_{jk} \bigwedge_{i\in I(k)} \left(\bigwedge_{h=1}^{\rho} (-1)^{1-\ell}\left(x_i + w_{ijk}^{\ell}\right)\right) \quad ; \quad \tau^j(\mathbf{x}) = p_j \bigvee_{k=1}^{K_j} \tau_k^j(\mathbf{x}), \tag{8.5}$$

are two distinct specific models. Unless otherwise mentioned, the lattice group $(\mathbb{R}, \wedge, \vee, +)$ will be employed when implementing equations 8.1 and 8.2 or 8.3 and 8.2. In contrast to standard ANNs currently in vogue, we allow both negative and positive synaptic weights as well as weights of value zero. The reason for this is that these values correspond to positive weights if one chooses the algebraically equivalent lattice group $(\mathbb{R}^+, \wedge, \vee, \times)$ (see Theorems 4.4 and 4.5). Having defined the computational model of an LBNN, it will be instructive to provide a few simple examples in order to illustrate several basic properties of LBNNs. The simplest examples are, of course, single-layer neural networks.

Example 8.1

1. The simplest examples are one input and one output neuron LBNNs. For instance, given the interval $[a,b] \subset \mathbb{R}$, the problem is to devise a network such that for any $x \in \mathbb{R}$, the output neuron can decide whether or not $x \in [a,b]$. Let N denote the input neuron and M the output neuron. In this case, there is no need for the two labels i and j. Furthermore, only one dendrite τ with two synaptic sites is necessary for this simple network, making the label k redundant. Let w_1^{ℓ} and w_2^{ℓ} denote the synaptic weight for the two synapses on d. Thus, equation 8.1 reduces to $\tau(x) = \bigotimes_{h=1}^{2} (-1)^{1-\ell}(x + w_h^{\ell})$. Letting $\otimes = \wedge$ results in $\tau(x) = (-1)^{1-\ell}(x + w_1^{\ell}) \wedge (-1)^{1-\ell}(x + w_2^{\ell})$. Finally, set $w_1^{\ell} = -a$ with $\ell = 1$ and $w_2^{\ell} = -b$ with $\ell = 0$. Then $\tau(x) = (x - a) \wedge (b - x)$. Using the hard-limiter activation function

$$f(\tau(\mathbf{x})) = \begin{cases} 1 \Leftrightarrow \tau(x) \geq 0 \\ 0 \Leftrightarrow \tau(x) < 0. \end{cases} \tag{8.6}$$

The network is shown in Figure 8.3a. Note that the problem of assigning the numeric values is self-evident when starting with the knowledge that $x \in [a,b] \Leftrightarrow (x - a) \wedge (b - x) \geq 0$.

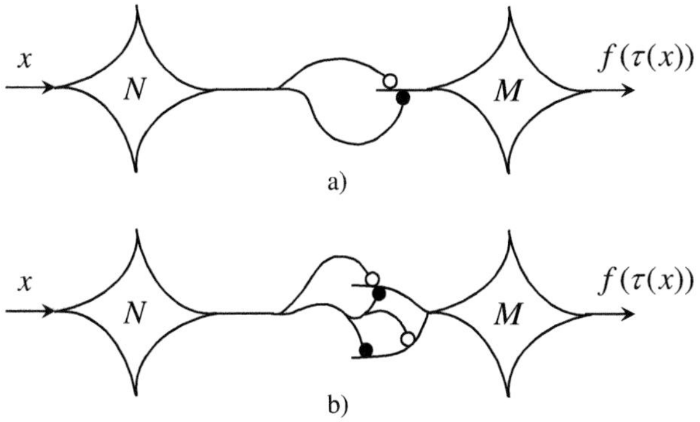

Figure 8.3 a) Sketch of the LNN constructed in Example 8.1(1), and b) depicts the LNN constructed in Example 8.1(2).

If instead the neuron M is asked to decide whether or not $x \in (a,b)$, then the same network can be used by simply changing the activation function to

$$f(\tau(\mathbf{x})) = \begin{cases} 1 \Leftrightarrow \tau(x) > 0 \\ 0 \Leftrightarrow \tau(x) \leq 0. \end{cases} \tag{8.7}$$

2. Suppose (a,b) and (c,d) are two disjoint open intervals and the problem is to construct the dendritic/synaptic makeup of the output neuron M so that for $x \in \mathbb{R}$, M can decide whether or not $x \in (a,b) \cup (c,d)$. In view of the LNN model constructed in the preceding example, all one needs to do is add another branch with two synaptic sites on the postsynaptic neuron M. Let τ_1 and τ_2 denote the two branches and let w_{11}^{ℓ}, w_{12}^{ℓ} and w_{21}^{ℓ}, w_{22}^{ℓ}, denote the respective synaptic weights. Observe that in this case the notation d_k and w_{kh} is being used since i and j are redundant. Mimicking Example 1, let $w_{11}^{\ell} = -a$ with $\ell = 1$, $w_{12}^{\ell} = -b$ with $\ell = 0$, and $w_{21}^{\ell} = -c$ with $\ell = 1$ and $w_{22}^{\ell} = -d$ with $\ell = 0$. Then $\tau_1(x) = (x-a) \wedge (b-x)$, $\tau_2(x) = (x-c) \wedge (d-x)$, and $\tau(x) = \tau_1(x) \vee \tau_2(x)$. A sketch of this LNN model is shown in Figure 8.3b. The final activation function is the one defined by equation 8.7.

3. Let $E \subset \mathbb{R}^2$ denote the line defined by the equation $\frac{1}{3}x_1 - x_2 = -\frac{4}{3}$ and suppose that the goal is to construct a construct an LNN that can decide whether or not a point $\mathbf{x} \in \mathbb{R}^2$ is a member of the half-space $\overline{E^+}$. Note that $\mathbf{x} \in \overline{E^+} \Leftrightarrow x_1 - 3x_2 + 4 \geq 0$. $\mathbf{x} \in X \Leftrightarrow x_1 - 3x_2 + 4 \geq 0$. Decomposing the variables and constants in this inequality provides the key in

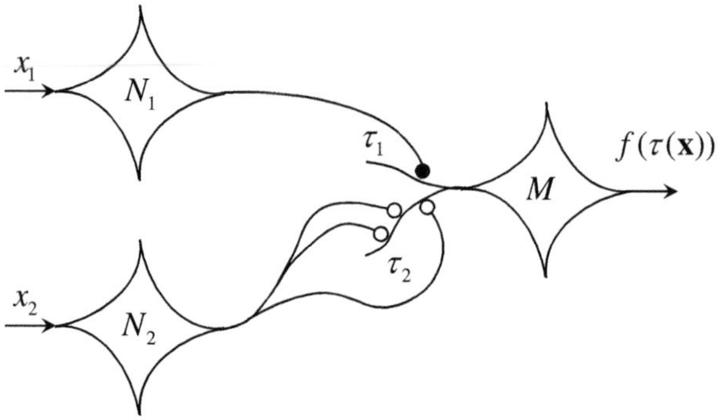

Figure 8.4 Sketch of the LNN that decides whether or not a point $\mathbf{x} \in \mathbb{R}^2$ belongs the half-plane $\overline{E^+} = \{\mathbf{x} \in \mathbb{R}^2 : x_1 - 3x_2 + 4 \geq 0\}$.

assembling the desired LNN. Explicitly, we have

$$\text{A)}\quad 0 \leq x_1 - 3x_2 + 4 = (x_1 + 1) - (x_2 - 1) - (x_2 - 1) - (x_2 - 1).$$

Using two dendritic branches τ_1 and τ_2, simply set $I(i) = \{i\}$ for $i = 1,2$, $h = 1$ for $k = 1$, and $h \in \{1,2,3\}$ for $k = 2$. Thus, if $i = 1$, then $w^\ell_{ikh} = w^\ell_{111}$ and if $i = 2$, then $w^\ell_{ikh} = w^\ell_{22h}$, where $h = 1,2,3$. Finally, set $w^\ell_{111} = w^1_{111} = 1$ and $w^\ell_{22h} = -1$ with $\ell = 0$ for $h = 1,2,3$. Then $\tau_1(\mathbf{x}) = x_1 + w^1_{111} = x_1 + 1$ and $\tau_2(\mathbf{x}) = \sum_{h=1}^3 (-1)^{1-\ell}(x_2 + w^\ell_{22h}) = -(x_2 - 1) - (x_2 - 1) - (x_2 - 1)$. It now follows that the LNN consisting of two input neurons N_1 and N_2 and one output neuron M with $\tau(\mathbf{x}) = \tau_1(\mathbf{x}) + \tau_2(\mathbf{x})$ satisfies inequality A) above. Applying the activation function $f(\tau(\mathbf{x}))$ defined in equation 8.6 completes the construction of the desired LNN shown in Figure 8.4.

The general formulas expressed in equations 8.1 and 8.2—as well as the lattice properties of associativity, absorption, and distributivity—enable different morphologies of LNNs for solving a specific problem. For instance, consider constructing an LNN that recognizes whether or not a point $\mathbf{x} \in \mathbb{R}^n$ is an element of the n-dimensional interval $[\mathbf{a}, \mathbf{b}] \subset \mathbb{R}^n$. Generalizing the approach discussed in Example 8.1(1), one needs n input neurons $N_1, N_2, \ldots, N_n$ so that for $\mathbf{x} \in \mathbb{R}^n$ the input to N_i is x_i. Following the example, let M denote the output neuron with one dendritic branch having $2n$ synaptic sites and set

$$\tau(\mathbf{x}) = \tau_1(\mathbf{x}) = \bigwedge_{i=1}^{n} \Big(\bigwedge_{h=1}^{2} (-1)^{1-\ell}(x_i + w^\ell_{ih}) \Big) = \bigwedge_{i=1}^{n} [(x_i - a_i) \wedge (b_i - x_i)], \qquad (8.8)$$

where $w_{i1}^{\ell} = w_{i1}^{1} = -a_i$ and $w_{i2}^{\ell} = w_{i2}^{0} = -b_i$ $\forall i \in \mathbb{N}_n$. An alternative route is to adorn M with a dendritic tree consisting of n branches $\tau_1, \ldots, \tau_n$, with each branch having two synaptic sites, and defining $\tau_i(\mathbf{x}) = \bigwedge_{h=1}^{2}(-1)^{1-\ell}(x_i + w_{ih}^{\ell}) = (x_i - a_i) \wedge (b_i - x_i)$. Then $\tau(\mathbf{x}) = \bigvee_{i=1}^{n} \tau_i(\mathbf{x})$.

Analogous to the single-layer feed-forward ANN or single-layer perceptron with one output neuron, the single-layer LNN with one output neuron also consists of a finite number of input neurons that are connected via axonal fibers to the output neuron. However, in contrast to the single-layer ANN, the output neuron of the single-layer LNN has a dendritic structure and performs the lattice computations embodied in equations 8.1 and 8.2. In fact, the biggest differences between traditional single-layer ANNs with one output neuron and a single-layer LNN with one output neuron are given by the following theorem:

Theorem 8.1 *Suppose d is an ℓ_p metric on $\mathbb{R}^n$. If $X \subset \mathbb{R}^n$ is compact and $\epsilon > 0$, then there exists a single-layer LNN with one output neuron that assigns every point of X to class C_1 and every point $\mathbf{x} \in \mathbb{R}^n$ to class C_0 whenever $d(\mathbf{x}, X) = \inf\{d(\mathbf{x}, \mathbf{y}) : \mathbf{y} \in X\} > \epsilon$.*

It follows from Example 2.8 and Section 2.2.2 that any two different ℓ_p metrics d_p and d_q generate topologically equivalent Euclidean space $\mathbb{R}^n$ since for $\mathbf{x} \in \mathbb{R}^n$ and any basic neighborhood $N_{p,r}(\mathbf{x})$ with $r \in \mathbb{R}^+$, there exists $t \in \mathbb{R}^+$ such that $N_{q,t}(\mathbf{x}) \subset N_{p,r}(\mathbf{x})$ and vice versa (see Figure2.1). Consequently, we may use any one ℓ_p in order to prove the theorem. For our objectives it will be convenient to use the ℓ_∞ metric $d(\mathbf{x}, \mathbf{y}) = \bigvee_{i=1}^{n} |x_i - y_i|$.

Proof. It follows from the paragraph preceding equation 4.37 that given a compact set $X \subset \mathbb{R}^n$, then there exists a smallest interval $[\mathbf{v}, \mathbf{u}]$ such that $X \subset [\mathbf{v}, \mathbf{u}]$. Thus, if $X = [\mathbf{v}, \mathbf{u}]$, then the LNN model defined in equation 8.8 and endowed with the activation function given in equation 8.6 proves the theorem for this special case.

For $X \neq [\mathbf{v}, \mathbf{u}]$, let $Y = [\mathbf{v}, \mathbf{u}] \setminus X$ and let $\bar{Y}$ denote the closure of Y in $\mathbb{R}^n$. Since $\bar{Y} \subset [\mathbf{v}, \mathbf{u}]$, $\bar{Y}$ is both closed and bounded and hence compact. For each $\mathbf{y} \in \bar{Y}$ define the neighborhood $N(\mathbf{y}) = \{\mathbf{x} \in \mathbb{R}^n : d(\mathbf{x}, \mathbf{y}) < \frac{\epsilon}{2}\}$. The collection $\mathfrak{N} = \{N(\mathbf{y}) : \mathbf{y} \in \bar{Y}\}$ is an open cover for $\bar{Y}$. Since $\bar{Y}$ is compact, there exists a finite subcollection $\mathfrak{N}_m = \{N(\mathbf{y}^k) : k = 1, \ldots, m\}$ of $\mathfrak{N}$ which also covers $\bar{Y}$. There are two specific cases that can occur in relationship to $\mathfrak{N}_m$, X, and ϵ. The first case happens when $N(\mathbf{y}^k) \cap X \neq \varnothing$ $\forall k \in \mathbb{N}_m$. In this case, any point $\mathbf{x} \in N(\mathbf{y}^k) \cap X$ still satisfies equation 8.8. If, on the other hand, $\mathbf{x} \in \mathbb{R}^n$ and

$d(\mathbf{x},X) > \epsilon$, then $\mathbf{x} \notin [\mathbf{v},\mathbf{u}]$. To prove this claim, suppose that $\mathbf{x} \in [\mathbf{v},\mathbf{u}]$. Then $\mathbf{x} \in Y \subset \bar{Y}$, which means that there must exist a member $N(\mathbf{y}^k) \in \mathfrak{N}_m$ such that $\mathbf{x} \in N(\mathbf{y}^k)$. Since by our assumption $N(\mathbf{y}^k) \cap X \neq \varnothing$, there must exist a point $\mathbf{z} \in X \cap N(\mathbf{y}^k)$. Therefore $d(\mathbf{x},X) \leq d(\mathbf{x},\mathbf{z}) \leq d(\mathbf{x},\mathbf{y}^k) + d(\mathbf{y}^k,\mathbf{z}) < \frac{\epsilon}{2} + \frac{\epsilon}{2} = \epsilon$. But this contradicts the fact that $d(\mathbf{x},X) > \epsilon$ and proves that $\mathbf{x} \notin [\mathbf{v},\mathbf{u}]$. It follows that for each $\mathbf{x} \notin [\mathbf{v},\mathbf{u}]$, there exists an index $i \in \mathbb{N}_n$ such that $x_i < v_i$ or $x_i > u_i$. Hence either $x_i - v_i = x_i + w_{i1}^1 < 0$ or $-(x_i - u_i) = -(x_i + w_{i2}^0) < 0$. Thus, $\tau(\mathbf{x}) < 0$, where $\tau(\mathbf{x})$ is defined in equation 8.6 and $f(\tau(\mathbf{x})) = 0$, where f is the activation function mentioned above. This proves the theorem if case one occurs.

The second case occurs if the first case does not happen. In this case, some elements of $\mathfrak{N}_m$ do not intersect X. Let $\{N(\mathbf{y}^{k_r}) : r = 1,\ldots,R-1\}$ denote the maximal sub-collection of $\mathfrak{N}_m$ for which $N(\mathbf{y}^{k_r}) \cap X = \varnothing$. For each $i \in \mathbb{N}_n$ and each $r = \{1,\ldots,R-1\}$ define the bounds

$$v_{ir} = y_i^{k_r} - \frac{\epsilon}{2} = \inf\{p_i(\mathbf{x}) : \mathbf{x} \in N(\mathbf{y}^{k_r})\}\ ;\ u_{ir} = y_i^{k_r} + \frac{\epsilon}{2} = \sup\{p_i(\mathbf{x}) : \mathbf{x} \in N(\mathbf{y}^{k_r})\}.$$

As before, let N_i denote the input neurons and let $d_1,\ldots,d_R$ denote the dendritic branches of the output neuron M. Define the LNN model so that each N_i has two axonal branches that synapse on d_r for each $r \in \mathbb{N}_R$, and define the weights of the two synaptic sites by setting

$$w_{ir2}^0 = \begin{cases} -u_i \Leftrightarrow r = 1 \\ -u_{ir-1} \Leftrightarrow r = 2,\ldots,R \end{cases} ;\ w_{ir1}^1 = \begin{cases} -v_i \Leftrightarrow r = 1 \\ -v_{ir-1} \Leftrightarrow r = 2,\ldots,R \end{cases}$$

where v_i and u_i correspond to the ith component of $\mathbf{v}$ and $\mathbf{u}$, respectively. Finally, set

$$p_r = \begin{cases} +1 \Leftrightarrow r = 1 \\ -1 \Leftrightarrow r \in \{2,3,\ldots,R\}. \end{cases}$$

We now show that the LNN model defined by

$$\tau(\mathbf{x}) = \bigwedge_{r=1}^{R} \tau_r(\mathbf{x}) \quad \text{where} \quad \tau_1(\mathbf{x}) = \bigwedge_{i=1}^{n} \Big(\bigwedge_{h=1}^{2} (-1)^{1-\ell}(x_i + w_{i1h}^{\ell})\Big) \quad \text{and}$$

$$\tau_{r+1}(\mathbf{x}) = p_{r+1} \bigwedge_{i=1}^{n} \Big(\bigwedge_{h=1}^{2} (-1)^{1-\ell}(x_i + w_{ir+1h}^{\ell})\Big) \quad \text{for} \quad r = 1,\ldots,R-1$$

satisfies the conclusion of the theorem.

Suppose that $\mathbf{x} \in X$. Then since $N(\mathbf{y}^{k_r}) \cap X = \varnothing \Rightarrow \mathbf{x} \notin N(\mathbf{y}^{k_r})$ for $r \in \{1,\ldots,R-1\}$, it follows that for any given r there must exist an index $i \in \mathbb{N}_n$ such that either $x_i < v_{ir}$ or $u_{ir} < x_i$. Hence, either

C. $(x_i - v_{ir}) = (x_i + w_{ir+11}^1) < 0$ or $-(x_i - u_{ir}) = -(x_i + w_{ir+12}^0) < 0$.

But this implies that $\wedge_{i=1}^{n}(\wedge_{h=1}^{2}(-1)^{1-\ell}(x_i + w_{ir+1h}^{\ell})) < 0$ for $r = 1,\ldots,R-1$. Since $p_{r+1} = -1$, we have $\tau_{r+1}(\mathbf{x}) > 0$ for $r = 1,\ldots,R-1$. Also, since $X \subset [\mathbf{v},\mathbf{u}]$, it follows that $\tau_1(\mathbf{x}) \geq 0$ and, hence, $\tau(\mathbf{x}) \geq 0$. Therefore $f(\tau(\mathbf{x})) = 1\ \forall \mathbf{x} \in X$, where f denotes the hard-limiter defined in Example 8.1(1).

Now suppose that $d(\mathbf{x}, X) > \epsilon$. Then two cases may occur, namely $\mathbf{x} \notin [\mathbf{v},\mathbf{u}]$ or $\mathbf{x} \in [\mathbf{v},\mathbf{u}]$. If $\mathbf{x} \notin [\mathbf{v},\mathbf{u}]$, then—as before—there must exist an index $i \in \mathbb{N}_n$ such that the ith coordinate of $\mathbf{x}$ satisfies one of the inequality of equation C. As a result, $\tau_1(\mathbf{x}) < 0$ and $\tau(\mathbf{x}) < 0$. Therefore $f(\tau(\mathbf{x})) = 0$.

If $d(\mathbf{x}, X) > \epsilon$ and $\mathbf{x} \in [\mathbf{v},\mathbf{u}]$, then since $\mathbf{x} \notin X$, $\mathbf{x} \in Y$. But this means that $\mathbf{x} \in N(\mathbf{y}^k)$ for some $k \in \mathbb{N}_m$. If $N(\mathbf{y}^k) \cap X \neq \varnothing$, then there exists a point $\mathbf{z} \in N(\mathbf{y}^k) \cap X$ so that $d(\mathbf{x}, X) \leq d(\mathbf{x},\mathbf{z}) \leq d(\mathbf{x},\mathbf{y}^k) + d(\mathbf{y}^k,\mathbf{z}) < (\frac{\epsilon}{2}) + \frac{\epsilon}{2}) = \epsilon$, contradicting the fact that $d(\mathbf{x}, X) > \epsilon$. Therefore $N(\mathbf{y}^k) \cap X = \varnothing$ and, consequently, $k = k_r$ for some $r \in \{1,\ldots,R-1\}$. Since $\mathbf{x} \in N(\mathbf{y}^{k_r})$, $v_{ir} < x_i < u_{ir}\ \forall i \in \mathbb{N}_n$. Equivalently, $\wedge_{i=1}^{n}(\wedge_{h=1}^{2}(-1)^{1-\ell}(x_i + w_{ir+1h}^{\ell})) > 0$, but $p_{r+1} = -1$, hence, $\tau_{r+1}(\mathbf{x}) < 0$. It follows that $\tau(\mathbf{x}) < 0$ and $f(\tau(\mathbf{x})) = 0$. Therefore $f(\tau(\mathbf{x})) = 0$ whenever $d(\mathbf{x}, X) > \epsilon$. □

According to Theorem 8.1 all points of X will be classified as belonging to class C_1 and all points outside the *banded* region of thickness ϵ will be classified as belonging to class C_0 (see also Figure 8.5). Points within the banded region may be misclassified. Accordingly, any compact set, whether it is convex or non-convex, connected or not connected, contains a finite or infinite number of points, can be approximated to any desired degree of accuracy $\epsilon > 0$ by a single-layer LNN with one output neuron.

Example 8.2 Consider the interval $[\mathbf{a},\mathbf{b}] \subset \mathbb{R}^2$, where $\mathbf{a} = \binom{2}{1}$ and $\mathbf{b} = \binom{6}{1.5}$, and the quarter disk C with center $\mathbf{c} = \binom{1}{5}$ defined by $C = \{\mathbf{x} \in \mathbb{R}^2 : d(\mathbf{c},\mathbf{x}) \leq 2, 1 \leq x_1 \leq 3, 3 \leq x_2 \leq 5\}$. The set $X = C \cup [\mathbf{a},\mathbf{b}]$ is a disconnected compact set in $\mathbb{R}^2$, with $X \subset [\mathbf{v},\mathbf{u}] = [\binom{1}{1},\binom{6}{5}]$. In this case, $X \cup Y \subset [\mathbf{v},\mathbf{u}] = X \cup \bar{Y}$ as indicated in Figure 8.5. Since the parameters $\mathbf{v}$ and $\mathbf{u}$ are known, then for any $\epsilon > 0$ one can cover the interval $[\mathbf{v},\mathbf{u}]$ in a systematic way with a finite number of ℓ_∞ neighborhoods of diameter $\frac{\epsilon}{2}$ such that no two open squares intersect, but the union of the closed squares cover the interval. For instance, if $\epsilon = \frac{1}{4}$, then for $s = 1,2,\ldots,20$ and $t = 1,2,\ldots,16$ set $\mathbf{x}(s,t) = (v_1 + \frac{2s-1}{8}, v_2 + \frac{2t-1}{8})'$, $Z = \{\mathbf{x}(s,t) : s \in \mathbb{N}_{20}$ and $t \in \mathbb{N}_{16}\}$, $r = \frac{1}{8}$, and let $N_r(\mathbf{x}(s,t))$ denote the ℓ_∞ neighborhood of radius r. Then

$$[\mathbf{v},\mathbf{u}] = \bigcup_{\mathbf{x}(s,t) \in Z} \bar{N}_r(\mathbf{x}(s,t)).$$

By definition of ℓ_∞ neighborhoods we have $\bar{N}_r(\mathbf{x}(s,t)) = [\mathbf{a}(s,t),\mathbf{b}(s,t)]$,

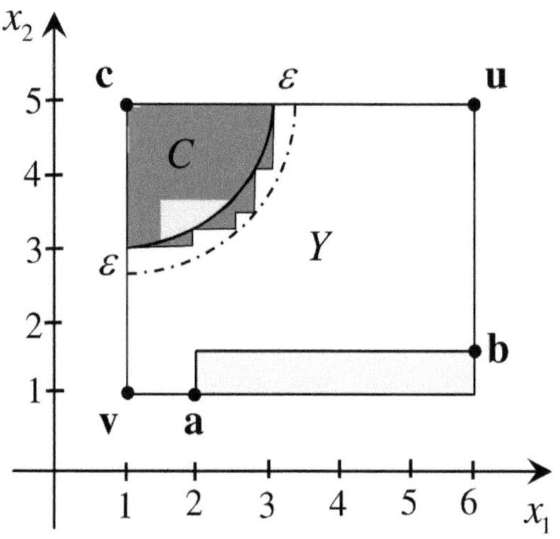

Figure 8.5 The set $X = C \cup [\mathbf{a}, \mathbf{b}]$. All points in the shaded region will be classified as belonging to X.

where $\mathbf{a}(s,t) = (v_1 + \frac{2s-1}{8} - r, v_2 + \frac{2t-1}{8} - r)' = (1 + \frac{2(s-1)}{8}, 1 + \frac{2(t-1)}{8})'$ and, similarly, $\mathbf{b}(s,t) = (1 + \frac{2s}{8}, 1 + \frac{2t}{8})'$ when adding r to the components of $\mathbf{x}(s,t)$. Setting $U = \{(s,t)) : N_r(\mathbf{x}(s,t)) \cap X \neq \varnothing\}$ and

$$\tilde{X} = \bigcup_{(s,t)\in U} \bar{N}_r(\mathbf{x}(s,t)) = \bigcup_{(s,t)\in U} [\mathbf{a}(s,t), \mathbf{b}(s,t)]$$

results in $X \subset \tilde{X} \subset [\mathbf{v}, \mathbf{u}]$ with $\tilde{X} \setminus X \subset Y$. Figure 8.5 illustrates these relationships. Also, since $\tilde{X}$ is the finite union of intervals and recalling the LNNs in Example 8.1 and equation 8.6, it becomes an easy task to construct a two-layer LNN with two input neurons and one output neuron that can decide whether or not $\mathbf{x} \in \tilde{X}\ \forall \mathbf{x} \in \mathbb{R}^2$. Any point $\mathbf{x} \in \tilde{X} \setminus X$ will be misclassified.

The set X in this example consists of two distinct compact components. In a myriad of applications, it is important to distinguish between various components by assigning them to different classes. More precisely, given a collection $X_1, \ldots, X_k$ of mutually disjoint compact subsets of $\mathbb{R}^n$, then the goal is to construct an LNN with n input neurons and k output neurons $M_1, \ldots, M_k$ that assigns each point $\mathbf{x} \in \mathbb{R}^n$ to class C_i whenever $\mathbf{x} \in X_i$. In order to prove that this goal is achievable with a single-layer LNN we need to employ the following distance function for compact subsets of $\mathbb{R}^n$:

Definition 8.1 If d is a metric on $\mathbb{R}^n$, then the *Hausdorff distance* (based on

d) between any two bounded subsets X and Y of $\mathbb{R}^n$ is denoted by $d_H(X,Y)$ and defined by

$$d_H(X,Y) = \sup\{d(\mathbf{x},Y) : \mathbf{x} \in X\} \vee \sup\{d(\mathbf{y},X) : \mathbf{y} \in Y\},$$

where $d(\mathbf{x},Y) = \inf\{d(\mathbf{x},\mathbf{y}) : \mathbf{y} \in Y\}$ and $d(\mathbf{y},X)$ are the distances from a point to a set as defined in Theorem 8.1.

It follows that $d_H(X,Y) \geq 0$ and $d_H(X,Y) = 0$ if and only if $X \cap Y \neq \varnothing$. The Hausdorff distance is key in the generalization of Theorem 8.1.

Theorem 8.2 *Suppose $X_1, X_2, \ldots, X_k$ are compact subsets of $\mathbb{R}^n$. If $X_i \cap X_j = \varnothing$ for all distinct pairs of integers $i, j \in \mathbb{N}_k$, then there exists a positive number $\delta \in \mathbb{R}$ such that for any number $\epsilon > 0$ with $\epsilon < \delta$, there exists a single-layer LNN with k output neurons that for $i \in \mathbb{N}_k$ assigns every point of X_i to class C_i and no point $\mathbf{x} \in \mathbb{R}^n$ to C_i if $d(\mathbf{x}, X_i) > \epsilon$.*

Proof. Since the proof of this theorem mimics the proof of Theorem 8.1, we only outline the proof and let the reader fill in the obvious details. In order to prove the generalization of Theorem 8.1, let $A = \{(i,j) : i,j \in \mathbb{N} \ni 1 \leq i < j \leq k\}$, $\rho = \bigvee_{(i,j)\in A} d_H(X_i, X_j)$, and $\delta = \frac{\rho}{2}$. Since $X_i \cap X_j = \varnothing \; \forall (i,j) \in A$, $\delta > 0$. The remainder of the proof follows in the steps of the proof of Theorem 8.1. To explicate, for each integer $j \in \{1, 2, \ldots, k\}$ let $[\mathbf{v}^j, \mathbf{u}^j]$ denote the smallest interval containing X_j and let $Y_j = [\mathbf{v}^j, \mathbf{u}^j] \setminus X_j$ and follow the steps in the proof of Theorem 8.1 in order to obtain the parameters

$$\tau^j(\mathbf{x}) = \bigwedge_{r=1}^{R_j} \tau_r^j(\mathbf{x}) \quad \text{where} \quad \tau_1^j(\mathbf{x}) = \bigwedge_{i=1}^{n} \Big(\bigwedge_{h=1}^{2} (-1)^{1-\ell}(x_i + w_{ij1h}^{\ell}) \Big) \quad \text{and}$$

$$\tau_{r+1}^j(\mathbf{x}) = p_{jr+1} \bigwedge_{i=1}^{n} \Big(\bigwedge_{h=1}^{2} (-1)^{1-\ell}(x_i + w_{ijr+1h}^{\ell}) \Big) \quad \text{for} \quad r = 1, \ldots, R_j - 1$$

that define the parameters of the output neuron M_j. Here the activation function f_j for M_j is the hard-limiter $f_j(\tau^j(\mathbf{x})) = 0$ whenever $d(\mathbf{x}, X_j) > \epsilon$ and $f_j(\tau^j(\mathbf{x})) = j$ if $\mathbf{x} \in X_j$. A point $\mathbf{x} \in \mathbb{R}^n$ will be said to belong to class C_0 whenever $f_1(\tau^1(\mathbf{x})) = f_2(\tau^2(\mathbf{x})) = \cdots = f_k(\tau^k(\mathbf{x})) = 0$ □.

The XOR, or *exlusive or*, problem is a classic problem in the development of ANNs. The domain of the problem is the set of binary numbers $\mathbb{Z}_2^2 = \{(0,0), (0,1), (1,0), (1,1)\}$ and the task is to construct an ANN that for $\mathbf{x} \in \mathbb{Z}_2^2$ can decide whether $\mathbf{x} \in X_1$ or $\mathbf{x} \in X_2$, where $X_1 = \{(1,0), (0,1)\}$ and

$X_0 = \{(0,0),(1,1)\}$. This simple problem served as an example that a single-layer perceptron can only separate different classes that are linearly separable [192]. This means that no supervised or machine learning algorithm can teach a single-layer perceptron to solve the XOR problem. In view of Theorem 8.2, this is not a problem for single-layer LNNs. More to the point, since $\mathbb{Z}_2^2 \subset \mathbb{R}^2$ one can simply follow the steps outlined in the proof of Theorem 8.2 in order to obtain a single-layer LNN with the sought after decision for each input $\mathbf{x} \in \mathbb{Z}_2^2$. Nevertheless, there are even more elementary methods for obtaining such LNNs.

Example 8.3 Given $\mathbf{x} = (x_1, x_2) \in \mathbb{Z}_2^2$ one needs two input neurons N_1 and N_2 whose input is $x_1 \in N_1$ and $x_2 \in N_2$, and one output neuron M such that

$$f(\tau(\mathbf{x})) = \begin{cases} 1 \Leftrightarrow \mathbf{x} \in \{(0,1),(1,0)\} \\ 0 \Leftrightarrow \mathbf{x} \in \{(0,0),(1,1)\} \end{cases}$$

It is obvious that the simple lattice algebra expression $(x_1 - x_2) \vee (x_2 - x_1)$ solves the XOR problem since for $\mathbf{x} \in X_1, (x_1 - x_2) \vee (x_2 - x_1) = 1$ and for $\mathbf{x} \in X_0$, $(x_1 - x_2) \vee (x_2 - x_1) = 0$. Knowing this fact, it easy to construct the desired LNN. According to the lattice expression, we want $\tau(\mathbf{x}) = (x_1 - x_2) \vee (x_2 - x_1)$ and the above activation function f representing the identity function. Next decompose the expression for $\tau(\mathbf{x})$ into two more basic elements $\tau_1(\mathbf{x}) = (x_1 - x_2)$ and $\tau_2(\mathbf{x}) = (x_2 - x_1)$, so that $\tau(\mathbf{x}) = \tau_1(\mathbf{x}) \vee \tau_2(\mathbf{x})$, which also implies that M must have two dendritic branches d_1 and d_2. The final task is to assign weights and synaptic sites to the dendrites. It follows from the definition of the functions τ_k for $k = 1,2$ that the weights need to be of equal strengths. That is, if $x_1 - x_2 = (-1)^{(1-\ell)}(x_1 + w_{11}^{\ell}) + (-1)^{(1-\ell)}(x_2 + w_{21}^{\ell})$ and $x_2 - x_1 = (-1)^{(1-\ell)}(x_1 + w_{12}^{\ell}) + (-1)^{(1-\ell)}(x_2 + w_{22}^{\ell})$, then $w_{1k}^{\ell} = w_{2k}^{\ell}$ with only ℓ depending on i and k. More precisely, choosing any number $r \in \mathbb{R}$ and setting $w_{ik}^{\ell} = r$, then $x_1 - x_2 = (x_1 + r) - (x_2 + r) = (x_1 + w_{11}^{1}) - (x_2 + w_{21}^{0})$ and $x_2 - x_1 = -(x_1 + w_{12}^{0}) + (x_2 + w_{22}^{1})$. Note that $\ell = 1$ if $i = k$ and $\ell = 0$ if $i \neq k$. It now follows that

$$\tau_k(\mathbf{x}) = p_k \sum_{i=1}^{2} (-1)^{1-\ell}(x_i + w_{ik}^{\ell}) \quad \text{and} \quad \tau(\mathbf{x}) = p \bigvee_{k=1}^{2} \tau_k(\mathbf{x}),$$

where $p_k = 1 = p$. This LNN corresponds to the model defined by equation 8.4. An alternative approach for generating a single-layer LNN model that solves the XOR problem is given in Example 4 of [221].

It follows from the definition of boxes (or intervals) determined by two points $\mathbf{v}, \mathbf{u} \in \mathbb{R}^n$ with $\mathbf{v} \leq \mathbf{u}$ that a point $\mathbf{v} \in \mathbb{R}^n$ can also be thought of as a 0-dimensional box $[\mathbf{v}, \mathbf{v}] = \{\mathbf{x} \in \mathbb{R}^n : v_i \leq x_i \leq v_i \text{ for } i = 1, \ldots, n\}$. Consequently, given a point $\mathbf{x} \in \mathbb{R}^n$, then

$$\mathbf{x} = \mathbf{v} \Leftrightarrow \bigwedge_{i=1}^{n} ((x_i - v_i) \wedge (v_i - x_i)) = 0. \tag{8.9}$$

Equation 8.9 helps in identifying specific points in $\mathbb{R}^n$ when using LNNs.

Example 8.4 Suppose the task is to construct an LNN that classifies an arbitrary point $\mathbf{x} \in \mathbb{R}^2$ as belonging to class C_1 if $\mathbf{x} \in X_0$, to class C_2 if $\mathbf{x} \in X_1$ and to class C_0 if $\mathbf{x} \in \mathbb{R}^2 \setminus (X_0 \cup X_1)$, where X_0 and X_1 are the sets defined in Example 8.3. Since the dimension $n = 2$ and the task is a three-class differential problem, the desired LNN requires two input neurons N_1 and N_2 and two output neurons M_1 and M_2. As before, the activation functions for the two output neurons will be a hard-limiter, namely

$$f_j(y) = \begin{cases} 1 \Leftrightarrow y \geq 0 \\ 0 \Leftrightarrow y < 0, \end{cases}$$

where $j = 1, 2$ and $y \in \mathbb{R}$. Thus, goal is to construct M_j, for $j = 1, 2$, so that $\mathbf{x}$ belongs to class $C_j \Leftrightarrow f_j(\tau^j(\mathbf{x})) = 1$.

Taking advantage of equation 8.9 we have for $\mathbf{x} \in \mathbb{R}^2$

$$\begin{aligned} 0 &\geq (x_1 \wedge -x_1) \wedge (x_2 \wedge -x_2) \\ &= [(x_1 + w^1_{111}) \wedge -(x_1 + w^0_{111})] \wedge [(x_2 + w^1_{211}) \wedge -(x_2 + w^0_{211})] \\ &= \bigwedge_{i=1}^{2} (\bigwedge_{h=1}^{2} (-1)^{1-\ell}(x_i + w^\ell_{i11h})) = \tau^1_1(\mathbf{x}), \end{aligned}$$

where $w^\ell_{i11} = r$ for $\ell = 0, 1$ and $r \in \mathbb{R}$. The number r can be chosen by the user. The weight $r = 0$ is the weight of choice as it represents direct signal transfer at the synapse and reduces computational cost. It follows from equation 8.9 that $\tau^1_1(\mathbf{x}) = 0 \Leftrightarrow \mathbf{x} = \mathbf{0}$. Similarly,

$$\begin{aligned} 0 &\geq [(x_1 - 1) \wedge (1 - x_1)] \wedge [(x_2 - 1) \wedge (1 - x_2)] \\ &= \bigwedge_{i=1}^{2} (\bigwedge_{h=1}^{2} (-1)^{1-\ell}(x_i + w^\ell_{i12h})) = \tau^1_2(\mathbf{x}), \end{aligned} \tag{8.10}$$

where $w^\ell_{i12} = -1$ for $\ell = 0, 1$ and $i = 1, 2$. Here $\tau^1_2(\mathbf{x}) = 0 \Leftrightarrow \mathbf{x} = (1, 1)$. Thus, if $\tau^1(\mathbf{x}) = \tau^1_1(\mathbf{x}) \vee \tau^1_2(\mathbf{x})$, then $\tau^1(\mathbf{x}) = 0 \Leftrightarrow \mathbf{x} \in X_0$. Finally, defining establishes the morphology and function of the output neuron M_1.

The construction of the output neuron M_2 is analogous. Here

$$0 \geq [(x_1 - 1) \wedge (1 - x_1)] \wedge [(x_2 \wedge -x_2)] = \bigwedge_{i=1}^{2} \Big(\bigwedge_{h=1}^{2} (-1)^{1-\ell}(x_i + w_{i21h}^{\ell}) \Big) = \tau_1^2(\mathbf{x}),$$

where $w_{121}^{\ell} = -1$ for $\ell = 0,1$ and $w_{221}^{\ell} = 0$ for $\ell = 0,1$. Furthermore, $\tau_1^2(\mathbf{x}) = 0 \Leftrightarrow \mathbf{x} = (1,0)$. In the same way set

$$0 \geq [(x_1 \wedge -x_1)] \wedge [(x_2 - 1) \wedge (1 - x_2)] = \bigwedge_{i=1}^{2} \Big(\bigwedge_{h=1}^{2} (-1)^{1-\ell}(x_i + w_{i22h}^{\ell}) \Big) = \tau_2^2(\mathbf{x}),$$

where $w_{122}^{\ell} = 0$ and $w_{221}^{\ell} = -1$ for $\ell = 0,1$. Here $\tau_2^2(\mathbf{x}) = 0 \Leftrightarrow \mathbf{x} = (0,1)$. Defining $\tau^2(\mathbf{x}) = \tau_1^2(\mathbf{x}) \vee \tau_2^2(\mathbf{x})$, then $\tau^2(\mathbf{x}) = 0 \Leftrightarrow \mathbf{x} \in X_1$. It now follows that $f_j(\tau^j(\mathbf{x})) = 1 \Leftrightarrow \mathbf{x} \in C_j$ for $j = 1,2$ and $\mathbf{x} \in C_0$ whenever $f_1(\tau^1(\mathbf{x})) = 0 = f_2(\tau^2(\mathbf{x}))$.

Theorems 8.1 and 8.2 have proven to be useful in the development of algorithms for constructing various LNNs for applications in pattern recognition and related tasks. Two approaches, referred to as the *elimination* and *merge* procedures, were early techniques for training single-layer LNNs [222, 224, 227]. However, a major problem encountered by these LNNs training algorithms is that many shapes cannot be modeled exactly and may require an unreasonable large number of synaptic sites for a close approximation. For instance, the triangle $T = \{\mathbf{x} \in \mathbb{R}^2 : x_1 - x_2 \geq 0 \text{ and } 0 \leq x_i \leq 2, i = 1,2\}$ shown in Figure 8.6 is bounded by only three lines, but its points cannot be classified exactly by either the elimination or merging techniques derived from the proof of Theorem 8.1. Using the elimination algorithm described in Chapter 9, then for $\epsilon = \frac{1}{2}$, the algorithm eliminates all points in the white region of the interval $[\mathbf{v}, \mathbf{u}]$, where $\mathbf{v} = (0,0)$ and $\mathbf{u} = (2,2)$. Similarly, the right-hand figure in the illustration shows the gray area obtained by merging four maximal rectangles containing only points of T, but not all points of T. The reason that these early elimination and merging methods are unable to exactly model the simple figure of a triangle is that both employ only intervals and their support hyperplanes in order to construct LNNs. In order to expand the capabilities of LNNs, it is imperative to include lines, planes, and hyperplanes whose orientations are not parallel to $E(\mathbf{e}^i)\ \forall i \in \mathbb{N}_n$.

Example 8.5

1. Let T be the triangle defined in the preceding paragraph. The vertices of the triangle determine three lines whose equations are given by $x_1 =$

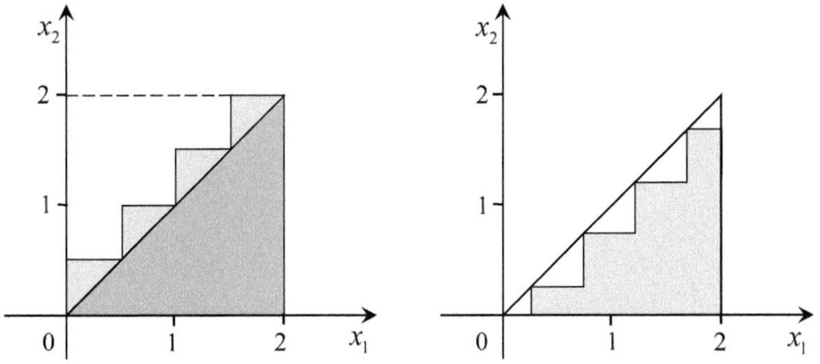

Figure 8.6 The shaded region in the left figure is generated when using elimination and the right-hand figure when using merging [236].

2, $x_2 = 0$, and $x_1 = x_2$. Thus, $\mathbf{x} \in T \Leftrightarrow x_1 \leq 2$ and $0 \leq x_2$ and $x_2 \leq x_1$. Equivalently, $0 \leq 2 - x_1$ and $0 \leq x_2$ and $0 \leq x_1 - x_2$. Therefore $0 \leq (2 - x_1) \wedge x_2 \wedge (x_1 - x_2) \Leftrightarrow \mathbf{x} \in T$. It is now easy to construct an LNN that can classify the set T exactly. The first term of the inequality can be expressed as $x_1 + w_{11}^{\ell}$, where $\ell = 0$ and $w_{11}^{0} = -2$, while the second term corresponds to $x_2 + w_{21}^{\ell}$ with $\ell = 1$ and $w_{21}^{1} = 0$. This means that there are two synaptic sites on a dendritic branch d_1. To combine the information generated by the two synapses, simply set $\tau_1(\mathbf{x}) = -(x_1 + w_{11}^{0}) \wedge (x_2 + w_{21}^{1})$.

In similar fashion, define $w_{12}^{1} = 0 = w_{22}^{0}$ and $\tau_2(\mathbf{x}) = \sum_{i=1}^{2}(x_i + w_{i2}^{\ell}) = (x_1 + w_{12}^{1}) - (x_2 + w_{22}^{0}) = (x_1 - x_2) \geq 0 \Leftrightarrow x_1 \geq x_2$. Finally, set $\tau(\mathbf{x}) = \tau_1(\mathbf{x}) \wedge \tau_2(\mathbf{x})$ and let f denote the hard-limiter activation function for the output neuron M defined in Example 8.4. Then $f(\tau(\mathbf{x})) = 1 \Leftrightarrow \tau(\mathbf{x}) \geq 0 \Leftrightarrow \mathbf{x} \in T$.

2. The classical 3-dimensional XOR problem assumes that the patterns under consideration are Boolean; i.e., the patterns are elements of $\mathbb{Z}_2^3$. For a point $\mathbf{x} = (x_1, x_2, x_3) \in \mathbb{Z}_2^3$, define its index ξ by setting $\xi = 4x_1 + 2x_2 + x_3 + 1$ so that $\mathbf{x}^1 = (0,0,0)$, $\mathbf{x}^2 = (0,0,1), \ldots, \mathbf{x}^8 = (1,1,1)$. The class pattern of $\mathbf{x}^{\xi} = (x_1^{\xi}, x_2^{\xi}, x_3^{\xi})$ for $\xi \in \{1,\ldots,8\}$ is defined by the addition $\oplus$ of its coordinates by setting $c_{\xi} = x_1^{\xi} \oplus x_2^{\xi} \oplus x_3^{\xi}$, where $\oplus$ denotes the binary addition of the group $(\mathbb{Z}_2^3, \oplus)$. Specifically, the class C_1 is given by $C_1 = \{\mathbf{x} \in \mathbb{Z}_2^3 : c_{\xi} = 1\} = \{\mathbf{x}^2, \mathbf{x}^3, \mathbf{x}^5, \mathbf{x}^8\}$, while $C_0 = \{\mathbf{x} \in \mathbb{Z}_2^3 : c_{\xi} = 0\} = \{\mathbf{x}^1, \mathbf{x}^4, \mathbf{x}^6, \mathbf{x}^7\}$. Figure 8.7(a) shows that the four elements of C_1 (black dots) corresponding to the vertices of four triangles with any two triangles sharing an edge. Each triangle is a subset of a unique

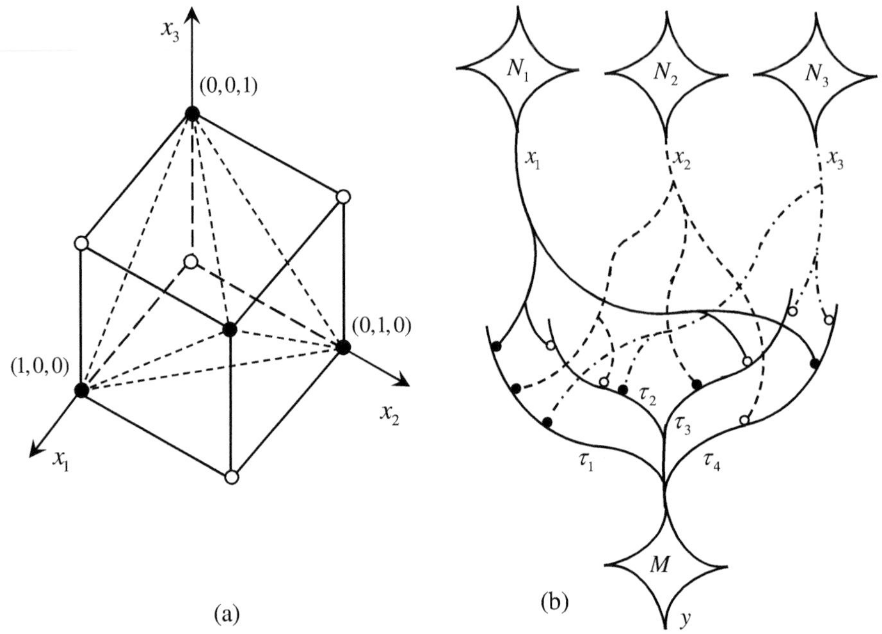

Figure 8.7 (a) The black vertices of the unit cube correspond to the elements of C_1. (b) A sketch of the LNN capable of solving the 3-dimensional XOR problem.

plane, where the planes are given by the equations

$$
\begin{aligned}
E_1(\mathbf{x}) &= \{\mathbf{x} : x_1 + x_2 + x_3 - 1 = 0\}, \\
E_2(\mathbf{x}) &= \{\mathbf{x} : -x_1 - x_2 + x_3 + 1 = 0\}, \\
E_3(\mathbf{x}) &= \{\mathbf{x} : -x_1 + x_2 - x_3 + 1 = 0\}, \\
E_4(\mathbf{x}) &= \{\mathbf{x} : x_1 - x_2 - x_3 + 1 = 0\}.
\end{aligned}
$$

The expressions $x_1 + x_2 + x_3 - 1$, $-x_1 - x_2 + x_3 + 1$, $-x_1 + x_2 - x_3 + 1$, and $x_1 - x_2 - x_3 + 1$ defining the planes provide a simple solution for solving the 3-dimensional XOR problem using a single-layer LNN with one output neuron M. Specifically, the four expressions can be rewritten as

$$
\begin{aligned}
&\tau_1(\mathbf{x}) = (x_1 + w_{11}^1) + (x_2 + w_{21}^1) + (x_3 + w_{31}^1);\ w_{11}^1 = 0 = w_{21}^1, w_{31}^1 = -1 \\
&\tau_2(\mathbf{x}) = -(x_1 + w_{12}^0) - (x_2 + w_{22}^0) + (x_3 + w_{32}^1);\ w_{12}^0 = 0 = w_{22}^0, w_{32}^1 = 1 \\
&\tau_3(\mathbf{x}) = -(x_1 + w_{13}^0) + (x_2 + w_{23}^1) - (x_3 + w_{33}^0);\ w_{13}^0 = 0 = w_{23}^1, w_{33}^0 = -1 \\
&\tau_4(\mathbf{x}) = (x_1 + w_{14}^1) - (x_2 + w_{24}^0) - (x_3 + w_{34}^0);\ w_{14}^1 = 0 = w_{24}^0, w_{34}^0 = -1.
\end{aligned}
$$

Thus, $\tau_k(\mathbf{x}) = \sum_{i=1}^{3}(-1)^{1-\ell}(x_i + w_{ik}^{\ell})$ for $k = 1,2,3,4$ with each branch having three synaptic sites. Finally, defining $\tau(\mathbf{x}) = \bigwedge_{k=1}^{4} \tau_k(\mathbf{x})$ results in $\tau(\mathbf{x}) = -1 \Leftrightarrow \mathbf{x} \in C_0$ and $\tau(\mathbf{x}) = 0 \Leftrightarrow \mathbf{x} \in C_1$. A graphical representation of this network is shown in Figure 8.7(b).

Note that the equation associated with $\tau_1(\mathbf{x})$ in Example 8.5(2) could just as well have been expressed as $\tau_1(\mathbf{x}) = (x_1 - \frac{1}{3}) + (x_2 - \frac{1}{3}) + (x_3 - \frac{1}{3})$ so that $\tau_1(\mathbf{x}) = \sum_{i=1}^{3}(x_i + w_{i1}^{1})$, where $w_{i1}^{1} = -\frac{1}{3}$. Similar distributions of fractions of the number 1 for the replacement of zero weights at synapses on $\tau_k(\mathbf{x})$ for $k = 2,3,4$ can be easily established. This avoids the zero weights, but increases the number of additions.

Another meaningful observation is the absence of spike trains in the examples of this section. More to the point, all input neurons N_i transfer the inputs x_i directly to the dendrites of the presynaptic neurons. For instance, the two inputs in x_1 and x_2 in Example 8.1(3) transfer to the two respective dendritic branches τ_1 and τ_2 of the output neuron M, with x_2 being transferred to three separate synaptic sites in order to express the polynomial $x_1 - 3x_2 + 4$. A simpler method would be to have N_2 send a spike train directly to one synapse using the expression $(11)x_2$, where the spike train represents the binary number $11_2(=3_{10})$ generated by the AIS of N_2 in response to the input x_2. Consequently, only one synapse on τ_2 with weight $w_{22}^{0} = -4$ suffices.

Exercises 8.2.1

1. Let $I_1 = [(0,-2),(2,3]]$, $I_2 = [(-2,0),(0,2)]$, $I_3 = [(-3,-2),(2,0)]$ and $I_4 = [(-1,-1),(1,1)]$ denote three intervals and let $X = (I_1 \cup I_2 \cup I_3) \setminus I_4$. Construct a biomimetic neural network with a single output neuron with output $f(\tau(\mathbf{x})) = 0$ if $\mathbf{x} \in \mathbb{R}^2 \setminus X$ and $f(\tau(\mathbf{x})) = 1 \, \forall \mathbf{x} \in X$.

2. Suppose that $X = \{\mathbf{x} \in \mathbb{R}^2 : x_1^2 + x_2^2 \leq 1$ and $0 \leq x_i$ for $i = 1,2\}$ and $\epsilon = 0.3$. Construct a single-layer LNN with one output neuron that assigns every point of X to class C_1 and every point $\mathbf{x} \in \mathbb{R}^2$ to class C_0 whenever $d(\mathbf{x},X) = \inf\{d(\mathbf{x},\mathbf{y}) : \mathbf{y} \in X\} > \epsilon$.

CHAPTER 9

Learning in Biomimetic Neural Networks

LEARNING rules for ANNs are dependent on specific data environments (inputs) and on the problem to be solved (the desired outputs); e.g. cluster detection, face recognition, etc. The rules for learning are expressed in terms of algorithms based on mathematical logic. The goal of these algorithms is to improve the performance and/or the training time of the network. The ANN is constructed before training begins. The number of input neurons (input nodes) are known, as they are data-dependent. The number of output neurons are known as well since they are derived from the problem to be solved. However, the number of hidden layers as well as the number of neurons in the different hidden layers needs to be approximated. Deep learning requires $k \geq 2$ hidden layers where k is either approximated using optimization methods or established by trial and error. The same holds for the number of nodes (neurons) in each of the hidden layers. This chapter provides several different types of LBNNs and learning methods for these networks. Since LBNNs are significantly different from the ANNs currently in vogue, it is not surprising that learning algorithms for LBNNs are drastically different from ANN learning methods.

9.1 LEARNING IN SINGLE-LAYER LBNNS

Initial learning rules were designed for single-layer LBNNs with one or more output neurons in order to solve one-class and multi-class problems. These rules were based on the proofs of Theorems 8.1 and 8.2. During the learning phase the output neurons grow new dendrites and synaptic sites, while the input neurons expand their axonal branches to terminate on the new dendritic

 DOI: 10.1201/9781003154242-9

synaptic sites. The algorithms always converge and have rapid convergence rate when compared to backpropagation learning in traditional perceptrons. Training can be realized in one of two main strategies, which differ in the way the separation surfaces in pattern space are determined. One strategy is based on elimination, whereas the other is based on merging. In the former approach, a hyperbox is initially constructed large enough to enclose all patterns belonging to the same class, possibly including foreign patterns from other classes. This large region is then carved to eliminate the foreign patterns. Training completes when all foreign patterns in the training set have been eliminated. The elimination is performed by computing the intersection of the regions recognized by the dendrites, as expressed by the equation $\tau^j(\mathbf{x}) = p_j \bigwedge_{k=1}^{K_j} \tau_k^j(x)$ for some neuron M_j.

The latter approach starts by creating small hyperboxes around individual patterns or small groups of patterns all belonging to the same class. Isolated boxes that are identified as being close according to a distance measure are then merged into larger regions that avoid including patterns from other classes. Training is completed after merging the hyperboxes for all patterns of the same class. The merging is performed by computing the union of the regions recognized by the dendrites. Thus, the total net value received by output neuron M_j is computed as in equation 8.4, namely $\tau^j(\mathbf{x}) = p_j \bigvee_{k=1}^{K_j} \tau_k^j(x)$.

The two strategies are equivalent in the sense that they are based on the same mathematical framework and they both result in closed separation surfaces around patterns. The equivalence follows from the equation $\bigvee_{k=1}^{K} a_k = -\bigwedge_{k=1}^{K}(-a_k)$ as proven in [227]. The major difference between the two approaches is in the shape of the separation surface that encloses the patterns of a class, and in the number of dendrites that are grown during training to recognize the region delimited by that separation surface. Since the elimination strategy involves removal of pieces from an originally large hyperbox, the resulting region is bigger than the one obtained with the merging strategy. The former approach is thus more general, while the latter is more specialized. This observation can guide the choice of the method for solving a particular problem.

Figure 9.1 illustrates two possible partitions of the pattern space $\mathbb{R}^2$ in terms of intersection (a) and, respectively, union (b), in order to recognize the solid circles (•) as one class C_1. In part (a), the C_1 region is determined as the intersection of three regions, each identified by a corresponding dendrite. The rectangular region marked D_1 is intersected with the complement of the regions marked D_2 and D_3 so that $C_1 = D_1 \setminus (D_2 \cup D_3)$. This means that we need three dendrites, one for each region plus one output neuron M

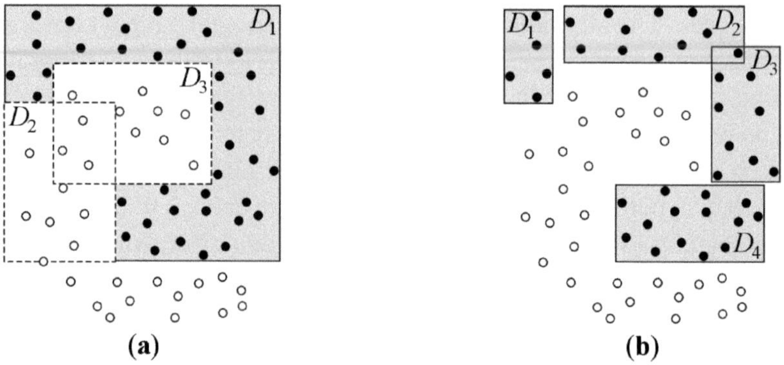

Figure 9.1 Two partitions of the pattern space $\mathbb{R}^2$ in terms of intersection (a) and union (b), respectively. The solid circles (•) belong to class C_1, which is recognized as the shaded area. Solid and dashed lines enclose regions learned by excitatory and, respectively, inhibitory dendrites [225].

for differentiating class $C_!$ from $\neg C_1$. In order to recognize all the solid circles as belonging to class C_1, we need one dendrite with excitatory output to recognizes the interior of an enclosed region and the other two dendrites with inhibitory output to recognizes the exterior of a delimited region. Since there is only one output neuron, we can simplify the expressions $\tau^j(\mathbf{x})$ and $\tau_k^j(\mathbf{x})$ to $\tau(\mathbf{x})$ and $\tau_k(\mathbf{x})$, respectively. In this example, the dendrite τ_1 will provide the excitatory output, while the dendrites τ_2 and τ_3 provide the inhibitory outputs. More precisely, using equation 8.5 to compute $\tau_k(\mathbf{x})$ results in $\tau_k(\mathbf{x}) = \bigwedge_{i=1}^{2}(-1)^{1-\ell}(x_i + w_{ik}^{\ell})$, where for $k = 1$, $\ell = 1$, while for $k = 2,3$, $\ell = 0$. Finally, using the equation $\tau(\mathbf{x}) = p \bigwedge_{k=1}^{3} \tau_k(x)$ with $p = 1$ and the hard-limiter f one obtains the output value $y_1 = f[\tau(\mathbf{x})] = f[\bigwedge_{k=1}^{3} \tau_k(x)]$ of M.

In part (b) the C_1 region is determined as the union of four regions $D_1, \ldots, D_4$, each identified by a corresponding dendritic branch $\tau_1, \ldots, \tau_4$. This time, all the dendritic branches have excitatory synapses. Thus, $\ell = 1$ for each synapse and $p_k = 1$ for $k = 1, \ldots, 4$. Again, employing eqn. 8.5 one obtains the output value $y = f[\tau(\mathbf{x})] = f[\bigvee_{k=1}^{4} \tau_k(\mathbf{x})]$. As noted above, the output value can be equivalently computed with minimum instead of maximum as $y = f[\tau(\mathbf{x})] = f[p^* \bigwedge_{k=1}^{4} p_k^* \tau_k(\mathbf{x})]$ by conjugating the responses $p^* = -p$ and $p_k^* = -p_k$.

9.1.1 Training Based on Elimination

A training algorithm that employs elimination is given below. In the training or learning set $T = \{(\mathbf{x}^\xi, c_\xi) : \xi = 1,\ldots,m\}$, the vector $\mathbf{x}^\xi = (x_1^\xi,\ldots,x_n^\xi) \in \mathbb{R}^n$ denotes an exemplar pattern and $c_\xi \in \{0,1\}$ stands for the corresponding class number in a *one-class* problem, i.e., $c_\xi = 1$ if $\mathbf{x}^\xi \in C_1$, and $c_\xi = 0$ if $\mathbf{x}^\xi \in C_0 = \neg C_1$. Numbered steps are prefixed by S and comments are provided within brackets.

Algorithm 9.1 (Single-Layer LBNN training by elimination)

S1. ***Let*** $k = 1$, $\Xi = \{1,\ldots,m\}$, $I = \{1,\ldots,n\}$, $L = \{0,1\}$; ***for*** $i \in I$ ***do***

$$w_{ik}^1 = -\bigwedge_{c_\xi=1} x_i^\xi \; ; \; w_{ik}^0 = -\bigvee_{c_\xi=1} x_i^\xi$$

[*Initialize parameters and auxiliary index sets; set weights for first dendrite (hyperbox enclosing class C_1); k is the dendrite counter.*]

S2. ***Let*** $p_k = (-1)^{sgn(k-1)}$; ***for*** $i \in I$, $\ell \in L$ ***do*** $r_{ik}^\ell = (-1)^{(1-\ell)}$; ***for*** $\xi \in \Xi$ ***do***

$$\tau_k(\mathbf{x}^\xi) = p_k \bigwedge_{i\in I} \bigwedge_{\ell \in L} r_{ik}^\ell (x_i^\xi + w_{ik}^\ell)$$

[*Compute response of current dendrite; sgn(x) is the signum function.*]

S3. ***For*** $\xi \in \Xi$ ***do*** $\tau(\mathbf{x}^\xi) = \bigwedge_{j=1}^k \tau_j(\mathbf{x}^\xi)$
[*Compute total response of output neuron M.*]

S4. ***If*** $f(\tau(\mathbf{x}^\xi)) = c_\xi \; \forall \xi \in \Xi$ ***then*** *let* $K = k$; ***for*** $k = 1,\ldots,K$ *and* ***for*** $i \in I, \ell \in L$ ***do*** *print network parameters:* k, w_{ik}^ℓ, r_{ik}^ℓ, p_k STOP
[*If with k generated dendrites learning is successful, output final weights and input-output synaptic responses of the completed network; K denotes the final number of dendrites grown in neuron M upon convergence (the algorithm ends here). If not, training continues by growing additional dendrites (next step).*]

S5. ***Let*** $k = k+1$, $I = I' = X = E = H = \varnothing$, *and* $D = C_1$
[*Add a new dendrite to M and initialize several auxiliary index sets; initially, set D gets all class 1 patterns.*]

S6. ***Let*** *choice*$(\mathbf{x}^\gamma) \in C_0$ *such that* $f(\tau(\mathbf{x}^\gamma)) = 1$
[*Select a misclassified pattern from class C_0; γ is the index of the misclassified pattern and function choice(x) is a random or sort selection mechanism.*]

S7. ***Let***

$$\mu = \bigwedge_{\xi \neq \gamma} \{ \bigvee_{i=1}^{n} |x_i^\gamma - x_i^\xi| : \mathbf{x}^\xi \in D \}$$

[*Compute the minimum Chebyshev distance from the selected misclassified pattern* $\mathbf{x}^\gamma$ *to all patterns in set D; since D is updated in step S12,* μ *changes during the generation of new terminals on the same dendrite.*]

S8. ***Let*** $I' = \{i : |x_i^\gamma - x_i^\xi| = \mu, \mathbf{x}^\xi \in D, \xi \neq \gamma\}$ *and*

$$X = \{(i, x_i^\xi) : |x_i^\gamma - x_i^\xi| = \mu, \mathbf{x}^\xi \in D, \xi \neq \gamma\}$$

[*Keep indices and coordinates of patterns in set D that are on the border of the hyperbox centered at* $\mathbf{x}^\gamma$.]

S9. ***For*** $(i, x_i^\xi) \in X$, ***if*** $x_i^\xi < x_i^\gamma$ ***then*** $w_{ik}^1 = -x_i^\xi$, $E = \{1\}$ *and*

if $x_i^\xi > x_i^\gamma$ ***then*** $w_{ik}^0 = -x_i^\xi$, $H = \{0\}$

[*Assign weights and input values for new axonal fibers in the current dendrite that provide correct classification for the misclassified pattern.*]

S10. ***Let*** $I = I \cup I'$; $L = E \cup H$

[*Update index sets I and L with only those input neurons and fibers needed to classify* $\mathbf{x}^\gamma$ *correctly.*]

S11. ***Let*** $D' = \{\mathbf{x}^\xi \in D : \forall i \in I, -w_{ik}^1 < x_i^\xi \text{ and } x_i^\xi < -w_{ik}^0\}$

[*Keep* C_1 *exemplars* $\mathbf{x}^\xi$ $(\xi \neq \gamma)$ *that do not belong to the recently created region in step S9; auxiliary set D′ is used for updating set D that is reduced during the creation of new possible axonal fibers on the current dendrite.*]

S12. ***If*** $D' = \varnothing$ ***then*** *return to S2* ***else*** *let* $D = D'$ *and loop to step S7*

[*Check if there is no need to wire more axonal fibers to the current dendrite. Going back to S2 means that the current dendrite is done, no more terminal fibers need to be wired in it but the neuron response is computed again to see if learning has been achieved; returning to S7 means that the current dendrite needs more fibers.*]

To illustrate the results of an implementation of the training algorithm based on elimination, we employed a data set from [310], where it was used to test a simulation of a *radial basis function network* (RBFN). The data set consists of two nonlinearly separable classes of 10 patterns each, where the class of interest, C_1, comprises the patterns depicted with solid circles (•). All patterns were used for both training and testing. Figure 9.2 compares the class C_1 regions learned by a single-layer LBNN with dendritic structures using the elimination-based algorithm (a) and, respectively, by a backpropagation multi-layer perceptron (b).

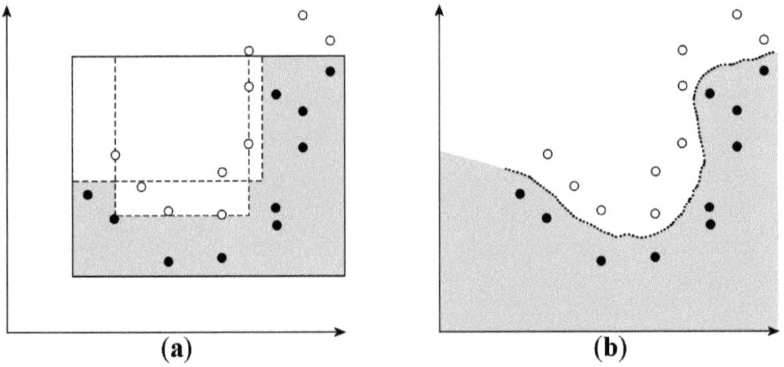

Figure 9.2 The closed class C_1 region (shaded) learned by a single-layer LBNN with dendritic structures using the elimination algorithm (a), in comparison to the open region learned by an MLP (b), both applied to the data set from [310]. During training, the LBNN grows only 3 dendrites, one excitatory and two inhibitory (dashed). Compare (a) to the output in Figure 9.3 of the merging version of the single-layer LBNN training algorithm [225].

The first step of the algorithm creates the first dendrite, which sends an excitatory message to the cell body of the output neuron if and only if a point of $\mathbb{R}^2$ is in the rectangle (solid lines) shown in Figure 9.2(a). This rectangle encloses the entire training set of points belonging to class C_1. Subsequent steps of the algorithm create two more dendrites having inhibitory responses. These dendrites will inhibit responses to points in the carved out region of the rectangle as indicated by the dashed lines in Figure 9.2(a). The only "visible" region for the output neuron will now be the dark shaded area of Figure 9.2(a).

The three dendrites grown in a single epoch during training of the single-layer LBNN are sufficient to partition the pattern space. In contrast, the multi-layer perceptron created the open surface in Figure 9.2(b) using 13 hidden units and 2000 epochs. The RBFN (Radial Basis Function NN) also required 13 basis functions in its hidden layer [310]. The separation surfaces drawn by the single-layer LBNN are closed, which is not the case for the multi-layer perceptron, and classification is 100% correct, as guaranteed by the Theorems 8.1 and 8.2. Additional comments and examples related to Algorithm 9.1 are discussed in [221].

Exercises 9.1.1

1. Use training based on elimination in order to solve the following prob-

lem. Suppose $A = \{\mathbf{x} \in \mathbb{R}^n : x_1^2 + x_2^2 = 1\}$ and $B = \{\mathbf{x} \in \mathbb{R}^n : x_1^2 + x_2^2 = 1\} = \frac{3}{4}$. The goal is to separate A from B, where every element $\mathbf{x} \in A$ belongs to class $c_\xi = 1$ and every element $\mathbf{x} \in B$ belongs to class $c_\xi = 0$. Making use of Algorithm 9.1, randomly choose m members from $X = A \cup B$ in order to obtain the training set $T = \{(\mathbf{x}^\xi, c_\xi) : \xi = 1, \ldots, m\}$. What is the smallest integer m that will separate A from B when using the elimination approach? Sketch the final single-layer neural network that separates A from B.

9.1.2 Training Based on Merging

A training algorithm based on merging for a BLNN is given below. The algorithm constructs and trains an BLNN with dendritic structures to recognize the patterns belonging to the class of interest C_1. The remaining patterns in the training set are labeled as belonging to class $C_0 = \neg C_1$. As in the preceding section, the input to the algorithm consists of a set of training patterns $T = \{(\mathbf{x}^\xi, c_\xi) : \xi = 1, \ldots, m\}$ with associated desired outputs y^ξ with $y^\xi = 1$ if and only if $\mathbf{x}^\xi \in C_1$, and $y^\xi = 0$ if $\mathbf{x}^\xi$ is not in class C_1. In the form provided below, the algorithm will construct small hyperboxes about training patterns belonging to C_1 and then merge these hyperboxes that are close to each other in the pattern space. For this reason, the number of dendrites of the output neuron M created by the algorithm will exceed the cardinality of the number of training patterns belonging to class C_1. Besides having one dendrite per pattern in class C_1, M may also have dendrites for some unique pattern pairs of patterns in C_1.

Algorithm 9.2 (Single-layer LBNN training by merging)

S1. ***Let*** *$k = 0$ and set all n-dimensional pattern pairs $\{\mathbf{x}^\xi, \mathbf{x}^\nu\}$ from C_1 as unmarked;* ***let***

$$d_{min} = \bigwedge_{\xi \neq \gamma} d(\mathbf{x}^\xi, \mathbf{x}^\gamma) = \bigwedge_{\xi \neq \gamma} \{\bigvee_{i=1}^{n} |x_i^\xi - x_i^\gamma| : \mathbf{x}^\xi \in C_1, \mathbf{x}^\gamma \in C_0\};$$

let *$\varepsilon = 0.5d_{min}$;* ***for*** *$i = 1, \ldots, n$ choose $\alpha_i^1, \alpha_i^0 \in (0, \varepsilon)$*

[k is a dendrite counter, the α's are $2n$ accuracy factors where d_{min} is the minimal Chebyshev interset distance between classes C_1 and C_0; ε is a tolerance parameter for merging two hyperboxes.]

S2. ***For*** *$\mathbf{x}^\xi \in C_1$* ***do*** *step S3*

S3. ***Let*** $k = k+1$; ***for*** $i = 1,\ldots,n$ ***do*** $w_{ik}^1 = -(x_i^\xi - \alpha_i^1)$, $w_{ik}^0 = -(x_i^\xi + \alpha_i^0)$
[Create a dendrite on the neural body of the output neuron M and for each input neuron N_i grow two axonal fibers on the kth dendrite by defining their weights.]

S4. ***For*** $\mathbf{x}^\xi \in C_1$ ***do*** *steps S5 to S10 [Outer loop.]*

S5. ***Mark*** *each unmarked pattern pair* $\{\mathbf{x}^\xi, \mathbf{x}^\nu\}$, *where* $\mathbf{x}^\xi$ *is fixed by step S4 and* ***do*** *steps S6 to S10*
[The pattern pair can be chosen in the order $\mathbf{x}^\nu$ *appears in the training set or at random.]*

S6. ***If*** $d(\mathbf{x}^\xi, \mathbf{x}^\nu) < d_{min} + \varepsilon$, ***do*** *steps S7 to S10* ***else*** *loop to step S5*

S7. ***Identify*** *a region* $\mathcal{R}$ *in pattern space that connects patterns* $\mathbf{x}^\xi$ *and* $\mathbf{x}^\nu$;
for $i = 1,\ldots,n$ *define a set of merging parameters* ε_i
such that $0 < \varepsilon_i \leq d_{min} + \varepsilon$
[This ensures a user defined merging size for each dimension. If $|x_i^\xi - x_i^\nu| \geq \varepsilon_i$, *region* $\mathcal{R}$ *will be bounded by the hyperplanes* x_i^ξ *and* x_i^ν *or by the hyperplanes* $0.5(x_i^\xi + x_i^\nu - \varepsilon_i)$ *and* $0.5(x_i^\xi + x_i^\nu + \varepsilon_i)$ *otherwise.]*

S8. ***If*** $\mathbf{x}^\gamma \in C_0$ *and* $\mathbf{x}^\gamma \notin \mathcal{R}$ ***then*** *do step S9* ***else*** *do step S10*
[$\mathcal{R}$ is the merged region determined in step S7.]

S9. ***Let*** $k = k+1$; ***for*** $i = 1,\ldots,n$ ***do***
if $|x_i^\xi - x_i^\gamma| \geq \varepsilon_i$
then $w_{ik}^1 = -\min\{x_i^\xi, x_i^\gamma\}$, $w_{ik}^0 = -\max\{x_i^\xi, x_i^\gamma\}$,
else $w_{ik}^1 = -0.5(x_i^\xi + x_i^\gamma - \varepsilon_i)$, $w_{ik}^0 = -0.5(x_i^\xi + x_i^\gamma + \varepsilon_i)$
[Create a new dendrite and axonal branches that recognize the new region.]

S10. ***If*** *there are unmarked pattern pairs remaining, loop to step S5;* ***if*** *all class* C_1 *pattern pairs* $\{\mathbf{x}^\xi, \mathbf{x}^\nu\}$ *with the fixed* $\mathbf{x}^\xi$ *have been marked,* ***then*** *loop to step S4; when outer loop exits,* ***let*** $K = k$ *and* STOP

According to this algorithm, the output neuron M will be endowed with K dendrites. If $|C_1|$ denotes the number of training patterns belonging to class C_1, then $K - |C_1|$ corresponds to the number of dendrites that recognize regions that connect a pattern pair, while $|C_1|$ of the dendrites recognize regions around individual class one patterns. We need to point out that this training algorithm as well as the previously mentioned algorithm starts with the creation of hyperboxes enclosing training points. In this sense, there is some similarity between this algorithms and those established in the fuzzy min-max neural

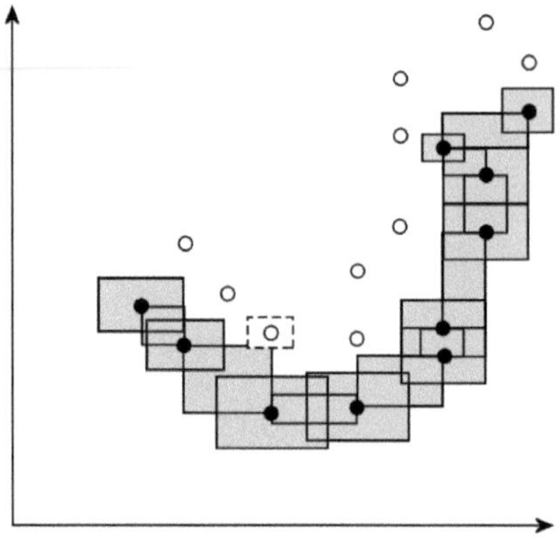

Figure 9.3 The class C_1 region (shaded) learned by a single-layer LBNN with dendritic structures using the merging-based algorithm, applied to the data set from [310]. During training, the LBNN grows 20 dendrites, 19 excitatory and 1 inhibitory (dashed). Compare to the results in Figure 9.2(a) obtained with the elimination version of the algorithm [225].

networks approach, which also uses hyperboxes [259, 260]. However, this is also where the similarity ends as all subsequent steps are completely different since the approach presented here does not rely on fuzzy set theory.

Figure 9.3 illustrates the results of an implementation of the single-layer LBNN training algorithm based on merging, applied to the same data set as in Figure. 9.2. Again, all patterns were used for both training and test. During training 19 excitatory dendrites are grown, 10 for regions around each pattern from class C_1 and 9 more to merge the individual regions. The separation surface is closed and recognition is 100% correct, as expected.

There is one more region in Figure 9.3, drawn in dashed line, corresponding to an inhibitory dendrite, and its presence is explained as follows. The merging algorithm outlined above creates regions that are sized to avoid touching patterns belonging to class C_0 or approaching them closer than a certain distance. In a more general version, larger hyper-boxes are allowed to be constructed. In this case, an additional step would identify foreign patterns that are approached or touched, and create inhibitory dendrites to eliminate those patterns. This more general approach is being used in the experiment of Figure 9.3 and explains the presence of the inhibitory dendrite whose region is depicted with dashed line.

Exercises 9.1.2

1. Use training based on merging in order to solve the same problem posed in Exercises 9.1.1.

9.1.3 Training for Multi-Class Recognition

For better clarity of description, the training algorithms described thus far were limited to a single non-zero class, which corresponds to a single output neuron. This subsection presents a straightforward generalization to multiple classes by invoking either one of the two procedures (elimination or merging) as a subroutine.

The generalized algorithm consists of a main loop that is iterated m times, where m represents the number of non-zero classes as well as the number of output neurons. Within the loop, the single-class procedure is invoked. Thus, one output neuron at a time is created and trained to classify the patterns belonging to its corresponding class. The algorithm proceeds as follows.

Algorithm 9.3 (Multi-class Single-layer LBNN training)

S1. ***For*** $j = 1, \ldots, m$ ***do*** *steps S2 through S4*
[*j is a non-zero class index.*]

S2. ***Generate*** *the output neuron* M_j
[*Create a new output neuron for each non-zero class.*]

S3. ***For*** *each pattern* $\mathbf{x}^{\xi}$ *of the training set* ***do***
    If $\mathbf{x}^{\xi}$ *is labeled as belonging to class* C_j
        then *temporarily reassign* $\mathbf{x}^{\xi}$ *as belonging to* C_1
        else *temporarily reassign* $\mathbf{x}^{\xi}$ *to class* C_0
[*The assignment is for this iteration only; original pattern labels are needed in subsequent iterations.*]

S4. ***Call*** *a single-class procedure to train output neuron* M_j *on the training set modified to contain patterns of only one non-zero class*
[*Algorithm 9.1 (elimination) or 9.2 (merging) can be used here.*]

Single and multi-class training sets used in applications are finite and mutually disjoint subsets of Euclidean space. Consequently, we have that if ρ denotes the minimal Hausdorff distance (based on the d_∞ metric) between the training sets, then $\rho > 0$. In accordance with Theorem 8.2, the tolerance factor ϵ guarantees the existence of a single-layer LBNN with multiple output neurons that will correctly identify each training class. However, a training set T_j is generally only a subset of C_j. Specifically, a training set $T_j \subset C_j$

may consist of only 60% of randomly chosen elements of C_j. It is therefore probable that the LBNN defined in terms of a training set will misclassify data points that are elements of $C_j \setminus T_j$.

A way of modifying the algorithm based on training sets is to draw regions of controlled size - based on minimum interset distance - in such a way that two regions that belong to different classes cannot touch. A similar idea was mentioned in the training algorithm based on merging. An alternative method is to take into account current information during training about the shape of the regions learned so far, and using this information when growing new dendrites and assigning synaptic weights. The algorithm discussed next provides more refined boundary shapes of regions and can be appended to the algorithms discussed thus far.

Exercises 9.1.3

1. Suppose that,

$$A = \{\begin{pmatrix}2.5\\3\end{pmatrix}, \begin{pmatrix}3\\3.5\end{pmatrix}, \begin{pmatrix}3.5\\3.5\end{pmatrix}, \begin{pmatrix}4\\4\end{pmatrix}, \begin{pmatrix}4.5\\4\end{pmatrix}, \begin{pmatrix}4.5\\4\end{pmatrix}, \begin{pmatrix}5\\3.5\end{pmatrix}, \begin{pmatrix}5.5\\2.5\end{pmatrix},$$
$$\begin{pmatrix}5.5\\2\end{pmatrix}, \begin{pmatrix}5.5\\1.5\end{pmatrix}, \begin{pmatrix}5.5\\1\end{pmatrix}, \begin{pmatrix}4.5\\1.5\end{pmatrix}\},$$
$$B = \{\begin{pmatrix}2.5\\1.5\end{pmatrix}, \begin{pmatrix}3\\2\end{pmatrix}, \begin{pmatrix}3.5\\2.5\end{pmatrix}, \begin{pmatrix}4\\3\end{pmatrix}, \begin{pmatrix}4.5\\3\end{pmatrix}, \begin{pmatrix}5\\2.5\end{pmatrix}, \begin{pmatrix}5\\2\end{pmatrix}, \begin{pmatrix}5\\1.5\end{pmatrix}\},$$
$$C = \{\begin{pmatrix}3\\3\end{pmatrix}, \begin{pmatrix}2.5\\2.5\end{pmatrix}, \begin{pmatrix}2\\2\end{pmatrix}, \begin{pmatrix}1.5\\1.5\end{pmatrix}, \begin{pmatrix}2\\1\end{pmatrix}, \begin{pmatrix}3\\1\end{pmatrix}, \begin{pmatrix}3.5\\1.5\end{pmatrix}, \begin{pmatrix}4\\2\end{pmatrix}\},$$

where A, B and C belong to class C_1, C_2, and C_3, respectively. Let $X = A \cup B \cup C$ and use Algorithm **9.3** in order to obtain a single-layer neural network that separate these three classes. Draw a sketch of the network.

9.1.4 Training Based on Dual Lattice Metrics

Algorithms 9.1 and 9.2 were respectively based on the removal of intervals containing points not belonging to a desired class or joining intervals containing only members of the desired class. These intervals where computed using the Chebyshev distance $d_\infty(\mathbf{x}, \mathbf{y})$ and represent a generalization of the closed N_∞ neighborhood discussed in Section 2.2.3. As noted in Section 8.2.1, it is imperative to include lines, planes, and hyperplanes whose orientations are not parallel to $E(\mathbf{e}^i) \forall i \in \mathbb{N}_n$. An interval is a special type of polytope, as any given interval $[\mathbf{a}, \mathbf{b}] \subset \mathbb{R}^n$ can also be viewed as the intersection of the half-spaces $\bigcap_{i=1}^{n}[\overline{E_{\mathbf{a}}^{+}}(\mathbf{e}^i) \cap \overline{E_{\mathbf{b}}^{-}}(\mathbf{e}^i)]$. Using the dot product, we also have that $\mathbf{x} \in \overline{E_{\mathbf{a}}^{+}}(\mathbf{e}^i) \cap \overline{E_{\mathbf{b}}^{-}}(\mathbf{e}^i) \Leftrightarrow E_{\mathbf{x}}(\mathbf{e}^i) \subset \overline{E_{\mathbf{a}}^{+}}(\mathbf{e}^i) \cap \overline{E_{\mathbf{b}}^{-}}(\mathbf{e}^i)$ since $\mathbf{e}^i \cdot \mathbf{a} = a_i \leq x_i = \mathbf{e}^i \cdot \mathbf{x}$ and $\mathbf{e}^i \cdot \mathbf{x} = x_i \leq b_i = \mathbf{e}^i \cdot \mathbf{b}\ \forall i \in \mathbb{N}_n$. Another type of polytope, known as the *cross polytope* or *hyperoctahedron* in dimension $n > 3$, represents the generalization of the closed N_1 neighborhood. In $\mathbb{R}^2$, they correspond to the intersections of the half-planes generated by two types of lines. Explicitly, let $E_1(\mathbf{x}) = x_1 + x_2$ and $E_2(\mathbf{x}) = x_1 - x_2$ and suppose that for $i = 1, 2$, $a_i, b_i \in \mathbb{R}$. If $a_i \leq b_i$, then the cross polytope in $\mathbb{R}^2$ is defined by

$$P^2 = \{\mathbf{x} \in \mathbb{R}^2 : a_i \leq E_i(\mathbf{x}) \leq b_i, i = 1,2\}. \tag{9.1}$$

Note that if $a_i = -1$ and $b_i = 1$, then $P^2 = \overline{N_{1,1}(\mathbf{0})}$.

If the dimension $n = 3$, then– in a likewise fashion–setting $E_1(\mathbf{x}) = x_1 + x_2 + x_3$, $E_2(\mathbf{x}) = x_1 + x_2 - x_3$, $E_3(\mathbf{x}) = x_1 - x_2 + x_3$, $E_4(\mathbf{x}) = x_1 - x_2 - x_3$, and $a_i \leq b_i$ for $i = 1,\ldots,4$, one obtains the cross polytope

$$P^3 = \{\mathbf{x} \in \mathbb{R}^3 : a_i \leq E_i(\mathbf{x}) \leq b_i, i = 1,\ldots,4\}. \tag{9.2}$$

Equations 9.1 and 9.2 readily generalize to any dimension $n > 3$. Observe that the vector $\mathbf{p}^1 = (1,1) = (1,0) + (0,1) = \mathbf{e}^1 + \mathbf{e}^2$ is perpendicular to the two lines $E_1(\mathbf{x}) = a_1$ and $E_1(\mathbf{x}) = b_1$ in eqnuation 9.1, while the vector $\mathbf{p}^2 = (1,-1) = (1,0) + (0,-1) = \mathbf{e}^1 - \mathbf{e}^2$ is perpendicular to the two lines $E_2(\mathbf{x}) = a_2$ and $E_2(\mathbf{x}) = b_2$. Similarly, the vector $\mathbf{p}^1 = (1,1,1) = \mathbf{e}^1 + \mathbf{e}^2 + \mathbf{e}^3$ is orthogonal to the two planes $E_1(\mathbf{x}) = a_1$ and $E_1(\mathbf{x}) = b_1$, the vector $\mathbf{p}^2 = (1,1,-1) = \mathbf{e}^1 + \mathbf{e}^2 - \mathbf{e}^3$ is orthogonal to the two planes $E_2(\mathbf{x}) = a_2$ and $E_2(\mathbf{x}) = b_2$, the vector $\mathbf{p}^3 = (1,-1,1) = \mathbf{e}^1 - \mathbf{e}^2 + \mathbf{e}^3$ is orthogonal to the two planes $E_3(\mathbf{x}) = a_3$ and $E_3(\mathbf{x}) = b_3$, and the vector the vector $\mathbf{p}^4 = (1,-1,-1) = \mathbf{e}^1 - \mathbf{e}^2 - \mathbf{e}^3$ is orthogonal to the two planes $E_4(\mathbf{x}) = a_4$ and $E_1(\mathbf{x}) = b_4$.

The vectors $\mathbf{p}^i$ are bipolar with components $p^i_j \in \{1,-1\}$, and they are key in defining the hyperplanes $E_i(\mathbf{x}) = a$ for $i \in \mathbb{N}_n$. Consider the set of integers $\mathbb{Z}_{2^n} = \{0, 1, \ldots, 2^n - 1\}$ modulo 2^n and the Cartesian product set $\mathbb{Z}_2^n = \{\mathbf{z} : z_j \in \mathbb{Z}_2$, for $j = 1, 2, \ldots, n\}$. Both sets have 2^n elements and if $\mathbf{z}^i$ denotes the elements of $\mathbb{Z}_2^n$ for $i = 1,2,\ldots,2^n$, then the function $\beta : \mathbb{Z}_{2^n} \to \mathbb{Z}_2^n$ defined by $\beta(i-1) = \mathbf{z}^i$ is a bijection. For $i = 1,\ldots, 2^n$ and $j \in \mathbb{N}_n$, set $\beta(i-1, j) = z^i_j$. Equivalently, the bit value of the jth coordinate of the binary number $\mathbf{z}^i$ representing the integer $(i-1) \in \mathbb{Z}_{2^n}$ is given by

$$\beta(i-1, j) = \mathrm{mod}\{(i-1)2^j, 2\} = z^i_j \in \{0,1\}. \tag{9.3}$$

For $i = 1,\ldots, 2^{n-1}$, the bipolar vectors $\mathbf{p}^i$ can now be defined by $p^i_j = 1$ if $z^i_j = 0$, and $p^i_j = -1$ if $z^i_j = 1$, where $j = 1,\ldots,n$. Thus, $\mathbf{p}^i =$

$((-1)^{\beta(i-1,1)}, \ldots, (-1)^{\beta(i-1,n)})$. If $E_i(\mathbf{x})$ is defined as the dot product $E_i(\mathbf{x}) = \mathbf{p}^i \cdot \mathbf{x}$, then setting $E_i(\mathbf{x}) = a$, where a is an arbitrary constant, results in a hyperplane with $\mathbf{p}^i$ orthogonal to this hyperplane. The n-dimensional cross polytope is now defined

$$P^n = \{\mathbf{x} \in \mathbb{R}^n : a_i \leq E_i(\mathbf{x}) \leq b_i, i = 1, \ldots, 2^{n-1}\}, \tag{9.4}$$

where $a_i, b_i \in \mathbb{R}$, $E_i(\mathbf{x}) = a_i$ and $E_i(\mathbf{x}) = b_i$ are parallel hyperplanes with $\mathbf{p}^i$ orthogonal to both.

Finding the smallest cross polygon containing a given data set $T = \{\mathbf{x}^\xi \in \mathbb{R}^n : \xi = 1, \ldots, r\}$ is equivalent to finding, for each $i \in \{1, 2, \ldots, 2^{n-1}\}$, the two constants $a_i \leq b_i$ such that $P^n_i = \{\mathbf{x} \in \mathbb{R}^n : a_i \leq E_i(\mathbf{x}) \leq b_i\}$ is the smallest region containing T for any pair of hyperplanes parallel to $E_i(\mathbf{x}) = b_i$. It follows that $P^n = \bigcap_{i=1}^{2^{n-1}} P^n_i$ is the desired cross polygon. The sought-after constants a_i and b_i can be quickly computed using the following equations:

$$a_i = \bigwedge_{\xi=1}^{r} E_i(\mathbf{x}^\xi) = \bigwedge_{\xi=1}^{r} (\mathbf{p}^i \cdot \mathbf{x}^\xi) \quad \text{and} \quad b_i = \bigvee_{\xi=1}^{r} E_i(\mathbf{x}^\xi) = \bigvee_{\xi=1}^{r} (\mathbf{p}^i \cdot \mathbf{x}^\xi) \tag{9.5}$$

Example 9.1 Let $T = \{\mathbf{x}^\xi \in \mathbb{R}^2 : \xi = 1, \ldots, 13\}$, where

$$\mathbf{x}^1 = \begin{pmatrix} 8 \\ 3 \end{pmatrix}, \mathbf{x}^2 = \begin{pmatrix} 10 \\ 3 \end{pmatrix}, \mathbf{x}^3 = \begin{pmatrix} 9 \\ 5 \end{pmatrix}, \mathbf{x}^4 = \begin{pmatrix} 9 \\ 6 \end{pmatrix}, \mathbf{x}^5 = \begin{pmatrix} 8 \\ 4 \end{pmatrix}, \mathbf{x}^6 = \begin{pmatrix} 7 \\ 5 \end{pmatrix}, \mathbf{x}^7 = \begin{pmatrix} 8 \\ 3 \end{pmatrix},$$

$$\mathbf{x}^8 = \begin{pmatrix} 6 \\ 6 \end{pmatrix}, \mathbf{x}^9 = \begin{pmatrix} 6 \\ 7 \end{pmatrix}, \mathbf{x}^{10} = \begin{pmatrix} 7 \\ 8 \end{pmatrix}, \mathbf{x}^{11} = \begin{pmatrix} 4 \\ 7 \end{pmatrix}, \mathbf{x}^{12} = \begin{pmatrix} 8 \\ 6 \end{pmatrix}, \mathbf{x}^{13} = \begin{pmatrix} 5 \\ 7 \end{pmatrix}.$$

Applying equation 9.5 results in the four lines

$$a_1 = \bigwedge_{\xi=1}^{13} (x_1^\xi + x_2^\xi) = 11, \quad b_1 = \bigvee_{\xi=1}^{13} (x_1^\xi + x_2^\xi) = 15,$$

$$a_2 = \bigwedge_{\xi=1}^{13} (x_1^\xi - x_2^\xi) = -3, \quad b_2 = \bigvee_{\xi=1}^{13} (x_1^\xi - x_2^\xi) = 7.$$

It follows that $\mathbf{x} \in P^2 \Leftrightarrow 11 \leq x_1 + x_2 \leq 15$ *and* $-3 \leq x_1 - x_2 \leq 7$. The first inequality is equivalent to $11 \leq x_1 + x_2$ *and* $x_1 + x_2 \leq 15 \Leftrightarrow 0 \leq x_1 + x_2 - 11$ *and* $0 \leq -x_1 - x_2 + 15$. Thus, $\mathbf{x}$ satisfies the inequality equation $11 \leq x_1 + x_2 \leq 15$ if and only if $\mathbf{x}$ satisfies the inequality $0 \leq (x_1 + x_2 - 11) \wedge (-x_1 - x_2 + 15)$. In a likewise fashion, one can rewrite the second inequality equation as $0 \leq (x_1 - x_2 + 3) \wedge (-x_1 + x_2 + 7)$. Consequently,

A) $\mathbf{x} \in P^2 \Leftrightarrow [(x_1 + x_2 - 11) \wedge (-x_1 - x_2 + 15)] \wedge [(x_1 - x_2 + 3) \wedge (-x_1 + x_2 + 7)] \geq 0.$

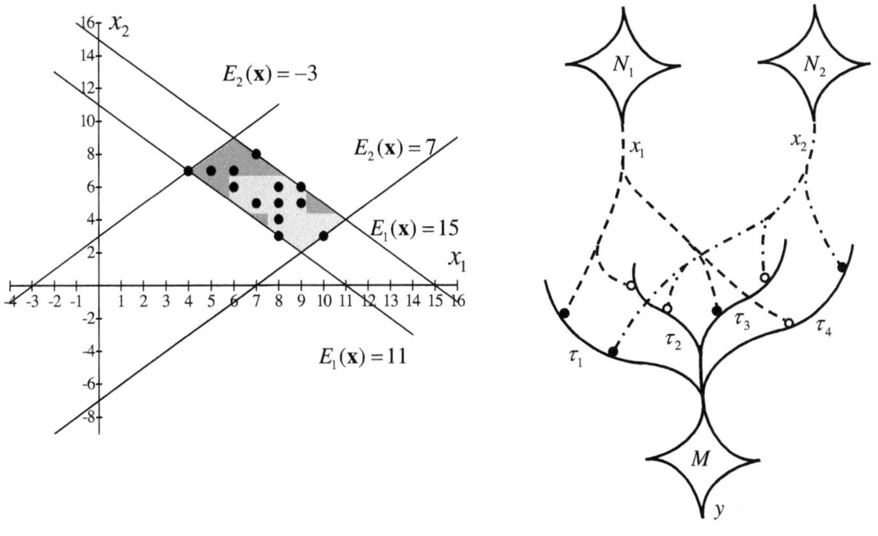

(a) (b)

Figure 9.4 (a) The boundary lines of P^2 containing the data set. (b) Sketch of the single-layer LNN capable of differentiating between the points of P^2 and the points of $\mathbb{R}^2 \setminus P^2$.

In lockstep with Example 8.5(2), this last inequality can be used to define a single-layer LNN with one output neuron M that classifies the elements of P^2 as belonging to class C_1 and all elements of $\mathbb{R}^2 \setminus P^2$ as belonging to class C_0. Specifically, the four planes in equation A. can be rewritten in terms of four dendritic branches–with each branch having two synaptic sites–as follows:

$$\tau_1(\mathbf{x}) = (x_1 + w_{11}^1) + (x_2 + w_{21}^1); \quad w_{11}^1 = 0,\ w_{21}^1 = -11,$$
$$\tau_2(\mathbf{x}) = -(x_1 + w_{12}^0) - (x_2 + w_{22}^0); \quad w_{12}^0 = 0,\ w_{22}^0 = -15,$$
$$\tau_3(\mathbf{x}) = (x_1 + w_{13}^1) - (x_2 + w_{23}^0); \quad w_{13}^1 = 0,\ w_{23}^0 = -3,$$
$$\tau_4(\mathbf{x}) = -(x_1 + w_{14}^0) + (x_2 + w_{24}^1); \quad w_{14}^0 = 0,\ w_{24}^1 = 7.$$

Equivalently, $\tau_k(\mathbf{x}) = \sum_{i=1}^{3}(-1)^{1-\ell}(x_i + w_{ik}^{\ell})$ for $k = 1, 2, 3, 4$. The *and* sign $\wedge$ in equation A. presupposes that $\tau(\mathbf{x}) = \bigwedge_{k=1}^{4} \tau_k(\mathbf{x})$. Utilizing the hard-limiter of equation 8.1 as the activation function for M results in $f[\tau(\mathbf{x}))] = 0 \Leftrightarrow \mathbf{x} \in \mathbb{R}^2 \setminus P^2$ and $f[\tau(\mathbf{x}))] = 1 \Leftrightarrow \mathbf{x} \in P^2$. Figure 9.4(b) depicts a geometric representation of this neural network.

The dual metric approach is derived from a simple observation: If

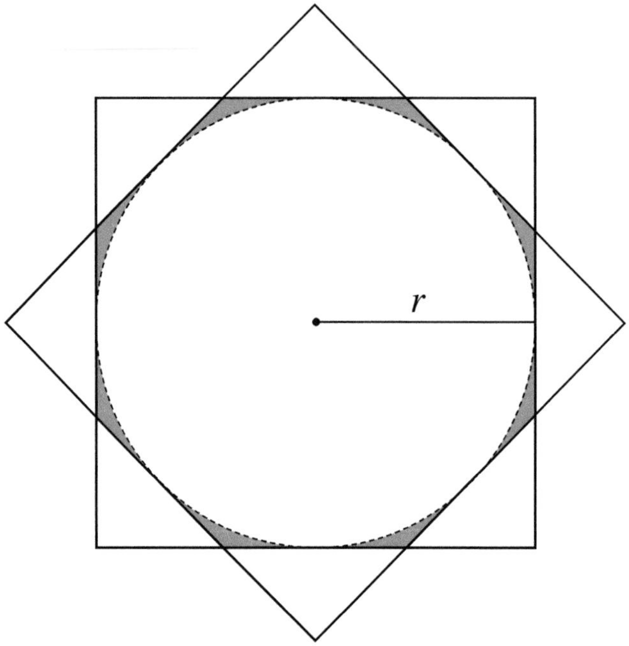

Figure 9.5 The error Area($P^2 \cap [\mathbf{v},\mathbf{i}]) - \pi r^2$, represented by the shaded regions, is drastically less than Area($[\mathbf{v},\mathbf{u}]) - \pi r^2 = r^2(4-\pi)$.

$P^2 \cap [\mathbf{v},\mathbf{u}]$ represents the intersection of the smallest cross polygon and the smallest 2-dimensional interval containing the triangular data set T illustrated in Figure 8.6 and discussed in Example 8.5(1.), then $T = P^2 \cap [\mathbf{v},\mathbf{u}]$. Compared to most data sets, the triangle provides an ideal setting for the dual (d_∞, d_1) metric approach. Nonetheless, the dual metric approach does reduce the error rate that occurs when using only one metric. For example, consider a data set that is densely packed in a circular region of radius r. Enclosing the set with smallest interval $[\mathbf{v},\mathbf{u}]$, an error of more than 27% in extraneous area occurs. In comparison, using the dual approach $P^2 \cap [\mathbf{v},\mathbf{u}]$ reduces the error to less 5.5% in extraneous area as indicated by the shaded region in Figure 9.5.

The dual metric approach has its own downside in that it requires the growth of more dendrites and additional synapses, thus increasing computational complexity. For instance, applying the dual metric approach on the data set T of Example 9.1 results in $[\mathbf{v},\mathbf{u}] = [(4,3)', (10,8)']$. Taking advantage of equation 8.8, simply attach an additional dendritic branch τ_5 with four

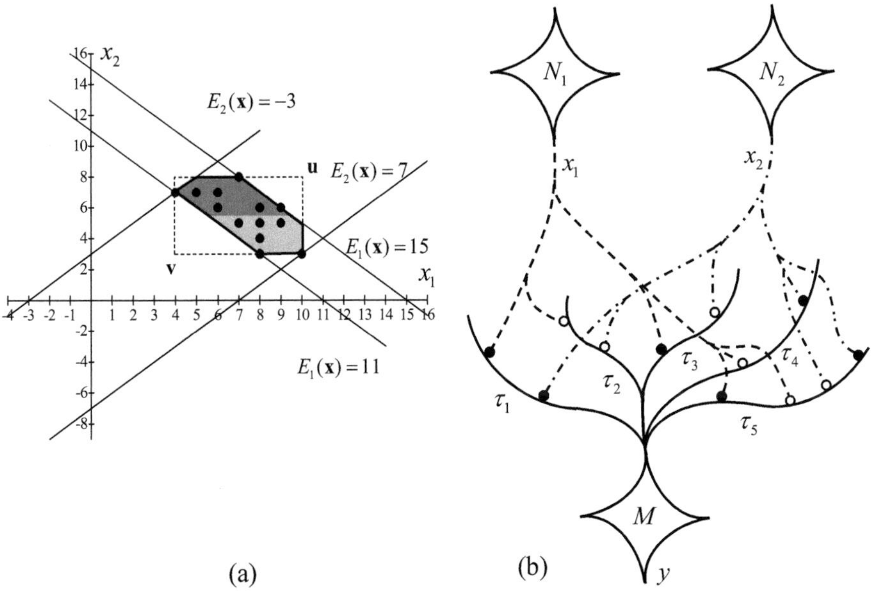

Figure 9.6 (a) The intersection $I = P^2 \cap [\mathbf{v}, \mathbf{u}]$ containing the data set T. (b) Sketch of the single-layer LBNN capable of differentiating between the points of I and the points of $\mathbb{R}^2 \setminus I$.

synaptic sites to the network shown in Figure. 9.4(b) by setting

$$\tau_5(\mathbf{x}) = \bigwedge_{i=1}^{2} \Big(\bigwedge_{h=1}^{2} (-1)^{1-\ell}(x_i + w_{i5h}^{\ell}) \Big),$$

where $w_{i51}^{\ell} = w_{i51}^{1} = -v_i$ and $w_{i52}^{\ell} = w_{i52}^{0} = -u_i$. In this case $w_{151}^{1} = -4$, $w_{251}^{1} = -3$, $w_{152}^{0} = -10$, and $w_{252}^{0} = -8$. The final computation by the output neuron is given by $\tau(\mathbf{x}) = \bigwedge_{k=1}^{5} \tau_k(\mathbf{x})$ and the corresponding LBNN is shown in Figure 9.6

More often than not, the various intersection and unions of hyperboxes and hyperoctahedrons, constructed during the learning process, provide for far more synaptic connections than are necessary for correct classification. For instance, consider the triangle T of Example 8.5(1). The single-layer LNN constructed in the example is capable of classifying all points of T and points of $\mathbb{R}^2 \setminus T$ correctly. The output neuron M has two dendritic branches with each branch having two synaptic sites. Applying the dual metric method instead results in a network with M having five branches and twelve synaptic

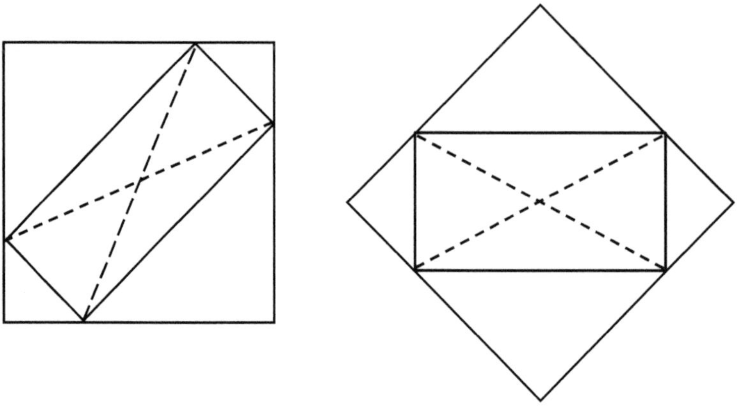

Figure 9.7 (a) The dual $[\mathbf{v},\mathbf{u}]$ of P^2 and (b) representing the dual P^2 of $[\mathbf{v},\mathbf{u}]$.

sites. Except for having different synaptic weights, the diagram of the LNN is identical to the one in Figure 9.4(b).

There is a simple method for reducing the number of synapses and dendritic branches when using the dual metric approach. The simplicity is due to the fact that a hyperbox is the dual of a cross polyhedron and vice versa. More precisely, if $P^n \subset \mathbb{R}^n$ and V denotes the set of vertices of P^n, then the polyhedron $[\mathbf{v},\mathbf{u}]$ is called the *dual* of P^n if and only if the vertices of P^n correspond to the faces of $[\mathbf{v},\mathbf{u}]$ or, equivalently, $P^n \cap \partial[\mathbf{v},\mathbf{u}] = V$ as shown in Figure 9.7(a). The polyhedron $[\mathbf{v},\mathbf{u}]$ is unique in that $\mathbf{v} = \bigwedge_{\mathbf{x}\in V} \mathbf{x}$ and $\mathbf{u} = \bigvee_{\mathbf{x}\in V} \mathbf{x}$.

Conversely, given some n-dimensional interval $[\mathbf{v},\mathbf{u}]$ with vertex set S, then $[\mathbf{v},\mathbf{u}] \subset P^n$, where $P^n = \{\mathbf{x} \in \mathbb{R}^n : a_i \le E_i(\mathbf{x}) \le b_i, i = 1,\ldots,2^{n-1}\}$, $a_i = \bigwedge_{\mathbf{x}\in S} E_i(\mathbf{x})$ and $b_i = \bigvee_{\mathbf{x}\in S} E_i(\mathbf{x})$. As illustrated in Figure 9.7(b), here again the vertices define the faces since $[\mathbf{v},\mathbf{u}] \cap \partial P^n = S$ and P^n is the dual of $[\mathbf{v},\mathbf{u}]$.

Suppose that $X \subset \mathbb{R}^n$ is a finite data set and $[\mathbf{v},\mathbf{u}]$ and P^n are smallest hyperbox and cross-polytope containing X. If $[\mathbf{v},\mathbf{u}]$ is the dual of P^n, then $[\mathbf{v},\mathbf{u}] \cap P^n = P^n$ and $[\mathbf{v},\mathbf{u}]$ can be eliminated from the dual metric approach. Similarly, P^n can be eliminated if it is the dual of $[\mathbf{v},\mathbf{u}]$. Unfortunately, these cases are extremely rare. In the typical case, neither of the two polytopes is a dual of the other. If this is the case, then for each $i \in \mathbb{N}_n$ the hyperplane $H_i \in \{E_{\mathbf{v}}(\mathbf{e}^i), E_{\mathbf{u}}(\mathbf{e}^i)\}$ has the property that $H_i \cap P^n \neq \varnothing$. It is also easy to prove that if $H_i \cap P^n = \{\mathbf{x}\}$, then $\mathbf{x}$ is a vertex of P^n. Likewise, if $H_i \in \{E_i(\mathbf{x}) = a_i, E_i(\mathbf{x}) = b_i\}$, then $H_i \cap [\mathbf{v},\mathbf{u}] \neq \varnothing$, and if $H_i \cap [\mathbf{v},\mathbf{u}] = \{\mathbf{x}\}$, then $\mathbf{x}$ is a vertex of $[\mathbf{v},\mathbf{u}]$.

Knowing the vertices of $[\mathbf{v},\mathbf{u}]$ and P^n is key in eliminating extraneous hyperplanes and, hence, extraneous dendritic branches and synaptic sites. The set of vertices of $[\mathbf{v},\mathbf{u}]$ is given by $S = \{\mathbf{x} \in \mathbb{R}^n : x_i \in \{v_i, u_i\} \forall i \in \mathbb{N}_n\}$. Thus, given a data set $X \subset \mathbb{R}^n$ and computing the vectors $\mathbf{v}$ and $\mathbf{u}$ also determines the $2^n = |S|$ vertices of $[\mathbf{v},\mathbf{u}]$. The ease of computing the elements of S is due to the fact that n of the $2n$ support hyperplanes contain $\mathbf{v}$ and n contain the vertex $\mathbf{u}$. As an aside, note that each face of $[\mathbf{v},\mathbf{u}]$ is an $(n-1)$-dimensional interval.

Although P^n has only $2n$ vertices, their determination is based on the various intersections of the 2^n support hyperplanes $E_i(\mathbf{x}) = a_i$ and $E_i(\mathbf{x}) = b_i$. Here each face $(E_i(\mathbf{x}) = c_i) \cap P^n$, where $c_i \in \{a_i, b_i\}$, is an $(n-1)$-dimensional simplex. To provide a simple systematic method for deriving the vertices of P^n, consider the case where $X \subset \mathbb{R}^2$ and let a_i, b_i are derived from equation 9.5. In this case P^2 has four distinct support lines as long as $(E_i(\mathbf{x}) = a_i) \cap (E_i(\mathbf{x}) = b_i) = \varnothing$, and four vertices corresponding to the intersections $(E_i(\mathbf{x}) = c_i) \cap (E_j(\mathbf{x}) = d_j)$, where $c_i \in \{a_i, b_i\}$, $d_j \in \{a_j, b_j\}$, and $i \neq j$.

Example 9.2 Consider the four lines $E_1(\mathbf{x}) = 11$, $E_1(\mathbf{x}) = 15$, $E_2(\mathbf{x}) = -3$, and $E_2(\mathbf{x}) = 7$ computed in Example 9.1. To find the four vertices terms of the intersections reduces to solving the following four sets of simple linear equations:

$$\begin{array}{llll} x_1 + x_2 = 11 & x_1 + x_2 = 11 & x_1 + x_2 = 15 & x_1 + x_2 = 15 \\ x_1 - x_2 = -3 & x_1 - x_2 = 7 & x_1 - x_2 = -3 & x_1 - x_2 = 7 \end{array}$$

The respective solutions are the vertices $\binom{4}{7}$, $\binom{9}{2}$, $\binom{6}{9}$, and $\binom{11}{4}$. Note that $(x_1 - x_2 = 7) \cap [\mathbf{v},\mathbf{u}] = (10,3)' = (u_1, v_2)'$ and $(x_1 = 4) \cap P^2 = (4,7)'$.

The method for obtaining the vertices of P^2 remains the same for any $n \geq 2$. For example if $n = 3$, then the six vertices of P^3 require the solutions of six sets of linear equations with each set consisting three linear equations. One straightforward method is to divide the six sets of linear equations into two classes, where the first class is given by

$$\begin{array}{lll} x_1 + x_2 + x_3 = a_1 & x_1 + x_2 + x_3 = a_1 & x_1 + x_2 + x_3 = a_1 \\ x_1 + x_2 - x_3 = a_2 & x_1 + x_2 - x_3 = a_2 & x_1 + x_2 - x_3 = b_2 \\ x_1 - x_2 + x_3 = a_3 & x_1 - x_2 + x_3 = b_3 & x_1 - x_2 + x_3 = b_3 \end{array}$$

and the second class by

$$\begin{array}{lll} x_1 + x_2 + x_3 = b_1 & x_1 + x_2 + x_3 = b_1 & x_1 + x_2 + x_3 = b_1 \\ x_1 + x_2 - x_3 = b_2 & x_1 + x_2 - x_3 = b_2 & x_1 + x_2 - x_3 = a_2 \\ x_1 - x_2 + x_3 = b_3 & x_1 - x_2 + x_3 = a_3 & x_1 - x_2 + x_3 = a_3 \end{array}$$

Let A and B denote the first class and second class, respectively. The three columns of equations of A and B are denoted as A_1, A_2, A_3 and B_1, B_2, B_3 in the order shown above. Considering the columns of equations for $n = 2$ in Example 9.6 and for $n = 3$ above, it becomes transparent that for $n \geq 2$, class A and class B each consist of n columns with each column consisting of n linear equations. More explicitly, for $i, j \in \{1, \ldots, n\}$

$$A_i = \begin{pmatrix} E_1(\mathbf{x}) = c_1 \\ \vdots \\ E_n(\mathbf{x}) = c_n \end{pmatrix} \quad \text{where} \quad c_j = \begin{cases} a_j \text{ if } j \leq n-(i-1) \\ b_j \text{ if } j > n-(i-1) \end{cases}, \tag{9.6}$$

and

$$B_i = \begin{pmatrix} E_1(\mathbf{x}) = c_1 \\ \vdots \\ E_n(\mathbf{x}) = c_n \end{pmatrix} \quad \text{where} \quad c_j = \begin{cases} b_j \text{ if } j \leq n-(i-1) \\ a_j \text{ if } j > n-(i-1) \end{cases}. \tag{9.7}$$

If $\mathfrak{N}$ represents the LBNN obtained from Algorithm 9.4 for a data set $X \subset \mathbb{R}^n$, then $\mathfrak{N}$ classifies all points in $P^n \cap [\mathbf{v}, \mathbf{u}]^n$ as belonging to one class C and all points in $\mathbb{R}^n \setminus P^n \cap [\mathbf{v}, \mathbf{u}]^n$ as not belonging to C. The dendritic branches and synaptic sites of the output neuron M are created from the support hyperplanes of the polyhedron $P^n \cap [\mathbf{v}, \mathbf{u}]^n$. As remarked earlier, this can result in encoding extraneous branches and synapses. Knowing the vertices of P^n and $[\mathbf{v}, \mathbf{u}]^n$ provides the means of eliminating the extraneous hyperplanes. The elimination is based on a simple rule: for $i = 1, \ldots, n$ let $c_i \in \{a_i, b_i\}$ and $\mathbf{c} \in \{\mathbf{v}, \mathbf{u}\}$ remove $E_i(\mathbf{x}) = c_i$ if $(E_i(\mathbf{x}) = c_i) \cap [\mathbf{v}, \mathbf{u}] = \{\mathbf{x}\}$ an remove $E_{\mathbf{c}}(\mathbf{e}^i)$ if $E_{\mathbf{c}}(\mathbf{e}^i) \cap P^n = \{\mathbf{x}\}$.

For an example, consider the intersection I shown in Figure 9.6(a). There the intersection $(x_1 - x_2 = 7) \cap [\mathbf{v}, \mathbf{u}]$ is the vertex $(10, 3)' = (u_1, v_2)'$ of $[\mathbf{v}, \mathbf{u}]$, while the intersection $(x_1 = 4) \cap P^2$ is the vertex $(4, 7)'$ of P^2. Considering the sketch 9.6(a), it is obvious that eliminating the lines $x_1 - x_2 = 7$ and $x_1 = 4$ does not affect the intersection, as the remaining lines are sufficient in supporting the structure of the polytope. Expressing mathematically this reduced support proves that it is indeed sufficient:

$$\{\mathbf{x} \in \mathbb{R}^2 : x_1 \leq u_1, v_2 \leq x_2 \leq u_2, 11 \leq x_1 + x_2 \leq 15, x_1 - x_2 \leq -3\} = P^2 \cap [\mathbf{v}, \mathbf{u}].$$

Elimination of the line $x_1 = 4$ reduces the number of synaptic sites on branch τ_5 of Figure 9.6(b) to three, while the elimination the line $E_2(\mathbf{x}) = 7$ eliminated all synaptic sites on branch τ_4 and, hence, the elimination of the branch.

Exercises 9.1.4

1. Suppose $X \subset \mathbb{R}^2$ denotes the set defined in Exercise 2 of **Exercises 8.2.1**. Using the dual lattice metrics approach, construct the convex polytope containing X.

2. Let P denote the polytope obtained in Exercise 1. Construct a single-layer LBNN with output neuron M such that $f[\tau(\mathbf{x}))] = 0 \Leftrightarrow \mathbf{x} \in \mathbb{R}^2 \setminus P$ and $f[\tau(\mathbf{x}))] = 1 \Leftrightarrow \mathbf{x} \in P$.

3. Let $X = \{\mathbf{x} \in \mathbb{R}^3 : 0 \le x_i \text{ for } i = 1,2,3 \text{ and } x_1^2 + x_2^2 + x_3^2 \le 1\}$. Using the dual lattice metrics approach, construct the convex polytope P containing X and the associated single-layer LBNN with output neuron M with output $f[\tau(\mathbf{x}))] = 1 \Leftrightarrow \mathbf{x} \in P$ and $f[\tau(\mathbf{x}))] = 0 \Leftrightarrow \mathbf{x} \in \mathbb{R}^3 \setminus P$.

4. Create a single-layer LBNN training algorithm based on the dual metric approach.

9.2 MULTI-LAYER LATTICE BIOMIMETIC NEURAL NETWORKS

Current artificial neural networks do not take into account the development of the various neurons and neural pathways during the prenatal period. Neural production in humans begins very early in the embryo; roughly 42 days after conception [36]. Stem cells generate the various types of neurons. When neurons are produced they migrate to different regions in the developing brain and make connections with other neurons to establish elementary neural networks. The brain must grow at a rate of approximately 216,000 neurons per minute in order to reach the 85 billion neurons at birth (section 8.1.2). These processes in the fetal brain are genetically controlled and do not require or depend on specific *learning* methods. Nevertheless, these elementary networks do carry out important tasks such as muscle movement and control of fetal heart beats.

In lockstep with the biological generation of elementary neural networks, the goal of this section concerns the construction of elementary LBNNs that are capable of solving various tasks. Different tasks require different networks and different types of neurons. Tasks such as associated memory, decision making, speech recognition, face recognition, etc, cannot be achieved by single-layer networks and require deeper networks.

9.2.1 Constructing a Multi-Layer DLAM

As discussed in Chapter 5, a special task for associative memories is the perfect recall of exemplar patterns that are distorted by different transformations or corrupted by random noise. This section introduces a two hidden layer LBNN associative memory, called a *dendritic lattice associative memory* (or DLAM), which provides for direct and fast association of perfect or imperfect input patterns with stored associated exemplar patterns without convergence problems. In this subsection the focus is on auto-associative memories.

The proposed DLAM consists of four layers of neurons: an input layer, two intermediate layers, and an output layer. The number of neurons in each layer is predetermined by the dimensionality of the pattern domain and the number of stored exemplar patterns. Explicitly, if $Q = \{\mathbf{q}^1, \ldots, \mathbf{q}^k\} \subset \mathbb{R}^n$, then the number of neurons in the input and output layers is n whereas each hidden layers consists of k neurons. The neurons in the first hidden layer will be denoted by $A_1, \ldots, A_k$, the second hidden layer neurons by $B_1, \ldots, B_k$, and in the output layer by $M_1, \ldots, M_n$. For a given input pattern $\mathbf{x} = (x_1, \ldots, x_n) \in \mathbb{R}^n$, the input neuron N_i assumes as its value the ith coordinate x_i of $\mathbf{x}$ and propagates this value through its axonal arborization to the dendrites of the first hidden-layer neurons. The dendritic tree of each neuron A_j has n branches $\tau_1^j, \ldots, \tau_n^j$ and each neuron N_i will have exactly two axonal fibers terminating on the synaptic sites located on the corresponding dendrite of A_j. The synaptic weights at these two synaptic sites are given by w_{ijkh}^{ℓ}, where $h = 1, 2$, $\ell = 0, 1$ and $k = i$. Thus, $w_{ijkh}^{\ell} = w_{ijih}^{\ell}$. The assumption that $\ell = 0, 1$ implies that there are two synaptic sites. This makes h superfluous so that $w_{ijih}^{\ell} = w_{iji}^{\ell} = w_{ij}^{\ell}$, where the last equality follows from the redundancy of the second i.

It is helpful to rename the synaptic weights for the A-layer neurons by setting $\alpha_{ij}^{\ell} = w_{ij}^{\ell}$ as illustrated in Figure 9.8. In order to store the exemplar patterns in the synaptic weights, set $\alpha_{ij}^{\ell} = -q_i^j$ for $\ell = 0, 1$. The total response or output of the branch τ_i^j to the received input at its synaptic sites can be computed by using the first equation of either equation. 8.4 or the first equation of equation. 8.5. Since $I(k) = I(i) = \{i\}$ and $p_{jk} = p_{ji}$, either of these choices reduce to $\tau_i^j(\mathbf{x}) = p_{ji} \bigwedge_{\ell=0}^{1} (-1)^{1-\ell} \left(x_i + \alpha_{ij}^{\ell}\right)$. Setting the postsynaptic dendrite response to $p_{ji} = 1$ results in

$$\tau_i^j(\mathbf{x}) = \bigwedge_{\ell=0}^{1} (-1)^{1-\ell} \left(x_i + \alpha_{ij}^{\ell}\right) = (q_i^j - x_i) \wedge (x_i - q_i^j). \tag{9.8}$$

In equation 9.8, $\tau_i^j(\mathbf{x}) = 0 \Leftrightarrow x_i = q_i^j$ and $\tau_i^j(\mathbf{x}) < 0 \Leftrightarrow x_i \neq q_i^j$. The value

$\tau_i^j(\mathbf{x})$ is passed to the cell body of A_j and its state is a function of the combined values received from its dendritic structure and is given by

$$\begin{aligned}\tau_A^j(\mathbf{x}) &= p_j \sum_{i=1}^{n} \tau_i^j(\mathbf{x}) = -\sum_{i=1}^{n} [(q_i^j - x_i) \wedge (x_i - q_i^j)] \\ &= \sum_{i=1}^{n} [(x_i - q_i^j) \vee (q_i^j - x_i)] = \sum_{i=1}^{n} |x_i - q_i^j|. \end{aligned} \tag{9.9}$$

Note that by setting $p_j = -1$ to inhibit the received input, the minimum of differences is changed to a maximum of differences. Thus, the *total response* τ_A^j of the A_j neuron provides the d_1 distance between the input pattern $\mathbf{x}$ and the j-th exemplar pattern $\mathbf{q}^j$, i.e., $\tau_A^j(\mathbf{x}) = d_1(\mathbf{x}, \mathbf{q}^j)$. Since pattern recall will be accomplished using the minimum distance obtained between the input pattern and all memorized exemplar patterns, the activation function used for the A-layer neurons is the identity function $f_A(\tau) = \tau$. Thus, the output of A_j, denoted by s_j^α, is equal to $\tau_A^j(\mathbf{x})$. In some cases it has proven advantageous to use the squared Euclidean distance $d_2^2(\mathbf{x}, \mathbf{q}^j)$ instead. In this case $\tau_A^j(\mathbf{x})$ is defined by

$$\begin{aligned}\tau_A^j(\mathbf{x}) &= p_j \sum_{i=1}^{n} \tau_i^j(\mathbf{x}) = -\sum_{i=1}^{n} [(q_i^j - x_i)^2 \wedge (x_i - q_i^j)^2] \\ &= \sum_{i=1}^{n} [(x_i - q_i^j)^2 \vee (q_i^j - x_i)^2] = \sum_{i=1}^{n} (x_i - q_i^j)^2. \end{aligned} \tag{9.10}$$

The output of the A-layer neurons serves as input to the B-layer neurons. Here, each neuron B_j has two dendrites τ_1^j and τ_2^j. The dendrite τ_1^j has only one synaptic site and the dendrite τ_2^j receives input from all the remaining neurons of the A-layer, i.e., $\{A_1, \ldots, A_k\} \setminus \{A_j\}$. Accordingly, τ_1^j has $k-1$ synaptic sites. As before, rename and simplify the weights at the synaptic sites by setting $\beta_{rj}^\ell = w_{rj}^\ell$, where $r = 1, \ldots, k$ refers to the presynaptic neuron A_r, $j = 1, \ldots, k$ refers to the postsynaptic neuron B_j, and the corresponding weight values of the synaptic target sites for A_r onto B_j are given by $\beta_{rj}^\ell = 0$. As shown in Figure 9.8, the weight of the single synapse on τ_{j1} is inhibitory and connects with a terminal axonal branch of A_j. All synapses on τ_{j2} are excitatory. Consequently, $\beta_{rj}^\ell = \beta_{rj}^0 \Leftrightarrow r = j$ and $\beta_{rj}^\ell = \beta_{rj}^1 \Leftrightarrow r \neq j$. Next, set $p_{j1} = 1 = p_{j2}$ so that the computation performed by the first dendrite of each B_j neuron is given by

$$\tau_1^j(\mathbf{s}^\alpha) = p_{j1}(-1)^{1-\ell}\left(s_j^\alpha + \beta_{jj}^\ell\right) = -s_j^\alpha. \tag{9.11}$$

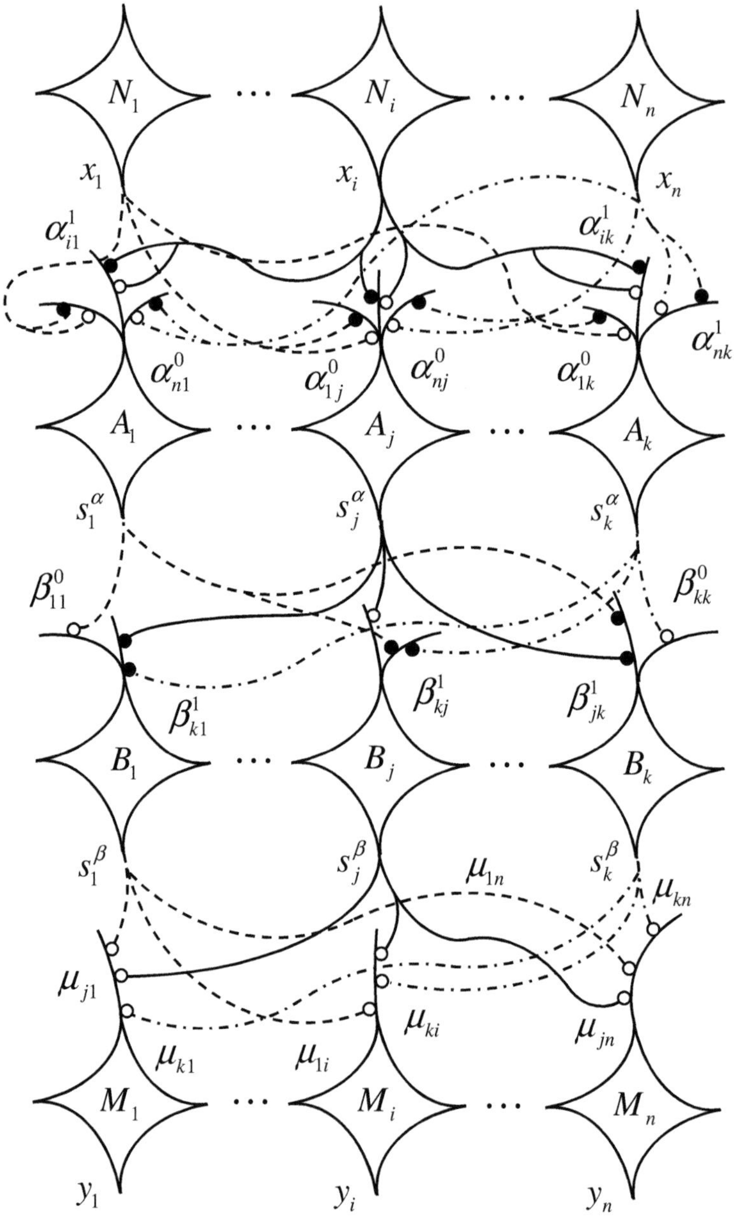

Figure 9.8 The network of the two-layer lattice auto-associative memory based on the dendritic model. The network is fully connected; all axonal branches from input neurons N_i synapse via two fibers on all dendrites of A_j-neurons, which in turn connect to all B_j-neurons via a single inhibitory synapse on one dendrite and multiple excitatory synopses on a second dendrite. Finally, the B-neurons synapse with output nodes M_i on a single dendrite having only inhibitory synapses.

In similar fashion, set $p_{j2} = 1$ in order to obtain

$$\tau_2^j(\mathbf{s}^\alpha) = p_{j2} \bigwedge_{r \neq j} (-1)^{1-\ell}\left(s_r^\alpha + \beta_{rj}^\ell\right) = \bigwedge_{r \neq j} s_r^\alpha. \tag{9.12}$$

The values $\tau_1^j(\mathbf{s}^\alpha)$ and $\tau_2^j(\mathbf{s}^\alpha)$ flow to the cell body of B_j and its state is a function of the combined values received from its dendritic structure and is given by

$$\begin{aligned} \tau_B^j(\mathbf{s}^\alpha) &= \sum_{k=1}^{2} \tau_k^j(\mathbf{s}^\alpha) = \tau_1^j(\mathbf{s}^\alpha) + \tau_2^j(\mathbf{s}^\alpha) \\ &= -s_j^\alpha + \bigwedge_{r \neq j} s_r^\alpha = \bigwedge_{r \neq j} d_1(\mathbf{x}, \mathbf{q}^r) - d_1(\mathbf{x}, \mathbf{q}^j). \end{aligned} \tag{9.13}$$

Note that if a distorted input vector $\mathbf{x}$ is nearest to the exemplar $\mathbf{q}^j$ than to the remaining exemplar patterns $\mathbf{q}^r$ for all $r \neq j$, then $\tau_B^j(\mathbf{s}^\alpha) > 0$ since $\bigwedge_{r \neq j} s_r^\alpha > s_j^\alpha$. Otherwise, if $\mathbf{x}$ is nearest another exemplar, say $\mathbf{q}^r$ with $r \neq j$, then $\tau_B^j(\mathbf{s}^\alpha) < 0$ since $\bigwedge_{r \neq j} s_r^\alpha = s_r^\alpha < s_j^\alpha$. In the first case, a firing signal will be passed from B_j to the output neurons; in the second case, no signal is passed from B_j to the output layer. Hence, the lattice-based hard-limiter defined by

$$f_B(\tau) = \begin{cases} 0 \Leftrightarrow \tau > 0 \\ -\infty \Leftrightarrow \tau \leq 0 \end{cases} \tag{9.14}$$

is used as the activation function for B_j to obtain the desired response. Therefore, the output of B_j given by $s_j^\beta = f_B[\tau_B^j(\mathbf{s}^\alpha)]$ serves as input to the output neurons. Here, each neuron M_i has a single dendrite d_i receiving inhibitory input from all k neurons in the B-layer. The weights of these synaptic sites are defined as $w_{ji}^0 = \mu_{ji} = q_i^j$ for $i = 1, \ldots, n$. To compute the recalled pattern note that $I(j) = \{1, \ldots, k\}$, and $\ell = 0$ for all synapses. Setting $p_{ji} = -1$ one obtains

$$y_i(\mathbf{s}^\beta) = p_{ji} \bigvee_{j=1}^{n} (-1)^{1-\ell}(s_j^\beta + w_{ji}^\ell) = -\bigwedge_{j=1}^{k} -(s_j^\beta + \mu_{ji}^0) = \bigvee_{j=1}^{k} (s_j^\beta + q_i^j). \tag{9.15}$$

The activation function for each output neuron M_i is again the identity function so that the total output $\mathbf{y}$ of the output layer is $\mathbf{y} = (y_1, \ldots, y_n)$.

Recall from perfect input. To verify the capabilities of the DLAM model, the first step is to show that the model provides perfect recall for perfect inputs. Let the input pattern $\mathbf{x}$ be one of the exemplars stored in the memory, i.e., let $\mathbf{x} = \mathbf{q}^j$ for some $j \in \{1,\ldots,k\}$. In this case it follows from eqn. 9.13 that

$$\tau_B^j(\mathbf{s}^\alpha) = \bigwedge_{r \neq j} d_1(\mathbf{q}^j, \mathbf{q}^r) - d_1(\mathbf{q}^j, \mathbf{q}^j) = \bigwedge_{r \neq j} d_1(\mathbf{q}^j, \mathbf{q}^r) > 0. \tag{9.16}$$

Since $d_1(\mathbf{q}^j, \mathbf{q}^r) > 0$ for each $r \neq j$, it follows that

$$s_j^\beta = f_B[\tau_B^j(\mathbf{s}^\alpha)] = 0. \tag{9.17}$$

Likewise, if $h \neq j$, then

$$\begin{aligned} \tau_B^h(\mathbf{s}^\alpha) &= \bigwedge_{r \neq h} d_1(\mathbf{q}^j, \mathbf{q}^r) - d_1(\mathbf{q}^j, \mathbf{q}^h) = d_1(\mathbf{q}^j, \mathbf{q}^j) \wedge \bigwedge_{r \neq h, j} d_1(\mathbf{q}^j, \mathbf{q}^r) - d_1(\mathbf{q}^j, \mathbf{q}^h) \\ &= 0 \wedge \bigwedge_{r \neq h, j} d_1(\mathbf{q}^j, \mathbf{q}^r) - d_1(\mathbf{q}^j, \mathbf{q}^h) = -d_1(\mathbf{q}^j, \mathbf{q}^h) \qquad (9.18) \\ &\Rightarrow s_h^\beta = f_B[\tau_B^h(\mathbf{s}^\alpha)] = -\infty \quad \forall h \neq j. \end{aligned}$$

With the outputs computed in equations 9.17 and 9.18, the proof is complete after evaluating equation 9.15 as follows:

$$\begin{aligned} y_i(\mathbf{s}^\beta) &= \bigvee_{j=1}^{k} (s_j^\beta + q_i^j) = (s_j^\beta + q_i^j) \vee \bigvee_{h \neq j} (s_h^\beta + q_i^h) \\ &= (0 + q_i^j) \vee \bigvee_{h \neq j} (-\infty + q_i^h) = q_i^j. \end{aligned} \tag{9.19}$$

Since the previous argument is true for all $i \in \{1,\ldots,n\}$, the recalled pattern $\mathbf{y}$ matches exactly the exemplar $\mathbf{q}^j$ given as input $\mathbf{x}$.

Recall from imperfect input. For a noisy or imperfect input two general situations may occur. An exemplar pattern can be distorted in such a way as to be still recognizable by the network in the sense that it will associate the input to the exemplar that most resembles it. A second possible scenario could happen if an exemplar pattern is affected by a large amount of distortion and consequently the network will not be able to establish the correct association by giving as output a different exemplar. Thus, for highly corrupted inputs the memory loses its capability to associate. We first show that the DLAM model can give perfect recall for imperfect input. Let the distorted input pattern $\widetilde{\mathbf{x}}$ be *closest* to one of the exemplars stored in the memory, i.e., let $\widetilde{\mathbf{x}} \approx \mathbf{q}^j$

for only one $j \in \{1,\ldots,k\}$. Then based on eqn. 9.12 results in the following approximation

$$\tau_B^j(\mathbf{s}^\alpha) = \bigwedge_{r \neq j} d_1(\widetilde{\mathbf{x}},\mathbf{q}^r) - d_1(\widetilde{\mathbf{x}},\mathbf{q}^j) \approx \bigwedge_{r \neq j} d_1(\widetilde{\mathbf{x}},\mathbf{q}^r) > 0 \tag{9.20}$$

$$\Rightarrow s_j^\beta = f_B[\tau_B^j(\mathbf{s}^\alpha)] = 0. \tag{9.21}$$

Equation 9.20 follows from the assumption that any distance $d_1(\widetilde{\mathbf{x}},\mathbf{q}^r) > d_1(\widetilde{\mathbf{x}},\mathbf{q}^j) \approx 0$ for each $r \neq j$. Similarly, for $h \neq j$ we have

$$\begin{aligned} \tau_B^h(\mathbf{s}^\alpha) &= \bigwedge_{r \neq h} d_1(\widetilde{\mathbf{x}},\mathbf{q}^r) - d_1(\widetilde{\mathbf{x}},\mathbf{q}^h) = d_1(\widetilde{\mathbf{x}},\mathbf{q}^j) \wedge \bigwedge_{r \neq h,j} d_1(\widetilde{\mathbf{x}},\mathbf{q}^r) - d_1(\widetilde{\mathbf{x}},\mathbf{q}^h) \\ &\approx 0 \wedge \bigwedge_{r \neq h,j} d_1(\widetilde{\mathbf{x}},\mathbf{q}^r) - d_1(\widetilde{\mathbf{x}},\mathbf{q}^h) = -d_1(\widetilde{\mathbf{x}},\mathbf{q}^h) \qquad (9.22) \\ \Rightarrow \quad & s_h^\beta = f_B[\tau_B^h(\mathbf{s}^\alpha)] = -\infty \quad \forall h \neq j. \end{aligned}$$

With the outputs computed using eqnuations 9.21 and 9.22, this synopsis ends with the evaluation of equation 9.15 in order to again obtain the output given by equation 9.19. Since the preceding argument is true for all $i \in \{1,\ldots,n\}$, the recalled vector $\mathbf{y} = \mathbf{x}$ is the pattern that most resembles the given input $\widetilde{\mathbf{x}}$. In the next section, the second scenario is illustrated by means of a visual example using a small number of grayscale images as exemplar patterns.

Storage and recall for image associations. Training a DLAM to recognize associations in the presence of imperfect input patterns is fairly straightforward. For a finite set of auto-associations, $\{(\mathbf{q}^\xi,\mathbf{q}^\xi) : \xi = 1,\ldots,k\} \subset \mathbb{R}^n \times \mathbb{R}^n$, the network's topology is as described earlier in this section. All weights are *preassigned* using the exemplar patterns and all that remains is to test the network using as imperfect inputs any imprinted exemplar pattern that has been transformed or corrupted a finite number of times, respectively, with different parameter values or with increasing noise levels to measure its recall capability or robustness to noise. We point out that the following computational experiment using "grayscale images" has been devised as a visual test to grasp how the network performs in the presence of imperfect input. Obviously, other potential applications are not restricted to images.

Example 9.3 In this example we consider recall from transformed images, where the set X of exemplar patterns consist of 10 grayscale images of size 128×128 pixels displayed in Figure 9.9. After scanning by rows, each image matrix $\mathbf{A}^\xi$ is converted into a vector $\mathbf{x}^\xi$ of dimension 16384; i.e. $x^\xi_{128(r-1)+c} =$

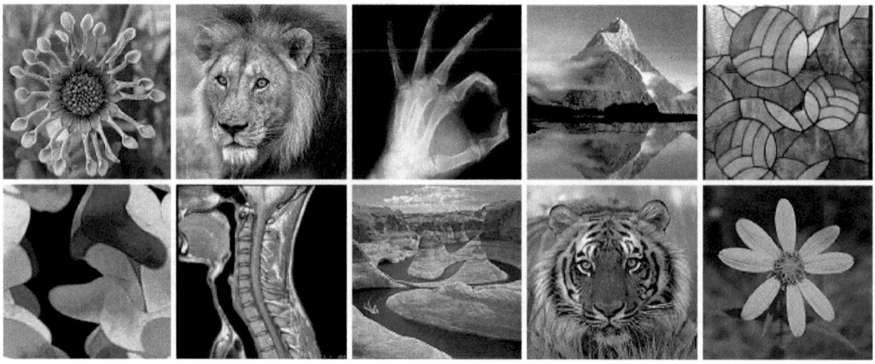

Figure 9.9 Set of 10 exemplar grayscale images of size 128×128 pixels used as example. 1st row, left to right: $\mathbf{x}^1$ to $\mathbf{x}^5$; 2nd row: $\mathbf{x}^6$ to $\mathbf{x}^{10}$ [234].

A^{ξ}_{rc} where $r, c = 1, \ldots, 128$ and $\xi = 1, \ldots, 10$. Hence, $X = \{\mathbf{x}^1, \mathbf{x}^2, \ldots, \mathbf{x}^9, \mathbf{x}^{10}\} \subset \mathbb{R}^{16384}$. For this example, each exemplar image $\mathbf{x}^{\xi}$ was changed by applying specific effects or transformations to produce imperfect inputs $\tilde{\mathbf{x}}^{\xi}$ that are very similar to its corresponding exemplar. The selected effects include, for example, opaque horizontal and vertical blinds, black pencil, soft illumination from different directions, texturization, and wind blowing from left or right. Other image transformations include morphological flat dilation and erosion, Gaussian and motion blur, and pixelation. Figure 9.10 displays some of the listed effects applied on image exemplar $\mathbf{x}^4$ ('mountain'). More specifically, a total of 33 digital effects and transformations were applied to each exemplar image in X. The response of the DLAM model outputs perfect recall for each imperfect input, a result that was expected from the explanations of equations. 9.21 and 9.22.

The next example examines the recall from noisy data.

Example 9.4 Suppose X is the set of images given in example 9.3. In this complementary example each exemplar image $\mathbf{x}^{\xi}$ is being corrupted by impulse, uniform and Gaussian noise. The production of imperfect inputs is arranged so that if the exemplar is contaminated with high levels of noise it tends to become more dissimilar from its original form, and the memory fails to perform the correct association. Incremental values of impulse noise level percentage p (%) with a step size of $\Delta p = 2\%$ was added to each exemplar image, e.g., Figure 9.11 shows some noisy instances of the exemplar image $\mathbf{x}^2$ corrupted with increasing percentages of impulse random noise. The uniform probability density function was evaluated with incremental values of

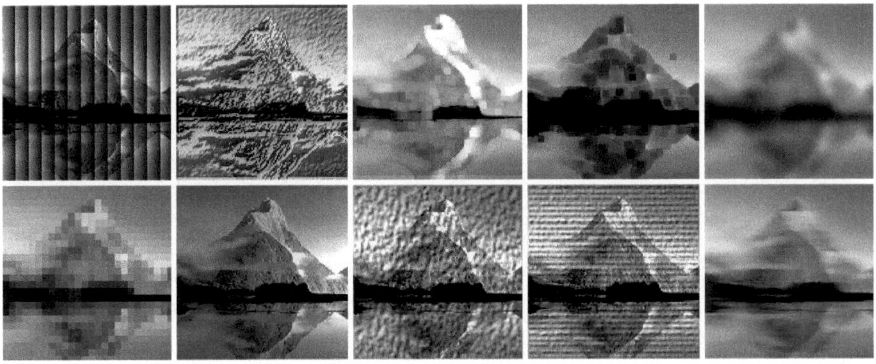

Figure 9.10 Exemplar image $\mathbf{x}^4$ ('mountain') subject to different effects and transformations. 1st row, left to right: opaque blinds (width=10 px, opacity=80), black pencil (detail=90, opacity=50), 5×5 morphological dilation, 5×5 morphological erosion, and Gaussian blur (radius=3). 2nd row: 5×5 pixelation, right illumination (intensity=65), gravel and stripes texturization (depth=1, angle=315°), and wind blowing from the right (velocity=10 px) [234].

interval length $\ell = b - a$ around a mean of 128, i.e., by setting its parameter values to $a = 128 - 2i$ and $b = 128 + 2i$, hence $\ell = 4i$ for $i = 1, \ldots, 64$. Similarly, the normal probability density function was computed by coupling its mean and standard deviation to the uniform density; thus, $\mu = (a+b)/2 = 128$ and $\sigma = (b-a)/2\sqrt{3} = 2i/\sqrt{3}$ for $i = 1, \ldots, 64$. Figure 9.12 displays some noisy instances also of exemplar image $\mathbf{x}^2$ contaminated with increasing length interval and standard deviation parameter values of uniform and Gaussian random noise respectively.

The DLAM's mean behavior is quantified in the presence of any type of random noise described by a probability density function (PDF) using as a measure the *average fraction of perfect recalls* (AFPR) which for brevity we denote by φ. Hence, for each noisy version $\widetilde{\mathbf{x}}^\xi$ of a given exemplar image $\mathbf{x}^\xi$ used as imperfect input, the output $\mathbf{y}_t(\widetilde{\mathbf{x}}^\xi)$ obtained with the DLAM model is verified for perfect recall and all hits are counted and averaged over a finite number T of trials. Specifically, the computed φ-quantities are given by, $\bar{\varphi}_I^\xi = \frac{1}{T}\sum_{t=1}^{T} d_1(\mathbf{y}_t(\widetilde{\mathbf{x}}_p^\xi), \mathbf{x}^\xi)$, $\bar{\varphi}_U^\xi = \frac{1}{T}\sum_{t=1}^{T} d_1(\mathbf{y}_t(\widetilde{\mathbf{x}}_{a,b}^\xi), \mathbf{x}^\xi)$, and $\bar{\varphi}_G^\xi = \frac{1}{T}\sum_{t=1}^{T} d_1(\mathbf{y}_t(\widetilde{\mathbf{x}}_{\mu,\sigma}^\xi), \mathbf{x}^\xi)$, where the subindex labels I, U, and G refer to the impulse, uniform, and Gaussian noise PDFs respectively, and $\xi = 1, \ldots, 10$, and $T = 50$ is the number of times an exemplar image $\mathbf{x}^\xi$ is corrupted with each type of random noise. In the case of impulse noise, $\bar{\varphi}_I^\xi = 1$ for every ξ for

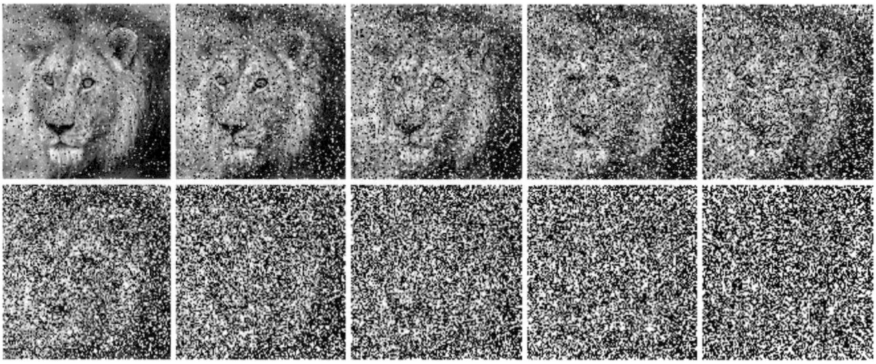

Figure 9.11 Exemplar image $\mathbf{x}^2$ ('lion') subject to different percentages of impulse random noise. 1st row, left to right: $p = 10\%, 20\%, 30\%, 40\%$, and 50%; 2nd row: $p = 60\%, 70\%, 80\%, 90\%$, and 100%. Perfect recall begins to fail when $p > 95\%$ [234].

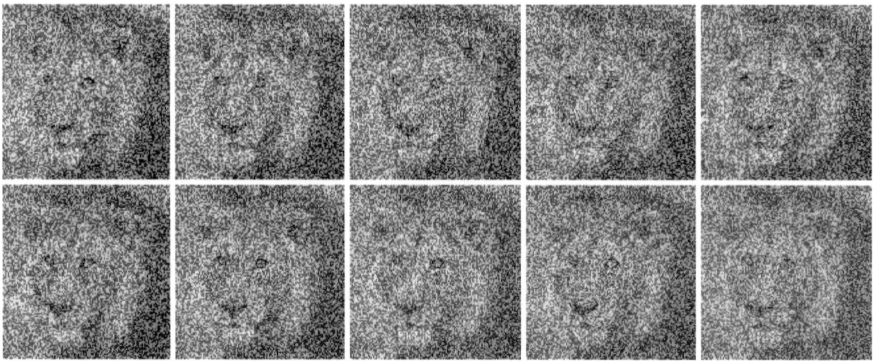

Figure 9.12 Exemplar image $\mathbf{x}^2$ ('lion') subject to different amounts of uniform and Gaussian random noise. 1st row, left to right: uniform interval length is $\ell = 16, 32, 48, 64$, and 80. 2nd row: Gaussian standard deviation is $\sigma = 4.62, 9.24, 13.86, 18.48$, and 23.1 [234].

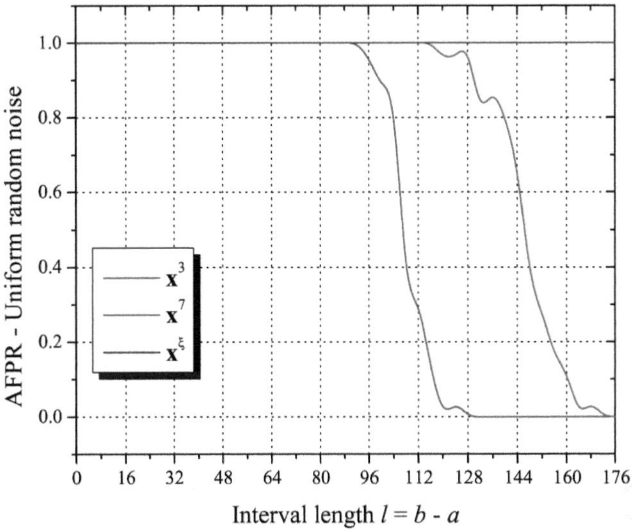

Figure 9.13 High rates of perfect recall hits occur for $\ell \leq 96$ for any ξ, otherwise non-perfect recall occurs (exemplar images $\mathbf{x}^3$ and $\mathbf{x}^7$); here, $\xi \neq 3, 7$ [234].

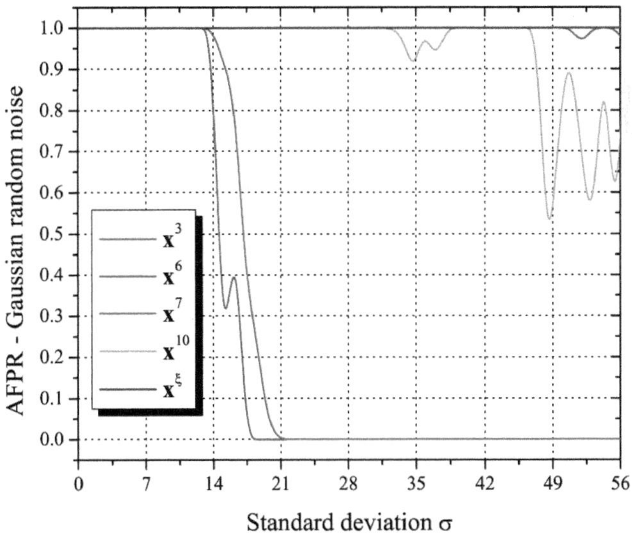

Figure 9.14 High rates of perfect recall hits occur for $\sigma \leq 13$ for any ξ, otherwise non-perfect recall occurs (mainly exemplar images $\mathbf{x}^3$ and $\mathbf{x}^7$); here, $\xi \neq 3, 6, 7, 10$ [234].

noise percentages up to 95% and for uniform and Gaussian noise, Figs. 9.13 and 9.14, respectively, give the overall response of the DLAM. The memory robustness can be seen by comparing, e.g., the parameter noise values shown in Fig. 9.12, with the corresponding values ℓ and σ on the horizontal axis of these two figures.

9.2.2 Learning for Pattern Recognition

The focus of this section is on hetero-associative memories that assign a class number, $c \in \{1,\ldots,m\} \subset \mathbb{Z}^+$, to a stored prototype pattern $\mathbf{q} \in \mathbb{R}^n$, where $\mathbf{q}$ is associated with c by some predefined relationship. More precisely, if $Q = \{\mathbf{q}^1,\ldots,\mathbf{q}^k\} \subset \mathbb{R}^n$ denotes the set of prototype pattern vectors, then the desired class association can be expressed as $\{(\mathbf{q}^j, c_j) : j = 1,\ldots,k \text{ and } c_j \in \mathbb{N}_m\}$. The goal is to store these pairs in some memory M such that for $j = 1,\ldots,k$, M assigns a class number c_j with a prototype pattern $\mathbf{q}^j$. As in Section 5.2.1, we use the notation $\mathbf{q}^j \to M \to c_j$ in order to express the association. Additionally, it is desirable for M to be able to assign the correct class number c_j to a non-stored or validation test pattern $\mathbf{x}$ known to belong to the same class as $\mathbf{q}^j$.

The utilization of LBNNs for multi-class pattern recognition was discussed in Section 9.1.3. There the methodology was based on either elimination or merging of interval neighborhoods. Furthermore, no hidden layers were required. A very different and simple approach to multi-class pattern recognition is the construction of a DLAM that is very similar to the DLAM shown in Figure 9.8. Explicitly, the proposed DLAM consists of four layers of neurons: an input layer, two intermediate layers, and an output layer. Assuming that, for any $j \in \{1,\ldots,k\}$, $\mathbf{q}^j \in \mathbb{R}^n$ and $c_j \in \{1,\ldots,m\}$, the number of neurons in the input layer is n and the output layer has a single neuron, whereas, the number of neurons in both intermediate layers is k. We denote the neurons in the input layer by $N_1,\ldots,N_n$, in the A-layer (first intermediate layer) by $A_1,\ldots,A_k$, in the B-layer (second intermediate layer) by $B_1,\ldots,B_k$, and in the output layer by M. As illustrated in Figure 9.15, with exception of the single output neuron M, the weights and synapses are the same as those shown in Figure 9.8 and discussed in Section 9.2.1. In fact, equations 9.8 through 9.13 define the dendritic branches and their associated synapses for the A-neurons as well as the B-neurons. The only difference of the proposed DLAM model is the single output neuron M and its synaptic weights. Here the single output neuron M has a single dendrite with k synaptic sites (j,ℓ) with weight $w_j^0 = c_j$ for $j = 1, \ldots, k$. The input received at (j,ℓ) is given by

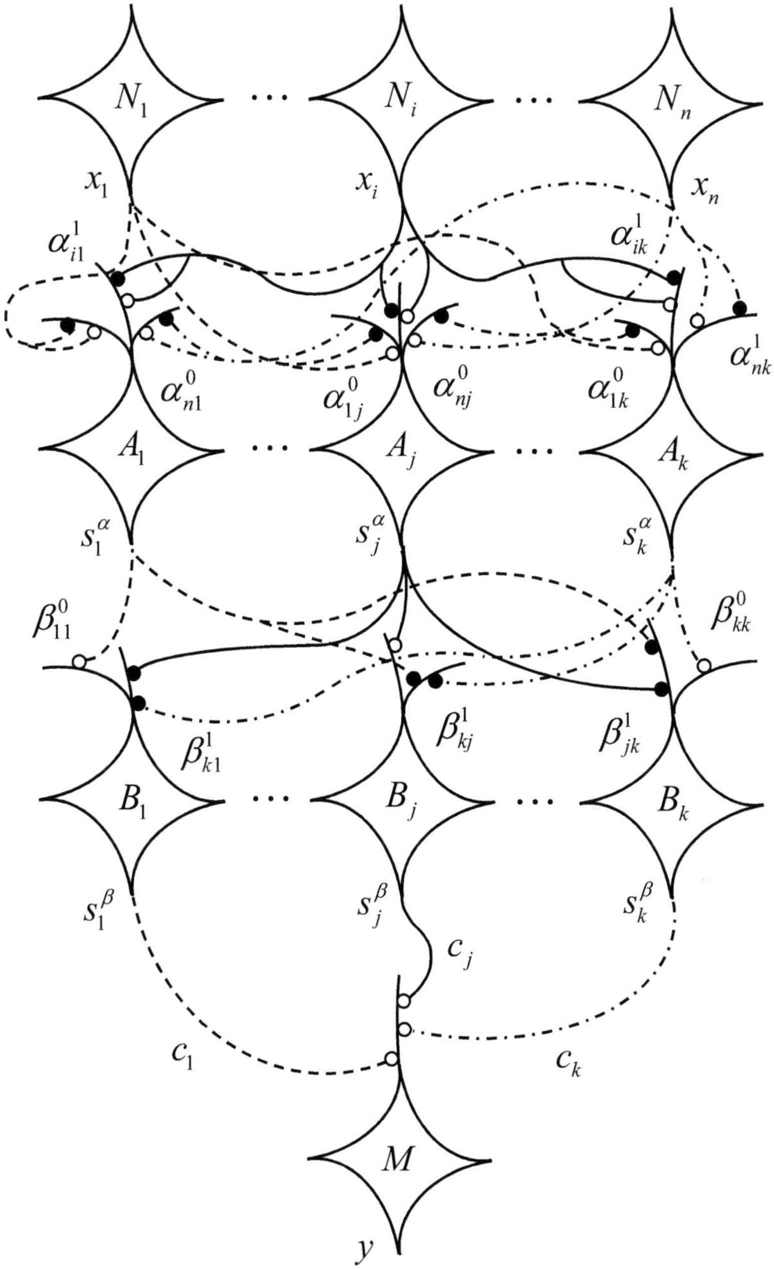

Figure 9.15 The topology of the two-layer lattice hetero-associative memory based on the dendritic model. With the exception of the output layer the network configuration is identical to the network illustrated in Fig.9.8. Here the B-neurons synapse with the output node M on a single dendrite.

$s_j^\beta = f_B[\tau_B^j(\mathbf{s}^\alpha)]$. This changes equation 9.15 to

$$y(\mathbf{s}^\beta) = -\bigwedge_{j=1}^{k}(-1)(s_j^\beta + w_j^0) = \bigvee_{j=1}^{k}(s_j^\beta + c_j). \tag{9.23}$$

The activation function for the output neuron M is again the identity function and the output y gives a class number.

Testing the DLAM network for correct classification when using prototype patterns as input is basically the same as testing the DLAM of Section 9.2.1 for recall from perfect input. As before, let the input pattern $\mathbf{x} = \mathbf{q}^j$ be one of the prototypes patterns stored in the memory and apply equations 9.16, 9.17, and 9.18. With the outputs computed in equations 9.17 and 9.18, the argument ends by evaluating equation 9.23 as follows:

$$\begin{aligned} y(\mathbf{s}^\beta) &= \bigvee_{j=1}^{k}(s_j^\beta + c_j) = (s_j^\beta + c_j) \vee \bigvee_{h\neq j}(s_h^\beta + c_h) \\ &= (0 + c_j) \vee \bigvee_{h\neq j}(-\infty + c_h) = c_j \vee (-\infty) = c_j. \end{aligned} \tag{9.24}$$

Thus, the class number $y = c_j$ is assigned to the prototype $\mathbf{q}^j$ given as input $\mathbf{x}$.

Classification of test patterns. For an input that is a non-stored or test pattern, two cases may occur. The test pattern is known to belong to one of the given m classes and will be correctly classified by the memory network in the sense that it will associate the input to the class pattern that most resembles a stored prototype pattern. A second possible scenario could happen if the test pattern is an unknown input and as such the network will not be able to establish the correct class association by giving as output an arbitrary class pattern. Thus, in the case of ambiguous inputs the memory loses its capability as a classifier. More specifically, we show that the dendritic model can give correct classification in the first case mentioned above. Let the test input pattern $\mathbf{x}$ be *closest* to one of the prototype patterns stored in the memory (same j-th class), i.e., let $\mathbf{x} \approx \mathbf{q}^j$ for only one $j \in \{1,\ldots,k\}$. Then, from equation 9.13,

$$\begin{aligned} \tau_B^j(\mathbf{s}^\alpha) &= \bigwedge_{r\neq j} d_1(\mathbf{x},\mathbf{q}^r) - d_1(\mathbf{x},\mathbf{q}^j) \approx \bigwedge_{r\neq j} d_1(\mathbf{x},\mathbf{q}^r) > 0 \\ &\Rightarrow s_j^\beta = f_B[\tau_B^j(\mathbf{s}^\alpha)] = 0. \end{aligned} \tag{9.25}$$

Equation 9.25 follows from the assumption that any distance $d_1(\mathbf{x},\mathbf{q}^r) > d_1(\mathbf{x},\mathbf{q}^j) \approx 0$ for each $r \neq j$. Similarly, for $h \neq j$ we have

$$\begin{aligned}
\tau_B^h(\mathbf{s}^\alpha) &= \bigwedge_{r\neq h} d_1(\mathbf{x},\mathbf{q}^r) - d_1(\mathbf{x},\mathbf{q}^h) = [d_1(\mathbf{x},\mathbf{q}^j) \wedge \bigwedge_{r\neq h,j} d_1(\mathbf{x},\mathbf{q}^r)] - d_1(\mathbf{x},\mathbf{q}^h) \\
&\approx [0 \wedge \bigwedge_{r\neq h,j} d_1(\mathbf{x},\mathbf{q}^r)] - d_1(\mathbf{x},\mathbf{q}^h) = -d_1(\mathbf{x},\mathbf{q}^h) \qquad (9.26) \\
&\Rightarrow \; s_h^\beta = f_B[\tau_B^h(\mathbf{s}^\alpha)] = -\infty \quad \forall h \neq j.
\end{aligned}$$

With the outputs computed in equations 9.25 and 9.26, the argument ends by evaluating equation 9.23 in order to obtain equation 9.24. Consequently, the recalled scalar $y = c_j$ is the class label corresponding to the given test input $\mathbf{x}$.

It is noteworthy to observe that the output vector $\mathbf{s}^\beta = (s_1^\beta,\ldots,s_k^\beta)$ from layer B, obtained by applying the hard-limiter defined in equation 9.14 to equation 9.13, provides numerical information equivalent to what a minimum distance classifier [76, 25] such as a 1-NN (1-Nearest Neighbor) or κ-NN (κ-Nearest Neighbor) algorithm computes. Note, for example, that the lattice operation expressed in equation 9.15 - computed after applying activation function (equation 9.14) - should not be confused with the majority vote procedure used with a κ-NN classifier and does not entail any search for a fixed number κ of nearest prototypes. Nonetheless, the proposed DLAM model can be categorized as a dynamic multiple 1-NN classifier that stores a variable number of prototypes of each class during training and due to its layered structure it computes distances in parallel.

Recognition Capability of DLAMs. Let $\{(\mathbf{q}^j, c_j) : j = 1,\ldots,k\} \subset \mathbb{R}^n \times \{1,\ldots,m\}$ be a set of hetero-associations, where $Q = \{\mathbf{q}^1, \ldots, \mathbf{q}^k\}$ denotes the data set and $c_j \in \{1, \ldots, m\}$ denotes the associated class index. For training the network, one starts with a reduced number of A and B neurons and adds new A and B neurons during training. More precisely, given the data set Q, select a family of prototype subsets, denoted by P_p, by randomly selecting predefined percentages $p\%$ of the total number k of samples in Q. We denote the complement of P_p with respect to Q by $Q_p = Q \setminus P_p$. In the four examples given below, the p percentages where considered in the range $\{10\%, 20\%, \ldots, 90\%\}$. Also, a finite number of runs were realized in order to compute the average fraction of hits for each selected percentage of all samples. Thus, if $|Q| = k$, $|P_p|$, $|Q_p|$, represent the set cardinality of data, prototype, and test sets, respectively, ρ denotes the number of runs, and μ_r denotes the number of misclassified test patterns in each run, then the *average fraction of hits* is given by

$$f_p^{\text{hits}} = 1 - \frac{\bar{\mu}}{k}, \quad \text{where} \quad \bar{\mu} = \frac{1}{\tau}\sum_{r=1}^{\rho} \mu_r, \tag{9.27}$$

and $k = |P_p| + |Q_p|$. In equation 9.25, we set $\rho = 20$ and use the same number of runs for every percentage p. If $p \in \{10\%, 20\%, \ldots, 90\%\}$, then for each run $r \in \{1, \ldots, 20\}$, create the set P_p by randomly choosing sample points. Thus, if P_p^r and P_p^s denote the set of randomly generated sample points at run r and run s, respectively, then $|P_p^r| = |P_p^s|$ but $P_p^r \neq P_p^s$. It follows that Q_p will have the same number of elements for each run with the same value of p, but $Q_p^r \neq Q_p^s$. Also, observe that a DLAM net trained with the prototype subset P_p can be tested either with the complete data set $Q = P_p \cup Q_p$ or with the partial set Q_p. However, since the DLAM is a perfect recall memory for all patterns in P_p as shown by equations 9.16 through 9.18, only Q_p is needed to test and measure network performance with the advantage of considerably reducing the amount of time required to compute f_p^{hits} for all selected values of p.

In the subsequent four examples of DLAM learning and classification of patterns, the example data sets are displayed in table format. Each table has seven columns: the 1st column gives the percentage p of sample points used to build the prototype and test subsets, the 2nd column provides the number of randomly selected prototype data patterns, and the 3rd column gives the number of test data patterns. The 4th column shows the average number of misclassified inputs obtained using the city block distance d_1. Similarly, the 5th column shows the average number of misclassified inputs obtained with the squared Euclidean distance d_2^2. The last two columns, 6th and 7th, give the corresponding average fraction of hits or correct classifications, respectively, for the d_1 and d_2^2 distances.

Example 9.5 This example uses an artificial data set consisting of 55 samples forming an 'X' shape in the plane, two features, the x and y coordinates, and two classes, $c \in \{1, 2\}$. The corresponding two-dimensional point set is shown in Figure 9.16 and Table 9.1 shows the numerical results.

Example 9.6 Here we use the 'Iris' data set [25, 86] with 150 samples where each sample is described by four flower features (sepal length, sepal width, petal length, petal width) and is equally distributed in 3 classes corresponding, respectively, to the subspecies of Iris setosa ($c = 1$), Iris versicolor ($c = 2$), and Iris virginica ($c = 3$). Table 9.2 displays the classification results.

Note that for the data sets 'X-shape' and 'Iris', a high average fraction of

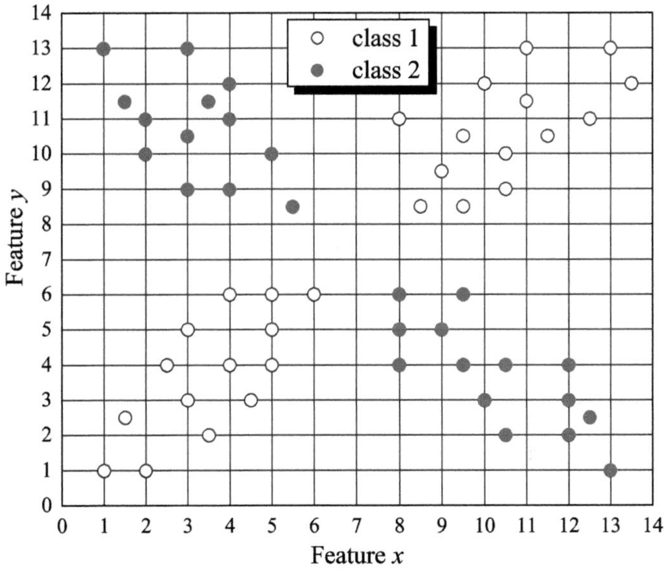

Figure 9.16 The 'X-shape' data set has 55 samples, described by 2 features (x and y coordinates), and is distributed in 2 classes. Class 1 has 28 points (blue circles) and class 2 has 27 points (red disks).

TABLE 9.1 DLAM classification performance for the 'X-shape' data set. Data set characteristics: $k = 55$ (samples), $n = 2$ (features), and $m = 2$ (classes)

p	$\|P_p\|$	$\|Q_p\|$	$\bar{\mu}_1$	$\bar{\mu}_2$	$f_p^{\text{hits}}(d_1)$	$f_p^{\text{hits}}(d_2^2)$
10%	6	49	12	8	0.769	0.837
20%	11	44	5	3	0.897	0.931
30%	17	38	1	0	0.965	0.984
40%	22	33	1	0	0.975	0.991
50%	28	27	0	0	0.987	0.999
60%	33	22	0	0	0.996	0.998
70%	39	16	0	0	0.999	1
80%	44	11	0	0	1	1
90%	50	5	0	0	1	1

TABLE 9.2 DLAM classification performance for the 'Iris' data set. Data set characteristics: $k = 150$ (samples), $n = 4$ (features), and $m = 3$ (classes)

p	$\lvert P_p\rvert$	$\lvert Q_p\rvert$	$\bar{\mu}_1$	$\bar{\mu}_2$	$f_p^{\text{hits}}(d_1)$	$f_p^{\text{hits}}(d_2^2)$
10%	15	135	12	11	0.915	0.920
20%	30	120	7	6	0.953	0.958
30%	45	105	5	4	0.963	0.967
40%	60	90	5	4	0.964	0.968
50%	75	75	3	3	0.974	0.978
60%	90	60	3	2	0.979	0.984
70%	105	45	2	2	0.983	0.986
80%	120	30	1	1	0.989	0.991
90%	135	15	0	0	0.996	0.996

hits such as $f_p^{\text{hits}} > 0.95$ is obtained for percentages p as low as 30% and that the use of the squared Euclidean distance gives slightly better results than the use of the city block distance. For instance, in the case of the 'Iris' data set, the DLAM net used as an individual classifier delivers similar performance against Linear or Quadratic Bayesian classifiers [320] for which, $f_{50}^{\text{hits}} = 0.953$ and $f_{50}^{\text{hits}} = 0.973$, respectively, or in comparison with an Edge-effect Fuzzy Support Vector Machine [165] whose $f_{60}^{\text{hits}} = 0.978$.

Example 9.7 The 'Column' data set in The 'Column' data set [86] has 310 patient samples. Each sample is specified by six biomechanical attributes derived from the shape and orientation of the pelvis and lumbar spine. Attribute one to six are numerical values of pelvic incidence, pelvic tilt, lumbar lordosis angle, sacral slope, pelvic radius, and grade of spondylolisthesis. Class one of patients diagnosed with disk hernia has 60 samples, class two of patients diagnosed with spondylolisthesis has 150 samples, and class three of normal patients has 100 samples. In this case, a high average fraction of hits occur for percentages p greater than 75% which is due to the presence of several interclass outliers. However, the DLAM response is still good if compared with other classifiers such as an SVM (Support Vector Machine), MLP (Multilayer Perceptron), and GRNN (General Regression Neural Network)[237] which give, correspondingly, $f_{80}^{\text{hits}} = 0.965$, $f_{80}^{\text{hits}} = 0.987$, and $f_{80}^{\text{hits}} = 0.965$ (with all outliers removed!). Table 9.3 lists the corresponding numerical results.

Example 9.8 The 'Wine' data set in [86] has 178 samples resulting of the chemical analysis of wines produced from three different cultivars (classes) of the same region in Italy. The features in each sample represent the quanti-

TABLE 9.3 DLAM classification performance for the 'Column' data set. Data set characteristics: $k = 310$ (samples), $n = 6$ (features), and $m = 3$ (classes)

p	$\|P_p\|$	$\|Q_p\|$	$\bar{\mu}_1$	$\bar{\mu}_2$	$f_p^{\text{hits}}(d_1)$	$f_p^{\text{hits}}(d_2^2)$
10%	31	279	66	65	0.786	0.790
20%	62	248	57	55	0.815	0.819
30%	93	217	47	46	0.848	0.852
40%	124	186	38	37	0.876	0.881
50%	155	155	31	29	0.889	0.904
60%	186	124	24	23	0.919	0.923
70%	217	93	19	17	0.937	0.943
80%	248	62	11	11	0.963	0.963
90%	279	31	5	5	0.983	0.983

ties of 13 constituents: alcohol, malic acid, ash, alcalinity of ash, magnesium, phenols, flavanoids, nonflavanoid phenols, proanthocyanins, color intensity, hue, diluted wines, and proline. Class one has 59 samples, class 2 has 71 samples, and class 3 has 48 samples. For this last example, since attribute 13 (proline) has a greater numerical range than the other attributes the DLAM model computes the d_1 and d_2^2 *partial distances* using the first 12 attributes. Thus, in equations 9.9 and 9.10, the subindex i goes from 1 to $n-1$. Table 9.4 shows the classification results. In this last example, a high average fraction of hits occur for percentages p greater than 70% and the DLAM performance is quite good if compared with other classifiers, based on the *leave one-out* technique, such as the 1-NN (1-Nearest Neighbor), LDA (Linear Discriminant Analysis), and QDA (Quadratic Discriminant Analysis) [3] which give, correspondingly, $f_p^{\text{hits}} = 0.961$, $f_p^{\text{hits}} = 0.989$, and $f_p^{\text{hits}} = 0.994$ where $p > 99\%$ and training must be repeated 178 times. Although not shown in Table 9.4, the DLAM net gives $f_{99}^{\text{hits}} \approx 1$, since almost all samples in the given data set are stored by the memory as prototype patterns. However, the DLAM model is outperformed by a short margin of misclassification error if compared to a FLNN classifier (Fuzzy Lattice Neural Network) that gives $f_{75}^{\text{hits}} = 0.997$ (leave-25%-out) [209].

9.2.3 Learning Based on Similarity Measures

. This section introduces an example and performance of learning methods for LBNNs based on the similarity measures. Recall that the mapping S defined in equation 4.28 is a similarity measure for any sublattice L of $\mathbb{R}^n$ that has a least element. Since the similarity measure is defined in terms of a val-

TABLE 9.4 DLAM classification performance for the 'Wine' data set. Data set characteristics: $k = 178$ (samples), $n = 13$ (features), and $m = 3$ (classes)

p	$\|P_p\|$	$\|Q_p\|$	$\bar{\mu}_1$	$\bar{\mu}_2$	$f_p^{\text{hits}}(d_1)$	$f_p^{\text{hits}}(d_2^2)$
10%	18	160	30	48	0.813	0.700
20%	36	142	19	27	0.869	0.815
30%	53	125	12	18	0.909	0.863
40%	71	107	8	12	0.929	0.896
50%	89	89	5	7	0.945	0.923
60%	107	71	3	5	0.960	0.935
70%	125	53	2	3	0.968	0.956
80%	142	36	1	2	0.978	0.972
90%	160	18	0	0	0.993	0.988

uation function, it is also called a *valuation* similarity measure. Likewise, the similarity measure defined in equation 4.27 is called a *norm* similarity measure. The application of learning algorithms based similarity measures are again within the realm of pattern recognition. Adopting the symbols used in the preceding sections, we let $Q \subset \mathbb{R}^n$ denote the data set consisting of k elements, $P_p \subset Q$ with $|P_p| < k$ and $Q_p = Q \setminus P_p$, where p denotes the percentage of the number of elements of P_p with respect to $k = 100^{\%}$.

Learning based on valuation similarity measure We employ the lattice $L = (\mathbb{R}^n_{[0,\infty)}, \vee, \wedge)$ in order to satisfy condition S_1 of Definition 4.4 and since the coordinates of the pattern vectors used in this section are non-negative. Since data sets are finite, data sets consisting of pattern vectors that are subsets of $\mathbb{R}^n$ have always an infimum $\mathbf{z}$ and a supremum $\mathbf{a}$. Thus, if $Q \subset \mathbb{R}^n$ is a data set whose pattern vectors have both negative and nonnegative coordinates, simply compute $\mathbf{z} = inf(Q) = \bigwedge_{\mathbf{q} \in Q} Q$ and $\mathbf{a} = sup(Q) = \bigvee_{\mathbf{q} \in Q} Q$. Note that the hyperbox $[\mathbf{z}, \mathbf{a}] = \{\mathbf{x} \in \mathbb{R}^n : z_i \leq x_i \leq a_i \text{ for } i = 1, 2, \ldots, n\}$ is a complete lattice. Setting $L = [\mathbf{z}, \mathbf{a}]$ and $\hat{\mathbf{x}} = \mathbf{x} - \mathbf{z}\ \forall \mathbf{x} \in L$, then $\hat{\mathbf{z}} = \mathbf{0}$ and $\hat{\mathbf{x}} \in \mathbb{R}^n_{[0,\infty)}\ \forall \mathbf{x} \in L$. Finally, define the mapping $s : L \to [0,1]$ by setting $s(\mathbf{x}, \mathbf{y}) = S(\hat{\mathbf{x}}, \hat{\mathbf{y}})$. It follows that $s(\mathbf{x}, \mathbf{z}) = S((\mathbf{x} - \mathbf{z}), \mathbf{0}) = 0$, which proves that condition S_1 is satisfied, and the remaining two conditions are just as easy to prove.

Suppose that $Q = \{\mathbf{q}^1, \mathbf{q}^2, \ldots, q^k\} \subset L$ is a data set consisting of prototype patterns, where each pattern $\mathbf{q}^j$ belongs to one of m different classes so that $R = \{(\mathbf{q}^j, c_j) : \mathbf{q}^j \in Q \text{ and } c_j \in \mathbb{N}_m\}$. As before, let $P_p = \{\mathbf{q}^{s_1}, \mathbf{q}^{s_2}, \ldots, \mathbf{q}^{s_t}\} \subset Q$ denote the training set obtained, where the elements $\mathbf{q}^{s_j}$ have been randomly chosen from $Q \subset \mathbb{R}^n_{[\mathbf{0},\infty)}$ and $t = |P_p|$. If Q consists of m classes, the user must assure that all m classes are represented in P_p. After selecting the training set P_p, precompute the values $v(\mathbf{q}^{s_j})$ for $j = 1, 2, \ldots, t$. These values will be stored at the synaptic sites of the LBNN. Knowing the dimension n and the size of the training set P_p, it is now an easy task to construct the network. As illustrated in Figure 9.17, the network has n input neurons denoted by $N_1, \ldots, N_n$, two hidden layer neurons, and a layer of output neurons. The first hidden layer neurons consist of two different types of neurons denoted by A_j and B_j, where $j = 1, 2, \ldots, t$. Each neuron A_j and B_j will have a single dendrite with each dendrite having n synaptic sites. For sake of simplicity, we denote the dendrite of A_j and of B_j by a^j and b^j, respectively. The second hidden layer has $t+1$ neurons denoted by C_j, where $j = 0, 1, \ldots, t$. Here C_0 has t dendrites, denoted by τ_j^0, with each dendrite having two synaptic sites for $j = 1, 2, \ldots, t$. For $j = 1, 2, \ldots, t$, each neuron C_j has only one dendrite, with each dendrite having two synaptic sites. The output layer is made up of t neurons, denoted by M_j for $j = 1, \ldots, t$, with each neuron M_j having a single dendrite τ^j and each dendrite having two synaptic sites.

Knowing the network's shell, the final step is describe the internal workings of the network. For a given $\mathbf{x} \in \mathbb{R}^n_{[0,\infty)}$, the input neuron N_i receives the input x_i and this information is sent to each of the neurons A_j and B_j. For each $i = 1, 2, \ldots, n$, the axonal arborization of N_i consists of $2t$ terminal branches with one terminal on each a^j and b^j. The synaptic weight α_{ij}^{ℓ} at the ith synapse on a^j is given by $\alpha_{ij}^{\ell} = q_i^{s_j}$ with $\ell = 1$. Each synapse on a^j at location (i, j) results in $x_i \vee q_i^{s_j}$ upon receiving the information x_i. The total response of the dendrite a^j is given by the summation $a^j(\mathbf{x}) = \sum_{i=1}^{n}(x_i \vee q_i^{s_j}) = v(\mathbf{x} \vee \mathbf{q}^{s_j})$.

In a similar fashion, the synaptic weight β_{ij}^{ℓ} at the ith synapse on b^j is given by $\beta_{ij}^{\ell} = q_i^{s_j}$ with $\ell = 1$. However, here each synapse on b^j at location (i, j) results in $x_i \wedge q_i^{s_j}$ upon receiving the information x_i, and each the neuron B_j computes $b^j(\mathbf{x}) = \sum_{i=1}^{n}(x_i \wedge q_i^{s_j}) = v(\mathbf{x} \wedge \mathbf{q}^{s_j})$. This information $a^j(\mathbf{x})$ and $b^j(\mathbf{x})$ travels through the soma towards its axon hillock of the respective neurons where the corresponding activation functions for neurons A_j and B_j, are given respectively by

$$f_j(\mathbf{x}) = \frac{v(\mathbf{q}^{s_j})}{v(\mathbf{x} \vee \mathbf{q}^{s_j})} \quad \text{and} \quad g_j(\mathbf{x}) = \frac{v(\mathbf{x} \wedge \mathbf{q}^{s_j})}{v(\mathbf{q}^{s_j})}. \tag{9.28}$$

The information $f_j(\mathbf{x})$ and $g_j(\mathbf{x})$ is being transferred via the axonal arborization of the first hidden layer neurons to the dendrites of the second layer neurons. The presynaptic neurons of C_0 are all the neurons of the first hidden layer. A terminal axonal fiber of A_j and one from B_j terminate on τ_j^0. The weight at each of the two synapses is $w_{aj0}^{\ell} = 0 = w_{bj0}^{\ell}$, where $\ell = 1$ and

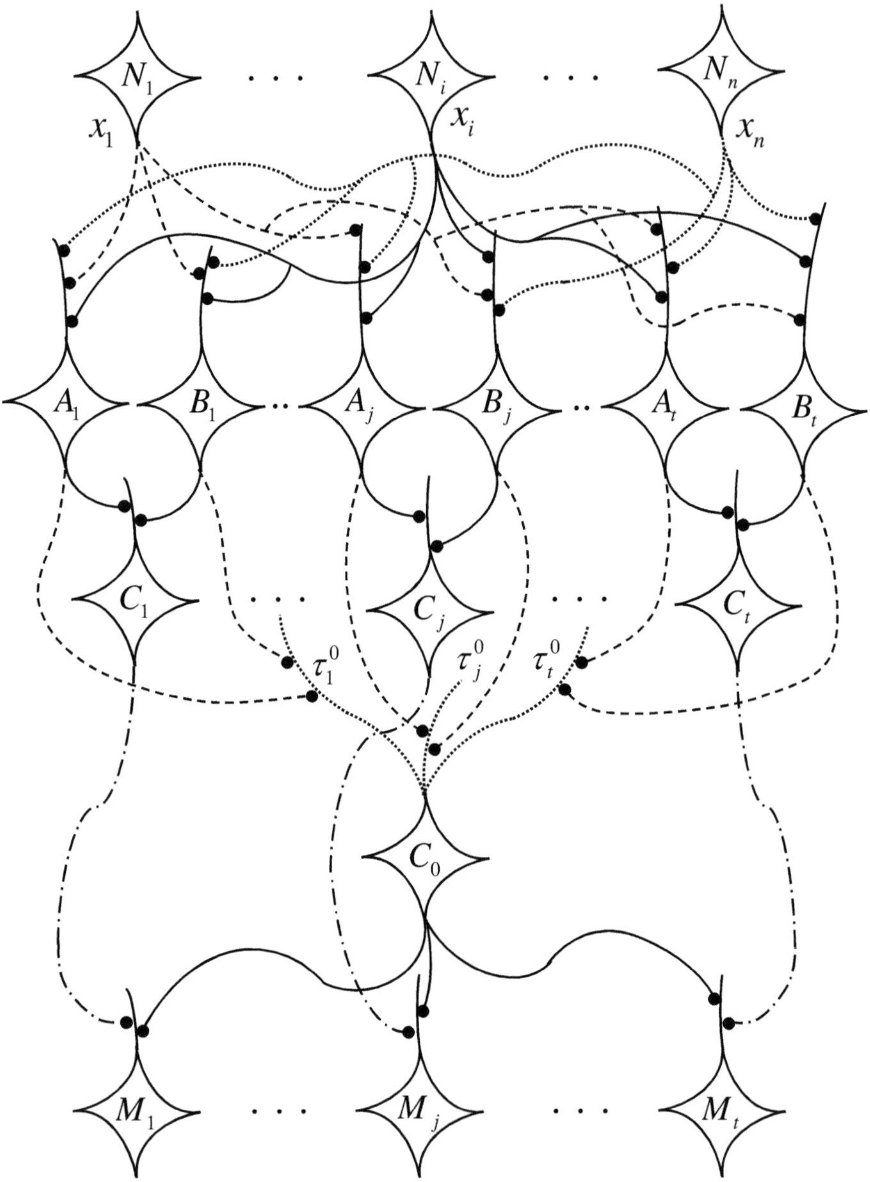

Figure 9.17 Diagram of the two hidden layer neural network for learning based on measures. The number of neurons in the two hidden layers will change during learning process.

aj, bj are address labels for the respective terminal axonal fibers from A_j and B_j. Thus, each synapse accepts the information $f_j(\mathbf{x})$ and $g_j(\mathbf{x})$. The total response of the dendrite is given by $\tau_j^0(\mathbf{x}) = f_j(\mathbf{x}) \wedge g_j(\mathbf{x})$. The total response of the neuron C_0 is given by

$$\tau^0(\mathbf{x}) = \bigvee_{j=1}^{t} \tau_j^0(\mathbf{x}) = \bigvee_{j=1}^{t} f_j(\mathbf{x}) \wedge g_j(\mathbf{x}) = \bigvee_{j=1}^{t} \left(\frac{v(\mathbf{q}^{s_j})}{v(\mathbf{x} \vee \mathbf{q}^{s_j})} \wedge \frac{v(\mathbf{x} \wedge \mathbf{q}^{s_j})}{v(\mathbf{q}^{s_j})}\right). \quad (9.29)$$

For $j = 1,2,\ldots,t$, the presynaptic neurons for the neuron C_j are the two neurons A_j and B_j. Denoting the single dendrite of C_j by τ^J, then a terminal axonal fiber of A_j and one from B_j terminate on τ^j. In lockstep with C_0, the weight at each of the two synapses is $w_{aj}^{\ell} = 0 = w_{bj}^{\ell}$, where $\ell = 1$ and aj, bj are address labels for the respective terminal axonal fibers from A_j and B_j. Again, the two synapses accept the information $f_j(\mathbf{x})$ and $g_j(\mathbf{x})$ and the response of the single dendrite is

$$\tau^j(\mathbf{x}) = f_j(\mathbf{x}) \wedge g_j(\mathbf{x}) = \frac{v(\mathbf{q}^{s_j})}{v(\mathbf{x} \vee \mathbf{q}^{s_j})} \wedge \frac{v(\mathbf{x} \wedge \mathbf{q}^{s_j})}{v(\mathbf{q}^{s_j})}. \quad (9.30)$$

The identity function $f(\mathbf{x}) = \mathbf{x}\ \forall\, j \in \{0,1,\ldots,t\}$ activates C_j. For the output layer, the presynaptic neurons for M_j are the two neurons C_j and C_0. As mentioned earlier, each output neuron M_j has one dendrite d^j with two synaptic regions, one for the terminal axonal bouton of C_j and one for C_0. The synaptic weight at the synapse of C_j on d^j is given by w_j^{ℓ}, where $\ell = 1$ and $w_j^1 = 0$, while the synaptic weight at the synapse of C_0 on d^j is given by w_0^{ℓ}, with $\ell = 0$ and $w_0^0 = 0$.

Because the activation function of C_j is the identity function, the input at the synapse with weight w_j^1 is $\tau^j(\mathbf{x})$, and since $w_j^1 = 0$, the synapse accepts the input. Likewise, the input from neuron C_0 at the synapse with weight w_0^0 is $\tau^0(\mathbf{x})$. However, because $\ell = 0$ the weight negates the input since $(-1)^{(1-\ell)}(\tau^0(\mathbf{x}) + w_0^0) = -\tau^0(\mathbf{x})$. The dendrite d^j adds the results of the synapses so the $d^j(\mathbf{x}) = \tau^j(\mathbf{x}) - \tau^0(\mathbf{x})$. This information flows to the hillock of M_j, and the activation function of M_j is the hard-limiter $f(d^j(\mathbf{x})) = 1 \Leftrightarrow d^j(\mathbf{x}) \geq 0$ and $f(d^j(\mathbf{x})) = 0$ if $d^j(\mathbf{x}) < 0$.

Since $\tau^j(\mathbf{x}) \leq \tau^0(\mathbf{x})$ for $j = 1, 2, \ldots, t$, it follows that $f(d^j(\mathbf{x})) = 1 \Leftrightarrow \tau^j(\mathbf{x}) = \tau^0(\mathbf{x})$. Thus, if $f(d^j(\mathbf{x})) = 1$ and $\forall k \in \{1, 2, \ldots, t\} \setminus \{j\}$ we have that $f(d^k(\mathbf{x})) = 0$, then we say that $\mathbf{x}$ belongs to class c_{s_j}; i.e. winner takes all. The case that $\mathbf{x}$ belongs to class c_j is enhanced if there are any other possible winners $f(d^k(\mathbf{x})) = 1$, with $c_{s_k} = c_{s_j}$. However, if there is another winner that is not a member of c_j, then repeat the steps with a new randomly obtained set P_p. If after several tries a single winner cannot be found, it becomes necessary to

TABLE 9.5 Similarity valuation classification performance for the 'X-shape' data set. Data set characteristics: $k = 55$ (samples), $n = 2$ (features), and $m = 2$ (classes)

p	$\lvert P_p \rvert$	$\lvert Q_p \rvert$	$\bar{\mu}_p$	f_{ps}^{hits}
10%	6	49	11	0.798
20%	11	44	4	0.916
30%	17	38	1	0.978
40%	22	33	0	0.989
50%	28	27	0	0.994
60%	33	22	0	0.998
70%	39	16	0	1
80%	44	11	0	1
90%	50	5	0	1

increase the percentage of points in P_p. The method described can be simplified by eliminating the neuron C_0 and using the C_j neurons as the output neurons. If there is one $\tau^j(\mathbf{x})$ such that $\tau^k(\mathbf{x}) < \tau^j(\mathbf{x})\ \forall\, s_k \in \{1,2,\ldots,m\} \setminus \{s_j\}$, then $\mathbf{x} \in c_{s_j}$, where c_{s_j} is the class of $\mathbf{q}^{s_j}$. If there are more than one winners where the other winner does not belong to class c_{s_j}, then repeat the steps with a new set P_p as described earlier.

Recognition capability of the BLNN based on the valuation similarity measure defined by equation 9.30, and applied to the data sets described earlier in Examples 9.5 through 9.8, follows an analogous rationale with respect to computational tests. Tables 9.5 through 9.8 show the classifications results corresponding to the 'X-shape', 'Iris', 'Column', and 'Wine' datasets. In each one of these tables, the first three columns give the same information as the tables in Examples 9.5 through 9.8. However, only two columns reporting the average value of misclassified patterns $\bar{\mu}_p$ and the average fraction of hits f_{ps}^{hits}, where s denotes the similarity measure being used. In relation to (9.27), we set $\rho = 50$ for the number of trial runs so that $r \in \{1,\ldots,50\}$ and the complete data set $Q = P_p \cup Q_p$ is used to test the BLNN for each percentage p of randomly generated training pattern subsets P_p.

Comparison between table entries for the datasets considered in the examples reveal that the classification performance of the BLNN based on metrics $f_p^{\text{hits}}(d_1)$, $f_p^{\text{hits}}(d_2^2)$, and the valuation similarity f_{ps}^{hits}, are all competitive and data-dependent. For the X-shape data, the scores for $f_p^{\text{hits}}(d_1)$ are slightly lower than those of f_{ps}^{hits} and $f_p^{\text{hits}}(d_2^2)$, and for $p \geq 60\%$, $f_{ps}^{\text{hits}} = f_p^{\text{hits}}(d_2^2)$.

For the Iris data set both f_{ps}^{hits} and $f_p^{\text{hits}}(d_2^2)$ are basically equal, and both are slightly better than $f_p^{\text{hits}}(d_1)$. In contrast, the scores of $f_p^{\text{hits}}(d_1)$ and f_{ps}^{hits}

TABLE 9.6 Similarity valuation classification performance for the 'Iris' data set. Data set characteristics: $k = 150$ (samples), $n = 4$ (features), and $m = 3$ (classes)

p	$\|P_p\|$	$\|Q_p\|$	$\bar{\mu}_p$	f_{ps}^{hits}
10%	15	135	11	0.921
20%	30	120	7	0.952
30%	45	105	5	0.960
40%	60	90	4	0.969
50%	75	75	3	0.975
60%	90	60	2	0.980
70%	105	45	1	0.987
80%	120	30	1	0.991
90%	135	15	0	0.997

TABLE 9.7 Similarity valuation classification performance for the 'Column' data set. Data set characteristics: $k = 310$ (samples), $n = 6$ (features), and $m = 3$ (classes)

p	$\|P_p\|$	$\|Q_p\|$	$\bar{\mu}_p$	f_{ps}^{hits}
10%	31	279	71	0.770
20%	62	248	56	0.818
30%	93	217	47	0.845
40%	124	186	39	0.872
50%	155	155	33	0.891
60%	186	124	25	0.917
70%	217	93	19	0.938
80%	248	62	12	0.958
90%	279	31	6	0.979

TABLE 9.8 Similarity valuation classification performance for the 'Wine' data set. Data set characteristics: $k = 178$ (samples), $n = 13$ (features), and $m = 3$ (classes)

p	$\|P_p\|$	$\|Q_p\|$	$\bar{\mu}_p$	f_{ps}^{hits}
10%	18	160	51	0.713
20%	36	142	42	0.764
30%	53	125	34	0.804
40%	71	107	28	0.843
50%	89	89	22	0.871
60%	107	71	17	0.901
70%	125	53	12	0.930
80%	142	36	7	0.956
90%	160	18	3	0.978

are very close when using the Column data set. Nonetheless, $f_p^{\text{hits}}(d_2^2)$ and $f_p^{\text{hits}}(d_1)$ are both better than f_{ps}^{hits}, with $f_p^{\text{hits}}(d_2^2)$ being the clear winner. In case of the Wine data set, the BLNN using the metric d_1 is the clear winner with $f_p^{\text{hits}}(d_1) >_p^{\text{hits}} (d_2^2)$ and $f_p^{\text{hits}}(d_1) > f_{ps}^{\text{hits}}$ for $p = 10\%$ through $p = 90\%$.

Exercises 9.2.3

1. Use the three similarity measures resulting from Exercise 3 of Exercises 4.2.1 for similarity learning and apply them to the four data sets employed in this section. Compare your results with those in Tables 9.5 through 9.8.

Epilogue

The work presented here is certainly not conclusive with respect to lattice algebra and its applications. We selected various lattices and theorems from Birkhoff's lattice theory that were needed for the mathematical foundation of lattice algebra. Thus, we had to forego discussing Borel lattices, σ-lattices, Boolean σ-algebras, von Neumann lattices, and many others. Considering applications, lattice algebra is at an extraordinary point in its development. Although the mathematical formulation of lattice theory and associated algebras began in the late 19th Century, their applications in other scientific fields started flourishing with the rise of the computer age. The number of researchers active in lattice-based applications has increased accordingly. These applications involve modeling methods, strategies, and procedures in a wide variety of scientific fields such as engineering, computer and information science, and applied mathematics. When surveying these widespread applications, one discovers a variety of different terminology and notations for the same objects, often causing some initial confusion.

It is our hope that researchers in lattice-based applications will continue expanding the field of lattice algebra. In order to have a coherent algebra, it is crucial to have consistent terminology and notation such as the two lattice-based matrix multiplications $\boxed{\vee}$ and $\boxed{\wedge}$. A *mea maxima culpa* for having introduced the term *morphological matrix products* for these two operations in 1992 [217]. The same holds for *morphological* memories and *morphological* neural network instead of *lattice* neural networks or *lattice-based* neural networks. Current morphological neural networks have little–if anything–to do with mathematical morphology. Another example is the name *tropical* neural networks, where the max-semring $(\mathbb{R}, \vee, +)$ and the min-semring $(\mathbb{R}, \wedge, +)$ are called *tropical* semirings [49, 329].

Another very interesting branch of lattice theory, that has not been discussed in this treatise, is Formal Concept Analysis (FCA). The main reason for this is that FCA evolved around the same time as the lattice algebra described in this book. Formal concept analysis is based on concept lattices and has proven to be an excellent tool for certain types of data analysis. It is a method which takes binary relations and produces a complete lattice. The current FCA theory is based on the pioneering efforts in the early 1980's by a research group led by Rudolf Wille, Bernhard Ganter, and Peter Baumeister at the Technical University of Darmstadt [313, 88, 89]. The motivation and goals of FCA and lattice algebra are the same, namely to expand lattice theory for utilization of real-world applications based on a solid mathematical foundation. There already exist examples where these two branches of lattice theory overlap [41]. We can foresee that in the future FCA could be added to an expanded version of lattice algebra by additional chapters devoted to FCA. This would be similar as in adding numerical linear algebra to linear algebra. It would also expand this book from a one-semester to a two-semester study.

Bibliography

[1] Abu-Mostafa, Y., St. Jacques, J. (1985). Information capacity of the Hopfield model. *IEEE Trans. on Information Theory*, **31**, 461–464.

[2] Adam, J.B., Smith, M.O., Johnson, P.E. (1986). Spectral mixture modeling: a new analysis of rock and soil types at the Viking Lander 1 site, *Journal of Geophysics Research*, **91**(B8), 8098–8112.

[3] Aeberhard, S., Coomans, D., de Vel, O. (1992). *Comparison of Classifiers in High Dimensional Settings*, Tech. Rep. no. 92-02. Dept. of Computer Science & Dept. of Mathematics and Statistics, James Cook University of North Queensland, Australia.

[4] Ajmal, N., Thomas, K.V. (1994). Fuzzy Lattices. *Information Sciences*, **79**, 271–291.

[5] Akayama, M.T., Kikuti, M. (2001). Recognition of character using morphological associative memory. Proceedings of XIV Brazilian Symposium on Computer Graphics and Image Processing, 400.

[6] Amit, D., Gutfreund, H., Sompolinsky, H. (1985). Storing infinite number of patterns in a spin-glass model neural network, *Physics Review Letters*, **55**(14), 1530–1533.

[7] Anderson, J.A. (1972). A simple neural network generating an interactive memory. *Mathematical Biosciences*, **14**, 197–220.

[8] Anderson, J.C., Gosink, L., Duchaineau, M.A., Joy, K.I. (2007). Feature identification and extraction in function fields, Proceedings Eurographics/IEEE-VGTC Symposium on Visualization, 195–201.

[9] Annovi, A., Bagliesi, M.G., et al. (2001). A pipeline of associative memory boards for track finding. *IEEE Trans. on Nuclear Science*, **48**(3), 595–600.

[10] Anton, H. (1977). *Elementary Linear Algebra*, John Wiley and Sons, New York, NY.

[11] Araki, S., Nomura, H., Wakami, N. (1993). Segmentation of thermal images using the fuzzy c-means algorithm. *IEEE Proc. Inter. Conf. on Fuzzy Systems*, Piscataway, NJ, 719–724.

[12] Arbib, M.A. (ed.) (1998). *The Handbook of Brain Theory and Neural Networks*. MIT Press, Boston, MA.

[13] Awcock, G.J., Thomas, R. (1996). *Applied Image Processing*, McGraw-Hill, New York, NY, 126–129.

[14] Backhouse, R.C., Carré, B. (1975). Regular algebra applied to path-finding problems. *Journal Institute Mathematics and Applications*, **15**, 161–186.

[15] Balcombe, J. (2016). *What a Fish Knows: The inner Lives of Our Underwater Cousins*. Farrrar, Straus and Giroux.

[16] Ballard, D.H., Brown, C.M. (1982). *Computer Vision*. Prentice Hall, Englewood Cliffs, NJ, 149–150.

[17] Barmpoutis, A., Ritter, G.X. (2007). Orthonormal basis lattice neural networks. *Studies in Computational Intelligence*, **67**, Springer-Verlag Berlin Heidelberg, 45–58.

[18] Bateson, C., Asner, G., Wessman, C. (2000). Endmember bundles: A new approach to incorporating endmember variability into mixture analysis. *IEEE Trans. Geoscience and Remote Sensing*, **38**(2), 1083–1094.

[19] Batten, L.M., Rentelspacher, A. (1993). *The Theory of Finite Linear Spaces*, Cambridge University Press, Cambridge, UK.

[20] Beham, M.P., Roomi, S.N.M. (2013). Review of face recognition methods. *International Journal of Pattern Recognition and Artificial Intelligence*, **27**(4), 1–35.

[21] Bekkers, J.M. (2003). Synaptic transmission: functional autapses in the cortex. *Current Biology*, **13**(11), 433–435.

[22] Benzaken, C. (1968). Structures algébra des cheminements. *Network and Switching Theory*. Academic Press, 40–57.

[23] Bezdeck, J.C. (1982). *Pattern recognition with fuzzy objective function algorithms*. Plenum Press, New York, NY.

[24] Bezdek, J.C., Pal, N.R. (1998). Some new indexes of cluster validity, *IEEE Trans. on Systems, Man, and Cybernetics*, **28**(3), 301–315.

[25] Bezdek, J.C., Keller, J., Krisnapuram, R., Pal, N.R. (1999). Cluster analysis for object data. In J.C. Bezdek, et al. (eds.), *Fuzzy Models and Algorithms for Pattern Recognition and Image Processing*. Kluver Academic, Dordrecht, Netherlands, 87–121.

[26] Birkhoff, G. (1933). On the combinations of subalgebras. *Proceedings Cambridge Philosophical Society*, **29**, 441–464.

[27] Birkhoff, G. (1933). On the structure of abstract algebras. *Proceedings Cambridge Philosophical Society*, **31**, 433-454.

[28] Birkhoff, G., von Neumann, J. (1933). On the logic of quantum mechanics. *Annals of Mathematics*, **37**, 823–843.

[29] Birkhoff, G. (1938). Partially ordered linear spaces. *Bulletin American Mathematical Society*, **44**, 186.

[30] Birkhoff, G. (1940). *Lattice Theory*, American Mathematical Society Colloquium Publications **25**, Providence, RI.

[31] Boardman, J. (1993). Automatic spectral unmixing of AVIRIS data using convex geometry concepts. Technical report N95-23848, *Center for the Study of Earth and Space*, University of Colorado, Boulder, CO, 11–14.

[32] Boi, L. (2006). From Riemannian geometry to Einstein's general relativity theory and beyond: Space time structures, geometrization and unification. *AIP Conference Proceedings*, **861**, 1066–1075.

[33] Branco, T., Häusser, M. (2010). The single dendritic branch as a fundamental functional unit in the nervous system. *Current Opinions in Neurology*, **20**, 494–502.

[34] Burnstock, G. (1976). Do some nerve cells release more than one transmitter? *Neuroscience*, **1**, 239–248.

[35] Burnstock, G. (2004). Cotransmission. *Current Opinion in Pharmacology*, **4**, 47–51.

[36] Bystrom, I., Blakemore, C., Rakic, P. (2008). Review: Development of the human cerebral cortex. Nature Reviews. *Neuroscience*, **9**(2), 110–22.

[37] Cantor, G. (1895). Beiträge zur Begründung der transfiniten Mengenlehre. *Mathematische Annalen*, **46**, 481–512.

[38] Zermelo, E. (ed.), (1932). *Georg Cantor: Gesammelte Abhandlungen*, Springer Verlag, Berlin, Germany.

[39] Čapek, K. (1920). *R.U.R.* A Play published by Aventinum, Prague, Szecholslovakia.

[40] Carathéodory, C. (1907). Über den Variablitätsbereich der Koeffizienten von Potenzreihen die gegebene Werte nicht annehmen. *Mathematische Annalen*, **64**(1), 95–115.

[41] Caro-Contreras, D.E., Mendez-Vazquez, A. (2013). Computing the concept lattice using dendritic neural networks. *The 10th International Conference on Concept Lattices and their Applications*, La Rochelle, France, 141–152.

[42] Celenk, M., Uijt de Haag, M. (1998). Optimal thresholding for color images. *Proceedings of SPIE, Nonlinear Image Processing IX*, **3304**, 250–259, San Jose, CA.

[43] Chakraborty, R., Vemuri, B.C. (1019). Statistics on the Stiefel manifold: Theory and applications. *Annals of Statistics*, **47**(1), 415–438.

[44] Chakraborty, R., Bouza, J., Manton, J., Vemuri, B.C. (2020). ManifoldNet: A deep neural network for manifold-valued data with applications. *IEEE Transactions on Pattern Analysis and Machine Intelligence,* Special Issue on Geometric Deep Learning. doi: 10.1109/TPAMI.2020.3003846.

[45] Chan, T.F., Vese, L.A. (2001). Active contours without edges. *IEEE Transactions on Image Processing*, **10**(2), 266–277.

[46] Chan, T-H., Chi, C-Y., Huang, Y-M., Ma, W-K. (2009). A convex analysis based minimum-volume enclosing simplex algorithm for hyperspectral unmixing. *IEEE Trans. on Signal Processing*, **57**(11), 4418–4432.

[47] Chang, C-I. (2003). *Hyperspectral Imaging Techniques for Spectral Detection and Classification*, Kluwer Academic/Plenum Publisher, New York, NY.

[48] Chang, C-I. (ed.), (2007). *Hyperspectral Data Exploitation: Theory and Applications*, Wiley-Interscience. John Wiley and Sons, Hoboken, NJ.

[49] Charisopoulos, V. and Maragos, P. (2018). A tropical approach to neural networks with piecewise linear activation. Preprint arxiv: 1805.08749, 2018 - arxiv.org.

[50] Chartier, S., Boukadom, M. (2006). A bidirectional heteroassociative memory for binary and grey-level patterns. *IEEE transaction on Neural Networks*, **17**(2), 385–396.

[51] Chartier, S., Boukadom, M. (2011). Encoding static and temporal patterns with a bidirectional heteroassociative memory. *Journal of Applied Mathematics*, 1–34.

[52] Chen, H., Lee,Y.C., Sun, G.Z., Lee, H.Y., Maxwell, T., Lee, C. (1986). Higher order correlation model for associative memories. In: Denker J.S. (ed.) *Neural Networks for Computing. AIP Proceedings*, **151**.

[53] Cheng, H.D., Jain, X.H., Sun, Y., Wang, J. Color image segmentation: advances and prospects. *Pattern Recognition*, **34**(12), 2259–2281.

[54] Chyzhyk, D., Graña, M. (2011). Optimal hyperbox shrinking in dendritic computing applied to Alzheimer's disease detection in MRI. *Advances in Intelligent and Soft Computing*, **87**, Springer-Berlin, 143–550.

[55] Clark, R.N., Swayze, G.A., Wise, R., Livo, E., Hoefen, T., Kokaly, R., Sutley, S.J. (2007). *USGS digital spectral library splib06a: U.S. Geological Survey*, Digital Data Series 231, http://speclab.cr.usgs.gov/spectral.lib06

[56] Cohen, P.J. (1964). The independence of the continuum hypothesis ii. *Proceedings of the National Academy of Sciences of the United States of America* **51**(1), 105–110.

[57] Craig, M.D. (1994). Minimum volume transforms for remotely sensed data. *IEEE Trans. Geoscience and Remote Sensing*, **32**(1), 99–109.

[58] Crespo, J., Schafer, R.W. (1994). The flat zone approach and color images. In: Serra, J. and Soille, P. (eds.) *Mathematical morphology and its applications to image processing*, 85–92. Kluwer Academic, Dordrecht, Netherlands.

[59] Critescu, R. (1977). *Topological Vector Spaces*, Nordhoff International Publishing, Leyden, Holland.

[60] Cripps, A., Nguyen, N. (2007). Fuzzy lattice reasoning classification using similarity measures. *Studies in Computational Intelligence*, **67**. Springer-Verlag Berlin Heidelberg, 263–284.

[61] Cuninghame-Green, R. (1960). Process synchronization in steelworks—a problem of feasibility, *Proceedings of the 2nd International Conference on Operations Research*, Banbury and Maitland, (eds.), 323–328, English University Press.

[62] Cuninghame-Green, R. (1962). Describing industrial processes with interference and approximating their steady-state behaviour, *Operations Research Quarterly*, **13**, 95–100.

[63] Cuninghame-Green, R. (1979). *Minimax Algebra: Lecture Notes in Economics and Mathematical Systems*, Springer-Verlag, New York, NY.

[64] Datta, A.K., Datta, M., Banerjee, P.K. (2015). *Face Detection and Recognition: Theory and Practice*. CRC Press, Taylor & Francis Group.

[65] Davidson, J.L. (1989). *Lattice Structures in the Image Algebra and Applications to Image Processing*, Ph.D. thesis, University of Florida, Gainesville, FL.

[66] Davidson, J.L. (1992). Lattice Structures in Image Algebra and Applications to Image Processing. In: Hawkes, P. (ed.), *Advances in Imaging and Electron Physics*, **84**, Academic Press, New York, NY, 61–130.

[67] Davies, D.L., Bouldin, D.W. (1979). A cluster separation measure. *IEEE Trans. on Pattern Analysis and Machine Intelligence*, **1**(4), 224–227.

[68] Dayan, P., Abott, L.F. (2006). *Theoretical Neuroscience: Computational and Mathematical Modeling of Neural Systems*, MIT Press, Boston, MA.

[69] Dedekind, R. (1897). *Über Zerlegungen von Zahlen durch ihre grössten gemainsamen Teiler*, Festschrift der Technischen Hochschule Braunschweig, Braunschweig, Germany.

[70] Dedekind, R. (1900). Über die von drei Moduln erzeugte Dualgruppe. *Mathematische Annalen*, **53**, 371–403.

[71] Dedekind, R. (1930). *Gesammelte Werke 2*, Göttinger Digitalizierungszentrum, Göttingen, Göttingen, Germany.

[72] Denker, J.S. (1986). Neural networks models of learning and adaption. *Physica*, **22**D, 216–222.

[73] Dobigeon, N., Tourneret, J.Y., Chang, C.I. (2007). Semi-supervised linear spectral unmixing using a hierarchical Bayesian model for hyperspectral imagery, IRIT-ENSEEHIT-TéSA, France, Tech. Report, 1–34.

[74] Drachman, D.A. (2005). Do we have a brain to spare? *Neurology*, **64**(12).

[75] Dubois, D., Karre, E., Mesiar, R., Prade, H. (2000). Fuzzy interval analysis. In Dubois D., Prade H.(eds.), *Fundamentals of Fuzzy Sets. The Handbook of Fuzzy Sets Series*, vol. **7**. Springer, Boston, MA, 483–581.

[76] Duda, R.O., Hart, P.E., Stork, D.G. (2000). *Pattern Classification 2nd Edition*, John Wiley and Sons, New York, NY.

[77] Dunn, J.C. (1974). A fuzzy relative of the ISODATA process and its use in detecting compact well-separated clusters. *Journal of Cybernetics*, **3**(3), 32–57.

[78] Editors of Scientific American Magazine (2001). *The Scientific American Book of the Brain.*

[79] Eccles, J.C. (1977). *The Understanding of the Brain.* McGraw-Hill, New York, NY.

[80] Elomaa, T., Koivistoinen, H. (2005). On autonomous K-means clustering. *Proc. 15th Int. Symposium on Methodologies for Intelligent Systems*, 228–236.

[81] Eggleston, H.G. (1963). *Convexity*, Cambridge University Press, Cambridge.

[82] Essaqote, H., Zahid, N., Haddaoui, I., Ettouhami, A. (2007). Color image segmentation based on new clustering algorithm and fuzzy eigenspace. *Research Journal of Applied Sciences*, **2**(8), 853–858.

[83] Farrand, W.H. (2005). Hyperspectral remote sensing of land and the atmosphere. B. Gunther and D. Steel (eds.) *Encyclopedia of Modern Optics*, **1**. Academic Press, San Diego, CA, 395–403.

[84] Feng, N., Qiu, Y., Wang, F., Sun, Y. (2006). A unified framework of morphological associative memories. *Lecture Notes in Control and Information Sciences*, **344**, 1–11.

[85] Finkbeiner, D.T. (1960). *Introduction to Matrices and Linear Transforms*, F.W.H. Freeman and Company, San Francisco, CA.

[86] Frank,A., Asuncion, A. (2010). *UCI Machine Learning Repository.* [http://archive.ics.uci.edu/ml]. University of California, School of Information & Computer Science, Irvine, CA.

[87] Freudenthal, H. (1936). Teilweise geordnete Moduln. *Proceedings Academy of Sciences Amsterdam*, **39**, 641–657.

[88] Ganter, B., Wille, R. (1999). *Formal Concept Analysis - Mathematical Foundation.* Springer Verlag Berlin.

[89] Ganter, B., Stumme, G., Wille, R. (eds.), (2005). *Formal Concept Analysis - Foundation and Appicatio.* Springer Verlag Berli-Heidelberg.

[90] Geng, X., Ji, L., Zhao, Y., Wang, F. (2013). A new endmember generation algorithm based on geometric optimization model for hyperspectral images. *IEEE Trans. on Geoscience and Remote Sensing*, **10**(4), 811–815.

[91] Geng, X., Sun, K., Ji, L., Zhao, Y., Tang, H. (2015). Optimizing the endmembers using volume invariant constrained model. *IEEE Trans. on Image Processing*, **24**(11), 3441–3449.

[92] Géraud, T., Strub, P-Y., Darbon, J. (2001). Color image segmentation based on automatic morphological clustering. *IEEE Proc., Inter. Conf. on Image Processing*, **3**, 70–73. Thessaloniki, Greece.

[93] Giffler, B. (1960). Mathematical solution of production planning and scheduling problems, *Technical Report: IBM ASDD*, **1060**, Yorktown Heights, NY.

[94] Gimenez-Martinez, V. (2000). A modified Hopfield auto-associative memory with improved capacity. *IEEE Trans. on Neural Networks*, **11**(4), 867–878.

[95] Goldie, A.W. (1950). The Jordan-Hölder Theorem for General Abstract Algebras. *Proceedings London Mathematics Society*, **2**(52), 107–131.

[96] Gonzalez, R.C., Woods, R.E. (2008). *Digital image processing*. Pearson Prentice-Hall, Upper Saddle River, NJ, 443–446.

[97] Gödel, K. (1940). *The Consistency of the Continuum Hypothesis*, Annals of Mathematics Studies, Princeton University Press, Princeton, RI.

[98] Graña, M., Raducanu, B. (2002). Increasing the robustness of hetero-associative morphological memories for practical applications. *Proceedings 6th International Symposium on Mathematical Morphology*, Sidney, Australia, 379–388.

[99] Graña, M., Gallego, J., Torrealdea, F.J., D'Anjou, A. (2003). On the application of associative morphological memories to hyperspectral image analysis. *Lecture Notes in Computer Sciences*, **2687**, 567–574.

[100] Graña, M., Sussner, P., Ritter, G.X. (2003). Associate morphological memories for endmember determination in spectral unmixing. *12th IEEE International Conference on Fuzzy Systems, FUZZ03*, **2**, 1285–1290.

[101] M. Graña, J.L. Jiménez, C. Hernández (2007). Lattice independence, autoassociative morphological memories and unsupervised segmentation of hyperspectral images. *Proceedings 10th Joint Conference on Information Sciences*, 1624–1631.

[102] M. Graña, I. Villaverde, J.O. Maldonado, C. Hernández (2009). Two lattice computing approaches for the unsupervised segmentation of hyperspectral images, *Neurocomputing*, **72**(10-12), 2111–2120.

[103] Graña, M., Savio, A.M., Garcia-Sebastian, M., Fernandez, E. (2009). A lattice approach to fMRI analysis. *Image and Vision Computing*, **28**(7), 1155–1161.

[104] Graña, M., Chyzhyc, D. (2012). Hybrid multivariate morphology using lattice auto-associative memories for resting-state fMRI network discovery. *12th International Conf. on Hybrid Intelligent Systems* (HIS), 537–542.

[105] Graña, M., Chyzhyc, D. (2015). Image understanding applications of lattice autoassociative memories. *IEEE Trans. on Neural Networks and Learning Systems*, **27**(9), 1920–1932.

[106] Grossberg, S. (1988). Nonlinear networks: principles, mechanisms, and architectures. *Neural Networks*, **1**, 17–61.

[107] Hadwiger, H., (1957). *Vorlesungen Über Inhalt, Oberfläche und Isoperimetrie*. Springer, Berlin.

[108] Ham, F.M., Kostanic, I. (1998). *Principles of neurocomputing for science and engineering*. McGraw-Hill, New York, NY.

[109] Haralick, R.M. and Shapiro, L.G. (1994). Glossary of Computer Vision Terms. Dougherty, E.R.(ed.), *Digital image processing methods*, Marcel Dekker, New York, NY, 439.

[110] Hausdorff, F., (1914). *Grundzüge der Mengenlehre*, Verlag von Veit und Co., Leipzig, Germany.

[111] Healey, G.E. (1990). Using physical color models in 3-d machine vision. *Proceedings of SPIE, Perceiving, Measuring and Using Color*, **1250**. San Diego, CA, 264–275.

[112] Heijmans, H.J.A.M., (1994). Morphological Image Operators In: Hawkes, P.(ed.), *Advances in Imaging and Electron Physics*, Supplement **84**, Academic Press, San Diego, CA, 61–130.

[113] Heinrich, B., (1999). *Mind of the Raven: Investigations and Adventures with Wolf-Birds*. HarperCollins Publishers.

[114] Hernandez, G., Zamora, E., Sossa, H. (2018). Morphological-Linear neural networks. *Fuzz IEEE*, 1–6.

[115] Hernandez G. et al. (2019). Using Morphological-Linear Neural Network for Upper Limb Movement Intention Recognition from EEG Signals. In: Carrasco-Ochoa J., Martnez-Trinidad J., Olvera-Lopez J., Salas J. (eds.) *Pattern Recognition*. MCPR 2019. Lecture Notes in Computer Science, vol **11524**. Springer, 389–397.

[116] Hilbert, D. (1899). *Grundlagen der Geometrie*, Teubner Verlag, Leipzig, Germany.

[117] Herculano-Houzel, S. (2012). The remarkable, yet not extraordinary, human brain as scaled-up primate brain and its associated cost. *Proc. of the National Academy of Sciences*, **109**, 10661–10668.

[118] Hole, K.J. Ahmad, S. (2019). Biologically driven artificial intelligence. *Computer*, **52**(8), 72–75. Publisher IEEE.

[119] Holmes, W.R., Rall, W. (1992). Electronic Models of Neuron Dendrites and Single Neuron Computation. In McKenna, T., Davis, J., Zornetzer, S.F. (eds.) *Single Neuron Computation*, Academic Press, New York, NY, 7–25.

[120] Hook, S.J. (1999). ASTER spectral library: Jet Propulsion Laboratory (http://speclib.jpl.nasa.gov), California Institute of Technology, Pasadena, California, CA.

[121] Hopfield, J.J. (1982). Neural networks and physical systems with emergent collective computational abilities, *Proc. of the National Academy of Sciences*, **79**, 2554–2558.

[122] Hopfield, J.J. (1984). Neurons with graded response have collective computational properties like those of two state neurons, *Proc. of the National Academy of Sciences*, **81**, 3088–3092.

[123] Hopfield, J.J. and Tank, D.W. (1986). Computing with neural circuits, *Science*, 233, 625–633.

[124] Hungerford, T.W. (1974). *Algebra*. Holt, Rinehart and Wiston Inc., New York, NY.

[125] Ikramov, Kh.D., Matin-far, M. (2006). Computer-algebra implementation of the least squares method on the nonnegative orthant, *J. of Mathematical Sciences*, **132**(2), 156–159.

[126] Iordache, M.D., Bioucas-Dias, J.M., Plaza, A. (2014). Collaborative sparse regression for hyperspectral unmixing, *IEEE Trans. Geoscience and Remote Sensing*, **52**(1), 341–354.

[127] Iudin, A.I. (1939). Solution de deux problemes de la théorie des espaces semi-ordonnés. *Doklady Akademii Nauk*, **23**, 418–422.

[128] Jain, R., Kasturi, R., Schunck, B.G. (1995). *Machine Vision*, McGraw-Hill, New York, NY, 73–76.

[129] Jensen, J.R. (2007). *Remote Sensing of the Environment: An Earth Resource Perspective*, 2nd ed, Prentice-Hall Series in Geographic Information Systems, Upper Saddle River, NJ.

[130] Kaburlasos, V.G., Pedridis, V. (1997). Fuzzy lattice neurocomputing (FLN): A novel connectionist scheme for versatile learning and decision making by clustering. *International J Computers and Their Applications*, **4**(3), 31–43.

[131] Kaburlasos, V.G. (2006). *Towards a Unified Modeling and Knowledge Representation based on Lattice Theory*. Studies in Computational Intelligence, **27**. Springer Verlag Berlin Heidelberg.

[132] Kaburlasos, V.G., Ritter, G.X. (eds.) (2007). *Computational Intelligence Based on Lattice Theory*. Studies in Computational Intelligence **67**. Springer Verlag Berlin Heidelberg.

[133] Kandel, E.R., Schwartz, J.H., Jessel, T.M. (2000). *Principles of Neural Systems*. McGraw-Hill, New York, NY.

[134] Kantorovich, L.V. (1937). Lineare halbgeordnete Räume. *Recueil Mathématiques (Sbornik)*, **2**(44), 121–168.

[135] Kantorovich, L.V. (1936). Einige Sätze über halbgeordnete Räume algemeiner Art. *C.R. Academy of Science, USSR*, **2**, 7–10.

[136] Kantorovich, L.V. (1936). The elements of the theory of functions of a real variable with values belonging to a semi ordered linear space. *C.R. Academy of Science, USSR*, **2**, 365–369.

[137] Kantorovich, L.V. (1937). Lineare halbgeordnete Räume. *Recueil Mathématiques (Sbornik)*, **2**(44), 121–168.

[138] Kawulok, M., Celebi, E., Smolka, B. (2016). *Advances ion Face Detection and Facial Image Analysis*. Springer Verlag Berlin Heidelberg.

[139] Keeler, J.D. (1886). Basins of attraction of neural of neural network models. In Denker J.S. (ed.) *Neural Networks for Computing. AIP Proceedings*, **151**.

[140] Keshava, N., Mustard, J.F. (2002). Spectral unmixing, *IEEE Signal Proc. Mag.*, **19**(1), 44–57.

[141] Keshava, N. (2003). A survey of spectral unmixing algorithms, *Lincoln Laboratory Journal*, **14**(1), 55–78.

[142] Kim, D-W., Lee, K.H., Lee, D. (2004). On cluster validity index for estimation of the optimal number of fuzzy clusters, *Pattern Recognition*, **37**, 2009–2025.

[143] Klaua, D. (1965). Über einen Ansatz zur mehrwertigen Mengenlehre. *Monatsberichte Deutsche Aka. der Wissensch. Berlin*, **7**, 859–876.

[144] Klinker, G.J., Schafer, S.A., Kanade, T. (1990). A physical approach to color image understanding. *International Journal of Computer Vision*, **4**(1), 7–38.

[145] Knuth, K.H. (2005). Lattice duality: The origin of probability and Entropy. *Neuron Computing*, **67**, 245–274.

[146] Knuth, K.H. (2007). Lattice Theory, Measures and Probabilities. *AIP Conference Proceedings*, **954**. American Institute of Physics, Melville NY., 22–36.

[147] Knuth, K.H. (2007). Valuation of Lattices: Fuzzification and its Implication. In Kaburlasos V.G., Ritter G.X. (eds.) *Computational Intelligence Based on Lattice Theory, Studies in Computational Intelligence*, **67**. Springer Verlag, Berlin Heidelberg New York, 309–324.

[148] Koch, C., Segev, I. (eds.) (1989). *Methods in Neuronal Modeling: From Synapses to Networks*. MIT Press, Cambridge, MA.

[149] Koch, C., Poggio, T. (1992). Multiplying with Synapses. In: McKenna, T., Davis, J., Zornetzer, S.F. (eds.) *Single Neuron Computation*, Academic Press, New York, 315–345.

[150] Koch, C. (1999). *Biophysics of Computation: Information Processing in Single Neurons*. Oxford University Press.

[151] Kohonen, T. (1972). Correlation matrix memory. *IEEE Trans. on Computers*, **21**, 353–359.

[152] Kohonen, T. (1987). *Self-Organization and Associative Memories*, 2nd ed., Springer-Verlag, Berlin.

[153] Kong, T.Y., Rosenfeld, A. (1989). Digital Topology: Introduction and Survey. *Computer Vision, Graphics, and Image Processing*, **48**, 357-393.

[154] Korenius, T., Laurikkala, J., Juhola, M. (2007). On principal component analysis, cosine, and Euclidean measures in information retrieval, *Information Sciences*, **177**(22), 4893–4905.

[155] Koschan, A., Abidi, M. (2008). *Digital color image processing*, 149–174. John Wiley & Sons, Hoboken, NJ.

[156] Kosko, B. (1987). Adaptive bidirectional associative memories. *IEEE 16th Workshop on Applied Images and Pattern Recognition*, Washington, D.C., 1–49.

[157] Kosko, B. (1987). Adaptive bidirectional associative memories. *IEEE Trans. Systems, Man, and Cybernetics, SMC 00*, 124–136.

[158] Kruse, F.A., Richardson, L.L., Ambrosia, V.G. (1997). Techniques developed for geological analysis of hyperspectral data applied to near-shore hyperspectral ocean data, *Proc. IV International Conference on Remote Sensing for Marine and Coastal Environments*, Ann Arbor, Michigan, Vol. **I**, 233–246.

[159] Kung, S.Y., Zhang, X. (2001). An associative memory approach to blind signal recovery for SMIO/MIMO systems. *Proceedings of the XI IEEE Signal Processing Society Workshop on Neural Networks for Signal Processing*, 343–362.

[160] Lawson, C.L., Hanson, R.J. (1974). *Solving Least Squares Problems*, Prentice-Hall, Englewood Cliffs, NJ.

[161] Lebanon, G. (2005). *Riemannian Geometry and Statistical Machine Learning*. Doctoral Thesis. School of Computer Science, Carnegie Mellon University. Pittsburgh, PA.

[162] Lee W.C.A., et al. (2006). Dynamic remodeling of dendritic arbors in GABAergic interneurons of adult visual cortex, *PLoS Biology*, **4**(2), 271–280.

[163] Leon, J. (1980). *Linear Algebra with Applications*. Macmillan Publishing Co. Inc. New York, NY.

[164] The Third Source (2010). A message of hope for education. *The Neuron*, in Media Gallery, *www.thethirdsource.org/media/charts-and-graphs*.

[165] Li, C.F., Xu, L., Wang, S.T. (2006). A comparative study on improved fuzzy support vector machines and Levenberg-Marquardt based BP network. *Intelligent Computing*, Springer LNCS, vol. **4113**, 73–82.

[166] Li, D., Ritter, G.X. (1990). Decomposition of Separable and symmetric convex templates. In *Image Algebra and Morphological Image Processing*, Proceedings of SPIE **1350**, 408–418.

[167] Li, D. (1992). Morphological template decomposition with max-polynomials. *Journal of Mathematical Image Processing and Vision*, **1**(3), 215–221.

[168] Li, W.J., Lee, T. (2001). Hopfield neural networks for affine invariant matching. *IEEE Trans. on Neural Networks*, **12**(6), 1400–1410.

[169] Lim, Y.W., Lee, S.U. (1990). On the color image segmentation algorithm based on the thresholding and fuzzy c-means techniques. *Pattern Recognition*, **23**(9), 935–952.

[170] Lin, T., Zha, H., Lee, S.U. (2006). Riemannian manifold learning for nonlinear dimensionality reduction. *ECCV 2006*, 45–55.

[171] Lippmann, R.P. (1987). An introduction to computing with neural nets. *IEEE Trans. on Acoust., Speech, Signal Processing*, ASSP-**4**, 4–22.

[172] Liu, J.,Yang, Y-H. (1994). Multiresolution color image segmentation. *IEEE Transactions on Pattern Analysis and Machine Intelligence*, **16**(7), 689–700.

[173] Lucchese, L., Mitra, S.K. (2001). Color image segmentation: A-state-of-the-art-survey. *Proc. Indian Nat. Sci. Acad. (INSA-A)*, **67**(2), 207–221.

[174] Ma, Y., Fu, Y. (2011). *Manifold Learning: Theory and Applications*. CRC Press, Taylor & Francis Group. Boca Raton, London, New York.

[175] Mac Lane, S. (1938). A lattice formulation for transcendence degrees and p-bases. *Duke Mathematics Journal*, **4**, 455–468.

[176] MacNeille, H. (1936). Extensions of partially ordered sets. *Proceedings National Academy of Science*, **22**(1), 45–50.

[177] MacNeille, H. (1937). Partially ordered sets. *Transactions of the American Mathematical Society*, **42**, 416–460.

[178] MacQueen, J. (1967). Some methods for classification and analysis of multivariate observations. In: *Proc. 5th Berkeley Symposium on Mathematics, Statistics, and Probabilities*, Vol. I, University of California, Berkeley, CA., 281–297.

[179] McElcie, R., Posner, E.C., Rodemich, R.R., Venkatesh, S.S. (1987). The capacity of the Hopfield associative memory. *IEEE Transaction of Information Theory*, **33**(4), 33–45.

[180] McKenna, T., Davis, J., Zornetzer, S.E. (eds.) (1992). *Single Neuron Computation.* Academic Press, San Diego, CA.

[181] Manning, H.P. (1914). *Geometry of Four Dimensions.* Macmillan Publishing Co. Inc. New York, NY.

[182] Marcus, G. (2018). *Deep learning: A critical appraisal.* Online available at https://arxiv.org/abs/1801.00631.

[183] Marois, R., Ivanoff, J. (2005). Capacity limits of information processing in the brain. *Trends in Cognitive Science*, **9**(6), 296–305.

[184] Matheron, G. (1975). *Random Sets and Integral Geometry.* Wiley, New York, NY.

[185] Matsuda, S. (2001). Theoretical limitations of a Hopfield Network for crossbar switching. *IEEE Trans. on Neural Networks*, **12**(3), 456–462.

[186] McCarthy, J., Hayes, P.J. (1981). *Readings in Artificial Intelligence.* Elsevier Publisher.

[187] Mel, B.W. (1993). Synaptic integration in excitable dendritic trees. *Journal of Neurophysiology*, **70**, 1086–1101.

[188] Mel, B.W. (1999). Why have Dendrites? A Computational Perspective. S.G., Spruston, N., Hausser, M.D. (eds.), *Dendrites.* Oxford University Press, 271–289.

[189] Meyer, F. (1992). Color image segmentation. *IEEE Proc., 4th Inter. Conf. on Image Processing and its Applications*, 303–306.

[190] Mihu, I.Z., Brad, R., Breazu, M. (2001). Specifications and FPGA implementation of a systolic Hopfield-type associative memory. *Proc. of the International Joint Conference on Neural Networks* (IJCNN'01), 228–233.

[191] Minkowsky, H. (1903). Volumen und Oberfläche. *Mathematische Annalen*, **57**, 447–495.

[192] Minsky M., Papert, S. (1969). *Perceptrons*, MIT Press, Cambridge, MA.

[193] Munehisa T., Kobayashi, M., Yamazaki, H. (2001). Cooperative updating in the Hopfield model. *IEEE Trans. on Neural Networks*, **12**(6). 1243–1251.

[194] D.S. Myers, Hyperspectral Endmember Detection Using Morphological Autoassociative Memories, M.S. Thesis, University of Florida, Gainesville, FL, 2005, 1–51.

[195] Myers, D.S. (2006). The synaptic morphological perceptron. In: Ritter G.X, Schmalz M.S., Barrera J., Astola J.T. (eds.). *Proc. SPIE Mathematics of Data/Image Pattern Recognition, Compression and Encryption*, **63150B**, 1–11.

[196] Myklebust, G., Steen, E. (1996). Neural networks in visualization of multispectral medical images, *Information Sciences*, **89**(3-4), 297–307.

[197] Nascimento, J.M.P., Bioncas-Dias, J.M. (2005). Vertex component analysis: A fast algorithm to unmix hyperspectral data. *IEEE Trans. Geoscience and Remote Sensing*, **43**(4), 898–910.

[198] Biochemistry of Neurotransmitters and Nerve Transmission. (2018) *URL: https://themedicalbiochemistrypage.org/nerves.php* themedicalbiochemistrypage.org, LLC.

[199] Nusbaum, M.P., et al. (2001). The roles of co-transmission in neural network modulation. *TRENDS in Neuroscience*, **24**(3), 146–154.

[200] Oh, H., Kothari, S.C. (1994). Adaption and relaxation method for learning in bidirectional associative memory. *IEEE Transactions on Neural Networks*, **5**, 573–583.

[201] Pakhira, M.K., Bandyopadhyay, S., Maulik, U. (2004). Validity index for crisp and fuzzy clusters, *Pattern Recognition*, **37**, 487–501.

[202] Pal, N.R., Pal, S.K. (1993). A review on image segmentation techniques. *Pattern Recognition*, **26**(9), 1277–1294.

[203] Palm, G. (1980). On associative memory. *Biological Cybernetics*. **36**, 19–31.

[204] Palm, G. (2013). Neural associative memory and sparse encoding. *Neural Networks*, **37**, 165-171.

[205] Palus, H., Kotyczka, T. (2001). Evaluation of colour image segmentation results. *Colour Image Processing Workshop*, Erlangen, Germany.

[206] Palus, H. (2006). Color image segmentation: selected techniques. Lukac, R. and Plataniotis, K.N. (eds.) *Color image processing: methods and applications*, 103–128. CRC Press, Boca Raton, FL.

[207] Park, S.H., Yun, I.D., Lee, S.U. (1998). Color image segmentation based on 3-D clustering: morphological approach. *Pattern Recognition*, **31**(8), 1061–1076.

[208] Peteanu, V. (1967). An algebra of the optimal path in networks. *Mathematica*, **9**, 335–342.

[209] Petridis, V. and Kaburlasos, V.G. (1998). Fuzzy lattice neural network (FLNN): a hybrid model of learning. *IEEE Transactions on Neural Networks*, **9**(5), 877–890.

[210] Pinsker, A.G. (1938). Sur l'extension des espaces semi-ordonnés. *C.R. Academy of Science, USSR*, **21**, 6–10.

[211] Plataniotis, K.N. and Venetsanopoulos, A.N. (2000). *Color image processing and applications*. Springer, Berlin, Germany, 237–273.

[212] Proceedings of the Second Symposium in Pure Mathematics (1959), Dilworth, R.P. (ed.) *Proceedings of Symposia in Pure Mathematics*, vol. **II**. American Mathematics Society.

[213] Raducanu, B., Graña, M., Albizuri, X.F. (2003). Morphological scale spaces and associative morphological memories: results on robustness and practical applications. *Journal of Mathematical Imaging and Vision* **19**(2), 851–867.

[214] Rall,W., Segev, I. (1987). Functional Possibilities for Synapses on Dendrites and Dendritic Spines. In: Edelman, G.M., Gall, E.E., Cowan, W.M. (eds.) *Synaptic Function*, John Wiley and Sons, New York, NY, 605–636.

[215] Richardson, L.L., Buison, D., Liu, C.J., Ambrosia, V.G. (1994). The detection of algal photosynthetic accessory pigments using AVIRIS spectral data, *Marine Technology Society Journal*, **28**, 10–21.

[216] Ritter, G.X., Li, D., Wilson, J.N. (1989). Image algebra and its relationship to neural networks. *SPIE Proceedings on Aerospace Pattern Recognition*, **1098**, 90–101.

[217] Ritter, G.X. (1992). Morphological matrix products and their applications in image processing. *Proceedings of the Canadian Conference on Electrical and Computer Engineering*, Toronto, Ontario, TA3.20.1–TA3.20.13.

[218] Ritter, G.X., Sussner, P., Díaz de León, J.L. (1998). Morphological associative memories. *IEEE Transactions on Neural Networks*, **9**(2), 281–293.

[219] Ritter, G.X., Díaz de León, J.L., Sussner, P. (1999). Morphological bidirectional associative memories. *Neural Networks*, **12**, 851–867.

[220] Ritter, G.X. Wilson, J.N. (2001). *Handbook of Computer Vision Algorithms in Image Algebra*, 2nd ed., CRC Press, Boca Raton.

[221] Ritter, G.X., Urcid, G. (2003). Lattice algebra approach to single-neuron computation. *IEEE Transactions on Neural Networks*, **14**(2), 282–295.

[222] Ritter, G.X., Iancu, L. Urcid, G. (2003). Morphological perceptrons with dendritic structures. *Proceedings of the IEEE International Conference on Fuzzy Systems*, **2**, 1296–1301.

[223] Ritter, G.X., Urcid, G., Iancu, L. (2003). Reconstruction of patterns from noisy inputs using morphological associative memories. *Journal of Mathematical Imaging and Vision*, **19**(2), 95–111.

[224] Ritter, G.X., Iancu, L. (2004). A morphological auto-associative memory based on dendritic computing. *Proceedings of the IEEE International Joint Conference on Neural Networks*, **2**, 915–920.

[225] Ritter G.X., Iancu L. (2005). A Lattice Algebraic Approach to Neural Computation. In: *Handbook of Geometric Computing*. Springer, Berlin, Heidelberg, 97–129.

[226] Ritter, G.X., Gader, P. (2006). Fixed Points of Lattice Transforms and Lattice Associative Memories. In: Hawkes, P.(ed.) *Advances in Imaging and Electron Physics*, **144**, Academic Press, San Diego, CA, 165–242.

[227] Ritter, G.X., Urcid, G. (2007). Learning in lattice neural networks that employ dendritic computing. *Studies in Computational Intelligence*, **67**, Springer-Verlag Berlin Heidelberg, 25–44.

[228] Ritter, G.X., Urcid, G. (2014). Lattice based dendritic computing: A biomimetic approach to ANNs. *Lecture Notes in Computer Science*, **8827**, Springer, Berlin/Heidelberg, 730–744.

[229] Ritter, G.X., Urcid, G. (2018). Extreme points of convex polytopes derived from lattice autoassociative memories. *Lecture Notes in Computer Science*, **10880**, Springer, Berlin/Heidelberg, 116–125.

[230] Ritter, G.X., Urcid, G., Schmalz, M.S. (2009). Autonomous single-pass endmember approximation using lattice auto-associative memories. *Neurocomputing*, **72**(10-12). 2101–2110.

[231] Ritter, G.X., Urcid, G. (2010). Lattice Algebra Approach to Endmember Determination in Hyperspectral Imagery. In: Hawkes, P. (ed.), *Advances in Imaging and Electron Physics*, **160**. Academic Press, San Diego, CA. 113–169.

[232] Ritter, G.X., Urcid, G. (2010). A lattice matrix method for hyperspectral image unmixing, *Information Sciences*, **181**(10), 1787–1803.

[233] Ritter, G.X., Urcid, G. (2011). A lattice matrix method for hyperspectral image unmixing. *Information Sciences*, **18**(10), 1787–1803.

[234] Ritter, G.X., Urcid, G., Valdiviezo, J.C. (2011). Grayscale image recall from imperfect inputs with a two layer dendritic lattice associative memory. *Third World Congress on Nature and Biologically Inspired Computing*, 261–266.

[235] Ritter, G.X., Urcid, G., Nievez-V, J.A. (2013). An autonomous endmember detection technique based on lattice algebra and data clustering. *Proceedings of WHISPERS 2013, 5th Workshop of Image and Signal Processing: Evolution in Remote Sensing*, Gainesville, FL, 1–4.

[236] Ritter, G.X., Urcid, G., Valdiviezo, J.C. (2014) Two lattice metrics dendritic computing for pattern recognition. *IEEE International Conference on Fuzzy Systems (FUZZ-IEEE)*, 45–52.

[237] Rocha, A.R., Barreto, G.A. (2009). On the application of ensembles of classifiers to the diagnosis of pathologies of the vertebral column:a comparative analysis. *IEEE Latin America Transactions*, **7**(4), 487–496.

[238] Rota, GC. (1997). *The Many Lives of Lattice Theory*. Notices of the AMS, American Mathematics Society, Providence, RI.

[239] Rudin, W. (1964). *Principles of Mathematical Analysis*. McGraw Hill Inc., New York, NY.

[240] Rudin, W. (1974). *Real and Complex Analysis*. McGraw Hill Inc., New York, NY.

[241] Said, S., Hajri, H., Bombrun, L., Vemuri, BC. (2019). Gaussian distributions on Riemannian symmetric spaces: statistical learning with structured covariance matrices. *IEEE Transactions on Information Theory*, **64**(2), 752–772.

[242] Savage, C. (1997). *Bird Brains: The Intelligence of Crows, Ravens, Magpies, and Jays*. Sierra Club Books.

[243] Schoute, P.H. (1902). *Mehrdimensionale Geometrie*. 1 Teil: Die linearen Räume. Göschen, Leipzig.

[244] Segev, I. (1998). Dendritic processing. Arbib, M.A. (ed.) *The Handbook of Brain Theory and Neural Networks*, MIT Press, Boston, MA, 282–289.

[245] Sejnowski, T.J., Qian, N. (1992). Synaptic integration by electrodiffusion in dendritic spines. McKenna, T., Davis, J., Zornetzer, S.F. (eds.) *Single Neuron Computation*. Academic Press, New York, NY.

[246] Serra, J. (1982). *Image Analysis and Mathematical Morphology*. Academic Press, London.

[247] Shafarenko, L., Petrou, H., Kittler, J. (1998). Histogram-based segmentation in a perceptually uniform color space. *IEEE Transactions on Image Processing*, **7**(9), 1354–1358.

[248] Shannon, C.E. (1938). A symbolic analysis of relay and switching circuits. *Trans. AIEE*, **57**(12), 713–723.

[249] Shannon, C.E. (1953). The lattice theory of information. *IEEE Transactions of Information Theory*, **1**(1), 105–107.

[250] Shannon, C.E. (1958). A note on a partial ordering for communication channels. *Information and Control*, **1**, 390–397.

[251] Shepherd, G.M. (1992). Canonical Neurons and their Computational Organization. In: McKenna, T., Davis, J., Zornetzer, S.F. (eds.) *Single Neuron Computation*, Academic Press, San Diego, CA, 27–55.

[252] Shi, H., Zhao, Y., Zhuang, X. (1998). A general model for bidirectional associative memories. *IEEE Transaction on Systems, Man, and Cybernetics*, **28**, 511–519.

[253] Shi, F-G. (2000). L-fuzzy relations and L-fuzzy subgroups. *Journal of Fuzzy Mathematics*, **8**(2), 494–499.

[254] Shi, F-G. (2000). L-fuzzy mappings of L-fuzzy sets. *Fuzzy Systems and Mathematics*, **14**(3), 16–24.

[255] Shi, J., Malik, J. (2000). Normalized cuts and image segmentation. *IEEE Transactions on Pattern Analysis and Machine Intelligence*, **22**(8), 888–905.

[256] Shimbel, A. (1954). Structure in communication nets. textitProceedings of the Symposium on Information Networks, 119–203, Polytechnic Institute of Brooklyn, Brooklyn, NY.

[257] Simon, H.A. (1955). A behavioral model of rational choice. *The Quarterly Journal of Economics*, 99–117.

[258] Simpson, P.K. (1990). Higher-ordered and intraconnected bidirectional associative memories. *IEEE Transaction on Systems, Man, and Cybernetics*, **20**, 637–652.

[259] Simpson, P.K. (1992). Fuzzy min-max neural networks-Part 1: Classification. *IEEE Trans. on Neural Networks*, **3**, 776–786.

[260] Simpson, P.K. (1993). Fuzzy min-max neural networks-Part 2: Clustering. *IEEE Trans. on Fuzzy Systems*, **1**, 32–45.

[261] Skarbek, W., Koschan, A. (1994). *Colour image segmentation: a survey*. Technical University of Berlin, TR. 94-32, 1–81.

[262] Sowmya, B., Sheelanari, B. (2009). Color image segmentation using soft computing techniques. *International Journal of Soft Computing Applications*, (**4**), 69–80.

[263] Srivastata, V., Sampath, S., Parker, D.J. (2014) Overcoming catastrophic interference in connectionist networks using Gram-Schmidt orthogonalization. *PLOS One* DOI:10.1371.

[264] Steinbuch, K. (1961). Die Lernmatrix, *Kybernetik*, vol.**1**, 36.

[265] Steinbuch, K. (1961). *Automat und Mensch*. Springer Verlag, Heidelberg.

[266] Steinbuch, K. (1963). *Automat und Mensch*, 2nd ed., Springer Verlag, Heidelberg.

[267] Steinbuch, K. (1965). *Automat und Mensch*, 3rd ed., Springer Verlag, Heidelberg.

[268] Steinbuch, K. (1972). *Automat und Mensch*, 4th ed., Springer Verlag, Heidelberg.

[269] Steinbuch, K., Piske, A.A.W. (1963). Learning matrices and their applications. *IEEE Trans. on Electronic Computers*, December 1963, 846–862.

[270] Stone, M.H. (1937). Topological representation of distributive lattices and Brouwerian logics. *Casopis Pestovani Matematiky Fysiky*, **69**, 1–25.

[271] Stone, M.H. (1937). Applications of the theory of Boolean rings to general topology. *Transactions American Mathematical Society*, **41**, 375–481.

[272] Sussner, P. (2000). Observations on morphological associative memories and the kernel method. *Neurocomputing*, **31**, 167–183.

[273] Sussner, P. (2001). A relationship between binary autoassociative morphological memories and fuzzy set theory. *Proceedings of the International Joint Conference on Neural Networks*, **4**, 2512–2517.

[274] Sussner, P. (2003). Generalizing operations of binary autoassiociative morphological memories using fuzzy set theory. *Journal of Mathematical Imaging and Vision*, **19**(2), 81–93.

[275] Sussner, P. (2003). A fuzzy autoassociative morphological memory. *Proceedings of the International Joint Conference on Neural networks*, Portland, OR, 326–331.

[276] Sussner, P., Valle, M.E. (2006). Implicative fuzzy associative memories. *IEEE Trans. on Fuzzy Systems*, **14**(6), 793–807.

[277] Sussner, P., Valle, M.E. (2006). Gray-scale morphological associative memories. *IEEE Trans. on Neural Networks*, **17**, 559–570.

[278] Szpilrajn, E. (1930). Sur l'extension de l'ordre partiel. *Fundamenta Mathematica*, **16**, 386–389.

[279] Tepavcevic, A., Trajkovski, G. (2001). L-fuzzy lattices: an introduction. *Fuzzy Sets and Systems*, **123**(2), 209–216.

[280] Thorndike, E.L. (1970). *Animal Intelligence: Experimental Studies.* Transaction Publishers.

[281] Trudeau, L.E., Gutirrez, R. (2007). On Cotransmission and Neurotransmitter Phenotype Plasticity. *Molecular Interventions* **7**(3), 138–146.

[282] Turing, A.M. (1950). Computing Machinery and Intelligence. *MIND, Quarterly Review of Psychology and Philosophy*, **LIX**(236), 433–460.

[283] Urcid, G., Ritter, G.X. (2003). Kernel computation in morphological associative memories for grayscale image recollection. Proceedings of the IASTED International Conference, August 13-15, 2003, Honolulu, HI. ACTA Press.

[284] Urcid, G., Ritter, G.X., Iancu, L. (2003). Kernel Computation in Morphological Bidirectional Associative Memories. In: Sanfeliu A., Ruiz-Shulcloper J. (eds.) *Progress in Pattern Recognition, Speech and Image Analysis* CIARP 2003. Lecture Notes in Computer Science, **2905**. Springer, Berlin, Heidelberg, 566–569.

[285] Urcid, G. (2005). Transformations of neural inputs in lattice dendrite computation. *Proc SPIE Math. Methods in Pattern and Image Analysis*, **5916**, 201–212.

[286] Urcid, G., Valdiviezo-N., J.C. (2009). Color image segmentation based on lattice auto-associative memories. *IASTED Proc., 13th Inter. Conference on Artificial Intelligence and Soft Computing*, Palma de Mallorca, Spain,166–173.

[287] Urcid, G., Valdiviezo-N., J.C., Ritter, G.X. (2010). Lattice associative memories for segmenting color images in different color spaces. *Lecture Notes in Artificial Intelligence*, **6077**(Part II), Springer, Berlin/Heidelberg, 359–366.

[288] Urcid, G., Valdiviezo, J.C., Ritter, G.X. (2009). Endmember search techniques based on lattice auto-associative memories: A case study of vegetation discrimination. *Proceedings of SPIE*, **7477**, 7477D-1–7477D-12.

[289] Urcid, G., Valdiviezo, J.C., Ritter, G.X. (2011). Lattice algebra approach to color image segmentation. *Journal of Mathematical Imaging and Vision*, **42**(2-3), 150–162.

[290] Urcid, G., Ritter, G.X. (2012). C-means Clustering of Lattice Auto-Associative Memories for Endmember Approximation. *Frontiers in Artificial Intelligence and Applications*, **243**, 2144–2148.

[291] Valente, R.A., Abrão, T. (2016). MIMO transmit scheme based on morphological perceptrons with competitive learning. *Neural Networks*, **80**, 9–18.

[292] Valentine, F.A. (1964). *Convex Sets*. McGraw-Hill, New York, NY.

[293] van der Waerden, B.L. (1964). *Modern Algebra*. Frederic Ungar Co., New York, NY.

[294] Vane, G., Green, R.O., Chrien, T.G., Enmark, H.T., Hansen, E.G., Porter, W.M. (1993). The Airborne Visible/Infrared Imaging Spectrometer (AVIRIS), *Remote Sensing Environment*, **44**, 127–143.

[295] Vazquez, R.A., Sossa, H. (2006). Associative memories applied to image categorization. *Lecture Notes in Computer Sciences*, **4225**, 549–558.

[296] Vazquez, R.A., Sossa, H. (2009). Behavior of Morphological associative memories with true-color image patterns. *Neurocomputing*, **72**, 225–244.

[297] Vazquez, R.A., Sossa, H. (2009). Morphological Hetero-Associative Memories Applied to Restore True-Color Patterns. In: Yu W., He H., Zhang N. (eds.) Advances in Neural Networks. *Lecture Notes in Computer Science*, **5553**. Springer, Berlin, Heidelberg, 520–529.

[298] Veblen, O. (1904). A system of axioms for geometry. *Transactions AMS*, **5**, 343–384.

[299] Veblen, O. (1911). *Foundations of Geometry*. Monographs of Modern Mathematics, J.W.A. Young, New York, NY.

[300] van der Loos, H., Glaser, E.M. (1972). Autapses in neocortex cerebri: synapses between pyramidal cell's axon and its own dendrites. *Brain Research*, **48**, 355–360.

[301] von Stieff, F. (2016). *Brain in Balance: Understanding the Genetics and Neurochemistry Behind Addiction and Sobriety*. Canyon Hill Publishing, San Francisco, CA.

[302] von Bartheld, C.S., Bahney, J., Herculano-Houzel, S. (2016). The search for true numbers of neurons and glial cells in the human brain: A review of 150 years of cell counting, quantification of neurons and glia in human brains. *Journal of Comparative Neurology*, **524**(18), 3865.

[303] Wallace, D.J. (1985). Memory and learning in a class of neural models. *Proc. Workshop on Lattice Gauge Theory*, Plenum Press, 313–330.

[304] Wallman, H. (1938). Lattices and topological spaces. *Annals of Mathematics*, **39**, 112–126.

[305] Wang, C.C., Don, H. (1995). An analysis of high-capability discrete exponential BAM. *IEEE Transaction on Neural Networks*, **6**, 492–496.

[306] Wang, J.F., Cruz, J.B., Mulligan, J.H. (1990). Two encoding strategies for bidirectional associative memory. *IEEE Transaction on Neural Networks*, **1**, 81–92.

[307] Wang, M., Wang, S.T., Wu, X.J. (2003). Initial results on fuzzy morphological associative memories (in Chinese). *Acta Electronica Sinica*, **31**, 690–693.

[308] Wang, Y.F., Cruz, J.B., Mulligan, J.H. (1991). Guarantied recall of all training pairs for bidirectional memory. *IEEE Transaction on Neural Networks*, **2**, 559–567.

[309] Wang, Z. (1996). A bidirectional associative memory based on optimal linear associative memory. *IEEE Transaction on Computers*, **45**, 1171–1179.

[310] Wasnikar, V.A., Kulkarni, A.D. (2000). Data mining with radial basis functions. In: Dagli C.H., et al (eds.) *Intelligent Engineering Systems Through Artificial Neural Networks*, ASME Press, New York, NY.

[311] Wechsler, H. (2009). *Face Recognition Methods: Applications: systems design, implementation and evaluation.* Springer Verlag Berlin Heidelberg.

[312] Wei, D.S., et al. (2001). Compartmentalized and binary behavior of terminal dendrites in hippocampal pyramidal neurons. *Science* **293**, 2272–2275.

[313] Wille, R. (1982). Restructuring lattice theory: An approach based on hierarchies of concepts. Rival I. (ed.), *Ordered Sets*. Dodrecht-Boston, 445–47.

[314] Williamson, J.H. (1962) *Lesbesgue Integration.* Holt, Rhinehart and Winston, New York, NY.

[315] Winter, M.E. (1999). N-FINDR: An algorithm for fast autonomous spectral endmember determination in hyperspectral data. *Proceedings SPIE*, **5**, 266–275.

[316] Winter, M.E. (2000). Comparison of approaches for determining endmembers in hyperspectral data, *IEEE Proc., Aerospace Conference 2000*, **3**, 305–313.

[317] Winter, M.E. (2007). Maximum Volume Transform for Endmember Spectra Determination. Chen I.C. (ed.) Chapter 7 of *Hyperspectral Data Exploitation - Theory and Application.* Wiley-Interscience, John Wiley and Sons, Hoboken, NJ.

[318] Wolfram, S. (2015). George Boole: A 200-Year View. @writings.stephenwolfram.com

[319] Woods, F.S. (1922). *Higher Geometry*. The Athenaeum Press, Boston, MA.

[320] Woods, K. (1997). Combination of multiple classifiers using local accuracy estimates. *IEEE Transactions on Pattern Analysis and Machine Intelligence* **19**(4), 405–410.

[321] Wu, K-L., Yang, M-S. (2005). A cluster validity index for fuzzy clustering, *Pattern Recognition Letters*, **26**, 1275–1291.

[322] Xie, X.L., Beni, G.A. (1991). A validity measure for fuzzy clustering. *IEEE Trans. on Pattern Analysis and Machine Intelligence*, **13**(8), 841–846.

[323] Xu, Z. (1994). Asymmetric bidirectional associative memories. *IEEE Transaction on Systems, Man, and Cybernetics*, **24**, 1558–1564.

[324] Yañez-Márquez, C. (2002). Associative Memories Based on Order Relations and Binary Operations. Ph.D. Thesis (in Spanish), IPN Center for Computing Research. Mexico City, Mexico.

[325] Yañez-Márquez, C., et al. (2007). Using alpha-beta memories to learn and recall RBG images. Symposium on Neural Networks, *Advances in Neural Networks - ISNN*, **20**, 828–833.

[326] Zahdeh, L.A. (1965). Fuzzy Sets. *Information and Control* **8**(3), 338–353.

[327] Zhang, C., Wang, P. (2000). A new method for color image segmentation based on intensity and hue clustering. *IEEE Proc., 15th Inter. Conf. on Pattern Recognition*, **3**, 613–616.

[328] Zhang, J-F. (2013). A novel definition of L-fuzzy lattice based on fuzzy sets. *The Scientific World Journal*, **2013**, Article ID 678586.

[329] Zhank, L., Naifzat, G., Lim, L.H. Tropical geometry and deep neural networks. *Proceeding of the 35th International Conference on Machine Learning/Proceedings of Machine Learning Research*, **80**, 5824–5832.

[330] Zhu, S.C., Yuille, A. (1996). Region competition: unifying snakes, region growing, and Bayes/MDL for multiband image segmentation. *IEEE Transactions on Pattern Analysis and Machine Intelligence*, **18**(9), 884-900.

[331] Zhuang, X., Huang, Y., Chen, S.S. (1993). Better learning for bidirectional associative memory. *Neural Networks*, **8**, 1131–1146.

[332] Zimmer, C. (2011). 100 trillion connections: New efforts probe and map the brain's detailed architecture. *Scientific American*, January 2011.

Index